Basic Theory of

# Fractional
# Differential
# Equations

Second Edition

Basic Theory of

# Fractional Differential Equations

## Second Edition

**Yong Zhou** *Xiangtan University, China*
**JinRong Wang** *Guizhou University, China*
**Lu Zhang** *Xiangtan University, China*

**World Scientific**

NEW JERSEY · LONDON · SINGAPORE · BEIJING · SHANGHAI · HONG KONG · TAIPEI · CHENNAI · TOKYO

*Published by*

World Scientific Publishing Co. Pte. Ltd.

5 Toh Tuck Link, Singapore 596224

*USA office:* 27 Warren Street, Suite 401-402, Hackensack, NJ 07601

*UK office:* 57 Shelton Street, Covent Garden, London WC2H 9HE

**Library of Congress Cataloging-in-Publication Data**

Names: Zhou, Yong, 1964–    | Wang, JinRong (Mathematics professor) | Zhang, Lu (Mathematics professor)

Title: Basic theory of fractional differential equations.

Description: 2nd edition / Yong Zhou (Xiangtan University, China), JinRong Wang
      (Guizhou University, China), Lu Zhang (Xiangtan University, China). |
      New Jersey : World Scientific, 2016. | Includes bibliographical references and index.

Identifiers: LCCN 2016032558 | ISBN 9789813148161 (hc : alk. paper)

Subjects: LCSH: Fractional differential equations. | Differential equations. | Fractional calculus.

Classification: LCC QA372 .Z47 2016 | DDC 515/.352--dc23

LC record available at https://lccn.loc.gov/2016032558

**British Library Cataloguing-in-Publication Data**

A catalogue record for this book is available from the British Library.

Desk Editors: V. Vishnu Mohan/Kwong Lai Fun

Typeset by Stallion Press
Email: enquiries@stallionpress.com

Printed in Singapore

# Preface to the Second Edition

This volume is the second edition of the monograph entitled Basic Theory of Fractional Differential Equations. In the second edition, some new topics have been added: Fractional impulsive differential equations, and fractional partial differential equations including fractional Navier-Stokes equations and fractional diffusion equations. The bibliography has also been updated and expanded.

The second edition of this book is divided into seven chapters. Chapter 1 introduces preliminary facts from fractional calculus, nonlinear analysis and semigroup theory. In Chapter 2, we present a unified framework to investigate the basic existence theory for discontinuous fractional functional differential equations with bounded delay, unbounded delay and infinite delay, respectively. Chapter 3 is devoted to the study of fractional differential equations in Banach spaces via measure of noncompactness method, topological degree method and Picard operator technique. In Chapter 4, we first present some techniques for the investigation of fractional evolution equations governed by $C_0$-semigroup, then we discuss fractional evolution equations with almost sectorial operators. Chapter 5 deals with initial boundary value problems of fractional impulsive differential equations including Langevin equations and evolution equations. In Chapter 6, by using critical point theory, we study existence and multiplicity of solutions for boundary value problems to fractional differential equations. And in the last chapter, we introduce the recent works on time-fractional partial differential equations including Navier-Stokes equations, Euler-Lagrange equations, diffusion equations and Schrödinger equations.

This book is useful to researchers, graduate or PhD students dealing with fractional calculus and applied analysis, differential equations and related areas of research.

We would like to thank Professors B. Ahmad, O.P. Agrawal, D. Baleanu, M. Benchohra, L. Bourdin, I. Vasundhara Devi, M. Fečkan, V. Kiryakova, F. Liu, J.A.T. Machado, M.M. Meerschaert, S. Momani, J.J. Nieto, V.E. Tarasov, J.J. Trujillo and M. Yamamoto for their support. We also wish to express our appreciation to our colleagues, Professors Z.B. Bai, Y.K. Chang, W.H. Deng, W. Jiang, Z.H. Liu, H.R. Sun, R.N. Wang and graduate students F. Jiao, Y.H. Lan and L. Peng for their help.

We acknowledge with gratitude the support of National Natural Science Foundation of China (11271309;11671339) and Training Object of High Level and Innovative Talents of Guizhou Province ([2016]4006).

Yong Zhou
*Xiangtan University, China*

JinRong Wang
*Guizhou University, China*

Lu Zhang
*Xiangtan University, China*

# Preface to the First Edition

The concept of fractional derivative appeared for the first time in a famous correspondence between G.A. de L'Hospital and G.W. Leibniz, in 1695. Many mathematicians have further developed this area and we can mention the studies of L. Euler (1730), J.L. Lagrange (1772), P.S. Laplace (1812), J.B.J. Fourier (1822), N.H. Abel (1823), J. Liouville (1832), B. Riemann (1847), H.L. Greer (1859), H. Holmgren (1865), A.K. Grünwald (1867), A.V. Letnikov (1868), N.Ya. Sonin (1869), H. Laurent (1884), P.A. Nekrassov (1888), A. Krug (1890), J. Hadamard (1892), O. Heaviside (1892), S. Pincherle (1902), G.H. Hardy and J.E. Littlewood (1917), H. Weyl (1919), P. Lévy (1923), A. Marchaud (1927), H.T. Davis (1924), A. Zygmund (1935), E.R. Love (1938), A. Erdélyi (1939), H. Kober (1940), D.V. Widder (1941), M. Riesz (1949) and W. Feller (1952). In the past sixty years, fractional calculus had played a very important role in various fields such as physics, chemistry, mechanics, electricity, biology, economics, control theory, signal and image processing, biophysics, blood flow phenomena, aerodynamics, fitting of experimental data, etc.

In the last decade, fractional calculus has been recognized as one of the best tools to describe long-memory processes. Such models are interesting for engineers and physicists but also for pure mathematicians. The most important among such models are those described by differential equations containing fractional-order derivatives. Their evolutions behave in a much more complex way than in the classical integer-order case and the study of the corresponding theory is a hugely demanding task. Although some results of qualitative analysis for fractional differential equations can be similarly obtained, many classical methods are hardly applicable directly to fractional differential equations. New theories and methods are thus required to be specifically developed, whose investigation becomes more challenging. Comparing with classical theory of differential equations, the researches on the theory of fractional differential equations are only on their initial stage of development.

This monograph is devoted to a rapidly developing area of the research for the qualitative theory of fractional differential equations. In particular, we are interested in the basic theory of fractional differential equations. Such basic theory should be the starting point for further research concerning the dynamics, control, numerical analysis and applications of fractional differential equations. The book is divided into six chapters. Chapter 1 introduces preliminary facts from fractional calculus, nonlinear analysis and semigroup theory. In Chapter 2, we present a unified framework to investigate the basic existence theory for discontinuous fractional functional differential equations with bounded delay, unbounded delay and infinite delay. Chapter 3 is devoted to the study of fractional differential equations in Banach spaces via measure of noncompactness method, topological degree method and Picard operator technique. In Chapter 4, we first present some techniques for the investigation of fractional evolution equations governed by $C_0$-semigroup, then we discuss fractional evolution equations with almost sectorial operators. In Chapter 5, by using critical point theory, we give a new approach to study boundary value problems of fractional differential equations. And in the last chapter, we present recent advances on theory for fractional partial differential equations including fractional Euler-Lagrange equations, time-fractional diffusion equations, fractional Hamiltonian systems and fractional Schrödinger equations.

The material in this monograph are based on the research work carried out by the author and other experts during the past four years. The book is self-contained and unified in presentation, and it provides the necessary background material required to go further into the subject and explore the rich research literature. Each chapter concludes with a section devoted to notes and bibliographical remarks and all abstract results are illustrated by examples. The tools used include many classical and modern nonlinear analysis methods. This book is useful for researchers and graduate students for research, seminars, and advanced graduate courses, in pure and applied mathematics, physics, mechanics, engineering, biology, and related disciplines.

I would like to thank Professors D. Baleanu, K. Balachandran, M. Benchohra, L. Bourdin, Y.Q. Chen, I. Vasundhara Devi, M. Fečkan, N.J. Ford, W. Jiang, V. Kiryakova, F. Liu, J.A.T. Machado, M.M. Meerschaert, S. Momani, G.M. N'Guérékata, J.J. Nieto, V.E. Tarasov, J.J. Trujillo, A.S. Vatsala and M. Yamamoto for their support. I also wish to express my appreciation to my colleagues, Professors Z.B. Bai, Y.K. Chang, H.R. Sun, J.R. Wang, R.N. Wang, S.Q. Zhang and my graduate students H.B. Gu, F. Jiao, Y.H. Lan and L. Zhang for their help. Finally, I thank the editorial assistance of World Scientific Publishing Co., especially Ms. L.F. Kwong.

I acknowledge with gratitude the support of National Natural Science Foundation of China (11271309, 10971173), the Specialized Research Fund for the Doctoral Program of Higher Education (20114301110001) and Key Projects of Hunan Provincial Natural Science Foundation of China (12JJ2001).

Yong Zhou
*Xiangtan, China*
October 2013

# Contents

*Preface to the Second Edition*                                    v

*Preface to the First Edition*                                    vii

1.  Preliminaries                                                   1

    1.1   Introduction . . . . . . . . . . . . . . . . . . . . . . . . .   1

    1.2   Some Notations, Concepts and Lemmas . . . . . . . . . . . . .   1

    1.3   Fractional Calculus . . . . . . . . . . . . . . . . . . . . .   3

        1.3.1   Definitions . . . . . . . . . . . . . . . . . . . . .   4

        1.3.2   Properties . . . . . . . . . . . . . . . . . . . . .   9

        1.3.3   Mittag-Leffler functions . . . . . . . . . . . . . .  12

    1.4   Some Results from Nonlinear Analysis . . . . . . . . . . . .  13

        1.4.1   Sobolev Spaces . . . . . . . . . . . . . . . . . . .  13

        1.4.2   Measure of Noncompactness . . . . . . . . . . . . .  14

        1.4.3   Topological Degree . . . . . . . . . . . . . . . . .  15

        1.4.4   Picard Operator . . . . . . . . . . . . . . . . . .  17

        1.4.5   Fixed Point Theorems . . . . . . . . . . . . . . . .  18

        1.4.6   Critical Point Theorems . . . . . . . . . . . . . . .  20

    1.5   Semigroups . . . . . . . . . . . . . . . . . . . . . . . .  22

        1.5.1   $C_0$-semigroup . . . . . . . . . . . . . . . . . . .  22

        1.5.2   Almost Sectorial Operators . . . . . . . . . . . . .  23

2.  Fractional Functional Differential Equations                   27

    2.1   Introduction . . . . . . . . . . . . . . . . . . . . . . . .  27

    2.2   Neutral Equations with Bounded Delay . . . . . . . . . . . .  28

        2.2.1   Introduction . . . . . . . . . . . . . . . . . . . .  28

        2.2.2   Existence and Uniqueness . . . . . . . . . . . . . .  28

        2.2.3   Extremal Solutions . . . . . . . . . . . . . . . . .  33

    2.3   $p$-Type Neutral Equations . . . . . . . . . . . . . . . . .  42

        2.3.1   Introduction . . . . . . . . . . . . . . . . . . . .  42

        2.3.2   Existence and Uniqueness . . . . . . . . . . . . . .  44

|       | 2.3.3 | Continuous Dependence . . . . . . . . . . . . . . . . . | 55 |
| 2.4 | Neutral Equations with Infinite Delay . . . . . . . . . . . . | 58 |
|       | 2.4.1 | Introduction . . . . . . . . . . . . . . . . . . . . | 58 |
|       | 2.4.2 | Existence and Uniqueness . . . . . . . . . . . . . . . | 60 |
|       | 2.4.3 | Continuation of Solutions . . . . . . . . . . . . . . | 67 |
| 2.5 | Iterative Functional Differential Equations . . . . . . . . . | 71 |
|       | 2.5.1 | Introduction . . . . . . . . . . . . . . . . . . . . | 71 |
|       | 2.5.2 | Existence . . . . . . . . . . . . . . . . . . | 72 |
|       | 2.5.3 | Data Dependence . . . . . . . . . . . . . . . . . | 78 |
|       | 2.5.4 | Examples and General Cases . . . . . . . . . . . . . | 79 |
| 2.6 | Notes and Remarks . . . . . . . . . . . . . . . . . . . . | 86 |

3. **Fractional Ordinary Differential Equations in Banach Spaces** — 87

| 3.1 | Introduction . . . . . . . . . . . . . . . . . . . . . . . . | 87 |
| 3.2 | Cauchy Problems via Measure of Noncompactness Method . . . . | 89 |
|       | 3.2.1 | Introduction . . . . . . . . . . . . . . . . . . . . | 89 |
|       | 3.2.2 | Existence . . . . . . . . . . . . . . . . . . . . . . | 89 |
| 3.3 | Cauchy Problems via Topological Degree Method . . . . . . . . | 98 |
|       | 3.3.1 | Introduction . . . . . . . . . . . . . . . . . . . . | 98 |
|       | 3.3.2 | Qualitative Analysis . . . . . . . . . . . . . . . . | 98 |
| 3.4 | Cauchy Problems via Picard Operators Technique . . . . . . . | 102 |
|       | 3.4.1 | Introduction . . . . . . . . . . . . . . . . . . . . | 102 |
|       | 3.4.2 | Results via Picard Operators . . . . . . . . . . . . | 102 |
|       | 3.4.3 | Results via Weakly Picard Operators . . . . . . . . . | 109 |
| 3.5 | Notes and Remarks . . . . . . . . . . . . . . . . . . . . | 113 |

4. **Fractional Abstract Evolution Equations** — 115

| 4.1 | Introduction . . . . . . . . . . . . . . . . . . . . . . . . | 115 |
| 4.2 | Evolution Equations with Riemann-Liouville Derivative . . . . | 116 |
|       | 4.2.1 | Introduction . . . . . . . . . . . . . . . . . . . . | 116 |
|       | 4.2.2 | Definition of Mild Solutions . . . . . . . . . . . . | 117 |
|       | 4.2.3 | Preliminary Lemmas . . . . . . . . . . . . . . . . . | 120 |
|       | 4.2.4 | Compact Semigroup Case . . . . . . . . . . . . . . . | 126 |
|       | 4.2.5 | Noncompact Semigroup Case . . . . . . . . . . . . . . | 131 |
| 4.3 | Evolution Equations with Caputo Derivative . . . . . . . . . | 134 |
|       | 4.3.1 | Introduction . . . . . . . . . . . . . . . . . . . . | 134 |
|       | 4.3.2 | Definition of Mild Solutions . . . . . . . . . . . . | 134 |
|       | 4.3.3 | Preliminary Lemmas . . . . . . . . . . . . . . . . . | 136 |
|       | 4.3.4 | Compact Semigroup Case . . . . . . . . . . . . . . . | 140 |
|       | 4.3.5 | Noncompact Semigroup Case . . . . . . . . . . . . . . | 143 |
| 4.4 | Nonlocal Problems for Evolution Equations . . . . . . . . . | 145 |
|       | 4.4.1 | Introduction . . . . . . . . . . . . . . . . . . . . | 145 |

|       | 4.4.2 | Definition of mild solutions | 145 |
|       | 4.4.3 | Existence | 147 |
| 4.5 | | Abstract Cauchy Problems with Almost Sectorial Operators | 153 |
|       | 4.5.1 | Introduction | 153 |
|       | 4.5.2 | Properties of Operators | 158 |
|       | 4.5.3 | Linear Problems | 164 |
|       | 4.5.4 | Nonlinear Problems | 169 |
|       | 4.5.5 | Applications | 177 |
| 4.6 | | Notes and Remarks | 179 |

**5. Fractional Impulsive Differential Equations** — 181

| 5.1 | | Introduction | 181 |
| 5.2 | | Impulsive Initial Value Problems | 182 |
|       | 5.2.1 | Introduction | 182 |
|       | 5.2.2 | Formula of Solutions | 182 |
|       | 5.2.3 | Existence | 185 |
| 5.3 | | Impulsive Boundary Value Problems | 190 |
|       | 5.3.1 | Introduction | 190 |
|       | 5.3.2 | Formula of Solutions | 190 |
|       | 5.3.3 | Existence | 193 |
| 5.4 | | Impulsive Langevin Equations | 197 |
|       | 5.4.1 | Introduction | 197 |
|       | 5.4.2 | Formula of Solutions | 198 |
|       | 5.4.3 | Existence | 206 |
| 5.5 | | Impulsive Evolution Equations | 213 |
|       | 5.5.1 | Introduction | 213 |
|       | 5.5.2 | Cauchy Problems | 214 |
|       | 5.5.3 | Nonlocal Problems | 216 |
| 5.6 | | Notes and Remarks | 222 |

**6. Fractional Boundary Value Problems** — 223

| 6.1 | | Introduction | 223 |
| 6.2 | | Solution for BVP with Left and Right Fractional Integrals | 223 |
|       | 6.2.1 | Introduction | 223 |
|       | 6.2.2 | Fractional Derivative Space | 226 |
|       | 6.2.3 | Variational Structure | 231 |
|       | 6.2.4 | Existence Under Ambrosetti-Rabinowitz Condition | 238 |
|       | 6.2.5 | Superquadratic Case | 243 |
|       | 6.2.6 | Asymptotically Quadratic Case | 247 |
| 6.3 | | Multiple Solutions for BVP with Parameters | 250 |
|       | 6.3.1 | Introduction | 250 |
|       | 6.3.2 | Existence | 251 |

6.4 Infinite Solutions for BVP with Left and Right Fractional
    Integrals . . . . . . . . . . . . . . . . . . . . . . . . . . . . 261
    6.4.1 Introduction . . . . . . . . . . . . . . . . . . . . 261
    6.4.2 Existence . . . . . . . . . . . . . . . . . . . . . . 262
6.5 Solutions for BVP with Left and Right Fractional Derivatives . . . 271
    6.5.1 Introduction . . . . . . . . . . . . . . . . . . . . 271
    6.5.2 Variational Structure . . . . . . . . . . . . . . . 272
    6.5.3 Existence of Weak Solutions . . . . . . . . . . . 275
    6.5.4 Existence of Solutions . . . . . . . . . . . . . . 279
6.6 Notes and Remarks . . . . . . . . . . . . . . . . . . . . . . 283

7. Fractional Partial Differential Equations                    285

7.1 Introduction . . . . . . . . . . . . . . . . . . . . . . . . . 285
7.2 Fractional Navier-Stokes Equations . . . . . . . . . . . . . 285
    7.2.1 Introduction . . . . . . . . . . . . . . . . . . . . 285
    7.2.2 Preliminaries . . . . . . . . . . . . . . . . . . . . 287
    7.2.3 Global Existence . . . . . . . . . . . . . . . . . . 290
    7.2.4 Local Existence . . . . . . . . . . . . . . . . . . 297
    7.2.5 Regularity . . . . . . . . . . . . . . . . . . . . . 301
7.3 Fractional Euler-Lagrange Equations . . . . . . . . . . . . 309
    7.3.1 Introduction . . . . . . . . . . . . . . . . . . . . 309
    7.3.2 Functional Spaces . . . . . . . . . . . . . . . . . 311
    7.3.3 Variational Structure . . . . . . . . . . . . . . . 314
    7.3.4 Existence of Weak Solution . . . . . . . . . . . . 317
7.4 Fractional Diffusion Equations . . . . . . . . . . . . . . . 321
    7.4.1 Introduction . . . . . . . . . . . . . . . . . . . . 321
    7.4.2 Preliminaries . . . . . . . . . . . . . . . . . . . . 324
    7.4.3 Existence and Regularity . . . . . . . . . . . . . 327
7.5 Fractional Schrödinger Equations . . . . . . . . . . . . . . 336
    7.5.1 Introduction . . . . . . . . . . . . . . . . . . . . 336
    7.5.2 Preliminaries . . . . . . . . . . . . . . . . . . . . 337
    7.5.3 Existence and Uniqueness . . . . . . . . . . . . . 340
7.6 Notes and Remarks . . . . . . . . . . . . . . . . . . . . . . 342

*Bibliography*                                                   343

*Index*                                                          363

# Chapter 1

# Preliminaries

## 1.1 Introduction

In this chapter, we introduce some notations and basic facts on fractional calculus, nonlinear analysis and semigroup which are needed throughout this book.

## 1.2 Some Notations, Concepts and Lemmas

As usual $\mathbb{N}$ denotes the set of positive integer numbers and $\mathbb{N}_0$ the set of nonnegative integer numbers. $\mathbb{R}$ denotes the real numbers, $\mathbb{R}_+$ denotes the set of nonnegative reals and $\mathbb{R}^+$ the set of positive reals. Let $\mathbb{C}$ be the set of complex numbers.

We recall that a vector space $X$ equipped with a norm $|\cdot|$ is called a normed vector space. A subset $E$ of a normed vector space $X$ is said to be bounded if there exists a number $K$ such that $|x| \leq K$ for all $x \in E$. A subset $E$ of a normed vector space $X$ is called convex if for any $x, y \in E$, $ax + (1 - a)y \in E$ for all $a \in [0, 1]$.

A sequence $\{x_n\}$ in a normed vector space $X$ is said to converge to the vector $x$ in $X$ if and only if the sequence $\{|x_n - x|\}$ converges to zero as $n \to \infty$. A sequence $\{x_n\}$ in a normed vector space $X$ is called a Cauchy sequence if for every $\varepsilon > 0$ there exists an $N = N(\varepsilon)$ such that for all $n, m \geq N(\varepsilon)$, $|x_n - x_m| < \varepsilon$. Clearly a convergent sequence is also a Cauchy sequence, but the converse may not be true. A space $X$ where every Cauchy sequence of elements of $X$ converges to an element of $X$ is called a complete space. A complete normed vector space is said to be a Banach space.

Let $E$ be a subset of a Banach space $X$. A point $x \in X$ is said to be a limit point of $E$ if there exists a sequence of vectors in $E$ which converges to $x$. We say a subset $E$ is closed if $E$ contains all of its limit points. The union of $E$ and its limit points is called the closure of $E$ and will be denoted by $\bar{E}$. Let $E$, $F$ be normed vector spaces, and $E$ be a subset of $X$. An operator $\mathscr{T} : E \to F$ is continuous at a point $x \in E$ if and only if for any $\varepsilon > 0$ there is a $\delta > 0$ such that $|\mathscr{T}x - \mathscr{T}y| < \varepsilon$ for all $y \in E$ with $|x - y| < \delta$. Further, $\mathscr{T}$ is continuous on $E$, or simply continuous, if it is continuous at all points of $E$.

We say that a subset $E$ of a Banach space $X$ is compact if every sequence of vectors in $E$ contains a subsequence which converges to a vector in $E$. We say that $E$ is relatively compact in $X$ if every sequence of vectors in $E$ contains a subsequence which converges to a vector in $X$, i.e., $E$ is relatively compact in $X$ if $\bar{E}$ is compact.

Let $J = [a, b]$ $(-\infty < a < b < \infty)$ be a finite interval of $\mathbb{R}$. We assume that $X$ is a Banach space with the norm $|\cdot|$. Denote $C(J, X)$ be the Banach space of all continuous functions from $J$ into $X$ with the norm

$$\|x\| = \sup_{t \in J} |x(t)|,$$

where $x \in C(J, X)$. $C^n(J, X)$ $(n \in \mathbb{N}_0)$ denotes the set of mappings having $n$ times continuously differentiable on $J$, $AC(J, X)$ is the space of functions which are absolutely continuous on $J$ and $AC^n(J, X)$ $(n \in \mathbb{N}_0)$ is the space of functions $f$ such that $f \in C^{n-1}(J, X)$ and $f^{(n-1)} \in AC(J, X)$. In particular, $AC^1(J, X) = AC(J, X)$. We also introduce the set of functions $PC(J, X) = \{x : J \to X \mid x$ is continuous at $t \in J \backslash \{t_1, t_2, \ldots, t_\delta\}$, and $x$ is continuous from left and has right hand limits at $t \in \{t_1, t_2, \ldots, t_\delta\}\}$ endowed with the norm

$$\|x\|_{PC} = \max \left\{ \sup_{t \in J} |x(t+0)|, \ \sup_{t \in J} |x(t-0)| \right\},$$

it is easy to see $(PC(J, X), \|\cdot\|_{PC})$ is a Banach space. Denote $PC^1(J, \mathbb{R}) \equiv \{x \in PC(J, \mathbb{R}) \mid \dot{x} \in PC(J, \mathbb{R})\}$. Set $\|x\|_{PC^1} = \|x\|_{PC} + \|\dot{x}\|_{PC}$. It can be seen that endowed with the norm $\|\cdot\|_{PC^1}$, $PC^1(J, \mathbb{R})$ is also a Banach space.

Let $1 \leq p \leq \infty$. $L^p(J, X)$ denotes the Banach space of all measurable functions $f : J \to X$. $L^p(J, X)$ is normed by

$$\|f\|_{L^p J} = \begin{cases} \left( \int_J |f(t)|^p dt \right)^{\frac{1}{p}}, & 1 \leq p < \infty, \\[2em] \inf_{\mu(\bar{J})=0} \left\{ \sup_{t \in J \backslash \bar{J}} |f(t)| \right\}, & p = \infty. \end{cases}$$

In particular, $L^1(J, X)$ is the Banach space of measurable functions $f : J \to X$ with the norm

$$\|f\|_{LJ} = \int_J |f(t)| dt,$$

and $L^\infty(J, X)$ is the Banach space of measurable functions $f : J \to X$ which are bounded, equipped with the norm

$$\|f\|_{L^\infty J} = \inf\{c > 0 \mid |f(t)| \leq c, \text{ a.e. } t \in J\}.$$

**Lemma 1.1.** *(Hölder inequality) Assume that $p, q \geq 1$, and $\frac{1}{p} + \frac{1}{q} = 1$. If $f \in L^p(J, X), g \in L^q(J, X)$, then for $1 \leq p \leq \infty$, $fg \in L^1(J, X)$ and*

$$\|fg\|_{LJ} \leq \|f\|_{L^p J} \|g\|_{L^q J}.$$

A family $F$ in $C(J, X)$ is called uniformly bounded if there exists a positive constant $K$ such that $|f(t)| \leq K$ for all $t \in J$ and all $f \in F$. Further, $F$ is called equicontinuous, if for every $\varepsilon > 0$ there exists a $\delta = \delta(\varepsilon) > 0$ such that $|f(t_1) - f(t_2)| < \varepsilon$ for all $t_1, t_2 \in J$ with $|t_1 - t_2| < \delta$ and all $f \in F$.

**Lemma 1.2.** *(Arzela-Ascoli theorem) If a family $F = \{f(t)\}$ in $C(J, \mathbb{R})$ is uniformly bounded and equicontinuous on $J$, then $F$ has a uniformly convergent subsequence $\{f_n(t)\}_{n=1}^{\infty}$. If a family $F = \{f(t)\}$ in $C(J, X)$ is uniformly bounded and equicontinuous on $J$, and for any $t^* \in J$, $\{f(t^*)\}$ is relatively compact, then $F$ has a uniformly convergent subsequence $\{f_n(t)\}_{n=1}^{\infty}$.*

Arzela-Ascoli theorem is the key to the following result: A subset $F$ in $C(J, \mathbb{R})$ is relatively compact if and only if it is uniformly bounded and equicontinuous on $J$.

**Lemma 1.3.** *(PC-type Arzela-Ascoli theorem) Let $X$ be a Banach space and $\mathcal{W} \subset PC(J, X)$. If the following conditions are satisfied:*

(i) *$\mathcal{W}$ is equicontinuous function subset of $PC(J, X)$;*
(ii) *$\mathcal{W}$ is equicontinuous in $(t_k, t_{k+1})$, $k = 0, 1, 2, \ldots, m$, where $t_0 = 0$, $t_{m+1} = T$;*
(iii) *$\mathcal{W}(t) = \{u(t) \mid u \in \mathcal{W}, t \in J \backslash \{t_1, \ldots, t_m\}\}$, $\mathcal{W}(t_k^+) = \{u(t_k^+) \mid u \in \mathcal{W}\}$ and $\mathcal{W}(t_k^-) \equiv \{u(t_k^-) \mid u \in \mathcal{W}\}$ is a relatively compact subsets of $X$.*

*Then $\mathcal{W}$ is a relatively compact subset of $PC(J, X)$.*

**Lemma 1.4.** *(Lebesgue dominated convergence theorem) Let $E$ be a measurable set and let $\{f_n\}$ be a sequence of measurable functions such that $\lim_{n \to \infty} f_n(x) = f(x)$ a.e. in $E$, and for every $n \in \mathbb{N}$, $|f_n(x)| \leq g(x)$ a.e. in $E$, where $g$ is integrable on $E$. Then*

$$\lim_{n \to \infty} \int_E f_n(x) dx = \int_E f(x) dx.$$

Finally, we state Bochner theorem.

**Lemma 1.5.** *(Bochner theorem) A measurable function $f : (a, b) \to X$ is Bochner integrable if $|f|$ is Lebesgue integrable.*

## 1.3 Fractional Calculus

The gamma function $\Gamma(z)$ is defined by

$$\Gamma(z) = \int_0^{\infty} t^{z-1} e^{-t} dt \quad (Re(z) > 0),$$

where $t^{z-1} = e^{(z-1)\log(t)}$. This integral is convergent for all complex $z \in \mathbb{C}$ $(Re(z) > 0)$.

For this function the reduction formula

$$\Gamma(z + 1) = z\Gamma(z) \quad (Re(z) > 0)$$

holds. In particular, if $z = n \in \mathbb{N}_0$, then

$$\Gamma(n+1) = n! \quad (n \in \mathbb{N}_0)$$

with (as usual) $0! = 1$.

Let us consider some of the starting points for a discussion of fractional calculus. One development begins with a generalization of repeated integration. Thus if $f$ is locally integrable on $(c, \infty)$, then the $n$-fold iterated integral is given by

$$_cD_t^{-n}f(t) = \int_c^t ds_1 \int_c^{s_1} ds_2 \cdots \int_c^{s_{n-1}} f(s_n)ds_n$$

$$= \frac{1}{(n-1)!} \int_c^t (t-s)^{n-1} f(s)ds$$

for almost all $t$ with $-\infty \leq c < t < \infty$ and $n \in \mathbb{N}$. Writing $(n-1)! = \Gamma(n)$, an immediate generalization is the integral of $f$ of fractional order $\alpha > 0$,

$$_cD_t^{-\alpha}f(t) = \frac{1}{\Gamma(\alpha)} \int_c^t (t-s)^{\alpha-1} f(s)ds \quad \text{(left hand)}$$

and similarly for $-\infty < t < d \leq \infty$

$$_tD_d^{-\alpha}f(t) = \frac{1}{\Gamma(\alpha)} \int_t^d (s-t)^{\alpha-1} f(s)ds \quad \text{(right hand)}$$

both being defined for suitable $f$.

A number of definitions for the fractional derivative has emerged over the years, we refer the reader to Diethelm, 2010; Hilfer, 2006; Kilbas, Srivastava and Trujillo, 2006; Miller and Ross, 1993; Podlubny, 1999. In this book, we restrict our attention to the use of the Riemann-Liouville and Caputo fractional derivatives. In this section, we introduce some basic definitions and properties of the fractional integrals and fractional derivatives which are used further in this book. The materials in this section are taken from Kilbas, Srivastava and Trujillo, 2006.

### 1.3.1  Definitions

**Definition 1.1.** (Left and right Riemann-Liouville fractional integrals) Let $J = [a, b]$ $(-\infty < a < b < \infty)$ be a finite interval of $\mathbb{R}$. The left and right Riemann-Liouville fractional integrals $_aD_t^{-\alpha}f(t)$ and $_tD_b^{-\alpha}f(t)$ of order $\alpha \in \mathbb{R}^+$, are defined by

$$_aD_t^{-\alpha}f(t) = \frac{1}{\Gamma(\alpha)} \int_a^t (t-s)^{\alpha-1} f(s)ds, \quad t > a, \quad \alpha > 0 \tag{1.1}$$

and

$$_tD_b^{-\alpha}f(t) = \frac{1}{\Gamma(\alpha)} \int_t^b (s-t)^{\alpha-1} f(s)ds, \quad t < b, \quad \alpha > 0, \tag{1.2}$$

respectively, provided the right-hand sides are pointwise defined on $[a, b]$. When $\alpha = n \in \mathbb{N}$, the definitions (1.1) and (1.2) coincide with the $n$-th integrals of the form

$$_aD_t^{-n}f(t) = \frac{1}{(n-1)!} \int_a^t (t-s)^{n-1} f(s)ds$$

and

$$_tD_b^{-n}f(t) = \frac{1}{(n-1)!} \int_t^b (s-t)^{n-1} f(s)ds.$$

**Definition 1.2.** (Left and right Riemann-Liouville fractional derivatives) The left and right Riemann-Liouville fractional derivatives $_aD_t^\alpha f(t)$ and $_tD_b^\alpha f(t)$ of order $\alpha \in \mathbb{R}_+$, are defined by

$$_aD_t^\alpha f(t) = \frac{d^n}{dt^n} {_aD_t^{-(n-\alpha)}} f(t)$$

$$= \frac{1}{\Gamma(n-\alpha)} \frac{d^n}{dt^n} \left( \int_a^t (t-s)^{n-\alpha-1} f(s)ds \right), \quad t > a$$

and

$$_tD_b^\alpha f(t) = (-1)^n \frac{d^n}{dt^n} {_tD_b^{-(n-\alpha)}} f(t)$$

$$= \frac{1}{\Gamma(n-\alpha)} (-1)^n \frac{d^n}{dt^n} \left( \int_t^b (s-t)^{n-\alpha-1} f(s)ds \right), \quad t < b,$$

respectively, where $n = [\alpha] + 1$, $[\alpha]$ means the integer part of $\alpha$. In particular, when $\alpha = n \in \mathbb{N}_0$, then

$$_aD_t^0 f(t) = {_tD_b^0} f(t) = f(t),$$

$$_aD_t^n f(t) = f^{(n)}(t) \quad \text{and} \quad _tD_b^n f(t) = (-1)^n f^{(n)}(t),$$

where $f^{(n)}(t)$ is the usual derivative of $f(t)$ of order $n$. If $0 < \alpha < 1$, then

$$_aD_t^\alpha f(t) = \frac{1}{\Gamma(1-\alpha)} \frac{d}{dt} \left( \int_a^t (t-s)^{-\alpha} f(s)ds \right), \quad t > a$$

and

$$_tD_b^\alpha f(t) = -\frac{1}{\Gamma(1-\alpha)} \frac{d}{dt} \left( \int_t^b (s-t)^{-\alpha} f(s)ds \right), \quad t < b.$$

**Remark 1.1.** If $f \in C([a,b], \mathbb{R}^N)$, it is obvious that Riemann-Liouville fractional integral of order $\alpha > 0$ exists on $[a, b]$. On the other hand, following Lemma 2.2 in Kilbas, Srivastava and Trujillo, 2006, we know that the Riemann-Liouville fractional derivative of order $\alpha \in [n-1, n)$ exists almost everywhere on $[a, b]$ if $f \in AC^n([a,b], \mathbb{R}^N)$.

The left and right Caputo fractional derivatives are defined via above Riemann-Liouville fractional derivatives.

**Definition 1.3.** (Left and right Caputo fractional derivatives) The left and right Caputo fractional derivatives ${}^{C}_{a}D^{\alpha}_{t}f(t)$ and ${}^{C}_{t}D^{\alpha}_{b}f(t)$ of order $\alpha \in \mathbb{R}_{+}$ are defined by

$$
{}^{C}_{a}D^{\alpha}_{t}f(t) = {}_{a}D^{\alpha}_{t}\left( f(t) - \sum_{k=0}^{n-1} \frac{f^{(k)}(a)}{k!}(t-a)^{k} \right)
$$

and

$$
{}^{C}_{t}D^{\alpha}_{b}f(t) = {}_{t}D^{\alpha}_{b}\left( f(t) - \sum_{k=0}^{n-1} \frac{f^{(k)}(b)}{k!}(b-t)^{k} \right),
$$

respectively, where

$$
n = [\alpha] + 1 \text{ for } \alpha \notin \mathbb{N}_{0}; \quad n = \alpha \text{ for } \alpha \in \mathbb{N}_{0}. \tag{1.3}
$$

In particular, when $0 < \alpha < 1$, then

$$
{}^{C}_{a}D^{\alpha}_{t}f(t) = {}_{a}D^{\alpha}_{t}(f(t) - f(a))
$$

and

$$
{}^{C}_{t}D^{\alpha}_{b}f(t) = {}_{t}D^{\alpha}_{b}(f(t) - f(b)).
$$

The Riemann-Liouville fractional derivative and the Caputo fractional derivative are connected with each other by the following relations.

**Proposition 1.1.**

(i) *If $\alpha \notin \mathbb{N}_{0}$ and $f(t)$ is a function for which the Caputo fractional derivatives ${}^{C}_{a}D^{\alpha}_{t}f(t)$ and ${}^{C}_{t}D^{\alpha}_{b}f(t)$ of order $\alpha \in \mathbb{R}^{+}$ exist together with the Riemann-Liouville fractional derivatives ${}_{a}D^{\alpha}_{t}f(t)$ and ${}_{t}D^{\alpha}_{b}f(t)$, then*

$$
{}^{C}_{a}D^{\alpha}_{t}f(t) = {}_{a}D^{\alpha}_{t}f(t) - \sum_{k=0}^{n-1} \frac{f^{(k)}(a)}{\Gamma(k-\alpha+1)}(t-a)^{k-\alpha}
$$

*and*

$$
{}^{C}_{t}D^{\alpha}_{b}f(t) = {}_{t}D^{\alpha}_{b}f(t) - \sum_{k=0}^{n-1} \frac{f^{(k)}(b)}{\Gamma(k-\alpha+1)}(b-t)^{k-\alpha},
$$

*where $n = [\alpha] + 1$. In particular, when $0 < \alpha < 1$, we have*

$$
{}^{C}_{a}D^{\alpha}_{t}f(t) = {}_{a}D^{\alpha}_{t}f(t) - \frac{f(a)}{\Gamma(1-\alpha)}(t-a)^{-\alpha}
$$

*and*

$$
{}^{C}_{t}D^{\alpha}_{b}f(t) = {}_{t}D^{\alpha}_{b}f(t) - \frac{f(b)}{\Gamma(1-\alpha)}(b-t)^{-\alpha}.
$$

(ii) *If $\alpha = n \in \mathbb{N}_0$ and the usual derivative $f^{(n)}(t)$ of order $n$ exists, then $_a^C D_t^n f(t)$ and $_t^C D_b^n f(t)$ are represented by*

$$_a^C D_t^n f(t) = f^{(n)}(t) \quad and \quad _t^C D_b^n f(t) = (-1)^n f^{(n)}(t). \tag{1.4}$$

**Proposition 1.2.** *Let $\alpha \in \mathbb{R}_+$ and let $n$ be given by (1.3). If $f \in AC^n([a,b], \mathbb{R}^N)$, then the Caputo fractional derivatives $_a^C D_t^\alpha f(t)$ and $_t^C D_b^\alpha f(t)$ exist almost everywhere on $[a,b]$.*

(i) *If $\alpha \notin \mathbb{N}_0$, $_a^C D_t^\alpha f(t)$ and $_t^C D_b^\alpha f(t)$ are represented by*

$$_a^C D_t^\alpha f(t) = \frac{1}{\Gamma(n-\alpha)} \left( \int_a^t (t-s)^{n-\alpha-1} f^{(n)}(s) ds \right)$$

*and*

$$_t^C D_b^\alpha f(t) = \frac{(-1)^n}{\Gamma(n-\alpha)} \left( \int_t^b (s-t)^{n-\alpha-1} f^{(n)}(s) ds \right),$$

*respectively, where $n = [\alpha] + 1$. In particular, when $0 < \alpha < 1$ and $f \in AC([a,b], \mathbb{R}^N)$,*

$$_a^C D_t^\alpha f(t) = \frac{1}{\Gamma(1-\alpha)} \left( \int_a^t (t-s)^{-\alpha} f'(s) ds \right) \tag{1.5}$$

*and*

$$_t^C D_b^\alpha f(t) = -\frac{1}{\Gamma(1-\alpha)} \left( \int_t^b (s-t)^{-\alpha} f'(s) ds \right). \tag{1.6}$$

(ii) *If $\alpha = n \in \mathbb{N}_0$ then $_a^C D_t^\alpha f(t)$ and $_t^C D_b^\alpha f(t)$ are represented by (1.4). In particular,*

$$_a^C D_t^0 f(t) = _t^C D_b^0 f(t) = f(t).$$

**Remark 1.2.** If $f$ is an abstract function with values in Banach space $X$, then integrals which appear in above definitions are taken in Bochner's sense.

The fractional integrals and derivatives, defined on a finite interval $[a,b]$ of $\mathbb{R}$, are naturally extended to whole axis $\mathbb{R}$.

**Definition 1.4.** (Left and right Liouville-Weyl fractional integrals on the real axis) The left and right Liouville-Weyl fractional integrals $_{-\infty}D_t^{-\alpha} f(t)$ and $_t D_{+\infty}^{-\alpha} f(t)$ of order $\alpha > 0$ on the whole axis $\mathbb{R}$ are defined by

$$_{-\infty}D_t^{-\alpha} f(t) = \frac{1}{\Gamma(\alpha)} \int_{-\infty}^t (t-s)^{\alpha-1} f(s) ds \tag{1.7}$$

*and*

$$_t D_{+\infty}^{-\alpha} f(t) = \frac{1}{\Gamma(\alpha)} \int_t^\infty (s-t)^{\alpha-1} f(s) ds,$$

respectively, where $t \in \mathbb{R}$ and $\alpha > 0$.

**Definition 1.5.** (Left and right Liouville-Weyl fractional derivatives on the real axis) The left and right Liouville-Weyl fractional derivatives $_{-\infty}D_t^{\alpha}f(t)$ and $_tD_{+\infty}^{\alpha}f(t)$ of order $\alpha$ on the whole axis $\mathbb{R}$ are defined by

$$_{-\infty}D_t^{\alpha}f(t) = \frac{d^n}{dt^n}(_{-\infty}D_t^{-(n-\alpha)}f(t))$$

$$= \frac{1}{\Gamma(n-\alpha)}\frac{d^n}{dt^n}\left(\int_{-\infty}^{t}(t-s)^{n-\alpha-1}f(s)ds\right)$$

and

$$_tD_{+\infty}^{\alpha}f(t) = (-1)^n\frac{d^n}{dt^n}(_tD_{+\infty}^{-(n-\alpha)}f(t))$$

$$= \frac{1}{\Gamma(n-\alpha)}(-1)^n\frac{d^n}{dt^n}\left(\int_{t}^{\infty}(s-t)^{n-\alpha-1}f(s)ds\right),$$

respectively, where $n = [\alpha]+1$, $\alpha \geq 0$ and $t \in \mathbb{R}$.

In particular, when $\alpha = n \in \mathbb{N}_0$, then

$$_{-\infty}D_t^0 f(t) = {}_tD_{+\infty}^0 f(t) = f(t),$$

$$_{-\infty}D_t^n f(t) = f^{(n)}(t) \quad \text{and} \quad {}_tD_{+\infty}^n f(t) = (-1)^n f^{(n)}(t),$$

where $f^{(n)}(t)$ is the usual derivative of $f(t)$ of order $n$. If $0 < \alpha < 1$ and $t \in \mathbb{R}$, then

$$_{-\infty}D_t^{\alpha}f(t) = \frac{1}{\Gamma(1-\alpha)}\frac{d}{dt}\left(\int_{-\infty}^{t}(t-s)^{-\alpha}f(s)ds\right)$$

$$= \frac{\alpha}{\Gamma(1-\alpha)}\int_{0}^{\infty}\frac{f(t)-f(t-s)}{s^{\alpha+1}}ds$$

and

$$_tD_{+\infty}^{\alpha}f(t) = -\frac{1}{\Gamma(1-\alpha)}\frac{d}{dt}\left(\int_{t}^{\infty}(s-t)^{-\alpha}f(s)ds\right)$$

$$= \frac{\alpha}{\Gamma(1-\alpha)}\int_{0}^{\infty}\frac{f(t)-f(t+s)}{s^{\alpha+1}}ds.$$

Formulas (1.5) and (1.6) can be used for the definition of the Caputo fractional derivatives on the whole axis $\mathbb{R}$.

**Definition 1.6.** (Left and right Caputo fractional derivatives on the real axis) The left and right Caputo fractional derivatives $_{-\infty}^C\mathbf{D}_t^{\alpha}f(t)$ and $_t^C\mathbf{D}_{+\infty}^{\alpha}f(t)$ of order $\alpha$ (with $\alpha > 0$ and $\alpha \notin \mathbb{N}$) on the whole axis $\mathbb{R}$ are defined by

$$_{-\infty}^C\mathbf{D}_t^{\alpha}f(t) = \frac{1}{\Gamma(n-\alpha)}\left(\int_{-\infty}^{t}(t-s)^{n-\alpha-1}f^{(n)}(s)ds\right) \tag{1.8}$$

and

$$_t^C\mathbf{D}_{+\infty}^{\alpha}f(t) = \frac{(-1)^n}{\Gamma(n-\alpha)}\left(\int_{t}^{\infty}(s-t)^{n-\alpha-1}f^{(n)}(s)ds\right), \tag{1.9}$$

respectively.

When $0 < \alpha < 1$, the relations (1.8) and (1.9) take the following forms

$$_{-\infty}^{C}\mathbf{D}_t^\alpha f(t) = \frac{1}{\Gamma(1-\alpha)}\left(\int_{-\infty}^{t}(t-s)^{-\alpha}f'(s)ds\right)$$

and

$$_t^C\mathbf{D}_{+\infty}^\alpha f(t) = -\frac{1}{\Gamma(1-\alpha)}\left(\int_t^\infty(s-t)^{-\alpha}f'(s)ds\right).$$

Now we present the Fourier transform properties of the fractional integral and fractional derivative operators.

**Definition 1.7.** The Fourier transform of a function $f(t)$ of real variable $t \in \mathbb{R}$ is defined by

$$\hat{f}(w) = \int_{-\infty}^\infty e^{-it\cdot w}f(t)dt \quad (w \in \mathbb{R}).$$

Let $f(t)$ be defined on $(-\infty, \infty)$ and $0 < \alpha < 1$. Then Fourier transform of Liouville-Weyl fractional integral and fractional differential operator satisfies

$$\widehat{_{-\infty}D_t^{-\alpha}f(t)}(w) = (iw)^{-\alpha}\hat{f}(w),$$

$$\widehat{_tD_\infty^{-\alpha}f(t)}(w) = (-iw)^{-\alpha}\hat{f}(w),$$

$$\widehat{_{-\infty}D_t^{\alpha}f(t)}(w) = (iw)^{\alpha}\hat{f}(w),$$

$$\widehat{_tD_\infty^{\alpha}f(t)}(w) = (-iw)^{\alpha}\hat{f}(w).$$

### 1.3.2 *Properties*

We present here some properties of the fractional integral and fractional derivative operators that will be useful throughout this book.

**Proposition 1.3.** *If $\alpha \geq 0$ and $\beta > 0$, then*

$$_aD_t^{-\alpha}(t-a)^{\beta-1} = \frac{\Gamma(\beta)}{\Gamma(\beta+\alpha)}(t-a)^{\beta+\alpha-1} \quad (\alpha > 0),$$

$$_aD_t^{\alpha}(t-a)^{\beta-1} = \frac{\Gamma(\beta)}{\Gamma(\beta-\alpha)}(t-a)^{\beta-\alpha-1} \quad (\alpha \geq 0)$$

*and*

$$_tD_b^{-\alpha}(b-t)^{\beta-1} = \frac{\Gamma(\beta)}{\Gamma(\beta+\alpha)}(b-t)^{\beta+\alpha-1} \quad (\alpha > 0),$$

$$_tD_b^{\alpha}(b-t)^{\beta-1} = \frac{\Gamma(\beta)}{\Gamma(\beta-\alpha)}(b-t)^{\beta-\alpha-1} \quad (\alpha \geq 0).$$

*In particular, if $\beta = 1$ and $\alpha \geq 0$, then the Riemann-Liouville fractional derivatives of a constant are, in general, not equal to zero:*

$$_aD_t^{\alpha}1 = \frac{(t-a)^{-\alpha}}{\Gamma(1-\alpha)}, \quad _tD_b^{\alpha}1 = \frac{(b-t)^{-\alpha}}{\Gamma(1-\alpha)}.$$

*On the other hand, for $j = 1, 2, \ldots, [\alpha] + 1$,*

$$_aD_t^{\alpha}(t-a)^{\alpha-j} = 0, \quad _tD_b^{\alpha}(b-t)^{\alpha-j} = 0.$$

The semigroup property of the fractional integral operators $_aD_t^{-\alpha}$ and $_tD_b^{-\alpha}$ are given by the following result.

**Proposition 1.4.** *If $\alpha > 0$ and $\beta > 0$, then the equations*

$$_aD_t^{-\alpha}(_aD_t^{-\beta}f(t)) = {_aD_t^{-\alpha-\beta}}f(t) \quad and \quad {_tD_b^{-\alpha}}(_tD_b^{-\beta}f(t)) = {_tD_b^{-\alpha-\beta}}f(t) \quad (1.10)$$

*are satisfied at almost every point $t \in [a, b]$ for $f \in L^p([a, b], \mathbb{R}^N)$ $(1 \le p < \infty)$. If $\alpha + \beta > 1$, then the relations in (1.10) hold at any point of $[a, b]$.*

**Proposition 1.5.**

(i) *If $\alpha > 0$ and $f \in L^p([a, b], \mathbb{R}^N)$ $(1 \le p \le \infty)$, then the following equalities*

$$_aD_t^{\alpha}(_aD_t^{-\alpha}f(t)) = f(t) \quad and \quad {_tD_b^{\alpha}}(_tD_b^{-\alpha}f(t)) = f(t) \quad (\alpha > 0)$$

  *hold almost everywhere on $[a, b]$.*

(ii) *If $\alpha > \beta > 0$, then, for $f \in L^p([a, b], \mathbb{R}^N)$ $(1 \le p \le \infty)$, the relations*

$$_aD_t^{\beta}(_aD_t^{-\alpha}f(t)) = {_aD_t^{-\alpha+\beta}}f(t) \quad and \quad {_tD_b^{\beta}}(_tD_b^{-\alpha}f(t)) = {_tD_b^{-\alpha+\beta}}f(t)$$

  *hold almost everywhere on $[a, b]$.*

  *In particular, when $\beta = k \in \mathbb{N}$ and $\alpha > k$, then*

$$_aD_t^{k}(_aD_t^{-\alpha}f(t)) = {_aD_t^{-\alpha+k}}f(t) \quad and \quad {_tD_b^{k}}(_tD_b^{-\alpha}f(t)) = (-1)^k {_tD_b^{-\alpha+k}}f(t).$$

To present the next property, we use the spaces of functions $_aD_t^{-\alpha}(L^p)$ and $_tD_b^{-\alpha}(L^p)$ defined for $\alpha > 0$ and $1 \le p \le \infty$ by

$$_aD_t^{-\alpha}(L^p) = \{f : f = {_aD_t^{-\alpha}}\varphi, \ \varphi \in L^p([a, b], \mathbb{R}^N)\}$$

and

$$_tD_b^{-\alpha}(L^p) = \{f : f = {_tD_b^{-\alpha}}\phi, \ \phi \in L^p([a, b], \mathbb{R}^N)\},$$

respectively. The composition of the fractional integral operator $_aD_t^{-\alpha}$ with the fractional derivative operator $_aD_t^{\alpha}$ is given by the following result.

**Proposition 1.6.** *Let $\alpha > 0$, $n = [\alpha] + 1$ and let $f_{n-\alpha}(t) = {_aD_t^{-(n-\alpha)}}f(t)$ be the fractional integral (1.1) of order $n - \alpha$.*

(i) *If $1 \le p \le \infty$ and $f \in {_aD_t^{-\alpha}}(L^p)$, then*

$$_aD_t^{-\alpha}(_aD_t^{\alpha}f(t)) = f(t).$$

(ii) *If $f \in L^1([a, b], \mathbb{R}^N)$ and $f_{n-\alpha} \in AC^n([a, b], \mathbb{R}^N)$, then the equality*

$$_aD_t^{-\alpha}(_aD_t^{\alpha}f(t)) = f(t) - \sum_{j=1}^{n} \frac{f_{n-\alpha}^{(n-j)}(a)}{\Gamma(\alpha - j + 1)}(t - a)^{\alpha-j},$$

  *holds almost everywhere on $[a, b]$.*

**Proposition 1.7.** *Let $\alpha > 0$ and $n = [\alpha] + 1$. Also let $g_{n-\alpha}(t) = {}_tD_b^{-(n-\alpha)}g(t)$ be the fractional integral (1.2) of order $n - \alpha$.*

**(i)** *If $1 \leq p \leq \infty$ and $g \in {}_tD_b^{-\alpha}(L^p)$, then*

$$
{}_tD_b^{-\alpha}({}_tD_b^{\alpha}g(t)) = g(t).
$$

**(ii)** *If $g \in L^1([a,b], \mathbb{R}^N)$ and $g_{n-\alpha} \in AC^n([a,b], \mathbb{R}^N)$, then the equality*

$$
{}_tD_b^{-\alpha}({}_tD_b^{\alpha}g(t)) = g(t) - \sum_{j=1}^{n} \frac{(-1)^{n-j}g_{n-\alpha}^{(n-j)}(a)}{\Gamma(\alpha - j + 1)}(b-t)^{\alpha-j},
$$

*holds almost everywhere on $[a,b]$.*

*In particular, if $0 < \alpha < 1$, then*

$$
{}_tD_b^{-\alpha}({}_tD_b^{\alpha}g(t)) = g(t) - \frac{g_{1-\alpha}(a)}{\Gamma(\alpha)}(b-t)^{\alpha-1},
$$

*where $g_{1-\alpha}(t) = {}_tD_b^{\alpha-1}g(t)$ while for $\alpha = n \in \mathbb{N}$, the following equality holds:*

$$
{}_tD_b^{-n}({}_tD_b^{n}g(t)) = g(t) - \sum_{k=0}^{n-1} \frac{(-1)^k g^{(k)}(a)}{k!}(b-t)^k.
$$

**Proposition 1.8.** *Let $\alpha > 0$ and let $y \in L^\infty([a,b], \mathbb{R}^N)$ or $y \in C([a,b], \mathbb{R}^N)$. Then*

$$
{}_a^C D_t^{\alpha}({}_aD_t^{-\alpha}y(t)) = y(t) \quad and \quad {}_t^C D_b^{\alpha}({}_tD_b^{-\alpha}y(t)) = y(t).
$$

**Proposition 1.9.** *Let $\alpha > 0$ and let $n$ be given by (1.3). If $y \in AC^n([a,b], \mathbb{R}^N)$ or $y \in C^n([a,b], \mathbb{R}^N)$, then*

$$
{}_aD_t^{-\alpha}({}_a^C D_t^{\alpha}y(t)) = y(t) - \sum_{k=0}^{n-1} \frac{y^{(k)}(a)}{k!}(t-a)^k
$$

*and*

$$
{}_tD_b^{-\alpha}({}_t^C D_b^{\alpha}y(t)) = y(t) - \sum_{k=0}^{n-1} \frac{(-1)^k y^{(k)}(b)}{k!}(b-t)^k.
$$

*In particular, if $0 < \alpha \leq 1$ and $y \in AC([a,b], \mathbb{R}^N)$ or $y \in C([a,b], \mathbb{R}^N)$, then*

$$
{}_aD_t^{-\alpha}({}_a^C D_t^{\alpha}y(t)) = y(t) - y(a) \quad and \quad {}_tD_b^{-\alpha}({}_t^C D_b^{\alpha}y(t)) = y(t) - y(b). \tag{1.11}
$$

On the other hand, we have the following property of fractional integration.

**Proposition 1.10.** *Let $\alpha > 0$, $p \geq 1$, $q \geq 1$, and $\frac{1}{p} + \frac{1}{q} \leq 1 + \alpha$ ($p \neq 1$ and $q \neq 1$ in the case when $\frac{1}{p} + \frac{1}{q} = 1 + \alpha$).*

**(i)** *If $\varphi \in L^p([a,b], \mathbb{R}^N)$ and $\psi \in L^q([a,b], \mathbb{R}^N)$, then*

$$
\int_a^b \varphi(t) \, {}_aD_t^{-\alpha}\psi(t)dt = \int_a^b \psi(t) \, {}_tD_b^{-\alpha}\varphi(t)dt. \tag{1.12}
$$

**(ii)** *If $f \in {}_tD_b^{-\alpha}(L^p)$ and $g \in {}_aD_t^{-\alpha}(L^q)$, then*

$$\int_a^b f(t) \, {}_aD_t^\alpha g(t)dt = \int_a^b g(t) \, {}_tD_b^\alpha f(t)dt. \tag{1.13}$$

Then applying Proposition 1.1, we can derive the integration by parts formula for the left and right Riemann-Liouville fractional derivatives looks as follows.

**Proposition 1.11.**

$$\int_a^b {}_aD_t^\alpha f(t) \cdot g(t)dt = \int_a^b {}_tD_b^\alpha g(t) \cdot f(t)dt, \quad 0 < \alpha \le 1,$$

*provided the boundary conditions*

$$f(a) = f(b) = 0, \quad f' \in L^\infty([a,b], \mathbb{R}^N), \quad g \in L^1([a,b], \mathbb{R}^N),$$

*or*

$$g(a) = g(b) = 0, \quad g' \in L^\infty([a,b], \mathbb{R}^N), \quad f \in L^1([a,b], \mathbb{R}^N)$$

*are fulfilled.*

**Remark 1.3.** If $f$, $g$ are abstract functions with values in Banach space $X$, then integrals which appear in above properties are taken in Bochner's sense.

### 1.3.3 *Mittag-Leffler functions*

**Definition 1.8.** (Miller and Ross, 1993; Podlubny, 1999) The generalized Mittag-Leffler function $E_{\alpha,\beta}$ is defined by

$$E_{\alpha,\beta}(z) := \sum_{k=0}^\infty \frac{z^k}{\Gamma(\alpha k + \beta)} = \frac{1}{2\pi i} \int_\Upsilon \frac{\lambda^{\alpha-\beta} e^\lambda}{\lambda^\alpha - z} d\lambda, \quad \alpha, \beta > 0, z \in \mathbb{C},$$

where $\Upsilon$ is a contour which starts and ends as $-\infty$ and encircles the disc $|\lambda| \le |z|^{1/\alpha}$ counter-clockwise.

If $0 < \alpha < 1$, $\beta > 0$, then the asymptotic expansion of $E_{\alpha,\beta}$ as $z \to \infty$ is given by

$$E_{\alpha,\beta}(z) = \begin{cases} \dfrac{1}{\alpha} z^{(1-\beta)/\alpha} \exp(z^{1/\alpha}) + \varepsilon_{\alpha,\beta}(z), & |\arg z| \le \dfrac{1}{2}\alpha\pi, \\[2mm] \varepsilon_{\alpha,\beta}(z), & |\arg(-z)| < \left(1 - \dfrac{1}{2}\alpha\right)\pi, \end{cases} \tag{1.14}$$

where

$$\varepsilon_{\alpha,\beta}(z) = -\sum_{n=1}^{N-1} \frac{z^{-n}}{\Gamma(\beta - \alpha n)} + O\left((|z|^{-N}\right), \quad \text{as } z \to \infty.$$

For short, set

$$E_\alpha(z) := E_{\alpha,1}(z), \quad e_\alpha(z) := E_{\alpha,\alpha}(z).$$

Then Mittag-Leffler functions have the following properties.

**Proposition 1.12.** *For $\alpha \in (0,1)$ and $t \in \mathbb{R}$,*

(i) $E_\alpha(t)$, $e_\alpha(t) > 0$;

(ii) $(E_\alpha(t))' = \frac{1}{\alpha} e_\alpha(t)$;

(iii) $\lim_{t\to-\infty} E_\alpha(t) = \lim_{t\to-\infty} e_\alpha(t) = 0$;

(iv) $_0^C D_t^\alpha E(\omega t^\alpha) = \omega E(\omega t^\alpha)$, $_0 D_t^{\alpha-1}(t^{\alpha-1} e_\alpha(\omega t^\alpha)) = E_\alpha(\omega t^\alpha)$, $\omega \in \mathbb{C}$.

**Definition 1.9.** (Mainardi, Paraddisi and Forenflo, 2000) The Wright function $M_\alpha$ is defined by

$$M_\alpha(z) := \sum_{n=0}^\infty \frac{(-z)^n}{n!\,\Gamma(-\alpha n + 1 - \alpha)}$$

$$= \frac{1}{\pi} \sum_{n=1}^\infty \frac{(-z)^n}{(n-1)!} \Gamma(n\alpha) \sin(n\pi\alpha), \quad z \in \mathbb{C}$$

with $0 < \alpha < 1$.

For $-1 < r < \infty, \lambda > 0$, the following results hold.

**Proposition 1.13.**

**(W1)** $M_\alpha(t) \geq 0$, $t > 0$;

**(W2)** $\int_0^\infty \frac{\alpha}{t^{\alpha+1}} M_\alpha(\frac{1}{t^\alpha}) e^{-\lambda t} dt = e^{-\lambda^\alpha}$;

**(W3)** $\int_0^\infty M_\alpha(t) t^r dt = \frac{\Gamma(1+r)}{\Gamma(1+\alpha r)}$;

**(W4)** $\int_0^\infty M_\alpha(t) e^{-zt} dt = E_\alpha(-z)$, $z \in \mathbb{C}$;

**(W5)** $\int_0^\infty \alpha t M_\alpha(t) e^{-zt} dt = e_\alpha(-z)$, $z \in \mathbb{C}$.

## 1.4 Some Results from Nonlinear Analysis

### 1.4.1 Sobolev Spaces

We refer to Cazenave and Haraux, 1998, for the definitions and results given below.

Consider an open subset $\Omega$ of $\mathbb{R}^N$. $\mathcal{D}(\Omega)$ is the space of $C^\infty$(real-valued or complex valued) functions with compact support in $\Omega$ and $\mathcal{D}'(\Omega)$ is the space of distributions on $\Omega$. A distribution $T \in \mathcal{D}'(\Omega)$ is said to belong to $L^p(\Omega)(1 \leq p \leq \infty)$ if there exists a function $f \in L^p(\Omega)$ such that

$$\langle T, \varphi \rangle = \int_\Omega f(x)\varphi(x) dx,$$

for all $\varphi \in \mathcal{D}(\Omega)$. In that case, it is well known that $f$ is unique. Let $m \in \mathbb{N}$ and let $p \in [1, \infty]$. Define

$$W^{m,p}(\Omega) = \{f \in L^p(\Omega) |\ D^\alpha f \in L^p(\Omega) \text{ for all } \alpha \in \mathbb{N}^N \text{ such that } |\alpha| \leq m\}.$$

$W^{m,p}(\Omega)$ is a Banach space when equipped with the norm

$$\|f\|_{W^{m,p}(\Omega)} = \sum_{|\alpha| \le m} \|D^\alpha f\|_{L^p},$$

for all $f \in W^{m,p}(\Omega)$. For all $m$, $p$ as above, we denote by $W_0^{m,p}(\Omega)$ the closure of $\mathcal{D}(\Omega)$ in $W^{m,p}(\Omega)$. If $p = 2$, one sets $W^{m,2}(\Omega) = H^m(\Omega)$, $W_0^{m,2}(\Omega) = H_0^m(\Omega)$ and one equips $H^m(\Omega)$ with the following equivalent norm:

$$\|f\|_{H^m} = \left( \sum_{|\alpha| \le m} \|D^\alpha u\|_{L^2}^2 \right)^{\frac{1}{2}}.$$

Then $H^m(\Omega)$ is a Hilbert space with the scalar product

$$\langle u, v \rangle_{H^m} = \sum_{|\alpha| \le m} \int_\Omega D^\alpha u \cdot D^\alpha v dx.$$

If $\Omega$ is bounded, there exists a constant $C(\Omega)$ such that

$$\|u\|_{L^2} \le C(\Omega) \|\nabla u\|_{L^2},$$

for all $u \in H_0^1(\Omega)$ (this is Poincaré inequality). It may be more convenient to equip $H_0^1(\Omega)$ with the following scalar product

$$\langle u, v \rangle = \int_\Omega \nabla u \cdot \nabla v dx,$$

which defines an equivalent norm to $\| \cdot \|_{H^1}$ on the closed space $H_0^1(\Omega)$.

### 1.4.2   Measure of Noncompactness

We recall here some definitions and properties of measure of noncompactness.

Assume that $X$ is a Banach space with the norm $|\cdot|$. The measure of noncompactness $\alpha$ is said to be:

(i) *Monotone* if for all bounded subsets $B_1$, $B_2$ of $X$, $B_1 \subseteq B_2$ implies $\alpha(B_1) \le \alpha(B_2)$;

(ii) *Nonsingular* if $\alpha(\{x\} \cup B) = \alpha(B)$ for every $x \in X$ and every nonempty subset $B \subseteq X$;

(iii) *Regular* $\alpha(B) = 0$ if and only if $B$ is relatively compact in $X$.

One of the most important examples of measure of noncompactness is the Hausdorff measure of noncompactness $\alpha$ defined on each bounded subset $B$ of $X$ by

$$\alpha(B) = \inf \left\{ \varepsilon > 0 : \ B \subset \bigcup_{j=1}^m B_\varepsilon(x_j) \text{ where } x_j \in X \right\},$$

where $B_\varepsilon(x_j)$ is a ball of radius $\le \varepsilon$ centered at $x_j$, $j = 1, 2, \ldots, m$. Without confusion, Kuratowski measure of noncompactness $\alpha_1$ defined on each bounded subset $B$ of $X$ by

$$\alpha_1(B) = \inf \left\{ \varepsilon > 0 : \ B \subset \bigcup_{j=1}^m M_j \text{ and } \text{diam}(M_j) \le \varepsilon \right\},$$

where the diameter of $M_j$ is defined by $\mathrm{diam}(M_j) = \sup\{|x - y| : x, y \in M_j\}$, $j = 1, 2, \ldots, m$.

It is well known that Hausdorff measure of noncompactness $\alpha$ and Kuratowski measure of noncompactness $\alpha_1$ enjoy the above properties (i)-(iii) and other properties. We refer the reader to Banaś and Goebel, 1980; Deimling, 1985; Heinz, 1983; Lakshmikantham and Leela, 1969.

(iv) $\alpha(B_1 + B_2) \le \alpha(B_1) + \alpha(B_2)$, where $B_1 + B_2 = \{x + y : x \in B_1, \ y \in B_2\}$;

(v) $\alpha(B_1 \cup B_2) \le \max\{\alpha(B_1), \alpha(B_2)\}$;

(vi) $\alpha(\lambda B) \le |\lambda|\alpha(B)$ for any $\lambda \in \mathbb{R}$.

In particular, the relationship of Hausdorff measure of noncompactness $\alpha$ and Kuratowski measure of noncompactness $\alpha_1$ is given by

(vii) $\alpha(B) \le \alpha_1(B) \le 2\alpha(B)$.

Let $J = [0, a], a \in \mathbb{R}^+$, For any $W \subset C(J, X)$, we define

$$\int_0^t W(s)ds = \left\{ \int_0^t u(s)ds : u \in W \right\}, \ \text{for } t \in [0, a],$$

where $W(s) = \{u(s) \in X : u \in W\}$.

We present here some useful properties.

**Proposition 1.14.** *If $W \subset C(J, X)$ is bounded and equicontinuous, then $\overline{co}W \subset C(J, X)$ is also bounded and equicontinuous.*

**Proposition 1.15.** *(Guo, Lakshmikantham and Liu, 1996) If $W \subset C(J, X)$ is bounded and equicontinuous, then $t \to \alpha(W(t))$ is continuous on $J$, and*

$$\alpha(W) = \max_{t \in J} \alpha(W(t)), \ \ \alpha\left( \int_0^t W(s)ds \right) \le \int_0^t \alpha(W(s))ds, \ \text{for } t \in [0, a].$$

**Proposition 1.16.** *(Mönch, 1980) Let $\{u_n\}_{n=1}^\infty$ be a sequence of Bochner integrable functions from $J$ into $X$ with $|u_n(t)| \le \tilde{m}(t)$ for almost all $t \in J$ and every $n \ge 1$, where $\tilde{m} \in L(J, \mathbb{R}^+)$, then the function $\psi(t) = \alpha(\{u_n(t)\}_{n=1}^\infty)$ belongs to $L(J, \mathbb{R}^+)$ and satisfies*

$$\alpha\left( \left\{ \int_0^t u_n(s)ds : n \ge 1 \right\} \right) \le 2 \int_0^t \psi(s)ds.$$

**Proposition 1.17.** *(Bothe, 1998) If $W$ is bounded, then for each $\varepsilon > 0$, there is a sequence $\{u_n\}_{n=1}^\infty \subset W$, such that*

$$\alpha(W) \le 2\alpha(\{u_n\}_{n=1}^\infty) + \varepsilon.$$

### 1.4.3 Topological Degree

For a minute description of the following notions we refer the reader to Banaś and Goebel, 1980; Deimling, 1985; Heinz, 1983; Lakshmikantham and Leela, 1969.

**Definition 1.10.** Consider $\Omega \subset X$ and $\mathscr{F} : \Omega \to X$ a continuous bounded mapping. We say that $\mathscr{F}$ is $\alpha$-Lipschitz if there exists $k \ge 0$ such that

$$\alpha(\mathscr{F}(B)) \le k\alpha(B) \ \ \forall B \subset \Omega \text{ bounded}.$$

If, in addition, $k < 1$, then we say that $\mathscr{F}$ is a strict $\alpha$-contraction. We say that $\mathscr{F}$ is $\alpha$-condensing if

$$\alpha(\mathscr{F}(B)) < \alpha(B) \quad \forall \, B \subset \Omega \text{ bounded with } \alpha(B) > 0.$$

In other words, $\alpha(\mathscr{F}(B)) \geq \alpha(B)$ implies $\alpha(B) = 0$. The class of all strict $\alpha$-contractions $\mathscr{F} : \Omega \to X$ is denoted by $SC_\alpha(\Omega)$ and the class of all $\alpha$-condensing mappings $\mathscr{F} : \Omega \to X$ is denoted by $C_\alpha(\Omega)$.

We remark that $SC_\alpha(\Omega) \subset C_\alpha(\Omega)$ and every $\mathscr{F} \in C_\alpha(\Omega)$ is $\alpha$-Lipschitz with constant $k = 1$. We also recall that $\mathscr{F} : \Omega \to X$ is Lipschitz if there exists $k > 0$ such that

$$|\mathscr{F}x - \mathscr{F}y| \leq k|x - y| \quad \forall \, x, y \in \Omega$$

and that $\mathscr{F}$ is a strict contraction if $k < 1$.

Next, we collect some properties of the applications defined above.

**Proposition 1.18.** *If $\mathscr{F}, \mathscr{G} : \Omega \to X$ are $\alpha$-Lipschitz mappings with constants $k$, $k'$, respectively, then $\mathscr{F} + \mathscr{G} : \Omega \to X$ is $\alpha$-Lipschitz with constant $k + k'$.*

**Proposition 1.19.** *If $\mathscr{F} : \Omega \to X$ is compact, then $\mathscr{F}$ is $\alpha$-Lipschitz with constant $k = 0$.*

**Proposition 1.20.** *If $\mathscr{F} : \Omega \to X$ is Lipschitz with constant $k$, then $\mathscr{F}$ is $\alpha$-Lipschitz with the same constant $k$.*

The theorem below asserts the existence and the basic properties of the topological degree for $\alpha$-condensing perturbations of the identity. For more details, see Isaia, 2006.

Let

$$\mathcal{T} = \left\{ (I - \mathscr{F}, \Omega, y) : \Omega \subset X \text{ open and bounded}, \mathscr{F} \in C_\alpha(\overline{\Omega}), y \in X \setminus (I - \mathscr{F})(\partial\Omega) \right\}$$

be the family of the admissible triplets.

**Theorem 1.1.** *There exists one degree function $D : \mathcal{T} \to \mathbb{N}_0$ which satisfies the properties:*

(i) *Normalization $D(I, \Omega, y) = 1$ for every $y \in \Omega$;*

(ii) *Additivity on domain: For every disjoint, open sets $\Omega_1, \Omega_2 \subset \Omega$ and every $y$ does not belong to $(I - \mathscr{F})(\overline{\Omega} \backslash (\Omega_1 \cup \Omega_2))$ we have*

$$D(I - \mathscr{F}, \Omega, y) = D(I - \mathscr{F}, \Omega_1, y) + D(I - \mathscr{F}, \Omega_2, y);$$

(iii) *Invariance under homotopy $D(I - H(t, \cdot), \Omega, y(t))$ is independent of $t \in [0, 1]$ for every continuous, bounded mapping $H : [0, 1] \times \overline{\Omega} \to X$ which satisfies*

$$\alpha(H([0, 1] \times B)) < \alpha(B) \quad \forall \, B \subset \overline{\Omega} \text{ with } \alpha(B) > 0$$

*and every continuous function $y : [0, 1] \to x$ which satisfies*

$$y(t) \neq x - H(t, x) \quad \forall \, t \in [0, 1], \, \forall \, x \in \partial\Omega;$$

(iv) *Existence* $D(I - \mathscr{F}, \Omega, y) \neq 0$ *implies* $y \in (I - \mathscr{F})(\Omega)$;

(v) *Excision* $D(I - \mathscr{F}, \Omega, y) = D(I - \mathscr{F}, \Omega_1, y)$ *for every open set* $\Omega_1 \subset \Omega$ *and every $y$ does not belong to* $(I - \mathscr{F})(\overline{\Omega} \backslash \Omega_1)$.

Having in hand a degree function defined on $\mathcal{T}$, we collect the usability of the a priori estimate method by means of this degree.

**Theorem 1.2.** *Let $\mathscr{F} : X \to X$ be $\alpha$-condensing and*

$$S = \{x \in X : \exists \, \lambda \in [0,1] \ \text{such that} \ x = \lambda \mathscr{F} x\}.$$

*If $S$ is a bounded set in $X$, so there exists $r > 0$ such that $S \subset B_r(0)$, then*

$$D(I - \lambda \mathscr{F}, B_r(0), 0) = 1 \quad \forall \, \lambda \in [0,1].$$

*Consequently, $\mathscr{F}$ has at least one fixed point and the set of the fixed points of $\mathscr{F}$ lies in $B_r(0)$.*

### 1.4.4 Picard Operator

Let $(X, d)$ be a metric space and $A : X \to X$ an operator. We shall use the following notations:

$P(X) = \{Y \subseteq X \mid Y \neq \varnothing\}$; $F_A = \{x \in X \mid A(x) = x\}$ — the fixed point set of $A$;
$I(A) = \{Y \in P(X) \mid A(Y) \subseteq Y\}$;
$O_A(x) = \{x, A(x), A^2(x), \ldots, A^n(x), \ldots\}$ — the $A$-orbit of $x \in X$;
$H : P(X) \times P(X) \to \mathbb{R}_+ \cup \{+\infty\}$;
$H(Y, Z) = - \max\{\sup_{a \in Y} \inf_{b \in Z} d(a,b), \ \sup_{b \in Z} \inf_{a \in Y} d(a,b)\}$
— the Pompeiu-Hausdorff functional on $P(X)$.

**Definition 1.11.** (Rus, 1987) Let $(X, d)$ be a metric space. An operator $A : X \to X$ is a Picard operator if there exists $x^* \in X$ such that $F_A = \{x^*\}$ and the sequence $\{A^n(x_0)\}_{n \in \mathbb{N}}$ converges to $x^*$ for all $x_0 \in X$.

**Definition 1.12.** (Rus, 1993) Let $(X, d)$ be a metric space. An operator $A : X \to X$ is a weak Picard operator if the sequence $\{A^n(x_0)\}_{n \in \mathbb{N}}$ converges for all $x_0 \in X$ and its limit (which may depend on $x_0$) is a fixed point of $A$.

If $A$ is a weak Picard operator, then we consider the operator

$$A^\infty : X \to X, \quad A^\infty(x) = \lim_{n \to \infty} A^n(x).$$

The following results are useful in what follows.

**Definition 1.13.** (Rus, 1979) Let $(Y, d)$ be a complete metric space and $A, B : Y \to Y$ two operators. Suppose that:

(i) $A$ is a contraction with contraction constant $\rho$ and $F_A = \{x_A^*\}$;
(ii) $B$ has fixed points and $x_B^* \in F_B$;
(iii) There exists $\eta > 0$ such that $d(A(x), B(x)) \leq \eta$, for all $x \in Y$.

Then $d(x_A^*, x_B^*) \leq \frac{\eta}{1-\rho}$.

**Definition 1.14.** (Rus and Mureşan, 2000) Let $(X, d)$ be a complete metric space and $A, B : X \to X$ two orbitally continuous operators. Assume that:

**(i)** there exists $\rho \in [0, 1)$ such that

$$d(A^2(x), A(x)) \leq \rho d(x, A(x)),$$
$$d(B^2(x), B(x)) \leq \rho d(x, B(x))$$

for all $x \in X$;

**(ii)** there exists $\eta > 0$ such that $d(A(x), B(x)) \leq \eta$ for all $x \in X$.

Then $H(F_A, F_B) \leq \frac{\eta}{1-\rho}$, where $H$ denotes the Pompeiu-Hausdorff functional.

**Theorem 1.3.** *(Rus, 1993) Let $(X, d)$ be a metric space. Then $A : X \to X$ is a weak Picard operator if and only if there exists a partition $X = \bigcup_{\lambda \in \Lambda} X_\lambda$ of $X$ such that*

**(i)** $X_\lambda \in I(A)$;
**(ii)** $A|_{X_\lambda} : X_\lambda \to X_\lambda$ *is a Picard operator, for all $\lambda \in \Lambda$.*

### 1.4.5   Fixed Point Theorems

In this subsection, we present some fixed point theorems which will be used in the following chapters.

**Theorem 1.4.** *(Banach contraction mapping principle) Let $(X, d)$ be a complete metric space, and $\mathscr{T} : \Omega \to \Omega$ a contraction mapping:*

$$d(\mathscr{T}x, \mathscr{T}y) \leq k d(x, y),$$

*where $0 < k < 1$, for each $x, y \in \Omega$. Then, there exists a unique fixed point $x$ of $\mathscr{T}$ in $\Omega$, i.e., $\mathscr{T}x = x$.*

**Theorem 1.5.** *(Schauder fixed point theorem). Let $X$ be a Banach space and $\Omega \subset X$ a convex, closed and bounded set. If $\mathscr{T} : \Omega \to \Omega$ is a continuous operator such that $\mathscr{T}\Omega \subset X$, $\mathscr{T}\Omega$ is relatively compact, then $\mathscr{T}$ has at least one fixed point in $\Omega$.*

**Theorem 1.6.** *(Schaefer fixed point theorem) Let $X$ be a Banach space and let $F : X \to X$ be a completely continuous mapping. Then either*

**(i)** *the equation $x = \lambda F x$ has a solution for $\lambda = 1$, or*
**(ii)** *the set $\{x \in X : x = \lambda F x \text{ for some } \lambda \in (0, 1)\}$ is unbounded.*

**Theorem 1.7.** *(Darbo-Sadovskii fixed point theorem) If $\Omega$ is bounded closed and convex subset of Banach space $X$, the continuous mapping $\mathscr{T} : \Omega \to \Omega$ is an $\alpha$-contraction, then the mapping $\mathscr{T}$ has at least one fixed point in $\Omega$.*

**Theorem 1.8.** *(Krasnoselskii fixed point theorem) Let $X$ be a Banach space, let $\Omega$ be a bounded closed convex subset of $X$ and let $\mathscr{S}$, $\mathscr{T}$ be mappings of $\Omega$ into $X$ such that $\mathscr{S}z + \mathscr{T}w \in \Omega$ for every pair $z, w \in \Omega$. If $\mathscr{S}$ is a contraction and $\mathscr{T}$ is completely continuous, then the equation $\mathscr{S}z + \mathscr{T}z = z$ has a solution on $\Omega$.*

**Theorem 1.9.** *(Nonlinear alternative of Leray-Schauder type) Let $C$ be a nonempty convex subset of $X$. Let $U$ be a nonempty open subset of $C$ with $0 \in U$ and $F : \overline{U} \to C$ be a compact and continuous operators. Then either*

**(i)** *$F$ has fixed points, or*
**(ii)** *there exist $y \in \partial U$ and $\lambda^* \in [0, 1]$ with $y = \lambda^* F(y)$.*

**Theorem 1.10.** *(O'Regan fixed point theorem) Let $U$ be an open set in a closed, convex set $C$ of $X$. Assume $0 \in U$, $T(\overline{U})$ is bounded and $T : \overline{U} \to C$ is given by $T = T_1 + T_2$, where $T_1 : \overline{U} \to X$ is completely continuous, and $T_2 : \overline{U} \to X$ is a nonlinear contraction. Then either*

**(i)** *$T$ has a fixed point in $\overline{U}$, or*
**(ii)** *there is a point $x \in \partial U$ and $\lambda \in (0, 1)$ with $x = \lambda T(x)$.*

A non-empty closed set $K$ in a Banach space $X$ is called a cone if:

**(i)** $K + K \subseteq K$;
**(ii)** $\lambda K \subseteq K$ for $\lambda \in \mathbb{R}, \lambda \geq 0$;
**(iii)** $\{-K\} \cap K = \{0\}$, where $0$ is the zero element of $X$.

We introduce an order relation "$\leq$" in $X$ as follows. Let $z, y \in X$. Then $z \leq y$ if and only if $y - z \in K$. A cone $K$ is called normal if the norm $\|\cdot\|_X$ is semi-monotone increasing on $K$, that is, there is a constant $N > 0$ such that $\|z\|_X \leq N\|y\|_X$ for all $z, y \in K$ with $z \leq y$. It is known that if the cone $K$ is normal in $X$, then every order-bounded set in $X$ is norm-bounded. Similarly, the cone $K$ in $X$ is called regular if every monotone increasing (resp. decreasing) order bounded sequence in $X$ converges in norm.

For any $a, b \in X, a \leq b$, the order interval $[a, b]$ is a set in $X$ given by

$$[a, b] = \{z \in X : a \leq z \leq b\}.$$

Let $X$ and $Y$ be two ordered Banach spaces. A mapping $\mathscr{T} : X \to Y$ is said to be nondecreasing or monotone increasing if $z \leq y$ implies $\mathscr{T}z \leq \mathscr{T}y$ for all $z, y \in [a, b]$.

**Theorem 1.11.** *(Hybrid fixed point theorem) (Dhage, 2006) Let $X$ be a Banach space and $A, B, C : X \to X$ be three monotone increasing operators such that*

**(i)** *$A$ is a contraction with contraction constant $k < 1$;*
**(ii)** *$B$ is completely continuous;*
**(iii)** *$C$ is totally bounded;*

**(iv)** *there exist elements $a$ and $b$ in $X$ such that $a \leq Aa + Ba + Ca$ and $b \geq Ab + Bb + Cb$ with $a \leq b$.*

*Further if the cone $K$ in $X$ is normal, then the operator equation $Az + Bz + Cz = z$ has a least and a greatest solution in $[a, b]$.*

### 1.4.6   Critical Point Theorems

Let $H$ be a real Banach space and $C^1(H, \mathbb{R}^N)$ denotes the set of functionals that are Fréchet differentiable and their Fréchet derivatives are continuous on $H$.

We need to use the critical point theorems to consider the fractional boundary value problems. For the reader's convenience, we state some necessary definitions and theorems and skip the proofs.

**Definition 1.15.** (Rabinowitz, 1986) Let $\psi \in C^1(H, \mathbb{R}^N)$. If any sequence $\{u_k\} \subset H$ for which $\{\psi(u_k)\}$ is bounded and $\psi'(u_k) \to 0$ as $k \to \infty$ possesses a convergent subsequence, then we say $\psi$ satisfies Palais-Smale condition (denoted by (PS) condition for short).

**Definition 1.16.** (Mawhin and Willem, 1989) Let $H$ be a real Banach space, $\psi : H \to \mathbb{R}$ is differentiable and $c \in \mathbb{R}$. We say that $\psi$ satisfies the $(PS)_c$ condition if the existence of a sequence $\{u_k\}$ in $H$ such that

$$\psi(u_k) \to c, \quad \psi'(u_k) \to 0$$

as $k \to \infty$, implies that $c$ is a critical value of $\psi$.

**Theorem 1.12.** *(Mawhin and Willem, 1989) Let $H$ be a real reflexive Banach space. If the functional $\psi : H \to \mathbb{R}^N$ is weakly lower semi-continuous and coercive, i.e., $\lim_{|z| \to \infty} \psi(z) = +\infty$, then there exists $z_0 \in H$ such that $\psi(z_0) = \inf_{z \in H} \psi(z)$. Moreover, if $\psi$ is also Fréchet differentiable on $H$, then $\psi'(z_0) = 0$.*

Let $B_r$ be the open ball in $H$ with the radius $r$ and centered at 0 and $\partial B_r$ denote its boundary.

**Theorem 1.13.** *(Mountain pass theorem) (Rabinowitz, 1986) Let $H$ be a real Banach space and $I \in C^1(H, \mathbb{R})$ satisfy (PS) condition . Suppose that $I$ satisfies the following conditions:*

**(i)** $I(0) = 0$;
**(ii)** *there exist constants $\rho, \beta > 0$ such that $I|_{\partial B_\rho(0)} \geq \beta$;*
**(iii)** *there exist $e \in H \setminus \overline{B}_\rho(0)$ such that $I(e) \leq 0$.*

*Then $I$ possesses a critical value $c \geq \beta$ given by*

$$c = \inf_{g \in \Gamma} \max_{s \in [0,1]} I(g(s)),$$

*where $B_\rho(0)$ is an open ball in $H$ of radius $\rho$ centered at 0, and*

$$\Gamma = \{g \in C([0, 1], H) : g(0) = 0, \ g(1) = e\}.$$

Let $X$ be a reflexive and separable Banach space, then there are $e_j \in X$ and $e_j^* \in X^*$ such that

$$X = \overline{\text{span}\{e_j : j = 1, 2, \ldots\}} \quad \text{and} \quad X^* = \overline{\text{span}\{e_j^* : j = 1, 2, \ldots\}},$$

and

$$\langle e_i^*, e_j \rangle = \begin{cases} 1, & i = j, \\ 0, & i \neq j. \end{cases}$$

For convenience, we write

$$X_j := \text{span}\{e_j\}, \quad Y_k := \bigoplus_{j=1}^{k} X_j, \quad Z_k := \overline{\bigoplus_{j=k}^{\infty} X_j}. \tag{1.15}$$

And let

$$B_k := \{u \in Y_k : |u| \leq \rho_k\}, \quad N_k := \{u \in Z_k : |u| = \gamma_k\}.$$

**Theorem 1.14.** *(Fountain theorem) (Bartsch, 1993) Suppose:*

**(H1)** $X$ *is a Banach space,* $\varphi \in C^1(X, \mathbb{R})$ *is an even functional, the subspace* $X_k, Y_k$ *and* $Z_k$ *are defined by* (1.15).

*If for every* $k \in \mathbb{N}$, *there exist* $\rho_k > r_k > 0$ *such that*

**(H2)** $a_k := \max\limits_{\substack{u \in Y_k \\ |u| = \rho_k}} \varphi(u) \leq 0;$

**(H3)** $b_k := \inf\limits_{\substack{u \in Z_k \\ |u| = r_k}} \varphi(u) \to \infty$, *as* $k \to \infty$;

**(H4)** $\varphi$ *satisfies the* $(PS)_c$ *condition for every* $c > 0$.

*Then* $\varphi$ *has an unbounded sequence of critical values.*

**Theorem 1.15.** *(Dual Fountain theorem) (Bartsch, 1993) Assume (H1) is satisfied, and there is a* $k_0 > 0$ *so as to for each* $k \geq k_0$, *there exist* $\rho_k > r_k > 0$ *such that*

**(H5)** $d_k := \inf\limits_{\substack{u \in Z_k \\ |u| \leq \rho_k}} \varphi(u) \to 0$, *as* $k \to \infty$;

**(H6)** $i_k := \max\limits_{\substack{u \in Y_k \\ |u| = r_k}} \varphi(u) < 0$;

**(H7)** $\inf\limits_{\substack{u \in Z_k \\ |u| = \rho_k}} \varphi(u) \geq 0$;

**(H8)** $\varphi$ *satisfies the* $(PS)_c^*$ *condition for every* $c \in [d_{k_0}, 0)$.

*Then* $\varphi$ *has a sequence of negative critical values converging to 0.*

**Remark 1.4.** $\varphi$ satisfies the $(PS)^*_c$ condition means that: if any sequence $\{u_{n_j}\} \subset X$ such that $n_j \to \infty, u_{n_j} \in Y_{n_j}, \varphi(u_{n_j}) \to c$ and $(\varphi|_{Y_{n_j}})'(u_{n_j}) \to 0$, then $\{u_{n_j}\}$ contains a subsequence converging to a critical point of $\varphi$. It is obvious that if $\varphi$ satisfies the $(PS)^*_c$ condition, then $\varphi$ satisfies the $(PS)_c$ condition.

Let $X$ be a nonempty set and $\Phi, \tilde{\Psi} : X \to \mathbb{R}$ be two functionals. For $r, r_1, r_2, r_3 \in \mathbb{R}$ with $r_1 < \sup_X \Phi, r_2 > \inf_X \Phi, r_2 > r_1$, and $r_3 > 0$, we define

$$\varphi(r) := \inf_{u \in \Phi^{-1}(-\infty,r)} \frac{\sup_{u \in \Phi^{-1}(-\infty,r)} \tilde{\Psi}(u) - \tilde{\Psi}(u)}{r - \Phi(u)}, \qquad (1.16)$$

$$\beta(r_1, r_2) := \inf_{u \in \Phi^{-1}(-\infty,r_1)} \sup_{v \in \Phi^{-1}[r_1,r_2)} \frac{\tilde{\Psi}(v) - \tilde{\Psi}(u)}{\Phi(v) - \Phi(u)}, \qquad (1.17)$$

$$\gamma(r_2, r_3) := \frac{\sup_{u \in \Phi^{-1}(-\infty,r_2+r_3)} \tilde{\Psi}(u)}{r_3}, \qquad (1.18)$$

$$\alpha(r_1, r_2, r_3) := \max\left\{\varphi(r_1), \varphi(r_2), \gamma(r_2, r_3)\right\}. \qquad (1.19)$$

**Lemma 1.6.** *(Averna and Bonanno, 2009; Bonanno and Candito, 2008) Let $X$ be a reflexive real Banach space, $\Phi : X \to \mathbb{R}$ be a convex, coercive, and continuously Gâteaux differentiable functional whose Gâteaux derivative admits a continuous inverse on $X^*$, $\tilde{\Psi} : X \to \mathbb{R}$ be a continuously Gâteaux differentiable functional whose Gâteaux derivative is compact, such that*

**(i)** $\inf_X \Phi = \Phi(0) = \tilde{\Psi}(0) = 0$;
**(ii)** *for every $u_1, u_2$ satisfying $\tilde{\Psi}(u_1) \geq 0$ and $\tilde{\Psi}(u_2) \geq 0$, one has*

$$\inf_{t \in [0,1]} \tilde{\Psi}(tu_1 + (1-t)u_2) \geq 0.$$

*Assume further that there exist three positive constants $r_1, r_2$ and $r_3$, with $r_1 < r_2$, such that*

**(iii)** $\alpha(r_1, r_2, r_3)) < \beta(r_1, r_2)$.

*Then, for each $\lambda \in \left(1/\beta(r_1, r_2), 1/\alpha(r_1, r_2, r_3)\right)$, the functional $\Phi - \lambda\tilde{\Psi}$ has three distinct critical points $u_1, u_2$ and $u_3$ such that $u_1 \in \Phi^{-1}(-\infty, r_1), u_2 \in \Phi^{-1}[r_1, r_2)$ and $u_3 \in \Phi^{-1}(-\infty, r_2 + r_3)$.*

## 1.5   Semigroups

### 1.5.1   $C_0$-semigroup

Let $X$ be a Banach space and $B(X)$ be the Banach space of linear bounded operators.

**Definition 1.17.** A semigroup is a one parameter family $\{T(t)\}_{t \geq 0} \subset B(X)$ satisfying the conditions:

(i) $T(t)T(s) = T(t + s)$, for $t, s \geq 0$;

(ii) $T(0) = I$.

Here $I$ denotes the identity operator in $X$.

**Definition 1.18.** A semigroup $\{T(t)\}_{t\geq 0}$ is uniformly continuous if

$$\lim_{t \to 0+} \|T(t) - T(0)\|_{B(X)} = 0,$$

that is if

$$\lim_{|t-s| \to 0} \|T(t) - T(s)\|_{B(X)} = 0.$$

**Definition 1.19.** We say that the semigroup $\{T(t)\}_{t\geq 0}$ is $C_0$-semigroup if the map $t \to T(t)x$ is strongly continuous, for each $x \in X$, i.e.

$$\lim_{t \to 0+} T(t)x = x, \quad \forall\, x \in X.$$

**Definition 1.20.** Let $T(t)$ be a $C_0$-semigroup defined on $X$. The infinitesimal generator $A$ of $T(t)$ is the linear operator defined by

$$A(x) = \lim_{t \to 0+} \frac{T(t)x - x}{t}, \quad \text{for } x \in D(A),$$

where $D(A) = \{x \in X : \lim_{t \to 0+} \frac{T(t)x-x}{t} \text{ exists in } X\}$.

### 1.5.2 *Almost Sectorial Operators*

We firstly introduce some special functions and classes of functions which will be used in the following, for more details, we refer to Markus, 2006; Periago and Straub, 2002.

Let $S_\mu^0$ with $0 < \mu < \pi$ be the open sector

$$\{z \in \mathbb{C}\backslash\{0\} : |\arg z| < \mu\}$$

and $S_\mu$ be its closure, that is

$$S_\mu = \{z \in \mathbb{C}\backslash\{0\} : |\arg z| \leq \mu\} \cup \{0\}.$$

We state the concept of almost sectorial operators as follows.

**Definition 1.21.** (Periago and Straub, 2002) Let $-1 < \gamma < 0$ and $0 < \omega < \pi/2$. By $\Theta_\omega^\gamma(X)$ we denote the family of all linear closed operators $A : D(A) \subset X \to X$ which satisfy:

(i) $\sigma(A) \subset S_\omega$;

(ii) for every $\omega < \mu < \pi$ there exists a constant $C_\mu$ such that

$$\|R(z; A)\|_{B(X)} \leq C_\mu |z|^\gamma, \quad \text{for all } z \in \mathbb{C} \setminus S_\mu,$$

where $R(z; A) = (zI - A)^{-1}$, $z \in \rho(A)$, which are bounded linear operators the resolvent of $A$. A linear operator $A$ will be called an almost sectorial operator on $X$ if $A \in \Theta_\omega^\gamma(X)$.

**Remark 1.5.** Let $A \in \Theta^{\gamma}_{\omega}(X)$. Then the definition implies that $0 \in \rho(A)$.

Set

$$\mathcal{F}^{\gamma}_0(S^0_{\mu}) = \bigcup_{s<0} \Psi^{\gamma}_s(S^0_{\mu}) \cup \Psi_0(S^0_{\mu}),$$

$$\mathcal{F}(S^0_{\mu}) = \{f \in \mathcal{H}(S^0_{\mu}) \mid \text{ there exist } k, n \in \mathbb{N} \text{ such that } f\psi^k_n \in \mathcal{F}_0(S^0_{\mu})\},$$

where

$$\mathcal{H}(S^0_{\mu}) = \{f : S^0_{\mu} \mapsto \mathbb{C} \mid f \text{ is holomorphic}\},$$
$$\mathcal{H}^{\infty}(S^0_{\mu}) = \{f \in \mathcal{H}(S^0_{\mu}) \mid f \text{ is bounded}\},$$
$$\varphi_0(z) = \frac{1}{1+z}, \quad \psi_n(z) := \frac{z}{(1+z)^n}, \quad z \in \mathbb{C}\backslash\{-1\}, n \in \mathbb{N} \cup \{0\},$$
$$\Psi_0(S^0_{\mu}) = \left\{f \in \mathcal{H}(S^0_{\mu}) \mid \sup_{z \in S^0_{\mu}} \left|\frac{f(z)}{\varphi_0(z)}\right| < \infty\right\},$$

and for each $s < 0$,

$$\Psi^{\gamma}_s(S^0_{\mu}) := \left\{f \in \mathcal{H}(S^0_{\mu}) \mid \sup_{z \in S^0_{\mu}} |\psi^s_n(z)f(z)| < \infty\right\},$$

where $n$ is the smallest integer such that $n \geq 2$ and $\gamma + 1 < -(n-1)s$.

Observe that the classes of functions introduced above satisfy the inclusions

$$\mathcal{F}^{\gamma}_0(S^0_{\mu}) \subset \mathcal{H}^{\infty}(S^0_{\mu}) \subset \mathcal{F}(S^0_{\mu}) \subset \mathcal{H}(S^0_{\mu}).$$

Moreover, taking $k, n \in \mathbb{N} \cup \{0\}$ with $n > k$, one easily sees that $\psi^k_n \in \mathcal{F}^{\gamma}_0(S^0_{\mu})$.

Assume that $A \in \Theta^{\gamma}_{\omega}(X)$ with $-1 < \gamma < 0$ and $0 < \omega < \pi/2$. Following Periago and Straub, 2002 (see also McIntosh, 1986; Cowling *et al.*, 1996), a closed linear operator $f \to f(A)$ can be constructed for every $f \in \mathcal{F}(S^0_{\mu})$ via an extended functional calculus. In the following we give a short overview to this construction.

For $f \in \mathcal{F}^{\gamma}_0(S^0_{\mu})$, via Dunford-Riesz integral, the operator $f(A)$ is defined by

$$f(A) = \frac{1}{2\pi i} \int_{\Gamma_{\theta}} f(z)R(z; A)dz, \tag{1.20}$$

where the integral contour $\Gamma_{\theta} := \{\mathbb{R}_+ e^{i\theta}\} \cup \{\mathbb{R}_+ e^{-i\theta}\}$, is oriented counter-clockwise and $\omega < \theta < \mu < \pi$. It follows that the integral is absolutely convergent and defines a bounded linear operator on $X$, and its value does not depend on the choice of $\theta$.

Notice in particular that for $k, n \in \mathbb{N} \cup \{0\}$ with $n > k$,

$$\psi^k_n(A) = A^k(A+1)^{-n}$$

and the operator $\psi^k_n(A)$ is injective. Notice also that if $f \in \mathcal{F}(S^0_{\mu})$, then there exist $k, n \in \mathbb{N}$ such that $f\psi^k_n \in \mathcal{F}^{\gamma}_0(S^0_{\mu})$. Hence, for $f \in \mathcal{F}(S^0_{\mu})$, one can define a closed linear operator, still denoted by $f(A)$,

$$D(f(A)) = \{x \in X \mid (f\psi^k_n)(A)x \in D(A^{(n-1)k})\},$$

$$f(A) = (\psi_n^k(A))^{-1}(f\psi_n^k)(A),$$

and the definition of $f(A)$ does not depend on the choice of $k$ and $n$. We emphasize that $f(A)$ is indeed an extension of the original and the triple $(\mathcal{F}_0^\gamma(S_\mu^0), \mathcal{F}(S_\mu^0), f(A))$ is called an *abstract functional calculus* on $X$ (see Markus, 2006).

With respect to this construction we collect some basic properties. For more details, we refer to Periago and Straub, 2002.

**Proposition 1.21.** *The following assertions hold.*

(i) $\alpha f(A) + \beta g(A) = (\alpha f + \beta g)(A)$, $(fg)(A) = f(A)g(A)$ *for all* $f, g \in \mathcal{F}_0^\gamma(S_\mu^0)$, $\alpha, \beta \in \mathbb{C}$;

(ii) $f(A)g(A) \subset (fg)(A)$ *for all* $f, g \in \mathcal{F}(S_\mu^0)$, *and*

(iii) $f(A)g(A) = (fg)(A)$, *provided that* $g(A)$ *is bounded or* $D((fg)(A)) \subset D(g(A))$.

Since for each $\beta \in \mathbb{C}$, $z^\beta \in \mathcal{F}(S_\mu^0)$ ($z \in \mathbb{C} \setminus (-\infty, 0]$, $0 < \mu < \pi$), one can define, via the triple $(\mathcal{F}_0^\gamma(S_\mu^0), \mathcal{F}(S_\mu^0), f(A))$, the complex powers of $A$ which are closed by

$$A^\beta = z^\beta(A), \quad \beta \in \mathbb{C},$$

However, in difference to the case of sectorial operators, having $0 \in \rho(A)$ does not imply that the complex powers $A^{-\beta}$ with $Re(\beta) > 0$, are bounded. The operator $A^{-\beta}$ belongs to $\mathcal{L}(X)$ whenever $Re(\beta) > 1+\gamma$. So, in this situation, the linear space $X^\beta := D(A^\beta)$, $\beta > 1 + \gamma$, endowed with the graph norm $|x|_\beta = |A^\beta x|$, $x \in X^\beta$, is a Banach space.

Next, we turn our attention to the semigroup associated with $A$. Since given $t \in S_{\frac{\pi}{2}-\omega}^0$, $e^{-tz} \in \mathcal{H}^\infty(S_\mu^0)$ satisfies the conditions (a) and (b) of Lemma 2.13 of Periago and Straub, 2002, the family

$$T(t) = e^{-tz}(A) = \frac{1}{2\pi i} \int_{\Gamma_\theta} e^{-tz} R(z; A)dz, \quad t \in S_{\frac{\pi}{2}-\omega}^0, \tag{1.21}$$

here $\omega < \theta < \mu < \frac{\pi}{2} - |\arg t|$, forms an analytic semigroup of growth order $1 + \gamma$.

**Remark 1.6.** From Periago and Straub, 2002, note that if $A \in \Theta_\omega^\gamma(X)$, then $A$ generates a semigroup $T(t)$ with a singular behavior at $t = 0$ in a sense, called semigroup of growth $1 + \gamma$. Moreover, the semigroup $T(t)$ is analytic in an open sector of the complex plane $\mathbb{C}$, but the strong continuity fails at $t = 0$ for data which are not sufficiently smooth.

For more properties on $T(t)$, please see the following proposition.

**Proposition 1.22.** *(Periago and Straub, 2002) Let* $A \in \Theta_\omega^\gamma(X)$ *with* $-1 < \gamma < 0$ *and* $0 < \omega < \pi/2$. *Then the following properties remain true.*

(i) $T(t)$ *is analytic in* $S_{\frac{\pi}{2}-\omega}^0$ *and*

$$\frac{d^n}{dt^n}T(t) = (-A)^n T(t), \quad \text{for all } t \in S_{\frac{\pi}{2}-\omega}^0;$$

**(ii)** *The functional equation $T(s+t) = T(s)T(t)$ for all $s$, $t \in S^0_{\frac{\pi}{2}-\omega}$ holds;*

**(iii)** *There exists a constant $C_0 = C_0(\gamma) > 0$ such that*

$$\|T(t)\|_{B(X)} \leq C_0 t^{-\gamma-1}, \quad \text{for all } t > 0;$$

**(iv)** *The range $R(T(t))$ of $T(t)$, $t \in S^0_{\frac{\pi}{2}-\omega}$ is contained in $D(A^\infty)$. Particularly, $R(T(t)) \subset D(A^\beta)$ for all $\beta \in \mathbb{C}$ with $Re(\beta) > 0$,*

$$A^\beta T(t)x = \frac{1}{2\pi i} \int_{\Gamma_\theta} z^\beta e^{-tz} R(z; A)x dz, \quad \text{for all } x \in X,$$

*and hence there exists a constant $C' = C'(\gamma, \beta) > 0$ such that*

$$\|A^\beta T(t)\|_{B(X)} \leq C' t^{-\gamma-Re(\beta)-1}, \quad \text{for all } t > 0;$$

**(v)** *If $\beta > 1+\gamma$, then $D(A^\beta) \subset \Sigma_T$, where $\Sigma_T$ is the continuity set of the semigroup $\{T(t)\}_{t \geq 0}$, that is,*

$$\Sigma_T = \left\{ x \in X \mid \lim_{t \to 0+} T(t)x = x \right\}.$$

**Remark 1.7.** We note that the condition (ii) of the proposition does not satisfy for $t = 0$ or $s = 0$.

Recall that semigroups of growth $1+\gamma$ were investigated earlier in deLaubenfels, 1994 and Toropova, 2003.

The relation between the resolvent operators of $A$ and the semigroup $T(t)$ is characterized by

**Proposition 1.23.** *(Periago and Straub, 2002) Let $A \in \Theta^\gamma_\omega(X)$ with $-1 < \gamma < 0$ and $0 < \omega < \pi/2$. Then for every $\lambda \in \mathbb{C}$ with $Re(\lambda) > 0$, one has*

$$R(\lambda; -A) = \int_0^\infty e^{-\lambda t} T(t) dt.$$

# Chapter 2

# Fractional Functional Differential Equations

## 2.1 Introduction

The main objective of this chapter is to present a unified framework to investigate the basic existence theory for a variety of fractional functional differential equations with applications. As far as we know, many complex processes in nature and technology are described by functional differential equations which are dominant nowadays because the functional components in equations allow one to consider prehistory or after-effect influence. Various classes of functional differential equations are of fundamental importance in many problems arising in bionomics, epidemiology, electronics, theory of neural networks, automatic control, etc. Quite long ago delay differential equations had shown their efficiency in the study of the behavior of real populations. One can show that even though the delay terms occurring in the equations are unbounded, the domain of the initial data (past history or memory) may be finite or infinite. Consequently, those two cases need to be discussed independently. Moreover, one can consider functional differential equations so that the delay terms also occur in the derivative of the unknown solution. Since the general formulation of such a problem is difficult to state, a special kind of equations called neutral functional differential equations has been introduced.

On the other hand, fractional calculus is one of the best tools to characterize long-memory processes and materials, anomalous diffusion, long-range interactions, long-term behaviors, power laws, allometric scaling laws, and so on. So the corresponding mathematical models are fractional differential equations. Their evolutions behave in a much more complicated way so to study the corresponding dynamics is much more difficult. Although the existence theorems for the fractional differential equations can be similarly obtained, not all the classical theory of differential equation can be directly applied to the fractional differential equations. Hence, a somewhat theoretical frame needs to be established.

In Section 2.2, we discuss the existence and uniqueness of solutions and the existence of extremal solutions of initial value problem for the fractional neutral differential equations with bounded delay. Section 2.3 is devoted to study of the basic existence theory for fractional $p$-type neutral differential equations with unbounded delay but finite memory. In Section 2.4, we present a unified treatment of fundamental existence theory of fractional neutral differential equations with infinite memory. In Section 2.5, we consider a fractional iterative functional differential equation with parameter. Some theorems to prove the existence of the iterative series solutions are presented under some nature conditions.

## 2.2   Neutral Equations with Bounded Delay

### 2.2.1   *Introduction*

Let $I_0 = [-\tau, 0]$, $\tau > 0$, $t_0 \geq 0$ and $I = [t_0, t_0 + \sigma]$, $\sigma > 0$ be two closed and bounded intervals in $\mathbb{R}$. Denote $J = [t_0 - \tau, t_0 + \sigma]$.

Let $\mathcal{C} = C(I_0, \mathbb{R}^n)$ be the space of continuous functions on $I_0$. For any element $\varphi \in \mathcal{C}$, define the norm

$$\|\varphi\|_* = \sup_{\theta \in I_0} |\varphi(\theta)|.$$

If $z \in C(J, \mathbb{R}^n)$, then for any $t \in I$ define $z_t \in \mathcal{C}$ by

$$z_t(\theta) = z(t + \theta), \quad \theta \in [-\tau, 0].$$

Consider the initial value problems (fractional IVP for short) of fractional neutral functional differential equations with bounded delay of the form

$$\begin{cases} {}^C_{t_0}D^\alpha_t(x(t) - k(t, x_t)) = F(t, x_t), & \text{a.e. } t \in (t_0, t_0 + a], \\ x_{t_0} = \varphi, \end{cases} \tag{2.1}$$

where ${}^C_{t_0}D^\alpha_t$ is Caputo fractional derivative of order $0 < \alpha < 1$, $F : I \times \mathcal{C} \to \mathbb{R}^n$ is a given function satisfying some assumptions that will be specified later, and $\varphi \in \mathcal{C}$.

In Subsection 2.2.2, we establish the existence and uniqueness theorems of fractional IVP (2.1). In Subsection 2.2.3, we discuss the existence of extremal solutions for fractional IVP (2.1). We firstly give the definitions of $L^{\frac{1}{\beta}}$-Carathéodory, $L^{\frac{1}{\gamma}}$-Chandrabhan and $L^{\frac{1}{\delta}}$-Lipschitz, where $\beta, \gamma, \delta$ are some given numbers. Next, we apply Hybrid fixed point theorem to prove the existence results of extremal solutions for fractional IVP (2.1) under $L^{\frac{1}{\beta}}$-Carathéodory, $L^{\frac{1}{\gamma}}$-Chandrabhan and $L^{\frac{1}{\delta}}$-Lipschitz conditions. We do not require the continuity of the nonlinearities involved in equation (2.1). In the end, we will present an example to illustrate our main results.

### 2.2.2   *Existence and Uniqueness*

Let $A(\sigma, \gamma) = \{x \in C([t_0 - \tau, t_0 + \sigma], \mathbb{R}^n) : x_{t_0} = \varphi, \sup_{t_0 \leq t \leq t_0 + \sigma} |x(t) - \varphi(0)| \leq \gamma\}$, where $\sigma, \gamma$ are positive constants.

Before stating and proving the main results, we introduce the following hypotheses:

**(H1)** $F(t, \varphi)$ is measurable with respect to $t$ on $I$;

**(H2)** $F(t, \varphi)$ is continuous with respect to $\varphi$ on $C(I_0, \mathbb{R}^n)$;

**(H3)** there exist $\alpha_1 \in (0, \alpha)$ and a real-valued function $m(t) \in L^{\frac{1}{\alpha_1}} I$ such that for any $x \in A(\sigma, \gamma)$, $|F(t, x_t)| \leq m(t)$, for $t \in I_0$;

**(H4)** for any $x \in A(\sigma, \gamma)$, $k(t, x_t) = k_1(t, x_t) + k_2(t, x_t)$;

**(H5)** $k_1$ is continuous and for any $x', x'' \in A(\sigma, \gamma)$, $t \in I$

$$|k_1(t, x'_t) - k_1(t, x''_t)| \leq l\|x' - x''\|, \quad \text{where} \quad l \in (0, 1);$$

**(H6)** $k_2$ is completely continuous and for any bounded set $\Lambda$ in $A(\sigma, \gamma)$, the set $\{t \to k_2(t, x_t) : x \in \Lambda\}$ is equicontinuous in $C(I, \mathbb{R}^n)$.

**Lemma 2.1.** *If there exist $\sigma \in (0, a)$ and $\gamma \in (0, \infty)$ such that (H1)-(H3) are satisfied, then for $t \in (t_0, t_0 + \sigma]$, fractional IVP (2.1) is equivalent to the following equation*

$$\begin{cases} x(t) = \varphi(0) - k(t_0, \varphi) + k(t, x_t) + \dfrac{1}{\Gamma(\alpha)} \displaystyle\int_{t_0}^t (t-s)^{\alpha-1} F(s, x_s) ds, \quad t \in I_0, \\ x_{t_0} = \varphi. \end{cases} \quad (2.2)$$

**Proof.** First, it is easy to obtain that $F(t, x_t)$ is Lebesgue measurable on $I$ according to conditions (H1) and (H2). A direct calculation gives that $(t-s)^{\alpha-1} \in L^{\frac{1}{1-\alpha_1}}([t_0, t], \mathbb{R})$, for $t \in I$. In the light of Hölder inequality and (H3), we obtain that $(t-s)^{\alpha-1} F(s, x_s)$ is Lebesgue integrable with respect to $s \in [t_0, t]$ for all $t \in I_0$ and $x \in A(\sigma, \gamma)$, and

$$\int_{t_0}^t |(t-s)^{\alpha-1} F(s, x_s)| ds \leq \|(t-s)^{\alpha-1}\|_{L^{\frac{1}{1-\alpha_1}}[t_0,t]} \|m\|_{L^{\frac{1}{\alpha_1}} I}. \quad (2.3)$$

According to Definitions 1.1 and 1.3, it is easy to see that if $x$ is a solution of the fractional IVP (2.1), then $x$ is a solution of equation (2.2).

On the other hand, if (2.2) is satisfied, then for every $t \in (t_0, t_0 + \sigma]$, we have

$$\begin{aligned} {}_{t_0}^C D_t^\alpha (x(t) - k(t, x_t)) &= {}_{t_0}^C D_t^\alpha \left( \varphi(0) - k(t_0, \varphi) + \frac{1}{\Gamma(\alpha)} \int_{t_0}^t (t-s)^{\alpha-1} F(s, x_s) ds \right) \\ &= {}_{t_0}^C D_t^\alpha \left( \frac{1}{\Gamma(\alpha)} \int_{t_0}^t (t-s)^{\alpha-1} F(s, x_s) ds \right) \\ &= {}_{t_0}^C D_t^\alpha ({}_{t_0} D_t^{-\alpha} F(t, x_t)) \\ &= {}_{t_0} D_t^\alpha ({}_{t_0} D_t^{-\alpha} F(t, x_t)) - {}_{t_0} D_t^{-\alpha} F(t, x_t)\big|_{t=t_0} \frac{(t-t_0)^{-\alpha}}{\Gamma(1-\alpha)} \\ &= F(t, x_t) - {}_{t_0} D_t^{-\alpha} F(t, x_t)\big|_{t=t_0} \frac{(t-t_0)^{-\alpha}}{\Gamma(1-\alpha)}. \end{aligned}$$

According to (2.3), we know that ${}_{t_0} D_t^{-\alpha} F(t, x_t)\big|_{t=t_0} = 0$, which means that ${}_{t_0}^C D_t^\alpha (x(t) - k(t, x_t)) = F(t, x_t)$, $t \in (t_0, t_0 + \sigma]$, and this completes the proof. $\square$

**Theorem 2.1.** *Assume that there exist $\sigma \in (0, a)$ and $\gamma \in (0, \infty)$ such that (H1)-(H6) are satisfied. Then the fractional IVP (2.1) has at least one solution on $[t_0, t_0 + \eta]$ for some positive number $\eta$.*

**Proof.** According to (H4), equation (2.2) is equivalent to the following equation

$$\begin{cases} x(t) = \varphi(0) - k_1(t_0, \varphi) - k_2(t_0, \varphi) + k_1(t, x_t) + k_2(t, x_t) \\ \qquad + \dfrac{1}{\Gamma(\alpha)} \displaystyle\int_{t_0}^t (t-s)^{\alpha-1} F(s, x_s) ds, \quad t \in I, \\ x_{t_0} = \varphi. \end{cases}$$

Let $\widetilde{\varphi} \in A(\sigma, \gamma)$ be defined as $\widetilde{\varphi}_{t_0} = \varphi$, $\widetilde{\varphi}(t_0 + t) = \varphi(0)$ for all $t \in [0, \sigma]$. If $x$ is a solution of the fractional IVP (2.1), let $x(t_0 + t) = \widetilde{\varphi}(t_0 + t) + y(t)$, $t \in [-\tau, \sigma]$, then we have $x_{t_0+t} = \widetilde{\varphi}_{t_0+t} + y_t$, $t \in [0, \sigma]$. Thus $y$ satisfies the equation

$$\begin{aligned} y(t) = &- k_1(t_0, \varphi) - k_2(t_0, \varphi) + k_1(t_0 + t, y_t + \widetilde{\varphi}_{t_0+t}) + k_2(t_0 + t, y_t + \widetilde{\varphi}_{t_0+t}) \\ &+ \dfrac{1}{\Gamma(\alpha)} \int_0^t (t-s)^{\alpha-1} F(t_0 + s, y_s + \widetilde{\varphi}_{t_0+s}) ds, \quad t \in [0, \sigma]. \end{aligned} \tag{2.4}$$

Since $k_1, k_2$ are continuous and $x_t$ is continuous in $t$, there exists $\sigma' > 0$, when $0 < t < \sigma'$,

$$|k_1(t_0 + t, y_t + \widetilde{\varphi}_{t_0+t}) - k_1(t_0, \varphi)| < \frac{\gamma}{3}, \tag{2.5}$$

and

$$|k_2(t_0 + t, y_t + \widetilde{\varphi}_{t_0+t}) - k_2(t_0, \varphi)| < \frac{\gamma}{3}. \tag{2.6}$$

Choose

$$\eta = \min \left\{ \sigma, \sigma', \left( \frac{\gamma \Gamma(\alpha)(1 + \beta)^{1-\alpha_1}}{3M} \right)^{\frac{1}{(1+\beta)(1-\alpha_1)}} \right\} \tag{2.7}$$

where $\beta = \frac{\alpha-1}{1-\alpha_1} \in (-1, 0)$ and $M = \|m\|_{L^{\frac{1}{\alpha_1}} I}$.

Define $E(\eta, \gamma)$ as follows

$$E(\eta, \gamma) = \{y \in C([-\tau, \eta], \mathbb{R}^n) : \; y(s) = 0 \;\text{ for } s \in [-\tau, 0] \;\text{ and } \|y\| \leq \gamma\}.$$

Then $E(\eta, \gamma)$ is a closed bounded and convex subset of $C([-\tau, \sigma], \mathbb{R}^n)$. On $E(\eta, \gamma)$ we define the operators $S$ and $U$ as follows

$$(Sy)(t) = \begin{cases} 0, & t \in [-\tau, 0], \\ - k_1(t_0, \varphi) + k_1(t_0 + t, y_t + \widetilde{\varphi}_{t_0+t}), & t \in [0, \eta], \end{cases}$$

$$(Uy)(t) = \begin{cases} 0, & t \in [-\tau, 0], \\ - k_2(t_0, \varphi) + k_2(t_0 + t, y_t + \widetilde{\varphi}_{t_0+t}) \\ \qquad + \dfrac{1}{\Gamma(\alpha)} \displaystyle\int_0^t (t-s)^{\alpha-1} F(t_0 + s, y_s + \widetilde{\varphi}_{t_0+s}) ds, & t \in [0, \eta]. \end{cases}$$

It is easy to see that the operator equation

$$y = Sy + Uy \tag{2.8}$$

has a solution $y \in E(\eta, \gamma)$ if and only if $y$ is a solution of equation (2.4). Thus $x(t_0 + t) = y(t) + \widetilde{\varphi}(t_0 + t)$ is a solution of equation (2.1) on $[0, \eta]$. Therefore, the existence of a solution of the fractional IVP (2.1) is equivalent that (2.8) has a fixed point in $E(\eta, \gamma)$.

Now we show that $S + U$ has a fixed point in $E(\eta, \gamma)$. The proof is divided into three steps.

**Claim I.** $Sz + Uy \in E(\eta, \gamma)$ for every pair $z, y \in E(\eta, \gamma)$.

In fact, for every pair $z, y \in E(\eta, \gamma)$, $Sz + Uy \in C([-\tau, \eta], \mathbb{R}^n)$. Also, it is obvious that $(Sz + Uy)(t) = 0$, $t \in [-\tau, 0]$.

Moreover, for $t \in [0, \eta]$, by (2.5)-(2.7) and the condition (H3), we have

$$|(Sz)(t) + (Uy)(t)|$$
$$\leq |-k_1(t_0, \varphi) + k_1(t_0 + t, z_t + \widetilde{\varphi}_{t_0+t})| + |-k_2(t_0, \varphi) + k_2(t_0 + t, y_t + \widetilde{\varphi}_{t_0+t})|$$
$$+ \frac{1}{\Gamma(\alpha)} \int_0^t |(t-s)^{\alpha-1} F(t_0 + s, y_s + \widetilde{\varphi}_{t_0+s})| ds$$
$$\leq \frac{2\gamma}{3} + \frac{1}{\Gamma(\alpha)} \left( \int_0^t (t-s)^{\frac{\alpha-1}{1-\alpha_1}} ds \right)^{1-\alpha_1} \left( \int_{t_0}^{t_0+t} (m(s))^{\frac{1}{\alpha_1}} ds \right)^{\alpha_1}$$
$$\leq \frac{2\gamma}{3} + \frac{1}{\Gamma(\alpha)} \left( \int_0^t (t-s)^{\frac{\alpha-1}{1-\alpha_1}} ds \right)^{1-\alpha_1} \left( \int_{t_0}^{t_0+\sigma} (m(s))^{\frac{1}{\alpha_1}} ds \right)^{\alpha_1}$$
$$\leq \frac{2\gamma}{3} + \frac{M\eta^{(1+\beta)(1-\alpha_1)}}{\Gamma(\alpha)(1+\beta)^{1-\alpha_1}}$$
$$\leq \gamma.$$

Therefore,

$$\|Sz + Uy\| = \sup_{t \in [0, \eta]} |(Sz)(t) + (Uy)(t)| \leq \gamma,$$

which means that $Sz + Uy \in E(\eta, \gamma)$ for any $z, y \in E(\eta, \gamma)$.

**Claim II.** $S$ is a contraction on $E(\eta, \gamma)$.

For any $y', y'' \in E(\eta, \gamma)$, $y'_t + \widetilde{\varphi}_{t_0+t}, y''_t + \widetilde{\varphi}_{t_0+t} \in A(\delta, \gamma)$. So by (H5), we get that

$$|(Sy')(t) - (Sy'')(t)| = |k_1(t_0 + t, y'_t + \widetilde{\varphi}_{t_0+t}) - k_1(t_0 + t, y''_t + \widetilde{\varphi}_{t_0+t})|$$
$$\leq l \|y' - y''\|,$$

which implies that

$$\|Sy' - Sy''\| \leq l \|y' - y''\|.$$

In view of $0 < l < 1$, $S$ is a contraction on $E(\eta, \gamma)$.

**Claim III.** $U$ is a completely continuous operator.

Let

$$(U_1 y)(t) = \begin{cases} 0, & t \in [-\tau, 0], \\ -k_2(t_0, \varphi) + k_2(t_0 + t, y_t + \widetilde{\varphi}_{t_0+t}), & t \in [0, \eta] \end{cases}$$

and

$$(U_2 y)(t) = \begin{cases} 0, & t \in [-\tau, 0], \\ \dfrac{1}{\Gamma(\alpha)} \displaystyle\int_0^t (t-s)^{\alpha-1} F(t_0 + s, y_s + \widetilde{\varphi}_{t_0+s}) ds, & t \in [0, \eta]. \end{cases}$$

Clearly, $U = U_1 + U_2$.

Since $k_2$ is completely continuous, $U_1$ is continuous and $\{U_1 y : y \in E(\eta, \gamma)\}$ is uniformly bounded. From the condition that the set $\{t \to k_2(t, x_t) : x \in \Lambda\}$ is equicontinuous for any bounded set $\Lambda$ in $A(\sigma, \gamma)$, we can conclude that $U_1$ is a completely continuous operator.

On the other hand, for any $t \in [0, \eta]$, we have

$$|(U_2 y)(t)| \leq \frac{1}{\Gamma(\alpha)} \int_0^t (t-s)^{\alpha-1} |F(t_0 + s, y_s + \widetilde{\varphi}_{t_0+s})| ds$$

$$\leq \frac{1}{\Gamma(\alpha)} \left( \int_0^t (t-s)^{\frac{\alpha-1}{1-\alpha_1}} ds \right)^{1-\alpha_1} \left( \int_{t_0}^{t_0+t} (m(s))^{\frac{1}{\alpha_1}} ds \right)^{\alpha_1}$$

$$\leq \frac{M\eta^{(1+\beta)(1-\alpha_1)}}{\Gamma(\alpha)(1+\beta)^{1-\alpha_1}}.$$

Hence, $\{U_2 y : y \in E(\eta, \gamma)\}$ is uniformly bounded.

Now, we will prove that $\{U_2 y : y \in E(\eta, \gamma)\}$ is equicontinuous. For any $0 \leq t_1 < t_2 \leq \eta$ and $y \in E(\eta, \gamma)$, we get that

$$|(U_2 y)(t_2) - (U_2 y)(t_1)|$$

$$= \left| \frac{1}{\Gamma(\alpha)} \int_0^{t_1} ((t_2-s)^{\alpha-1} - (t_1-s)^{\alpha-1}) F(t_0 + s, y_s + \widetilde{\varphi}_{t_0+s}) ds \right.$$

$$\left. + \frac{1}{\Gamma(\alpha)} \int_{t_1}^{t_2} (t_2-s)^{\alpha-1} F(t_0 + s, y_s + \widetilde{\varphi}_{t_0+s}) ds \right|$$

$$\leq \frac{1}{\Gamma(\alpha)} \int_0^{t_1} ((t_1-s)^{\alpha-1} - (t_2-s)^{\alpha-1}) |F(t_0 + s, y_s + \widetilde{\varphi}_{t_0+s})| ds$$

$$+ \frac{1}{\Gamma(\alpha)} \int_{t_1}^{t_2} (t_2-s)^{\alpha-1} |F(t_0 + s, y_s + \widetilde{\varphi}_{t_0+s})| ds$$

$$\leq \frac{M}{\Gamma(\alpha)} \left( \int_0^{t_1} ((t_1-s)^{\alpha-1} - (t_2-s)^{\alpha-1})^{\frac{1}{1-\alpha_1}} ds \right)^{1-\alpha_1}$$

$$+ \frac{M}{\Gamma(\alpha)} \left( \int_{t_1}^{t_2} ((t_2-s)^{\alpha-1})^{\frac{1}{1-\alpha_1}} ds \right)^{1-\alpha_1}$$

$$\leq \frac{M}{\Gamma(\alpha)} \left( \int_0^{t_1} (t_1-s)^{\beta} - (t_2-s)^{\beta} ds \right)^{1-\alpha_1} + \frac{M}{\Gamma(\alpha)} \left( \int_{t_1}^{t_2} (t_2-s)^{\beta} ds \right)^{1-\alpha_1}$$

$$\leq \frac{M}{\Gamma(\alpha)(1+\beta)^{1-\alpha_1}} (t_1^{1+\beta} - t_2^{1+\beta} + (t_2-t_1)^{1+\beta})^{1-\alpha_1}$$

$$+ \frac{M}{\Gamma(\alpha)(1+\beta)^{1-\alpha_1}} (t_2-t_1)^{(1+\beta)(1-\alpha_1)}$$

$$\leq \frac{2M}{\Gamma(\alpha)(1+\beta)^{1-\alpha_1}} (t_2-t_1)^{(1+\beta)(1-\alpha_1)},$$

which means that $\{U_2 y : y \in E(\eta, \gamma)\}$ is equicontinuous. Moreover, it is clear that $U_2$ is continuous. So $U_2$ is a completely continuous operator. Then $U = U_1 + U_2$ is a completely continuous operator.

Therefore, Krasnoselskii fixed point theorem shows that $S + U$ has a fixed point on $E(\eta, \gamma)$, and hence the fractional IVP (2.1) has a solution $x(t) = \varphi(0) + y(t - t_0)$ for all $t \in [t_0, t_0 + \eta]$. $\qquad\square$

In the case where $k_1 \equiv 0$, we get the following result.

**Corollary 2.1.** *Assume that there exist $\sigma \in (0, a)$ and $\gamma \in (0, \infty)$ such that (H1)-(H3) hold and*

**(H5)′** *$k$ is continuous and for any $x', x'' \in A(\sigma, \gamma)$, $t \in I$*

$$|k(t, x_t') - k(t, x_t'')| \le l\|x' - x''\|, \quad \text{where} \quad l \in (0, 1).$$

*Then fractional IVP (2.1) has at least one solution on $[t_0, t_0 + \eta]$ for some positive number $\eta$.*

In the case where $k_2 \equiv 0$, we have the following result.

**Corollary 2.2.** *Assume that there exist $\sigma \in (0, a)$ and $\gamma \in (0, \infty)$ such that (H1)-(H3) hold and*

**(H6)′** *$k$ is completely continuous and for any bounded set $\Lambda$ in $A(\sigma, \gamma)$, the set $\{t \to k(t, x_t) : x \in \Lambda\}$ is equicontinuous on $C(I, \mathbb{R}^n)$.*

*Then fractional IVP (2.1) has at least one solution on $[t_0, t_0 + \eta]$ for some positive number $\eta$.*

### 2.2.3  Extremal Solutions

Define the order relation " $\le$ " by the cone $K$ in $C(J, \mathbb{R}^n)$, given by

$$K = \{z \in C(J, \mathbb{R}^n) \mid z(t) \ge 0 \ \text{ for all } \ t \in J\}.$$

Clearly, the cone $K$ is normal in $C(J, \mathbb{R}^n)$. Note that the order relation " $\le$ " in $C(J, \mathbb{R}^n)$ also induces the order relation in the space $\mathcal{C}$ which we also denote by " $\le$ " itself when there is no confusion.

We give the following definitions in the sequel.

**Definition 2.1.** A mapping $f : I \times \mathcal{C} \to \mathbb{R}^n$ is called $L^{\frac{1}{\delta}}$-Lipschitz if

(i) $t \mapsto f(t, z)$ is Lebesgue measurable for each $z \in \mathcal{C}$;

(ii) there exist a constant $\delta \in [0, \alpha)$ and a function $l \in L^{\frac{1}{\delta}}(I, \mathbb{R}_+)$ such that

$$|f(t, z) - f(t, y)| \le l(t)\|z - y\|_*, \quad \text{a.e.} \ t \in I$$

for all $z, y \in \mathcal{C}$.

**Definition 2.2.** A mapping $g : I \times \mathcal{C} \to \mathbb{R}^n$ is said to be Carathéodory if

(i) $t \mapsto g(t, z)$ is Lebesgue measurable for each $z \in \mathcal{C}$;

(ii) $z \mapsto g(t, z)$ is continuous almost everywhere for $t \in I$.

Furthermore, a Carathéodory function $g(t, z)$ is called $L^{\frac{1}{\beta}}$-Carathéodory if

**(iii)** for each real number $r > 0$, there exist a constant $\beta \in [0, \alpha)$ and a function $m_r \in L^{\frac{1}{\beta}}(I, \mathbb{R}_+)$ such that

$$|g(t, z)| \leq m_r(t), \quad \text{a.e. } t \in I$$

for all $z \in \mathcal{C}$ with $\|z\|_* \leq r$.

**Definition 2.3.** A mapping $h : I \times \mathcal{C} \to \mathbb{R}^n$ is said to be Chandrabhan if

**(i)** $t \mapsto h(t, z)$ is Lebesgue measurable for each $z \in \mathcal{C}$;
**(ii)** $z \mapsto h(t, z)$ is nondecreasing almost everywhere for $t \in I$.

Furthermore, a Chandrabhan function $h(t, z)$ is called $L^{\frac{1}{\gamma}}$-Chandrabhan if

**(iii)** for each real number $r > 0$, there exist a constant $\gamma \in [0, \alpha)$ and a function $w_r \in L^{\frac{1}{\gamma}}(I, \mathbb{R}_+)$ such that

$$|h(t, z)| \leq w_r(t), \quad \text{a.e. } t \in I$$

for all $z \in \mathcal{C}$ with $\|z\|_* \leq r$.

**Definition 2.4.** A function $x \in C(J, \mathbb{R}^n)$ is called a solution of fractional IVP (2.1) on $J$ if

**(i)** the function $[x(t) - k(t, x_t)]$ is absolutely continuous on $I$;
**(ii)** $x_{t_0} = \varphi$, and
**(iii)** $x$ satisfies the equation in (2.1).

**Definition 2.5.** A function $a \in C(J, \mathbb{R}^n)$ is called a lower solution of fractional IVP (2.1) on $J$ if the function $[a(t) - k(t, a_t)]$ is absolutely continuous on $I$, and

$$\begin{cases} {}^{C}_{t_0}D_t^{\alpha}(a(t) - k(t, a_t)) \leq F(t, a_t), \quad \text{a.e. } t \in (t_0, t_0 + \sigma], \\ a_{t_0} \leq \varphi. \end{cases}$$

Again, a function $b \in C(J, \mathbb{R}^n)$ is called an upper solution of fractional IVP (2.1) on $J$ if the function $[b(t) - k(t, b_t)]$ is absolutely continuous on $I$, and $[b(t) - k(t, b_t)]$ is absolutely continuous on $I$, and

$$\begin{cases} {}^{C}_{t_0}D_t^{\alpha}(b(t) - k(t, b_t)) \geq F(t, b_t), \quad \text{a.e. } t \in (t_0, t_0 + \sigma], \\ b_{t_0} \geq \varphi. \end{cases}$$

Finally, a function $x \in C(J, \mathbb{R}^n)$ is a solution of fractional IVP (2.1) on $J$ if it is a lower as well as a upper solution of fractional IVP (2.1) on $J$.

**Definition 2.6.** A solution $x_M$ of fractional IVP (2.1) is said to be maximal if for any other solution $x$ to fractional IVP (2.1), one has $x(t) \leq x_M(t)$ for all $t \in J$. Again, a solution $x_m$ of fractional IVP (2.1) is said to be minimal if $x_m(t) \leq x(t)$ for all $t \in J$, where $x$ is any solution for fractional IVP (2.1) on $J$.

We need the following hypotheses in the sequel.

**(F1)** $F(t, z_t) = f(t, z_t) + g(t, z_t) + h(t, z_t)$, where $f, g, h : I \times C \to \mathbb{R}^n$;

**(F2)** fractional IVP (2.1) has a lower solution $a$ and an upper solution $b$ with $a \leq b$;

**(k0)** $k(t, z)$ is continuous with respect to $t$ on $I$ for any $z \in C$;

**(k1)** $|k(t, z) - k(t, y)| \leq k_0 \|z - y\|_*$, for $z, y \in C$, $t \in I$, where $k_0 > 0$;

**(k2)** $k(t, z)$ is nondecreasing with respect to $z$ for any $z \in C$ and almost all $t \in I$;

**(f1)** $f$ is $L^{\frac{1}{\delta}}$-Lipschitz, and there exists $\eta \in [0, \alpha)$ such that $|f(t, 0)| \in L^{\frac{1}{\eta}}(I, \mathbb{R}_+)$;

**(f2)** $f(t, z)$ is nondecreasing with respect to $z$ for any $z \in C$ and almost all $t \in I$;

**(g1)** $g$ is $L^{\frac{1}{\beta}}$-Carathéodory;

**(g2)** $g(t, z)$ is nondecreasing with respect to $z$ for any $z \in C$ and almost all $t \in I$;

**(h1)** $h$ is $L^{\frac{1}{\gamma}}$-Chandrabhan.

For any positive constant $r$, let $B_r = \{z \in C(J, \mathbb{R}^n) : \|z\| \leq r\}$. Set

$$q_1 = \frac{\alpha - 1}{1 - \delta} \in (-1, 0), \quad L = \|l\|_{L^{\frac{1}{\delta}} I}$$

and

$$q_2 = \frac{\alpha - 1}{1 - \beta} \in (-1, 0), \quad M_r = \|m_r\|_{L^{\frac{1}{\beta}} I}.$$

In order to prove our main results, we need the following lemma.

**Lemma 2.2.** *Assume that the hypotheses (F1), (f1), (g1) and (h1) hold. $x \in C(J, \mathbb{R}^n)$ is a solution for fractional IVP (2.1) on $J$ if and only if $x$ satisfies the following relation*

$$\begin{cases} x(t) = \varphi(0) + k(t, x_t) - k(t_0, \varphi) + \dfrac{1}{\Gamma(\alpha)} \displaystyle\int_{t_0}^{t} (t - s)^{\alpha - 1} F(s, x_s) ds, & \text{for } t \in I, \\ x(t_0 + \theta) = \varphi(\theta), & \text{for } \theta \in I_0. \end{cases}$$
$$(2.9)$$

**Proof.** For any positive constant $r$ and $x \in B_r$, since $x_t$ is continuous in $t$, according to $(g_1)$ and Definition 2.2(i)-(ii), $g(t, x_t)$ is a measurable function on $I$. Direct calculation gives that $(t - s)^{\alpha - 1} \in L^{\frac{1}{1 - \beta}}[t_0, t]$, for $t \in I$ and $\beta \in [0, \alpha)$. By using Lemma 1.1 (Hölder inequality) and Definition 2.2 (iii), for $t \in I$, we obtain that

$$\int_{t_0}^{t} |(t - s)^{\alpha - 1} g(s, x_s)| ds \leq \left( \int_{t_0}^{t} (t - s)^{\frac{\alpha - 1}{1 - \beta}} ds \right)^{1 - \beta} \|m_r\|_{L^{\frac{1}{\beta}}[t_0, t]}$$

$$= \left( \int_{t_0}^{t} (t - s)^{q_2} ds \right)^{1 - \beta} \|m_r\|_{L^{\frac{1}{\beta}}[t_0, t]}$$

$$\leq \frac{M_r}{(1 + q_2)^{1 - \beta}} \sigma^{(1 + q_2)(1 - \beta)},$$

which means that $(t - s)^{\alpha - 1} g(s, x_s)$ is Lebesgue integrable with respect to $s \in [t_0, t]$ for all $t \in I$ and $x \in B_r$.

According to (f1), for $t \in I$ and $x \in B_r$, we get that

$$|f(t, x_t)| \leq l(t)\|x_t\|_* + |f(t, 0)| \leq l(t)r + |f(t, 0)|.$$

Using the similar argument and noting that (f1) and (h1), we can get that $(t - s)^{\alpha-1}f(s, x_s)$ and $(t-s)^{\alpha-1}h(s, x_s)$ are Lebesgue integrable with respect to $s \in [t_0, t]$ for all $t \in I$ and $x \in B_r$. Thus, according to (F1), we get that $(t - s)^{\alpha-1}F(s, x_s)$ is Lebesgue integrable with respect to $s \in [t_0, t]$ for all $t \in I$ and $x \in B_r$.

Let $G(\theta, s) = (t - \theta)^{-\alpha}|\theta - s|^{\alpha-1}m_r(s)$. Since $G(\theta, s)$ is a nonnegative, measurable function on $D = [t_0, t] \times [t_0, t]$ for $t \in I$, we have

$$\int_{t_0}^t \left( \int_{t_0}^t G(\theta, s)ds \right) d\theta = \int_D G(\theta, s)dsd\theta = \int_{t_0}^t \left( \int_{t_0}^t G(\theta, s)d\theta \right) ds$$

and

$$\int_D G(\theta, s)dsd\theta = \int_{t_0}^t \left( \int_{t_0}^t G(\theta, s)ds \right) d\theta$$

$$= \int_{t_0}^t (t - \theta)^{-\alpha} \left( \int_{t_0}^t |\theta - s|^{\alpha-1}m_r(s)ds \right) d\theta$$

$$= \int_{t_0}^t (t - \theta)^{-\alpha} \left( \int_{t_0}^\theta (\theta - s)^{\alpha-1}m_r(s)ds \right) d\theta$$

$$+ \int_{t_0}^t (t - \theta)^{-\alpha} \left( \int_\theta^t (s - \theta)^{\alpha-1}m_r(s)ds \right) d\theta$$

$$\leq \frac{2M_r}{(1 + q_2)^{1-\beta}} \sigma^{(1+q_2)(1-\beta)} \int_{t_0}^t (t - \theta)^{-\alpha} d\theta$$

$$\leq \frac{2M_r}{(1 - \alpha)(1 + q_2)^{1-\beta}} \sigma^{(1+q_2)(1-\beta)+1-\alpha}.$$

Therefore, $G_1(\theta, s) = (t - \theta)^{-\alpha}(\theta - s)^{\alpha-1}g(s, x_s)$ is a Lebesgue integrable function on $D = (t_0, t) \times (t_0, t)$, then we have

$$\int_{t_0}^t d\theta \int_{t_0}^\theta G_1(\theta, s)ds = \int_{t_0}^t ds \int_s^t G_1(\theta, s)d\theta.$$

We now prove that

$$_{t_0}D_t^\alpha (_{t_0}D_t^{-\alpha}F(t, x_t)) = F(t, x_t), \quad \text{for } t \in (t_0, t_0 + \sigma].$$

Indeed, we have

$$_{t_0}D_t^\alpha (_{t_0}D_t^{-\alpha}g(t, x_t)) = \frac{1}{\Gamma(1 - \alpha)\Gamma(\alpha)} \frac{d}{dt} \int_{t_0}^t (t - \theta)^{-\alpha} \left( \int_{t_0}^\theta (\theta - s)^{\alpha-1}g(s, x_s)ds \right) d\theta$$

$$= \frac{1}{\Gamma(1 - \alpha)\Gamma(\alpha)} \frac{d}{dt} \int_{t_0}^t d\theta \int_{t_0}^\theta G_1(\theta, s)ds$$

$$= \frac{1}{\Gamma(1 - \alpha)\Gamma(\alpha)} \frac{d}{dt} \int_{t_0}^t ds \int_s^t G_1(\theta, s)d\theta$$

$$= \frac{1}{\Gamma(1-\alpha)\Gamma(\alpha)} \frac{d}{dt} \int_{t_0}^{t} g(s, x_s)ds \int_{s}^{t} (t-\theta)^{-\alpha}(\theta-s)^{\alpha-1}d\theta$$

$$= \frac{d}{dt} \int_{t_0}^{t} g(s, x_s)ds$$

$$= g(t, x_t) \quad \text{for } t \in (t_0, t_0 + \sigma].$$

Similarly, we can get

$$_{t_0}D_t^\alpha({}_{t_0}D_t^{-\alpha}f(t, x_t)) = f(t, x_t), \quad _{t_0}D_t^\alpha({}_{t_0}D_t^{-\alpha}h(t, x_t)) = h(t, x_t), \quad \text{for } t \in (t_0, t_0 + \sigma],$$

which implies

$$_{t_0}D_t^\alpha({}_{t_0}D_t^{-\alpha}F(t, x_t)) = F(t, x_t), \quad \text{for } t \in (t_0, t_0 + \sigma].$$

If $x$ satisfies the relation (2.9), then we get that $x(t) - k(t, x_t)$ is absolutely continuous on $I$. In fact, for any disjoint family of open intervals $\{(a_i, b_i)\}_{1 \le i \le n}$ on $I$ with $\sum_{i=1}^{n}(b_i - a_i) \to 0$, we have

$$\sum_{i=1}^{n} \frac{1}{\Gamma(\alpha)} \left| \int_{t_0}^{b_i} (b_i - s)^{\alpha-1}g(s, x_s)ds - \int_{t_0}^{a_i} (a_i - s)^{\alpha-1}g(s, x_s)ds \right|$$

$$\le \sum_{i=1}^{n} \frac{1}{\Gamma(\alpha)} \left| \int_{a_i}^{b_i} (b_i - s)^{\alpha-1}g(s, x_s)ds \right|$$

$$+ \sum_{i=1}^{n} \frac{1}{\Gamma(\alpha)} \left| \int_{t_0}^{a_i} (b_i - s)^{\alpha-1}g(s, x_s)ds - \int_{t_0}^{a_i} (a_i - s)^{\alpha-1}g(s, x_s)ds \right|$$

$$\le \sum_{i=1}^{n} \frac{1}{\Gamma(\alpha)} \int_{a_i}^{b_i} (b_i - s)^{\alpha-1}m_r(s)ds$$

$$+ \sum_{i=1}^{n} \frac{1}{\Gamma(\alpha)} \int_{t_0}^{a_i} ((a_i - s)^{\alpha-1} - (b_i - s)^{\alpha-1})m_r(s)ds$$

$$\le \sum_{i=1}^{n} \frac{1}{\Gamma(\alpha)} \left( \int_{a_i}^{b_i} (b_i - s)^{\frac{\alpha-1}{1-\beta}}ds \right)^{1-\beta} \|m_r\|_{L^{\frac{1}{\beta}}I}$$

$$+ \sum_{i=1}^{n} \frac{1}{\Gamma(\alpha)} \left( \int_{t_0}^{a_i} ((a_i - s)^{\frac{\alpha-1}{1-\beta}} - (b_i - s)^{\frac{\alpha-1}{1-\beta}})ds \right)^{1-\beta} \|m_r\|_{L^{\frac{1}{\beta}}I}$$

$$\le \sum_{i=1}^{n} \frac{(b_i - a_i)^{(1+q_2)(1-\beta)}}{\Gamma(\alpha)(1+q_2)^{1-\beta}} \|m_r\|_{L^{\frac{1}{\beta}}I}$$

$$+ \sum_{i=1}^{n} \frac{(a_i^{1+q_2} - b_i^{1+q_2} + (b_i - a_i)^{1+q_2})^{1-\beta}}{\Gamma(\alpha)(1+q_2)^{1-\beta}} \|m_r\|_{L^{\frac{1}{\beta}}I}$$

$$\le 2\sum_{i=1}^{n} \frac{(b_i - a_i)^{(1+q_2)(1-\beta)}}{\Gamma(\alpha)(1+q_2)^{1-\beta}} M_r$$

$$\to 0.$$

Using the similar method, as $\sum_{i=1}^{n}(b_i - a_i) \to 0$, we can get that

$$\sum_{i=1}^{n} \frac{1}{\Gamma(\alpha)} \left| \int_{t_0}^{b_i} (b_i - s)^{\alpha-1} f(s, x_s) ds - \int_{t_0}^{a_i} (a_i - s)^{\alpha-1} f(s, x_s) ds \right| \to 0$$

and

$$\sum_{i=1}^{n} \frac{1}{\Gamma(\alpha)} \left| \int_{t_0}^{b_i} (b_i - s)^{\alpha-1} h(s, x_s) ds - \int_{t_0}^{a_i} (a_i - s)^{\alpha-1} h(s, x_s) ds \right| \to 0.$$

Hence, $\sum_{i=1}^{n} \|x(b_i) - k(b_i, x_{b_i}) - x(a_i) + k(a_i, x_{a_i})\| \to 0$, as $\sum_{i=1}^{n}(b_i - a_i) \to 0$. Therefore, $x(t) - k(t, x_t)$ is absolutely continuous on $I$ which implies that $x(t) - k(t, x_t)$ is differentiable for almost all $t \in I$. According to the argument above, for almost all $t \in (t_0, t_0 + \sigma]$, we have

$$
\begin{aligned}
{}^{C}_{t_0}D^{\alpha}_t (x(t) - k(t, x_t)) &= {}^{C}_{t_0}D^{\alpha}_t \left( \varphi(0) - k(t_0, \varphi) + \frac{1}{\Gamma(\alpha)} \int_{t_0}^{t} (t-s)^{\alpha-1} F(s, x_s) ds \right) \\
&= {}^{C}_{t_0}D^{\alpha}_t \left( \frac{1}{\Gamma(\alpha)} \int_{t_0}^{t} (t-s)^{\alpha-1} F(s, x_s) ds \right) \\
&= {}^{C}_{t_0}D^{\alpha}_t ({}_{t_0}D^{-\alpha}_t F(t, x_t)) \\
&= {}_{t_0}D^{\alpha}_t ({}_{t_0}D^{-\alpha}_t F(t, x_t)) - {}_{t_0}D^{-\alpha}_t F(t, x_t)\big|_{t=t_0} \frac{(t-t_0)^{-\alpha}}{\Gamma(1-\alpha)} \\
&= F(t, x_t) - {}_{t_0}D^{-\alpha}_t F(t, x_t)\big|_{t=t_0} \frac{(t-t_0)^{-\alpha}}{\Gamma(1-\alpha)}.
\end{aligned}
$$

Since $(t-s)^{\alpha-1} F(s, x_s)$ is Lebesgue integrable with respect to $s \in [t_0, t]$ for all $t \in I$, we know that ${}_{t_0}D^{-\alpha}_t F(t, x_t)\big|_{t=t_0} = 0$, which means that ${}^{C}_{t_0}D^{q}_t x(t) = F(t, x_t)$, a.e. $t \in (t_0, t_0 + \sigma]$. Hence, $x \in C(J, \mathbb{R}^n)$ is a solution of fractional IVP (2.1). On the other hand, it is obvious that if $x \in C(J, \mathbb{R}^n)$ is a solution of fractional IVP (2.1), then $x$ satisfies the relation (2.9), and this completes the proof. $\square$

**Theorem 2.2.** *Assume that the hypotheses (F1), (F2), (k0)-(k2), (f1), (f2), (g1), (g2) and (h1) hold. Then fractional IVP (2.1) has a minimal and a maximal solution in $[a, b]$ defined on $J$ provided that*

$$k_0 + \frac{L\sigma^{(1+q_1)(1-\delta)}}{\Gamma(\alpha)(1+q_1)^{1-\delta}} < 1. \tag{2.10}$$

**Proof.** Define three operators $A$, $B$ and $C$ on $C(J, \mathbb{R}^n)$ as follows

$$
\begin{cases}
(Ax)(t) = k(t, x_t) - k(t_0, \varphi) + \dfrac{1}{\Gamma(\alpha)} \displaystyle\int_{t_0}^{t} (t-s)^{\alpha-1} f(s, x_s) ds, & \text{for } t \in I, \\
(Ax)(t_0 + \theta) = 0, & \text{for } \theta \in I_0,
\end{cases}
$$

$$
\begin{cases}
(Bx)(t) = \varphi(0) + \dfrac{1}{\Gamma(\alpha)} \displaystyle\int_{t_0}^{t} (t-s)^{\alpha-1} g(s, x_s) ds, & \text{for } t \in I, \\
(Bx)(t_0 + \theta) = \varphi(\theta), & \text{for } \theta \in I_0,
\end{cases}
$$

and

$$\begin{cases} (Cx)(t) = \dfrac{1}{\Gamma(\alpha)} \displaystyle\int_{t_0}^{t} (t-s)^{\alpha-1} h(s, x_s) ds, & \text{for } t \in I, \\[2mm] (Cx)(t_0 + \theta) = 0, & \text{for } \theta \in I_0, \end{cases}$$

where $x \in C(J, \mathbb{R}^n)$.

Obviously, $Ax + Bx + Cx \in C(J, \mathbb{R}^n)$ for every $x \in C(J, \mathbb{R}^n)$. From Lemma 2.2, we get that fractional IVP (2.1) is equivalent to the operator equation $(Ax)(t) + (Bx)(t) + (Cx)(t) = x(t)$ for $t \in J$. Now we show that the operator equation $Ax + Bx + Cx = x$ has a least and a greatest solution in $[a, b]$. The proof is divided into three steps.

**Claim I.** $A$ is a contraction in $C(J, \mathbb{R}^n)$.

For any $x, y \in C(J, \mathbb{R}^n)$ and $t \in I$, according to (k1) and (f1), we have

$$|(Ax)(t) - (Ay)(t)| \leq |k(t, x_t) - k(t, y_t)| + \frac{1}{\Gamma(\alpha)} \int_{t_0}^{t} (t-s)^{\alpha-1} |f(s, x_s) - f(s, y_s)| ds$$

$$\leq k_0 \|x_t - y_t\|_* + \frac{1}{\Gamma(\alpha)} \int_{t_0}^{t} (t-s)^{\alpha-1} l(s) \|x_s - y_s\|_* \, ds$$

$$\leq k_0 \|x - y\| + \frac{1}{\Gamma(\alpha)} \left( \int_{t_0}^{t} (t-s)^{\frac{\alpha-1}{1-\delta}} ds \right)^{1-\delta} \|l\|_{L^{\frac{1}{\delta}}[t_0, t]} \|x - y\|$$

$$\leq k_0 \|x - y\| + \frac{L \sigma^{(1+q_1)(1-\delta)}}{\Gamma(\alpha)(1 + q_1)^{1-\delta}} \|x - y\|$$

$$= \left( k_0 + \frac{L \sigma^{(1+q_1)(1-\delta)}}{\Gamma(\alpha)(1 + q_1)^{1-\delta}} \right) \|x - y\|,$$

which implies that $\|Ax - Ay\| \leq (k_0 + \frac{L \sigma^{(1+q_1)(1-\delta)}}{\Gamma(\alpha)(1+q_1)^{1-\delta}}) \|x - y\|$. Therefore, $A$ is a contraction in $C(J, \mathbb{R}^n)$ according to (2.10).

**Claim II.** $B$ is a completely continuous operator and $C$ is a totally bounded operator.

For any $x \in C(J, \mathbb{R}^n)$, we can choose a positive constant $r$ such that $\|x\| \leq r$. Firstly, we will prove that $B$ is continuous on $B_r$. For $x^n, x \in B_r, n = 1, 2, \ldots$ with $\lim_{n \to \infty} \|x^n - x\| = 0$, we get

$$\lim_{n \to \infty} x_s^n = x_s, \quad \text{for } s \in I.$$

Thus, by (g1) and Definition 2.2(ii), and noting that $x_s$ is continuous with respect to $s$ on $I$, we have

$$\lim_{n \to \infty} g(s, x_s^n) = g(s, x_s), \quad \text{a.e } s \in I.$$

On the other hand, noting that $|g(s, x_s^n) - g(s, x_s)| \leq 2m_r(s)$, by Lebesgue dominated convergence theorem, we have

$$|(Bx^n)(t) - (Bx)(t)| \leq \frac{1}{\Gamma(\alpha)} \int_{t_0}^{t} (t-s)^{\alpha-1} |g(s, x_s^n) - g(s, x_s)| ds \to 0, \quad \text{as } n \to \infty,$$

which implies

$$\|Bx^n - Bx\| \to 0 \quad \text{as} \quad n \to \infty.$$

This means that $B$ is continuous.

Next, we will show that for any positive constant $r$, $\{Bx : x \in B_r\}$ is relatively compact. It suffices to show that the family of functions $\{Bx : x \in B_r\}$ is uniformly bounded and equicontinuous.

For any $x \in B_r$ and $t \in I$, we have

$$|(Bx)(t)| \leq |\varphi(0)| + \frac{1}{\Gamma(\alpha)} \int_{t_0}^{t} (t-s)^{\alpha-1} |g(s, x_s)| ds$$

$$\leq |\varphi(0)| + \frac{1}{\Gamma(\alpha)} \left( \int_0^t (t-s)^{\frac{\alpha-1}{1-\beta}} ds \right)^{1-\beta} \|m_r\|_{L^{\frac{1}{\beta}}[t_0, t]}$$

$$\leq |\varphi(0)| + \frac{M_r}{\Gamma(\alpha)} \left( \int_0^t (t-s)^{q_2} ds \right)^{1-\beta}$$

$$\leq |\varphi(0)| + \frac{M_r \sigma^{(1+q_2)(1-\beta)}}{\Gamma(\alpha)(1+q_2)^{1-\beta}}.$$

For $\theta \in I_0$, we have $|(Bx)(t_0 + \theta)| = |\varphi(\theta)|$. Thus $\{Bx : x \in B_r\}$ is uniformly bounded. In the following, we will show that $\{Bx : x \in B_r\}$ is a family of equicontinuous functions.

For any $x \in B_r$ and $t_0 \leq t_1 < t_2 \leq t_0 + \sigma$, we get

$$|(Bx)(t_2) - (Bx)(t_1)|$$

$$= \frac{1}{\Gamma(\alpha)} \left| \int_{t_0}^{t_1} ((t_2-s)^{\alpha-1} - (t_1-s)^{\alpha-1}) g(s, x_s) ds + \int_{t_1}^{t_2} (t_2-s)^{\alpha-1} g(s, x_s) ds \right|$$

$$\leq \frac{1}{\Gamma(\alpha)} \int_{t_0}^{t_1} |((t_2-s)^{\alpha-1} - (t_1-s)^{\alpha-1}) g(s, x_s)| ds$$

$$+ \frac{1}{\Gamma(\alpha)} \int_{t_1}^{t_2} |(t_2-s)^{\alpha-1} g(s, x_s)| ds$$

$$\leq \frac{1}{\Gamma(\alpha)} \int_{t_0}^{t_1} ((t_1-s)^{\alpha-1} - (t_2-s)^{\alpha-1}) m_r(s) ds + \frac{1}{\Gamma(\alpha)} \int_{t_1}^{t_2} (t_2-s)^{\alpha-1} m_r(s) ds$$

$$\leq \frac{1}{\Gamma(\alpha)} \left( \int_{t_0}^{t_1} ((t_1-s)^{\alpha-1} - (t_2-s)^{\alpha-1})^{\frac{1}{1-\beta}} ds \right)^{1-\beta} \|m_r\|_{L^{\frac{1}{\beta}}[t_0, t_1]}$$

$$+ \frac{1}{\Gamma(\alpha)} \left( \int_{t_1}^{t_2} ((t_2-s)^{\alpha-1})^{\frac{1}{1-\beta}} ds \right)^{1-\beta} \|m_r\|_{L^{\frac{1}{\beta}}[t_1, t_2]}$$

$$\leq \frac{M_r}{\Gamma(\alpha)} \left( \int_{t_0}^{t_1} (t_1-s)^{q_2} - (t_2-s)^{q_2} ds \right)^{1-\beta} + \frac{M_r}{\Gamma(\alpha)} \left( \int_{t_1}^{t_2} (t_2-s)^{q_2} ds \right)^{1-\beta}$$

$$\leq \frac{M_r}{\Gamma(\alpha)(1+q_2)^{1-\beta}} ((t_1-t_0)^{1+q_2} - (t_2-t_0)^{1+q_2} + (t_2-t_1)^{1+q_2})^{1-\beta}$$

$$+ \frac{M_r}{\Gamma(\alpha)(1+q_2)^{1-\beta}} (t_2-t_1)^{(1+q_2)(1-\beta)}$$

$$\leq \frac{2M_r}{\Gamma(\alpha)(1+q_2)^{1-\beta}}(t_2-t_1)^{(1+q_2)(1-\beta)}.$$

As $t_2 - t_1 \to 0$, the right-hand side of the above inequality tends to zero independently of $x \in B_r$. In view of the continuity of $\varphi$, we can get that $\{Bx \ : \ x \in B_r\}$ is a family of equicontinuous functions. Therefore, $\{Bx \ : \ x \in B_r\}$ is relatively compact by Arzela-Ascoli theorem.

Using the similar argument, we can get that $\{Cx \ : \ x \in B_r\}$ is also relatively compact, which means that $C$ is totally bounded.

**Claim III.** $A, B$ and $C$ are three monotone increasing operators.

Since $x, y \in C(J, \mathbb{R}^n)$ with $x \leq y$ implies that $x_t \leq y_t$ for $t \in I$, according to (k2) and (f2), we have

$$(Ax)(t) = k(t, x_t) - k(t_0, \varphi) + \frac{1}{\Gamma(\alpha)} \int_{t_0}^t (t-s)^{\alpha-1} f(s, x_s) ds$$

$$\leq k(t, y_t) - k(t_0, \varphi) + \frac{1}{\Gamma(\alpha)} \int_{t_0}^t (t-s)^{\alpha-1} f(s, y_s) ds$$

$$= (Ay)(t).$$

Hence $A$ is a monotone increasing operator. Similarly, we can conclude that $B$ and $C$ are also monotone increasing operators according to (g2), (h1) and Definition 2.3 (ii).

Clearly, $K$ is a normal cone. From (F2) and Definition 2.5, we have that $a \leq Aa + Ba + Ca$ and $b \geq Ab + Bb + Cb$ with $a \leq b$. Thus the operators $A, B$ and $C$ satisfy all the conditions of Theorem 1.11 and hence the operator equation $Ax + Bx + Cx = x$ has a least and a greatest solution in $[a, b]$. Therefore, fractional IVP (2.1) has a minimal and a maximal solution on $J$. $\qquad\square$

**Example 2.1.** Consider the following IVP of scalar discontinuous fractional functional differential equation

$$\begin{cases} {}^C_0 D_t^{\frac{1}{2}} x(t) = F(t, x_t) \\ \qquad = f(t) + \zeta(t)x(t) + \dfrac{1}{t^{1/3}}x(t-1) + \zeta(t)h(x(t)), \quad \text{a.e. } t \in (0, \sigma], \\ x(\theta) = 0, \qquad\qquad\qquad\qquad\qquad\qquad\qquad \theta \in [-1, 0], \end{cases}$$

$$(2.11)$$

where $0 < \sigma \leq (\frac{1}{2\Gamma(3/2)})^2 = \frac{1}{\pi}$ and we take functions $f(t)$, $\zeta(t)$ and $h(x(t))$ as follows

$$f(t) = \begin{cases} t, & 0 \leq t \leq \dfrac{\sigma}{2}, \\ 0, & \dfrac{\sigma}{2} < t \leq \sigma, \end{cases} \qquad \zeta(t) = \begin{cases} 0, & 0 \leq t \leq \dfrac{\sigma}{2}, \\ 1, & \dfrac{\sigma}{2} < t \leq \sigma, \end{cases}$$

and

$$h(x(t)) = \begin{cases} x(t), & x(t) \geq 0, \\ x(t) - 1, & x(t) < 0. \end{cases}$$

Evidently, the function

$$F(t, \varphi) = f(t, \varphi) + g(t, \varphi) + h(t, \varphi), \quad \varphi \in C([-1, 0], \mathbb{R}),$$

where

$$f(t, \varphi) = f(t) + \zeta(t)\varphi(0), \quad g(t, \varphi) = \frac{1}{t^{1/3}}\varphi(-1) \quad \text{and} \quad h(t, \varphi) = \zeta(t)h(\varphi(0)).$$

One can easily check that $a(t) = 0$ is a lower solution of fractional IVP (2.11). On the other hand, let

$$b(t) = \begin{cases} t, & t \in [0, \sigma], \\ 0, & t \in [-1, 0]. \end{cases}$$

Then, $b \in C([-1, \sigma], \mathbb{R})$ is a upper solution of fractional IVP (2.11). In fact, direct calculation gives that

$$^C_0 D_t^{\frac{1}{2}} b(t) = \frac{t^{\frac{1}{2}}}{\Gamma(\frac{3}{2})} \geq 2t \geq F(t, b_t) = \begin{cases} t, & 0 < t \leq \frac{\sigma}{2}, \\ 2t, & \frac{\sigma}{2} < t \leq \sigma, \end{cases} \quad \text{for } t \in (0, \sigma].$$

Moreover, noting that $\Gamma(\frac{3}{2}) = \frac{\sqrt{\pi}}{2}$, it is easy to verify that conditions (k0)-(k2), (f1)-(f2), (g1)-(g2), (h1) and (2.10) are satisfied. Therefore, Theorem 2.2 allows us to conclude that fractional IVP (2.11) has a minimal and a maximal solution in $[0, b]$ defined on $[-1, \sigma]$.

## 2.3 *p*-Type Neutral Equations

### 2.3.1 *Introduction*

Let $\mathcal{C} = C([-1, 0], \mathbb{R}^n)$ denote the space of continuous functions on $[-1, 0]$. For any element $\varphi \in \mathcal{C}$, define the norm $\|\varphi\|_* = \sup_{\theta \in [-1, 0]} |\varphi(\theta)|$.

Consider the IVP of fractional $p$-type neutral functional differential equations of the form

$$^C_{t_0} D_t^q g(t, x_t) = f(t, x_t), \tag{2.12}$$

$$x_{t_0} = \varphi, \qquad (t_0, \varphi) \in \Omega, \tag{2.13}$$

where $^C_{t_0} D_t^q$ is Caputo fractional derivative of order $0 < q < 1$, $\Omega$ is an open subset of $[0, \infty) \times \mathcal{C}$ and $g, f : \Omega \to \mathbb{R}^n$ are given functionals satisfying some assumptions that will be specified later. $x_t \in \mathcal{C}$ is defined by $x_t(\theta) = x(p(t, \theta))$, where $-1 \leq \theta \leq 0$, $p(t, \theta)$ is a $p$-function.

**Definition 2.7.** (Lakshmikantham, Wen and Zhang, 1994) A function $p \in C(J \times [-1, 0], \mathbb{R})$ is called a $p$-function if it has the following properties:

**(i)** $p(t, 0) = t$;
**(ii)** $p(t, -1)$ is a nondecreasing function of $t$;
**(iii)** there exists a $\sigma \geq -\infty$ such that $p(t, \theta)$ is an increasing function for $\theta$ for each $t \in (\sigma, \infty)$;

**(iv)** $p(t,0) - p(t,-1) > 0$ for $t \in (\sigma, \infty)$.

In the following, we suppose $t \in (\sigma, \infty)$.

**Definition 2.8.** (Lakshmikantham, Wen and Zhang, 1994) Let $t_0 \geq 0, A > 0$ and $x \in C([p(t_0, -1), t_0 + A], \mathbb{R}^n)$. For any $t \in [t_0, t_0 + A]$, we define $x_t$ by

$$x_t(\theta) = x(p(t, \theta)), \quad -1 \leq \theta \leq 0,$$

so that $x_t \in \mathcal{C} = C([-1, 0], \mathbb{R}^n)$.

Note that the frequently used symbol "$x_t$" (in Hale, 1977; Lakshmikantham, 2008; Lakshmikantham, Wen and Zhang, 1994, $x_t(\theta) = x(t + \theta)$, where $-\tau \leq \theta \leq 0, r > 0, r = $ const) in the theory of functional differential equations with bounded delay is a partial case of the above definition. Indeed, in this case we can put $p(t, \theta) = t + r\theta, \ \theta \in [-1, 0]$.

**Definition 2.9.** A function $x$ is said to be a solution of fractional IVP (2.12)-(2.13) on $[p(t_0, -1), t_0 + \alpha]$, if there are $t_0 \geq 0, \alpha > 0$, such that

**(i)** $x \in C([p(t_0, -1), t_0 + \alpha], \mathbb{R}^n)$ and $(t, x_t) \in \Omega$, for $t \in [t_0, t_0 + \alpha]$;
**(ii)** $x_{t_0} = \varphi$;
**(iii)** $g(t, x_t)$ is differentiable and (2.12) holds almost everywhere on $[t_0, t_0 + \alpha]$.

We need the following lemma relative to $p$-function before we proceed further, which is taken from Lakshmikantham, Wen and Zhang, 1994.

**Lemma 2.3.** *(Lakshmikantham, Wen and Zhang, 1994) Suppose that $p(t, \theta)$ is a $p$-function. For $A > 0$, $\tau \in (\sigma, \infty)$ ($\tau$ may be $\sigma$ if $\sigma > -\infty$), let $x \in C([p(\tau, -1), \tau + A], \mathbb{R}^n)$ and $\varphi \in C([-1, 0], \mathbb{R}^n)$. Then we have*

**(i)** *$x_t$ is continuous in $t$ on $[\tau, \tau + A]$ and $\tilde{p}(t, \theta) = p(\tau + t, \theta) - \tau$ is also a $p$-function;*
**(ii)** *if $p(\tau + t, -1) < \tau$ for $t > 0$, then there exists $-1 < s(\tau, t) < 0$ such that $p(\tau + t, s(\tau, t)) = \tau$ and*

$$\begin{cases} p(\tau + t, -1) \leq p(\tau + t, \theta) \leq \tau, & \text{for } -1 \leq \theta \leq s(\tau, t), \\ \tau \leq p(\tau + t, \theta) \leq \tau + t, & \text{for } s(\tau, t) \leq \theta \leq 0. \end{cases}$$

*Moreover, $s \to 0$ uniformly in $\tau$ as $t \to 0$;*
**(iii)** *there exists a function $\eta \in C([p(\tau, -1), \tau], \mathbb{R}^n)$ such that*

$$\eta(p(\tau, \theta)) = \varphi(\theta) \quad \text{for } -1 \leq \theta \leq 0.$$

It is well known that a neutral functional differential equation (NFDE for short) is one in which the derivatives of the past history or derivatives of functionals of the past history are involved as well as the present state of the system. In other words, in order to guarantee that equation (2.12) is NFDE, the coefficient of $x(t)$ that is contained in $g(t, x_t)$ cannot be equal to zero. Then we need introduce the concept of atomic.

Let $g \in C(\mathbb{R}^+ \times \mathcal{C}, \mathbb{R}^n)$ and $g(t, \varphi)$ be linear in $\varphi$. Then Riesz representation theorem shows that there exists an $n \times n$ matrix function $\eta(t, \theta)$ of bounded variation such that

$$g(t, \varphi) = \int_{-\gamma}^{0} [d_\theta \eta(t, \theta)] \varphi(\theta).$$

For $t_0 \geq 0$ and $\theta_0 \in (-\gamma, 0)$, if

$$\det[\eta(t_0, \theta_0^+) - \eta(t_0, \theta_0^-)] \neq 0,$$

then we say that $g(t, \varphi)$ is atomic at $\theta_0$ for $t_0$. Similarly, one can define $g(t, \varphi)$ to be atomic at the endpoints $-r$ and $0$ for $t_0$. If for every $t \geq 0$, $g(t, \varphi)$ is atomic at $\theta_0$ for $t$, then we say that $g(t, \varphi)$ is atomic at $\theta_0$ for $\mathbb{R}^+$. If $g(t, \varphi)$ is not linear in $\varphi$, suppose that $g(t, \varphi)$ has a Fréchet derivative with respect to $\varphi$, then $g'_\varphi(t, \varphi) \psi \in \mathbb{R}^n$ for $(t, \varphi) \in \mathbb{R}^+ \times \mathcal{C}$ and $\psi \in \mathcal{C}$, where $g'_\psi$ denote the Fréchet derivative of $g$ with respect to $\varphi$. Then $g'_\varphi(t, \varphi)$ is a linear mapping from $\mathcal{C}$ into $\mathbb{R}^n$ and therefore

$$g'_\varphi(t, \varphi) \psi = \int_{-\gamma}^{0} [d_\theta \mu(t, \varphi, \theta)] \psi(\theta),$$

where $\mu(t, \varphi, \theta)$ is a matrix function of bounded variation. As before, if $\det[\mu(t_0, \varphi_0, \theta_0^+) - \mu(t_0, \varphi_0, \theta_0^-)] \neq 0$, for $t_0 \geq 0$, then we say that, the nonlinear $g(t, \varphi)$ is atomic at $\theta_0$ for $(t_0, \varphi_0)$. If $g(t, \varphi)$ is atomic at $\theta_0$, for every $(t, \varphi)$, then we say that $g(t, \varphi)$ is atomic at $\theta_0$ for $\mathbb{R}^+ \times \mathcal{C}$.

For a detailed discussion on atomic concept we refer the reader to the books Hale, 1977; Lakshmikantham, Wen and Zhang, 1994.

**Lemma 2.4.** *(Hale, 1977; Lakshmikantham, Wen and Zhang, 1994) Suppose that $g(t, \varphi)$ is atomic at zero on $\Omega$. Then there are a continuous $n \times n$ matrix function $A(t, \varphi)$ with $\det A(t, \varphi) \neq 0$ on $\Omega$ and a functional $L(t, \varphi, \psi)$ which is linear in $\psi$ such that*

$$g'_\varphi(t, \varphi) \psi = A(t, \varphi) \psi(0) + L(t, \varphi, \psi).$$

*Moreover, there exists a continuous function $\gamma : \Omega \times [0, 1] \to \mathbb{R}^+$ with $\gamma(t, \varphi, 0) = 0$ such that for every $s \in [0, 1]$ and $\psi$ with $(t, \psi) \in \Omega$, $\psi(\theta) = 0$ for $-1 \leq \theta \leq -s$,*

$$|L(t, \varphi, \psi)| \leq \gamma(t, \varphi, s) \|\psi\|_*.$$

In Subsection 2.3.2, we discuss various criteria on existence and uniqueness of solutions for fractional IVP (2.12)-(2.12). Subsection 2.3.3 is devoted to the continuous dependence on data for solutions.

### 2.3.2 *Existence and Uniqueness*

Assume that the functional $f : \Omega \to \mathbb{R}^n$ satisfies the following conditions.

**(H1)** $f(t, \varphi)$ is Lebesgue measurable with respect to $t$ for any $(t, \varphi) \in \Omega$;

**(H2)** $f(t, \varphi)$ is continuous with respect to $\varphi$ for any $(t, \varphi) \in \Omega$;

**(H3)** there exist a constant $q_1 \in (0, q)$ and a $L^{\frac{1}{q_1}}$-integrable function $m$ such that $|f(t, \varphi)| \leq m(t)$ for any $(t, \varphi) \in \Omega$.

For each $(t_0, \varphi) \in \Omega$, let $\tilde{p}(t, \theta) = p(t_0 + t, \theta) - t_0$. Define the function $\eta \in C([\tilde{p}(0, -1), \infty), \mathbb{R}^n)$ by

$$\begin{cases} \eta(\tilde{p}(0, \theta)) = \varphi(\theta), & \text{for } \theta \in [-1, 0], \\ \eta(t) = \varphi(0), & \text{for } t \in [0, \infty). \end{cases}$$

Let $x \in C([p(t_0, -1), t_0 + \alpha], \mathbb{R}^n)$, $\alpha < A$ and let

$$x(t_0 + t) = \eta(t) + z(t) \quad \text{for } \tilde{p}(0, -1) \leq t \leq \alpha. \tag{2.14}$$

**Lemma 2.5.** *$x(t)$ is a solution of fractional IVP (2.12)-(2.13) on $[p(t_0, -1), t_0 + \alpha]$ if and only if $z(t)$ satisfies the relation*

$$\begin{cases} g(t_0 + t, \tilde{\eta}_t + \tilde{z}_t) - g(t_0, \varphi) = \dfrac{1}{\Gamma(q)} \displaystyle\int_0^t (t - s)^{q-1} f(t_0 + s, \tilde{\eta}_s + \tilde{z}_s) ds, \quad t \in [0, \alpha], \\ \tilde{z}_0 = 0, \end{cases} \tag{2.15}$$

*where $\tilde{\eta}_t(\theta) = \eta(\tilde{p}(t, \theta))$, $\tilde{z}_t(\theta) = z(\tilde{p}(t, \theta))$, for $-1 \leq \theta \leq 0$.*

**Proof.** Since $x_t$ is continuous in $t$, $x_t$ is a measurable function, therefore according to conditions (H1) and (H2), $f(t, x_t)$ is Lebesgue measurable on $[t_0, t_0 + \alpha]$. Direct calculation gives that $(t - s)^{q-1} \in L^{\frac{1}{1-q_1}}[t_0, t]$, for $t \in [t_0, t_0 + \alpha]$ and $q_1 \in (0, q)$. In light of Hölder inequality, we obtain that $(t - s)^{q-1} f(s, x_s)$ is Lebesgue integrable with respect to $s \in [t_0, t]$ for all $t \in [t_0, t_0 + \alpha]$, and

$$\int_{t_0}^t |(t - s)^{q-1} f(s, x_s)| ds \leq \|(t - s)^{q-1}\|_{L^{\frac{1}{1-q_1}}[t_0, t]} \|m\|_{L^{\frac{1}{q_1}}[t_0, t_0 + \alpha]}.$$

Hence $x(t)$ is the solution of fractional IVP (2.12)-(2.13) if and only if it satisfies the relation

$$\begin{cases} g(t, x_t) - g(t_0, x_{t_0}) = \dfrac{1}{\Gamma(q)} \displaystyle\int_{t_0}^t (t - u)^{q-1} f(u, x_u) du, & \text{for } t \in [t_0, t_0 + \alpha], \\ x_{t_0} = \varphi, \end{cases}$$

or setting $u = t_0 + s$,

$$\begin{cases} g(t_0 + t, x_{t_0 + t}) - g(t_0, x_{t_0}) = \dfrac{1}{\Gamma(q)} \displaystyle\int_0^t (t - s)^{q-1} f(t_0 + s, x_{t_0 + s}) ds, & \text{for } t \in [0, \alpha], \\ x_{t_0} = \varphi. \end{cases} \tag{2.16}$$

In view of (2.14), we have

$$\begin{aligned} x_{t_0 + t}(\theta) = x(p(t_0 + t, \theta)) &= x(\tilde{p}(t, \theta) + t_0) \\ &= \eta(\tilde{p}(t, \theta)) + z(\tilde{p}(t, \theta)) \\ &= \tilde{\eta}_t(\theta) + \tilde{z}_t(\theta), \quad \text{for } t \in [0, \alpha]. \end{aligned}$$

In particular $x_{t_0}(\theta) = \tilde{\eta}_0(\theta) + \tilde{z}_0(\theta)$. Hence $x_{t_0} = \varphi$ if and only if $\tilde{z}_0 = 0$ according to $\tilde{\eta} = \varphi$. It is clear that $x(t)$ satisfies (2.16) if and only if $z(t)$ satisfies (2.15). $\quad\square$

For any $\sigma, \xi > 0$, let

$$E(\sigma, \xi) = \{z \in C([\tilde{p}(0, -1), \sigma], \mathbb{R}^n) : \tilde{z}_0 = 0, \ \|\tilde{z}_t\|_* \leq \xi \ \text{for} \ t \in [0, \sigma]\},$$

which is a bounded closed convex subset of the Banach space $C([\tilde{p}(0, -1), \sigma], \mathbb{R}^n)$ endowed with supremum norm $\|\cdot\|$.

**Lemma 2.6.** *Suppose $\Omega \subseteq R \times C$ is open, $W \subset \Omega$ is compact. For any a neighborhood $V' \subset \Omega$ of $W$, there is a neighborhood $V'' \subset V'$ of $W$ and there exist positive numbers $\delta$ and $\xi$ such that $(t_0 + t, \tilde{\eta}_t + \lambda \tilde{z}_t) \in V'$ with $0 \leq \lambda \leq 1$ for any $(t_0, \varphi) \in V'', t \in [0, \sigma]$ and $z \in E(\sigma, \xi)$.*

The proof of Lemma 2.6 is similar to that of (iii) of Lemma 2.1.8 in Lakshmikantham, Wen and Zhang, 1994, thus it is omitted.

Suppose $g$ is atomic at 0 on $\Omega$. Define two operators $S$ and $T$ on $E(\alpha, \beta)$ as follows

$$\begin{cases} (Sz)(t) = 0, & \text{for } t \in [\tilde{p}(0, -1), 0], \\ A(t_0 + t, \tilde{\eta}_t)(Sz)(t) = g(t_0, \varphi) - g(t_0 + t, \tilde{\eta}_t + \tilde{z}_t) \\ \qquad\qquad + g'_\varphi(t_0 + t, \tilde{\eta}_t)\tilde{z}_t - L(t_0 + t, \tilde{\eta}_t, \tilde{z}_t), & \text{for } t \in [0, \alpha] \end{cases} \tag{2.17}$$

and

$$\begin{cases} (Tz)(t) = 0, & \text{for } t \in [\tilde{p}(0, -1), 0], \\ A(t_0 + t, \tilde{\eta}_t)(Tz)(t) \\ \qquad = \dfrac{1}{\Gamma(q)} \displaystyle\int_0^t (t - s)^{q-1} f(t_0 + s, \tilde{\eta}_s + \tilde{z}_s)ds, & \text{for } t \in [0, \alpha], \end{cases} \tag{2.18}$$

where $A(t_0 + t, \tilde{\eta}_t)$, $L(t_0 + t, \tilde{\eta}_t, \tilde{z}_t)$ are functions described in Lemma 2.4.

It is clear that the operator equation

$$z = Sz + Tz \tag{2.19}$$

has a solution $z \in E(\alpha, \beta)$ if and only if $z$ is a solution of (2.15). Therefore the existence of a solution of fractional IVP (2.12)-(2.13) is equivalent to determining $\alpha, \beta > 0$ such that $S + T$ has a fixed point on $E(\alpha, \beta)$.

We are now in a position to prove the following existence results, and the proof is based on Krasnoselskii fixed point theorem.

**Theorem 2.3.** *Suppose $g : \Omega \to \mathbb{R}^n$ is continuous together with its first Fréchet derivative with respect to the second argument, and $g$ is atomic at 0 on $\Omega$. $f : \Omega \to \mathbb{R}^n$ satisfies conditions (H1)-(H3). $W \subset \Omega$ is a compact set. Then there exist a neighborhood $V \subset \Omega$ of $W$ and a constant $\alpha > 0$ such that for any $(t_0, \varphi) \in V$, fractional IVP (2.12)-(2.13) has a solution which exists on $[p(t_0, -1), t_0 + \alpha]$.*

**Proof.** As we have mentioned above, we only need to discuss operator equation (2.19). For any $(t, \varphi) \in \Omega$, the property of the matrix function $A(t, \varphi)$ which is nonsingular and continuous on $\Omega$ implies that its inverse matrix $A^{-1}(t, \varphi)$ exists and is continuous on $\Omega$. Let $V_0 \subset \Omega$ be the neighborhood of $W$, suppose that there is an $M > 0$ such that

$$|A^{-1}(t^0, \varphi)| \leq M, \quad \text{for every } (t^0, \varphi) \in V_0. \tag{2.20}$$

Note the complete continuity of the function $(m(t))^{\frac{1}{q_1}}$, hence, for a given positive number $N$, there must exist a number $\alpha_0 > 0$ satisfying

$$\left( \int_{t_0}^{t_0+\alpha_0} (m(s))^{\frac{1}{q_1}} ds \right)^{q_1} \leq N. \tag{2.21}$$

Due to the continuity of functions $\gamma$ and $g'_\varphi$ described in Lemma 2.4, there exist a neighborhood $V_1 \subset \Omega$ of $W$ and constants $h_1 > 0$, $h_2 \in (0, 1]$ such that

$$|\gamma(t_0 + t, \tilde{\eta}_t, -s)| = |\gamma(t_0 + t, \tilde{\eta}_t, -s) - \gamma(t_0 + t, \tilde{\eta}_t, 0)| < \frac{1}{4M}, \tag{2.22}$$

$$|g'_\varphi(t_0 + t, \tilde{\eta}_t + \psi) - g'_\varphi(t_0 + t, \tilde{\eta}_t)| < \frac{1}{8M}, \tag{2.23}$$

whenever $(t_0 + t, \tilde{\eta}_t)$, $(t_0 + t, \tilde{\eta}_t + \psi) \in V_1$ and $\|\psi\|_* < h_1$, $-s \in [0, h_2]$.

Let $V_2 = V_0 \cap V_1$. According to Lemma 2.6, we can find a neighborhood $V \subset V_2$ of $W$ and positive numbers $\alpha_1$ and $\beta$ with $\alpha_1 < \alpha_0$ and $\beta \leq h_1$ such that $(t_0 + t, \tilde{\eta}_t + \lambda \tilde{z}_t) \in V_2$ with $0 \leq \lambda \leq 1$ for any $(t_0, \varphi) \in V$, $t \in [0, \alpha_1]$ and $z \in E(\alpha_1, \beta)$. Let

$$h(t_0 + t, \tilde{\eta}_t, \tilde{z}_t) = g(t_0 + t, \tilde{\eta}_t + \tilde{z}_t) - g(t_0 + t, \tilde{\eta}_t) - g'_\varphi(t_0 + t, \tilde{\eta}_t)\tilde{z}_t.$$

Then we have

$$|h(t_0 + t, \tilde{\eta}_t, \tilde{z}_t)| = \left| \left( \int_0^1 g'_\varphi(t_0 + t, \tilde{\eta}_t + \lambda \tilde{z}_t) d\lambda - g'_\varphi(t_0 + t, \tilde{\eta}_t) \right) \tilde{z}_t \right|$$

$$\leq \left| \int_0^1 g'_\varphi(t_0 + t, \tilde{\eta}_t + \lambda \tilde{z}_t) - g'_\varphi(t_0 + t, \tilde{\eta}_t) d\lambda \right| \| \tilde{z}_t \|_*. \tag{2.24}$$

According to (2.20), (2.23) and (2.24), for any $(t_0, \varphi) \in V$, we have

$$|A^{-1}(t_0 + t, \tilde{\eta}_t) \, h(t_0 + t, \tilde{\eta}_t, \tilde{z}_t)| \leq \frac{\beta}{8}. \tag{2.25}$$

On the other hand, for any $z, w \in E(\alpha_1, \beta)$ and $t \in [0, \alpha_1]$

$$\|\lambda \tilde{z}_t + (1 - \lambda)\tilde{w}_t\|_* \leq \|\lambda \tilde{z}_t\|_* + \|(1 - \lambda)\tilde{w}_t\|_* \leq \lambda\beta + (1 - \lambda)\beta = \beta,$$

thus, $(t_0 + t, \tilde{\eta}_t + \lambda \tilde{z}_t + (1 - \lambda)\tilde{w}_t) \in V_2$, and

$$|h(t_0 + t, \tilde{\eta}_t, \tilde{z}_t) - h(t_0 + t, \tilde{\eta}_t, \tilde{w}_t)|$$
$$= |g(t_0 + t, \tilde{\eta}_t + \tilde{z}_t) - g(t_0 + t, \tilde{\eta}_t + \tilde{w}_t) - g'_\varphi(t_0 + t, \tilde{\eta}_t)(\tilde{z}_t - \tilde{w}_t)|$$
$$= \left| \left( \int_0^1 g'_\varphi(t_0 + t, \tilde{\eta}_t + \tilde{w}_t + \lambda(\tilde{z}_t - \tilde{w}_t)) d\lambda - g'_\varphi(t_0 + t, \tilde{\eta}_t) \right)(\tilde{z}_t - \tilde{w}_t) \right|$$
$$\leq \left| \int_0^1 (g'_\varphi(t_0 + t, \tilde{\eta}_t + \lambda \tilde{z}_t + (1 - \lambda)\tilde{w}_t) - g'_\varphi(t_0 + t, \tilde{\eta}_t)) d\lambda \right| \|\tilde{z}_t - \tilde{w}_t\|_*. \tag{2.26}$$

From (2.20), (2.23) and (2.26), we have

$$|A^{-1}(t_0 + t, \tilde{\eta}_t)[h(t_0 + t, \tilde{\eta}_t, \tilde{z}_t) - h(t_0 + t, \tilde{\eta}_t, \tilde{w}_t)]| \leq \frac{1}{8}\|\tilde{z}_t - \tilde{w}_t\|_*. \qquad (2.27)$$

By (ii) of Lemma 2.3, we can also choose $\alpha_2 < \alpha_1$ such that for $t \in [0, \alpha_2]$, $-s(0, t) \in [0, h_2]$. From (2.20) and (2.22), we have

$$\begin{aligned} &|A^{-1}(t_0 + t, \tilde{\eta}_t)||L(t_0 + t, \tilde{\eta}_t, \tilde{z}_t)| \\ &\leq |A^{-1}(t_0 + t, \tilde{\eta}_t)|\gamma(t_0 + t, \tilde{\eta}_t, -s(0, t))\|\tilde{z}_t\|_* \\ &\leq \frac{1}{4}\|\tilde{z}_t\|_*, \end{aligned} \qquad (2.28)$$

whenever $t \in [0, \alpha_2]$ and $z \in E(\alpha_2, \beta)$.

Now consider the expression $g(t_0, \varphi) - g(t_0 + t, \tilde{\eta}_t)$. Since $g$ is continuous in $\Omega$ and noting the facts that $\tilde{\eta}_t$ is continuous in $t$ and $\tilde{\eta}_0 = \varphi$, there exists a constant $\alpha_3 < \alpha_2$ such that

$$|g(t_0, \varphi) - g(t_0 + t, \tilde{\eta}_t)| < \frac{\beta}{8M}, \qquad (2.29)$$

whenever $t \in [0, \alpha_3]$.

Set

$$\alpha = \min\left\{\alpha_3, (1 + b)^{\frac{1}{1+b}}\left(\frac{\Gamma(q)\beta}{2MN}\right)^{\frac{1}{(1-q_1)(1+b)}}\right\}, \qquad (2.30)$$

where $b = \frac{q-1}{1-q_1} \in (-1, 0)$.

Now we show that for any $(t_0, \varphi) \in V$, $S + T$ has a fixed point on $E(\alpha, \beta)$, where $S$ and $T$ are defined as in (2.17) and (2.18) respectively. The proof is divided into three steps.

**Claim I.** $Sz + Tw \in E(\alpha, \beta)$ whenever $z, w \in E(\alpha, \beta)$.

Obviously, for every pair $z, w \in E(\alpha, \beta)$, $(Sz)(t)$ and $(Tw)(t)$ are continuous in $t \in [0, \alpha]$. From (2.25), (2.28) and (2.29), for $t \in [0, \alpha]$, we have

$$\begin{aligned} |(Sz)(t)| \leq &|A^{-1}(t_0 + t, \tilde{\eta}_t)|\left\{|g(t_0, \varphi) - g(t_0 + t, \tilde{\eta}_t)| + |L(t_0 + t, \tilde{\eta}_t, \tilde{z}_t)|\right. \\ &\left. + |g(t_0 + t, \tilde{\eta}_t) - g(t_0 + t, \tilde{\eta}_t + \tilde{z}_t) + g'_\varphi(t_0 + t, \tilde{\eta}_t)\tilde{z}_t|\right\} \\ \leq &\frac{\beta}{2}. \end{aligned}$$

For $t \in [0, \alpha]$, by using (2.20), (2.21), (2.30) and Hölder inequality, we have

$$\begin{aligned} |(Tw)(t)| &\leq |A^{-1}(t_0 + t, \tilde{\eta}_t)|\frac{1}{\Gamma(q)}\left|\int_0^t (t - s)^{q-1}f(t_0 + s, \tilde{\eta}_s + \tilde{w}_s)ds\right| \\ &\leq \frac{M}{\Gamma(q)}\left(\int_0^t ((t-s)^{q-1})^{\frac{1}{1-q_1}}ds\right)^{1-q_1}\left(\int_{t_0}^{t_0+\alpha}(m(s))^{\frac{1}{q_1}}ds\right)^{q_1} \\ &\leq \frac{MN}{\Gamma(q)}\left(\frac{1}{1+b}\alpha^{1+b}\right)^{1-q_1} \\ &\leq \frac{\beta}{2}. \end{aligned} \qquad (2.31)$$

Thus $|(Sz)(t) + (Tw)(t)| \leq \beta$ i.e. $Sz + Tw \in E(\alpha, \beta)$, whenever $z, w \in E(\alpha, \beta)$.

**Claim II.** $S$ is a contraction mapping from $E(\alpha, \beta)$ into itself whose contraction constant is independent of $(t_0, \varphi) \in V$.

For any $z, w \in E(\alpha, \beta)$, $\tilde{w}_0 - \tilde{z}_0 = 0$. Hence (ii) of Lemma 2.3 and Lemma 2.4 are applicable to $\tilde{w}_t - \tilde{z}_t$. For every pair $z, w \in E(\alpha, \beta)$, from (2.27), (2.28) and noting the fact that

$$
\begin{aligned}
\sup_{0 \leq t \leq \alpha} \|\tilde{z}_t - \tilde{w}_t\|_* &= \sup_{0 \leq t \leq \alpha} \sup_{-1 \leq \theta \leq 0} |z(\tilde{p}(t, \theta)) - w(\tilde{p}(t, \theta))| \\
&= \sup_{0 \leq t \leq \alpha} \sup_{\tilde{p}(t,-1) \leq s \leq t} |z(s) - w(s)| \\
&= \sup_{\tilde{p}(0,-1) \leq s \leq \alpha} |z(s) - w(s)| \\
&= \|z - w\|,
\end{aligned}
$$

we have

$$
\begin{aligned}
\|Sz - Sw\| &= \sup_{\tilde{p}(0,-1) \leq t \leq \alpha} |(Sz)(t) - (Sw)(t)| \\
&= \sup_{0 \leq t \leq \alpha} |(Sz)(t) - (Sw)(t)| \\
&\leq \sup_{0 \leq t \leq \alpha} \{|A^{-1}(t_0 + t, \tilde{\eta}_t)|(|L(t_0 + t, \tilde{\eta}_t, \tilde{w}_t - \tilde{z}_t)| \\
&\quad + |h(t_0 + t, \tilde{\eta}_t, \tilde{z}_t) - h(t_0 + t, \tilde{\eta}_t, \tilde{w}_t)|)\} \\
&\leq \left(\frac{1}{8} + \frac{1}{4}\right) \sup_{0 \leq t \leq \alpha} \|\tilde{z}_t - \tilde{w}_t\|_* \\
&\leq \frac{3}{8}\|z - w\|.
\end{aligned}
$$

Therefore $S$ is a contraction mapping from $E(\alpha, \beta)$ into itself whose contraction constant is independent of $(t_0, \varphi) \in V$.

**Claim III.** Now we show that $T$ is a completely continuous operator.

For any $z \in E(\alpha, \beta)$ and $0 \leq t_1 < t_2 \leq \alpha$, we get

$$
\begin{aligned}
&|(Tz)(t_2) - (Tz)(t_1)| \\
&= \left| A^{-1}(t_0 + t_2, \tilde{\eta}_{t_2}) \frac{1}{\Gamma(q)} \int_0^{t_2} (t_2 - s)^{q-1} f(t_0 + s, \tilde{\eta}_s + \tilde{z}_s) ds \right. \\
&\quad \left. - A^{-1}(t_0 + t_1, \tilde{\eta}_{t_1}) \frac{1}{\Gamma(q)} \int_0^{t_1} (t_1 - s)^{q-1} f(t_0 + s, \tilde{\eta}_s + \tilde{z}_s) ds \right| \\
&= \left| A^{-1}(t_0 + t_2, \tilde{\eta}_{t_2}) \frac{1}{\Gamma(q)} \int_{t_1}^{t_2} (t_2 - s)^{q-1} f(t_0 + s, \tilde{\eta}_s + \tilde{z}_s) ds \right. \\
&\quad + A^{-1}(t_0 + t_2, \tilde{\eta}_{t_2}) \frac{1}{\Gamma(q)} \int_0^{t_1} (t_2 - s)^{q-1} f(t_0 + s, \tilde{\eta}_s + \tilde{z}_s) ds \\
&\quad \left. - A^{-1}(t_0 + t_2, \tilde{\eta}_{t_2}) \frac{1}{\Gamma(q)} \int_0^{t_1} (t_1 - s)^{q-1} f(t_0 + s, \tilde{\eta}_s + \tilde{z}_s) ds \right.
\end{aligned}
$$

$$+ A^{-1}(t_0 + t_2, \tilde{\eta}_{t_2}) \frac{1}{\Gamma(q)} \int_0^{t_1} (t_1 - s)^{q-1} f(t_0 + s, \tilde{\eta}_s + \tilde{z}_s) ds$$

$$- A^{-1}(t_0 + t_1, \tilde{\eta}_{t_1}) \frac{1}{\Gamma(q)} \int_0^{t_1} (t_1 - s)^{q-1} f(t_0 + s, \tilde{\eta}_s + \tilde{z}_s) ds \Bigg|$$

$$\leq \frac{|A^{-1}(t_0 + t_2, \tilde{\eta}_{t_2})|}{\Gamma(q)} \left| \int_{t_1}^{t_2} (t_2 - s)^{q-1} f(t_0 + s, \tilde{\eta}_s + \tilde{z}_s) ds \right|$$

$$+ \frac{|A^{-1}(t_0 + t_2, \tilde{\eta}_{t_2})|}{\Gamma(q)} \left| \int_0^{t_1} [(t_2 - s)^{q-1} - (t_1 - s)^{q-1}] f(t_0 + s, \tilde{\eta}_s + \tilde{z}_s) ds \right|$$

$$+ \frac{|A^{-1}(t_0 + t_2, \tilde{\eta}_{t_2}) - A^{-1}(t_0 + t_1, \tilde{\eta}_{t_1})|}{\Gamma(q)} \left| \int_0^{t_1} (t_1 - s)^{q-1} f(t_0 + s, \tilde{\eta}_s + \tilde{z}_s) ds \right|$$

$$= \frac{|A^{-1}(t_0 + t_2, \tilde{\eta}_{t_2})|}{\Gamma(q)} (I_1 + I_2) + \frac{|A^{-1}(t_0 + t_2, \tilde{\eta}_{t_2}) - A^{-1}(t_0 + t_1, \tilde{\eta}_{t_1})|}{\Gamma(q)} I_3,$$

where

$$I_1 = \left| \int_{t_1}^{t_2} (t_2 - s)^{q-1} f(t_0 + s, \tilde{\eta}_s + \tilde{z}_s) ds \right|,$$

$$I_2 = \left| \int_0^{t_1} \left( (t_2 - s)^{q-1} - (t_1 - s)^{q-1} \right) f(t_0 + s, \tilde{\eta}_s + \tilde{z}_s) ds \right|,$$

$$I_3 = \left| \int_0^{t_1} (t_1 - s)^{q-1} f(t_0 + s, \tilde{\eta}_s + \tilde{z}_s) ds \right|.$$

By using analogous argument performed in (2.31), we can conclude that

$$I_1 \leq \frac{N}{(1+b)^{1-q_1}} (t_2 - t_1)^{(1+b)(1-q_1)},$$

$$I_3 \leq \frac{N}{(1+b)^{1-q_1}} \left( t_1^{1+b} \right)^{1-q_1},$$

and

$$I_2 \leq \left( \int_0^{t_1} \left| (t_2 - s)^{q-1} - (t_1 - s)^{q-1} \right|^{\frac{1}{1-q_1}} ds \right)^{1-q_1} \left( \int_{t_0}^{t_0+t_1} |f(s, x_s)|^{\frac{1}{q_1}} ds \right)^{q_1}$$

$$\leq N \left( \int_0^{t_1} (t_1 - s)^b - (t_2 - s)^b ds \right)^{1-q_1}$$

$$= \frac{N}{(1+b)^{1-q_1}} \left( t_1^{1+b} - t_2^{1+b} + (t_2 - t_1)^{1+b} \right)^{1-q_1}$$

$$\leq \frac{N}{(1+b)^{1-q_1}} (t_2 - t_1)^{(1+b)(1-q_1)},$$

where $b = \frac{q-1}{1-q_1} \in (-1, 0)$. Therefore

$$|(Tz)(t_2) - (Tz)(t_1)|$$

$$\leq \frac{|A^{-1}(t_0 + t_2, \tilde{\eta}_{t_2})|}{\Gamma(q)} \frac{2N}{(1+b)^{1-q_1}} (t_2 - t_1)^{(1+b)(1-q_1)}$$

$$+ \frac{|A^{-1}(t_0 + t_2, \tilde{\eta}_{t_2}) - A^{-1}(t_0 + t_1, \tilde{\eta}_{t_1})|}{\Gamma(q)} \frac{N}{(1+b)^{1-q_1}} \left(t_1^{1+b}\right)^{1-q_1}.$$

Since $A^{-1}(t_0 + t, \tilde{\eta}_t)$ is continuous in $t \in [0, \alpha]$, then $\{Tz; z \in E(\alpha, \beta)\}$ is equicontinuous. In addition, $T$ is continuous from the condition (H2) and $\{Tz; z \in E(\alpha, \beta)\}$ is uniformly bounded from (2.31), thus $T$ is a completely continuous operator by Arzela-Ascoli theorem.

Therefore, by Theorem 1.7, for every $(t_0, \varphi) \in V$, $S + T$ has a fixed point on $E(\alpha, \beta)$. Hence, fractional IVP (2.12)-(2.13) has a solution defined on $[p(t_0, -1), t_0 + \alpha]$. $\qquad \square$

**Corollary 2.3.** *Suppose that $(t_0, \varphi) \in \Omega$ is given, $g, f$ are defined as in Theorem 2.3. Then there exists a solution of fractional IVP (2.12)-(2.13).*

**Corollary 2.4.** *Suppose that $\Omega, f$ are defined as in Theorem 2.3. If $(t_0, \varphi) \in \Omega$ is given, then the fractional IVP relative to fractional p-type retarded differential equations of the form*

$$\begin{cases} {}^{C}_{t_0}D^q_t x(t) = f(t, x_t), \\ x_{t_0} = \varphi, \end{cases}$$

*has a solution.*

The following existence and uniqueness result for fractional IVP (2.12)-(2.13) is based on Banach contraction mapping principle.

**Theorem 2.4.** *Suppose $(t_0, \varphi) \in \Omega$ is given, $g$ is defined as in Theorem 2.3. $f : \Omega \to \mathbb{R}^n$ satisfies the condition (H3) and*

**(H4)** *$f(t, x_t)$ is measurable for every $(t, x_t) \in \Omega$;*

**(H5)** *let $A > 0$, there exists a nonnegative function $\ell : [0, A] \to [0, \infty)$ continuous at $t = 0$ and $\ell(0) = 0$ such that for any $(t, x_t), (t, y_t) \in \Omega, t \in [t_0, t_0 + A]$, we have*

$$\left| \int_{t_0}^{t} (t - s)^{q-1}(f(s, x_s) - f(s, y_s))ds \right| \leq \ell(t - t_0) \sup_{t_0 \leq s \leq t} \|x_s - y_s\|_*.$$

*Then fractional IVP (2.12)-(2.13) has a unique solution.*

**Proof.** According to the argument of Theorem 2.3, it suffices to prove that $S + T$ has a unique fixed point on $E(\alpha, \beta)$, where $\alpha, \beta > 0$ sufficiently small. Now, choose $\alpha \in (0, A]$ such that (2.30) holds and

$$c = \frac{3}{8} + \sup_{0 \leq s \leq \alpha} \frac{|A^{-1}(t_0 + s, \tilde{\eta}_s)| \, |\ell(s)|}{\Gamma(q)} < 1.$$

Obviously, $S + T$ is a mapping from $E(\alpha, \beta)$ into itself. Using the same argument as that of Theorem 2.3, for any $z, w \in E(\alpha, \beta), t \in [0, \alpha]$, we get

$$|(Sz)(t) - (Sw)(t)| \leq \frac{3}{8}\|z - w\|,$$

and

$$|(Tz)(t) - (Tw)(t)| \leq \frac{|A^{-1}(t_0 + t, \tilde{\eta}_t)|}{\Gamma(q)} \left| \int_0^t (t-s)^{q-1} f(t_0 + s, \tilde{\eta}_s + \tilde{z}_s) ds \right.$$

$$\left. - \int_0^t (t-s)^{q-1} f(t_0 + s, \tilde{\eta}_s + \tilde{w}_s) ds \right|$$

$$\leq \frac{|A^{-1}(t_0 + t, \tilde{\eta}_t)|}{\Gamma(q)} |\ell(t)| \sup_{0 \leq s \leq t} \|\tilde{z}_s - \tilde{w}_s\|_*$$

$$\leq \frac{\sup_{0 \leq s \leq \alpha} |A^{-1}(t_0 + s, \tilde{\eta}_s)| |\ell(s)|}{\Gamma(q)} \|z - w\|.$$

Therefore

$$|(S+T)z(t) - (S+T)w(t)| \leq \left( \frac{3}{8} + \sup_{0 \leq s \leq \alpha} \frac{|A^{-1}(t_0 + s, \tilde{\eta}_s)| |\ell(s)|}{\Gamma(q)} \right) \|z - w\|$$

$$= c\|z - w\|.$$

Hence, we have

$$\|(S+T)z - (S+T)w\| \leq c\|z - w\|,$$

where $c < 1$. By applying Theorem 1.4, we know that $S + T$ has a unique fixed point on $E(\alpha, \beta)$. The proof is complete. $\square$

**Corollary 2.5.** *Suppose the condition (H5) of Theorem 2.4 is replaced by the following condition:*

**(H5)**′ *let $A > 0$, there exist $q_2 \in (0, q)$ and a real-valued function $\ell_1 \in L^{\frac{1}{q_2}}[t_0, t_0 + A]$ such that for any $(t, x_t)$, $(t, y_t) \in \Omega$, $t \in [t_0, t_0 + A]$, we have*

$$|f(t, x_t) - f(t, y_t)| \leq \ell_1(t) \sup_{t_0 \leq s \leq t} \|x_s - y_s\|_*.$$

*Then the result of Theorem 2.4 holds.*

**Proof.** It suffices to prove that the condition (H5) of Theorem 2.4 holds. Note that $\ell_1 \in L^{\frac{1}{q_2}}[t_0, t_0 + A]$, let $K = \|\ell_1\|_{L^{\frac{1}{q_2}}[t_0, t_0 + A]}$. Then for any $(t, x_t)$, $(t, y_t) \in \Omega$ we have

$$\left| \int_{t_0}^t (t-s)^{q-1} (f(s, x_s) - f(s, y_s)) ds \right|$$

$$\leq \int_{t_0}^t (t-s)^{q-1} |f(s, x_s) - f(s, y_s)| ds$$

$$\leq \int_{t_0}^t (t-s)^{q-1} \ell_1(s) ds \sup_{t_0 \leq s \leq t} \|x_s - y_s\|_*$$

$$\leq \frac{K}{(1+b_1)^{1-q_2}} (t-t_0)^{(1+b_1)(1-q_2)} \sup_{t_0 \leq s \leq t} \|x_s - y_s\|_*,$$

where $b_1 = \frac{q-1}{1-q_2} \in (-1,0)$. Let

$$\ell(t - t_0) = \frac{K}{(1 + b_1)^{1-q_2}} (t - t_0)^{(1+b_1)(1-q_2)}.$$

Obviously, $\ell : [0, A] \to [0, \infty)$ continuous at $t = 0$ and $\ell(0) = 0$. Then the condition (H5) of Theorem 2.4 holds. $\qquad\square$

The next result is concerned with the uniqueness of solutions.

**Theorem 2.5.** *Suppose that $g$ is defined as in Theorem 2.3 and the condition (H5)′ of Corollary 2.5 holds. If $x$ is a solution of fractional IVP (2.12)-(2.13), then $x$ is unique.*

**Proof.** Suppose (for contradiction) $x$ and $y$ are the solutions of fractional IVP (2.12)-(2.13) on $[p(t_0, -1), t_0 + A]$ with $x \neq y$, let

$$t_1 = \inf\{t \in [t_0, t_0 + A] : x(t) \neq y(t)\}.$$

Then $t_0 \leq t_1 < t_0 + A$ and

$$x(t) = y(t) \quad \text{for } p(t_0, -1) \leq t < t_1,$$

which implies that

$$x_t(\theta) = x(p(t, \theta)) = y(p(t, \theta)) = y_t(\theta), \ t_0 \leq t < t_1, \ -1 \leq \theta \leq 0. \qquad (2.32)$$

Choose $\alpha > 0$ such that $t_1 + \alpha < t_0 + A$. According to (i) of Definition 2.9, we have

$$\{(t, x_t), \ t_1 \leq t \leq t_1 + \alpha\} \cup \{(t, y_t), \ t_1 \leq t \leq t_1 + \alpha\} \subset \Omega.$$

On the one hand, $x$ and $y$ satisfy (2.12)-(2.13) on $[t_0, t_0 + A]$, thus from (2.32) and the condition (H5)′, for $t \in [t_0, t_1 + \alpha]$, we have

$$
\begin{aligned}
|g(t, x_t) - g(t, y_t)| &\leq \frac{1}{\Gamma(q)} \left| \int_{t_0}^{t} (t - s)^{q-1} (f(s, x_s) - f(s, y_s)) ds \right| \\
&= \frac{1}{\Gamma(q)} \left| \int_{t_1}^{t} (t - s)^{q-1} (f(s, x_s) - f(s, y_s)) ds \right| \\
&\leq \frac{1}{\Gamma(q)} \int_{t_1}^{t} (t - s)^{q-1} \ell_1(s) \, ds \sup_{t_0 \leq s \leq t} \|x_s - y_s\|_* \\
&\leq \frac{K}{\Gamma(q)(1 + b_1)^{1-q_2}} \alpha^{(1+b_1)(1-q_2)} \sup_{t_1 \leq s \leq t_1 + \alpha} \|x_s - y_s\|_*, \quad (2.33)
\end{aligned}
$$

where $b_1 = \frac{q-1}{1-q_2} \in (-1,0), K = \|\ell_1\|_{L^{\frac{1}{q_2}}[t_0, \, t_0+A]}$.

On the other hand, since $g(t, \varphi)$ is continuously differentiable in $\varphi$, we have

$$g(t, x_t) - g(t, y_t) = g'_\varphi(t, y_t)(x_t - y_t) + k\|x_t - y_t\|_* \qquad (2.34)$$

with $k \to 0$ as $\|x_t - y_t\|_* \to 0$.

By the hypothesis that $g(t, \varphi)$ is atomic at $0$ on $\Omega$, there exist a nonsingular continuous matrix function $A(t, y_t)$ and a function $L(t, y_t, \psi)$ which is linear in $\psi$ such that

$$g'_\varphi(t, y_t)\psi = A(t, y_t)\psi(0) + L(t, y_t, \psi). \qquad (2.35)$$

Moreover, there is a positive real-valued continuous function $\gamma(t, y_t, -s)$ such that for every $s \in [-1, 0]$,

$$|L(t, y_t, \psi)| \le \gamma(t, y_t, -s)\|\psi\|_* \qquad (2.36)$$

if $\psi(\theta) = 0$ for $-1 \le \theta \le s$.

Hence for every $t \in [t_1, t_1 + \alpha]$, by (ii) of Lemma 2.3, there is $s(t_1, t - t_1) \in [-1, 0]$ with $s(t_1, t - t_1) \to 0$ as $t \to t_1$ such that

$$|L(t, y_t, x_t - y_t)| \le \gamma(t, y_t, \ s(t_1, t - t_1))\|x_t - y_t\|_*.$$

From (2.34)-(2.36), it follows that

$$g(t, x_t) - g(t, y_t) = A(t, y_t)(x(t) - y(t)) + L(t, y_t, x_t - y_t) + k\|x_t - y_t\|_*,$$

therefore

$$\begin{aligned} |x(t) - y(t)| \le |A^{-1}(t, y_t)|[ &|g(t, x_t) - g(t, y_t)| \\ &+ \gamma(t, y_t, -s(t_1, t - t_1))\|x_t - y_t\|_* + k\|x_t - y_t\|_*]. \end{aligned}$$

Let $M_1 = \max\{|A^{-1}(t, y_t)| : t_1 \le t \le t_1 + \alpha\}$. Then by relation (2.33), for $t \in [t_1, t_1 + \alpha]$, we have

$$|x(t) - y(t)| \le c_1 \sup_{t_1 \le s \le t_1 + \alpha} \|x_s - y_s\|_*,$$

where $c_1 = M_1\left(\frac{K}{\Gamma(q)(1+b_1)^{1-q_2}} \alpha^{(1+b_1)(1-q_2)} + \gamma(t, y_t, -s(t_1, t - t_1)) + k\right)$.

Noting that

$$\begin{aligned} \sup_{t_1 \le s \le t_1 + \alpha} \|x_s - y_s\|_* &= \sup_{t_1 \le s \le t_1 + \alpha} \ \sup_{-1 \le \theta \le 0} |x(p(s, \theta)) - y(p(s, \theta))| \\ &= \sup_{t_1 \le s \le t_1 + \alpha} \ \sup_{p(s, -1) \le \rho \le s} |x(\rho) - y(\rho)| \\ &= \sup_{p(t_1, -1) \le s \le t_1 + \alpha} |x(s) - y(s)|, \end{aligned}$$

we have

$$\sup_{p(t_1, -1) \le s \le t_1 + \alpha} |x(s) - y(s)| \le c_1 \sup_{p(t_1, -1) \le s \le t_1 + \alpha} |x(s) - y(s)|.$$

Choose $\alpha$ so small that $c_1 < 1$. Thus

$$\sup_{p(t_1, -1) \le s \le t_1 + \alpha} |x(s) - y(s)| = 0, \quad \text{i.e. } x(t) \equiv y(t), \quad \text{for } t_1 \le t \le t_1 + \alpha,$$

contradicting the definition of $t_1$. $\qquad\qquad \square$

### 2.3.3 *Continuous Dependence*

The following lemma is introduced in Lakshmikantham, Wen and Zhang, 1994. However, for the sake of completeness, we outline its proof here.

**Lemma 2.7.** *Assume* $x \in C([p(0,-1), A], \mathbb{R}^n)$. *Then for every* $t \in [0, A]$

$$\|x_t\| \leq \sup_{0 \leq s \leq t} |x(s)| + \|x_0\|.$$

**Proof.** By definition, $\|x_0\| = \sup_{-1 \leq \theta \leq 0} |x(p(0, \theta))|$. If $p(t, -1) \geq 0$, then

$$0 \leq p(t, \theta) \leq t \quad \text{for } -1 \leq \theta \leq 0.$$

Thus,

$$\sup_{-1 \leq \theta \leq 0} |x(p(t, \theta))| \leq \sup_{0 \leq s \leq t} |x(s)| \leq \sup_{0 \leq s \leq t} |x(s)| + \|x_0\|.$$

If $p(t, -1) < 0$, then by Lemma 2.3, there exists an $s \in [-1, 0]$ such that

$$p(t, -1) \leq p(t, \theta) \leq p(0, \theta) \quad \text{for } -1 \leq \theta \leq s,$$

while

$$0 \leq p(t, \theta) \leq t \quad \text{for } s \leq \theta \leq 0.$$

Hence

$$\sup_{-1 \leq \theta \leq 0} |x(p(t, \theta))| \leq \sup_{-1 \leq \theta \leq s} |x(p(t, \theta))| + \sup_{s \leq \theta \leq 0} |x(p(t, \theta))|$$

$$\leq \sup_{-1 \leq \theta \leq 0} |x(p(0, \theta))| + \sup_{s \leq \theta \leq 0} |x(p(t, \theta))|$$

$$= \|x_0\| + \sup_{0 \leq s \leq t} |x(s)|,$$

completing the proof. $\square$

We can now prove the following result on continuous dependence.

**Theorem 2.6.** *Let* $(t_0, \varphi) \in \Omega$ *be given. Suppose that the solution* $x = x(t_0, \varphi)$ *of* (2.12) *through* $(t_0, \varphi)$ *defined on* $[t_0, A]$ *is unique. Then for every* $\epsilon > 0$, *there exists a* $\delta(\epsilon) > 0$ *such that* $(s, \psi) \in \Omega$, $|s - t_0| < \delta$ *and* $\|\psi - \varphi\| < \varphi$ *imply*

$$\|x_t(s, \psi) - x_t(t_0, \varphi)\| < \epsilon \quad \text{for all } t \in [\sigma, A],$$

*where* $x(s, \psi)$ *is the solution of* (2.12) *through* $(s, \psi)$ *and* $\sigma = \max\{s, t_0\}$.

**Proof.** In order to prove the theorem, it is enough to show that if $\{(t_k, \varphi^k)\} \subset \Omega$, with $t_k \to t_0$ and $\varphi^k \to \varphi$ as $k \to \infty$, then there is a natural number $N$ such that each solution $x^k = x(t^k, \varphi^k)$ with $k \geq N$ of (2.12) through $(t^k, \varphi^k)$ exists on $[p(t_k, -1), A]$ and $x^k(t) \to x(t)$ uniformly on $[p(\sigma, -1), A]$, where $\sigma = \sup\{t_0, t^k : k \geq N\}$.

Since $x_t(t_0, \varphi)$ is continuous in $t \in [t_0, A]$, the set $W = \{(t, x_t(t_0, \varphi)) : t \in [t_0, A]\}$ is compact in $\Omega$. By Theorem 2.3, there exist a neighborhood $V$ of $W$ and number

$\alpha > 0$ such that for any $(s, \psi) \in V$, there is a solution $x(s, \psi)$ of (2.12) through $(s, \psi)$ which exists at least on $[s, s + \alpha]$. Without loss of generality, we let $V = V(W, r)$, choose $N$ so large that $|t_k - t_0| < \frac{r}{2}$ and $\|\varphi^k - \varphi\| < \frac{r}{2}$, so that $(t_k, \varphi^k) \in V$ for $k \geq N$. Thus $x^k = x(t_k, \varphi^k)$ exists at least on $[t_k, t_k + \alpha]$. For convenience, we shall denote $\varphi = \varphi^0$, $x = x^0$ and $x^k = x(t_k, \varphi^k)$, $k = 0, 1, \ldots$.

Let $p_k(t, \theta) = p(t_k + t, \theta) - t_k$. Define $\eta^k$, $y^k$ the same way as in Lemma 2.5. Recalling the proof of Lemma 2.5, we see that $y^k$ satisfy:

$$g(t_k + t, \eta_t^k + y_t^k) - g(t_k, \varphi^k) = \frac{1}{\Gamma(q)} \int_0^t (t - s)^{q-1} f(t_k + s, \eta_s^k + y_s^k) ds, \ t \in [0, \alpha] \ (2.37)$$

if and only if $x^k$ is the solution of (2.12) on $[p(t_k, -1), t_k + \alpha]$, where $\eta_t^k = \eta^k(p_k(t, \theta))$, $y_t^k = y^k(p_k(t, \theta))$.

Set $\bar{y}^k = y^k|_{[0,\alpha]}$, the restriction of $y^k$ to $[0, \alpha]$. Let $\Lambda = \{\bar{y}^k : k = 0, 1, 2, \ldots\}$. For every $z_k = (t_k, \varphi^k)$, define operators $S(z_k) : \Lambda \to C([0, \alpha], \mathbb{R}^n)$ and $T(z_k) : \Lambda \to C([0, \alpha], \mathbb{R}^n)$ as follows:

$$S(z_k)z(t) = A^{-1}(t_k + t, \eta_t^k)[g(t_k, \varphi^k) - g(t_k + t, \eta_t^k + z_t)$$
$$+ g'_\varphi(t_k + t, \eta_t^k)z_t - L(t_k + t, \eta_t^k + z_t)], \quad 0 \leq t \leq \alpha,$$

and

$$T(z_k)z(t) = A^{-1}(t_k + t, \eta_t^k) \frac{1}{\Gamma(q)} \int_0^t (t - s)^{q-1} f(t_k + s, \eta_s^k + z_s) ds, \quad 0 \leq t \leq \alpha,$$

where $z_t(\theta) = z(\bar{p}_k(t, \theta))$ with $z_0 = 0$.

It is easy to see that $\{T(z_k)\bar{y}^k\}$ is compact in $C([0, \alpha], \mathbb{R}^n)$. Recalling Theorem 2.3, we see that there exists a constant $\gamma \in [0, 1)$ which is independent to $z_k$ such that

$$\|Sz - Sy\| \leq \gamma \|z - y\| \quad \text{for any} \quad z, y \in \Lambda. \tag{2.38}$$

Let $\{z_k : k = 0, 1, 2, \ldots\} = \bar{\Lambda}$. Denote the Kuratowskii measure of $A \subset C([0, \alpha], \mathbb{R}^n)$ by $\alpha(A)$. Then (2.38) implies that

$$\alpha \left( \bigcup_{z_k \in \bar{\Lambda}} S(z_k)(\Lambda) \right) \leq \gamma \alpha(\Lambda).$$

Let $R = S + T$. Thus $\bar{y}^k = R(z_k)\bar{y}_k$. By the well-known properties of Kuratowskii measure $\alpha$, we immediately obtain that

$$\alpha(\Lambda) = \alpha(\{R(z_k)\bar{y}_k\}) \leq \alpha(\{S(z_k)\bar{y}^k\}) + \alpha(\{T(z_k)\bar{y}^k\})$$

$$= \alpha(\{S(z_k)\bar{y}^k\}) \leq \alpha \left( \bigcup_{z_k \in \bar{\Lambda}} S(z_k)(\Lambda) \right) \leq \gamma \alpha(\Lambda).$$

This means that $\alpha(\Lambda) = 0$ which implies $\Lambda$ is relatively compact in $C([0, \alpha], \mathbb{R}^n)$. Hence there exists a subsequence of $\Lambda$, say $\{\bar{y}^{k_i}\}$, which converges uniformly on $[0, \alpha]$. Assume that

$$\bar{y}^{k_i}(t) \to \bar{y}^*(t) \quad \text{uniformly on} \quad [0, \alpha].$$

Define a function $y^* : [p_0(0, -1), \alpha] \to \mathbb{R}^n$ by

$$\begin{cases} y^*(t) = \bar{y}^*(t), & \text{for } 0 \leq t \leq \alpha, \\ y_0^* = 0, \end{cases}$$

where $p_0$ is such a $p$-function that $p_0(t, \theta) = p(t + t_0, \theta) - t_0$. Let $\delta = \inf\{p_k(0, -1) : k = 0, 1, 2, \ldots\}$ and $\hat{y}^k$ denote the extension of $y^k$ to $[\delta, \alpha]$ which is defined by

$$\begin{cases} \hat{y}^{k_i}(t) = y^{k_i}(t), & \text{for } 0 \leq t \leq \alpha, \\ \hat{y}^{k_i}(t) = 0, & \text{for } \delta \leq t \leq 0. \end{cases}$$

Obviously, $\{\hat{y}^{k_i}(t)\}$ converges uniformly on $[\delta, \alpha]$ as $k_i \to \infty$. Consequently, $\{\hat{y}^{k_i}\}$ is a relatively compact set.

We claim that $y_t^{k_i} \to y_t^*$ uniformly in $t \in [0, \alpha]$. In fact,

$$|p_{k_i}(t, \theta) - p_0(t, \theta)| = |(p(t_{k_i} + t, \theta) - t_{k_i}) - (p(t_0 + t, \theta) - t_0)|$$
$$\leq |p(t_{k_i} + t, \theta) - p(t_0 + t, \theta)| + |t_0 - t_{k_i}|.$$

Hence, for every $\mu > 0$ there exists a number $L$ such that

$$|p_{k_i}(t, \theta) - p_0(t, \theta)| < \mu \quad \text{whenever} \quad k_i \geq L.$$

We have the inequality

$$\begin{aligned} \|y_t^{k_i} - y_t^*\| &= \sup_{-1 \leq \theta \leq 0} |\hat{y}^{k_i}(p_{k_i}(t, \theta)) - y^*(p_0(t, \theta))| \\ &= \sup_{-1 \leq \theta \leq 0} |\hat{y}^{k_i}(p_{k_i}(t, \theta)) - \hat{y}^{k_i}(p_0(t, \theta)) + \hat{y}^{k_i}(p_0(t, \theta)) - y^*(p_0(t, \theta))| \\ &\leq \sup_{-1 \leq \theta \leq 0} |\hat{y}^{k_i}(p_{k_i}(t, \theta)) - \hat{y}^{k_i}(p_0(t, \theta))| \\ &\quad + \sup_{-1 \leq \theta \leq 0} |\hat{y}^{k_i}(p_0(t, \theta)) - y^*(p_0(t, \theta))|. \end{aligned}$$

By Lemma 2.7, we get

$$\begin{aligned} \sup_{-1 \leq \theta \leq 0} |\hat{y}^{k_i}(p_0(t, \theta)) - y^*(p_0(t, \theta))| &\leq \sup_{0 \leq \theta \leq \alpha} |y^{k_i}(t) - y^*(t)| + \|\hat{y}_0^{k_i} - y_0^*\| \\ &= \sup_{0 \leq \theta \leq \alpha} |y^{k_i}(t) - y^*(t)|. \end{aligned}$$

For every $\epsilon > 0$ there exists a number $L_1$ such that

$$\sup_{0 \leq \theta \leq \alpha} |y^{k_i}(t) - y^*(t)| < \frac{\epsilon}{2} \quad \text{for} \quad k_i \geq L_1,$$

by the definition of $y^*$. On the other hand, since $\{\hat{y}^{k_i}\}$ is an equi-continuous set, for the given $\epsilon$, there exists a $\mu > 0$ such that

$$|\hat{y}^{k_i}(t) - \hat{y}^{k_i}(\tau)| < \frac{\epsilon}{2} \quad \text{for} \quad |t - \tau| < \mu. \tag{2.39}$$

We can choose $L \geq L_1$ so that $|p_{k_i}(t, \theta) - p_0(t, \theta)| < \mu$. Thus (2.39) holds as long as $k_i \geq L$. Furthermore,

$$\|y_t^{k_i} - y_t\| < \epsilon \quad \text{whenever} \quad k_i \geq L,$$

which is just our claim. A similar argument shows that $\eta_t^{k_i} \to \eta_t$ uniformly in $t \in [0, \alpha]$. The limiting process upon (2.37) yields

$$
\begin{cases}
g(t_0 + t, \eta_t + y_t^*) - g(t_0, \varphi) = \dfrac{1}{\Gamma(q)} \displaystyle\int_0^t (t-s)^{q-1} f(t_0 + s, \eta_s + y_s^*) ds, & 0 \le t \le \alpha, \\
y_0^* = 0,
\end{cases}
$$

which demonstrates that $y^*$ as well as $y^0$ is a solution of the fractional IVP

$$
\begin{cases}
g(t_0 + t, \eta_t + y_t^*) - g(t_0, \varphi) = \dfrac{1}{\Gamma(q)} \displaystyle\int_0^t (t-s)^{q-1} f(t_0 + s, \eta_s + y_s) ds, & 0 \le t \le \alpha, \\
y_0 = 0.
\end{cases}
$$

The hypothesis that $x(t_0, \varphi)$ is unique, that is, $y^0$ is unique implies that $y^* = y^0$. Thus $y^{k_i}(t) \to y^0(t)$ uniformly on $[0, \alpha]$. The verified fact that every subsequence of sequence $\{y^k\}$ has a convergent subsequence with a same limit $y^0$ implies that the entire sequence $\{y^k\}$ converges to $y^0$. Translating these remarks back into $x^k$, we have indeed obtained the result stated in this theorem for the interval $[p(\sigma, -1), \sigma + \alpha]$.

Let $b = \sigma + \alpha$. If $b < A$, $(b, x_b) \in W$, we can choose $N_1 \ge N$ such that $(b, x_b^k) \in V$ as long as $k \ge N_1$. By Theorem 2.3, for every point $(b, x_b^k)$ the solution $x^k(b, x_b^k)$ exists at least on $[p(b, -1), b + \alpha]$. The above argument can be adapted to this interval which yields the assertion that $x^k(b, x_b^k)(t) \to x^0(t_0, \varphi)(t)$ uniformly on the same interval. The conclusion stated in theorem can be verified by successive steps of finite intervals of length $\alpha$. Hence the proof is completed. $\qquad\square$

## 2.4  Neutral Equations with Infinite Delay

### 2.4.1  *Introduction*

In Section 2.4, we consider the initial value problem of fractional neutral functional differential equations with infinite delay of the form

$$
{}_{t_0}^C D_t^q g(t, x_t) = f(t, x_t), \qquad t \in [t_0, \infty), \tag{2.40}
$$

$$
x_{t_0} = \varphi, \qquad\qquad (t_0, \varphi) \in [0, \infty) \times \Omega, \tag{2.41}
$$

where ${}_{t_0}^C D_t^q$ is Caputo fractional derivative of order $0 < q < 1$, $\Omega$ is an open subset of $B$ and $g, f : [t_0, \infty) \times \Omega \to \mathbb{R}^n$ are given functionals satisfying some assumptions that will be specified later. $B$ is called a phase space that will be defined later.

If $x : (-\infty, A) \to \mathbb{R}^n, A \in (0, \infty)$, then for any $t \in [0, A)$ define $x_t$ by $x_t(\theta) = x(t + \theta)$, for $\theta \in (-\infty, 0]$.

Denote by $BC(J, \mathbb{R}^n)$ the Banach space of all continuous and bounded functions from $J$ into $\mathbb{R}^n$ with the norm $\| \cdot \|$.

To describe fractional neutral functional differential equations with infinite delay, we need to discuss a phase space $B$ in a convenient way. We shall provide a general description of phase spaces of neutral differential equations with infinite delay which is taken from Lakshmikantham, Wen and Zhang, 1994.

Let $B$ be a real vector space either

(i) of continuous functions that map $(-\infty, 0]$ to $\mathbb{R}^n$ with $\varphi = \psi$ if $\varphi(s) = \psi(s)$ on $(-\infty, 0]$ or

(ii) of measurable functions that map $(-\infty, 0]$ to $\mathbb{R}^n$ with $\varphi = \psi$ (or $\varphi$ is equivalent to $\psi$) in $B$ if $\varphi(s) = \psi(s)$ almost everywhere on $(-\infty, 0]$, and $\varphi(0) = \psi(0)$.

Let $B$ be endowed with a norm $\| \cdot \|_B$ such that $B$ is complete with respect to $\| \cdot \|_B$. Thus $B$ equipped with norm $\| \cdot \|_B$ is a Banach space. We denote this space by $(B, \| \cdot \|_B)$ or simply by $B$, whenever no confusion arises.

Let $0 \leq a < A$. If $x : (-\infty, A) \to \mathbb{R}^n$ is given such that $x_a \in B$ and $x \in [a, A) \to \mathbb{R}^n$ is continuous, then $x_t \in B$ for all $t \in [a, A)$.

This is a very weak condition that the common admissible phase spaces and $BC$ satisfy. For more details of the phase spaces, we refer the reader to Hino, Murakami and Naito, 1991; Lakshmikantham, Wen and Zhang, 1994.

**Definition 2.10.** A function $x : (-\infty, t_0 + \sigma) \to \mathbb{R}^n (t_0 \in [0, \infty), \sigma > 0)$ is said to be a solution of fractional IVP (2.40)-(2.41) through $(t_0, \varphi)$ on $[t_0, t_0 + \sigma)$, if

(i) $x_{t_0} = \varphi$;
(ii) $x$ is continuous on $[t_0, t_0 + \sigma)$;
(iii) $g(t, x_t)$ is absolutely continuous on $[t_0, t_0 + \sigma)$;
(iv) (2.40) holds almost everywhere on $[t_0, t_0 + \sigma)$.

Let $\Omega \subseteq B$ be an open set such that for any $(t_0, \varphi) \in [0, \infty) \times \Omega$, there exist constants $\sigma_1, \gamma_1 > 0$ so that $x_t \in \Omega$ provided that $x \in A(t_0, \varphi, \sigma_1, \gamma_1)$ and $t \in [t_0, t_0 + \sigma_1]$, where $A(t_0, \varphi, \sigma_1, \gamma_1)$ is defined as

$$A(t_0, \varphi, \sigma_1, \gamma_1) = \left\{ x : (-\infty, t_0 + \sigma_1] \to \mathbb{R}^n, x_{t_0} = \varphi, \sup_{t_0 \leq t \leq t_0 + \sigma_1} |x(t) - \varphi(0)| \leq \gamma_1 \right\}.$$

In order to guarantee that equation (2.40) is NFDE, the coefficient of $x(t)$ that is contained in $g(t, x_t)$ cannot be equal to zero. Then we need to introduce the generalized atomic concept.

**Definition 2.11.** (Lakshmikantham, Wen and Zhang, 1994) The functional $g : [0, \infty) \times \Omega \to \mathbb{R}^n$ is said to be generalized atomic on $\Omega$, if

$$g(t, \varphi) - g(t, \psi) = K(t, \varphi, \psi)(\varphi(0) - \psi(0)) + L(t, \varphi, \psi)$$

where $(t, \varphi, \psi) \in [0, \infty) \times \Omega \times \Omega$, $K : [0, \infty) \times \Omega \times \Omega \to \mathbb{R}^{n \times n}$ and $L : [0, \infty) \times \Omega \times \Omega \to \mathbb{R}^n$ satisfy

(i) $\det K(t, \varphi, \varphi) \neq 0$ for all $(t, \varphi) \in [0, \infty) \times \Omega$;

**(ii)** for any $(t_0, \varphi) \in [0, \infty) \times \Omega$, there exist constants $\delta_1, \gamma_1 > 0$, and $k_1, k_2 > 0$, with $2k_2 + k_1 < 1$ such that for all $x, y \in A(t_0, \varphi, \sigma_1, \gamma_1)$, $g(t, x_t)$, $K(t, x_t, y_t)$ and $L(t, x_t, y_t)$ are continuous in $t \in [t_0, t_0 + \sigma_1]$, and

$$|K^{-1}(t_0, \varphi, \varphi)L(t, x_t, y_t)| \leq k_1 \sup_{t_0 \leq s \leq t} |x(s) - y(s)|,$$

$$|K^{-1}(t_0, \varphi, \varphi)K(t, x_t, y_t) - I| \leq k_2,$$

where $I$ is the $n \times n$ unit matrix.

For a detailed discussion on the atomic concept we refer the reader to the books Hale, 1977; Lakshmikantham, Wen and Zhang, 1994.

In Subsection 2.4.2, we shall discuss existence and uniqueness of solutions for fractional IVP (2.40)-(2.41) on a class of comparatively comprehensive phase spaces. We establish various criteria on existence and uniqueness of solutions for fractional IVP (2.40)-(2.41). In Subsection 2.4.2, we proceed to consider the continuation of solutions.

### 2.4.2 *Existence and Uniqueness*

The following existence result for fractional IVP (2.40)-(2.41) is based on Krasnoselskii fixed point theorem.

**Theorem 2.7.** *Assume that $g$ is generalized atomic on $\Omega$, and that for any $(t_0, \varphi) \in [0, \infty) \times \Omega$, there exist constants $\sigma_1, \gamma_1 \in (0, \infty)$, $q_1 \in (0, q)$ and a real-valued function $m(t) \in L^{\frac{1}{q_1}}[t_0, t_0 + \sigma_1]$ such that*

**(H1)** *for any $x \in A(t_0, \varphi, \sigma_1, \gamma_1)$, $f(t, x_t)$ is measurable;*
**(H2)** *for any $x \in A(t_0, \varphi, \sigma_1, \gamma_1)$, $|f(t, x_t)| \leq m(t)$, for $t \in [t_0, t_0 + \sigma_1]$;*
**(H3)** *$f(t, \phi)$ is continuous with respect to $\phi$ on $\Omega$.*

*Then fractional IVP (2.40)-(2.41) has a solution.*

**Proof.** We know that $f(t, x_t)$ is Lebesgue measurable in $[t_0, t_0 + \sigma_1]$ according to conditions (H1). Direct calculation gives that $(t - s)^{q-1} \in L^{\frac{1}{1-q_1}}[t_0, t]$, for $t \in [t_0, t_0 + \sigma_1]$. In light of Hölder inequality and the condition (H2), we obtain that $(t-s)^{q-1}f(s, x_s)$ is Lebesgue integrable with respect to $s \in [t_0, t]$ for all $t \in [t_0, t_0 + \sigma_1]$, and

$$\int_{t_0}^{t} |(t-s)^{q-1}f(s, x_s)|ds \leq \|(t-s)^{q-1}\|_{L^{\frac{1}{1-q_1}}[t_0, t]} \|m\|_{L^{\frac{1}{q_1}}[t_0, t_0 + \sigma_1]}. \tag{2.42}$$

According to Definition 2.10, fractional IVP (2.40)-(2.41) is equivalent to the following equation

$$g(t, x_t) = g(t_0, \varphi) + \frac{1}{\Gamma(q)} \int_{t_0}^{t} (t-s)^{q-1}f(s, x_s)ds \quad \text{for } t \in [t_0, t_0 + \sigma_1]. \tag{2.43}$$

Let $\hat{\varphi} \in A(t_0, \varphi, \sigma_1, \gamma_1)$ be defined as $\hat{\varphi}_{t_0} = \varphi$, $\hat{\varphi}(t_0 + t) = \varphi(0)$ for all $t \in [0, \sigma_1]$. If $x$ is a solution of fractional IVP (2.40)-(2.41), let $x(t_0 + t) = \hat{\varphi}(t_0 + t) + z(t)$, $t \in$

$(-\infty, \sigma_1]$, then we have $x_{t_0+t} = \hat{\varphi}_{t_0+t} + z_t,\ t \in [0, \sigma_1]$. Thus (2.43) implies that $z$ satisfies the equation

$$g(t_0 + t, \hat{\varphi}_{t_0+t} + z_t) = g(t_0, \varphi) + \frac{1}{\Gamma(q)} \int_0^t (t-s)^{q-1} f(t_0 + s, \hat{\varphi}_{t_0+s} + z_s) ds, \quad (2.44)$$

for $0 \le t \le \sigma_1$.

Since $g$ is generalized atomic on $\Omega$, there exist positive constant $\alpha > 1$ and a positive function $\sigma_2(\gamma)$ defined in $(0, \gamma_1]$, such that for any $\gamma \in (0, \gamma_1]$, when $0 \le t \le \sigma_2(\gamma)$, we have

$$\alpha(2k_2 + k_1) < 1, \quad (2.45)$$

$$|K^{-1}(t_0, \varphi, \varphi)K(t_0 + t, x_{t_0+t}, y_{t_0+t}) - I| \le k_2, \quad (2.46)$$

$$|I - K^{-1}(t_0 + t, \hat{\varphi}_{t_0+t}, \hat{\varphi}_{t_0+t})K(t_0, \varphi, \varphi)| \le \min\{\alpha k_2, \alpha - 1\}, \quad (2.47)$$

$$|K^{-1}(t_0 + t, \hat{\varphi}_{t_0+t}, \hat{\varphi}_{t_0+t})||g(t_0 + t, \hat{\varphi}_{t_0+t}) - g(t_0, \varphi)| \le \frac{1 - \alpha(2k_2 + k_1)}{2}\gamma. \quad (2.48)$$

Note the completely continuity of the function $(m(t))^{\frac{1}{q_1}}$. Hence, for a given positive number $M$, there must exist a number $h > 0$, satisfying

$$\int_{t_0}^{t_0+h} (m(s))^{\frac{1}{q_1}} ds \le M.$$

For a given $\gamma \in (0, \gamma_1]$, choose

$$\sigma = \min\left\{\sigma_1, \sigma_2(\gamma), h, (1 + \beta)^{\frac{1}{1+\beta}} \left(\frac{(1 - \alpha(2k_2 + k_1))\Gamma(q)\gamma}{2\alpha|K^{-1}(t_0, \varphi, \varphi)|M^{q_1}}\right)^{\frac{1}{(1-q_1)(1+\beta)}}\right\}, \quad (2.49)$$

where $\beta = \frac{q-1}{1-q_1} \in (-1, 0)$.

For any $(t_0, \varphi) \in [0, \infty) \times \Omega$, define $E(\sigma, \gamma)$ as follows:

$$E(\sigma, \gamma) = \{z : (-\infty, \sigma) \to \mathbb{R}^n \text{ is continuous}; z(s) = 0 \text{ for } s \in (-\infty, 0] \text{ and } \|z\| \le \gamma\}$$

where $\|z\| = \sup_{0 \le s \le \sigma} |z(t)|$. Then $E(\sigma, \gamma)$ is a closed bounded and convex subset of Banach space $BC((-\infty, \sigma_1], \mathbb{R}^n)$.

Now, on $E(\sigma, \gamma)$ define two operators $S$ and $U$ as follows:

$$(Sz)(t) = \begin{cases} 0, & t \in (-\infty, 0], \\ K^{-1}(t_0 + t, \hat{\varphi}_{t_0+t}, \hat{\varphi}_{t_0+t})[-g(t_0 + t, \hat{\varphi}_{t_0+t} + z_t) \\ \quad + g(t_0, \varphi) + K(t_0 + t, \hat{\varphi}_{t_0+t}, \hat{\varphi}_{t_0+t})z(t)], & t \in [0, \sigma], \end{cases}$$

and

$$(Uz)(t) = \begin{cases} 0, & t \in (-\infty, 0], \\ K^{-1}(t_0 + t, \hat{\varphi}_{t_0+t}, \hat{\varphi}_{t_0+t}) \\ \quad \times \frac{1}{\Gamma(q)} \int_0^t (t-s)^{q-1} f(t_0 + s, \hat{\varphi}_{t_0+s} + z_s) ds, & t \in [0, \sigma], \end{cases}$$

where $z \in E(\sigma, \gamma)$.

It is easy to see that the operator equation

$$z = Sz + Uz \qquad (2.50)$$

has a solution $z \in E(\sigma, \gamma)$ if and only if $z$ is a solution of equation (2.44). Thus, $x_{t+t_0} = \hat{\varphi}_{t_0+t} + z_t$ is a solution of equation (2.40) on $[0, \sigma]$. Therefore, the existence of a solution of fractional IVP (2.40)-(2.41) is equivalent to determining $\sigma, \gamma > 0$ such that (2.50) has a fixed point in $E(\sigma, \gamma)$.

Now we show that $S + U$ has a fixed point in $E(\sigma, \gamma)$. The proof is divided into three steps.

**Claim I.** $Sz + Uw \in E(\sigma, \gamma)$ for every pair $z, w \in E(\sigma, \gamma)$.

Obviously, for every pair $z, w \in E(\sigma, \gamma)$, $(Sz)(t)$ and $(Uw)(t)$ are continuous in $t \in [0, \sigma]$, and for $t \in [0, \sigma]$, by using the Hölder inequality and (2.47), we have

$$|(Uw)(t)| \leq |K^{-1}(t_0 + t, \hat{\varphi}_{t_0+t}, \hat{\varphi}_{t_0+t}) \, K(t_0, \varphi, \varphi) \, K^{-1}(t_0, \varphi, \varphi)|$$
$$\times \frac{1}{\Gamma(q)} \left| \int_0^t (t-s)^{q-1} f(t_0 + s, \hat{\varphi}_{t_0+s} + w_s) ds \right|$$
$$\leq \alpha |K^{-1}(t_0, \varphi, \varphi)| \frac{1}{\Gamma(q)} \left( \int_0^t ((t-s)^{q-1})^{\frac{1}{1-q_1}} ds \right)^{1-q_1}$$
$$\times \left( \int_{t_0}^{t_0+\sigma} (m(s))^{\frac{1}{q_1}} ds \right)^{q_1}$$
$$\leq \alpha |K^{-1}(t_0, \varphi, \varphi)| \frac{M^{q_1}}{\Gamma(q)} \left( \frac{1}{1+\beta} \sigma^{1+\beta} \right)^{1-q_1}$$
$$\leq \frac{1 - \alpha(2k_2 + k_1)}{2} \gamma, \qquad (2.51)$$

where $\beta = \frac{q-1}{1-q_1} \in (-1, 0)$, and

$$|(Sz)(t)| = \left| K^{-1}(t_0 + t, \hat{\varphi}_{t_0+t}, \hat{\varphi}_{t_0+t})[-g(t_0 + t, \hat{\varphi}_{t_0+t} + z_t) + g(t_0 + t, \hat{\varphi}_{t_0+t}) \right.$$
$$\left. - g(t_0 + t, \hat{\varphi}_{t_0+t}) + g(t_0, \varphi) + K(t_0 + t, \hat{\varphi}_{t_0+t}, \hat{\varphi}_{t_0+t})z(t)] \right|$$
$$= \left| K^{-1}(t_0 + t, \hat{\varphi}_{t_0+t}, \hat{\varphi}_{t_0+t})[-K(t_0 + t, \hat{\varphi}_{t_0+t} + z_t, \hat{\varphi}_{t_0+t})z(t) \right.$$
$$- L(t_0 + t, \hat{\varphi}_{t_0+t} + z_t, \hat{\varphi}_{t_0+t}) - g(t_0 + t, \hat{\varphi}_{t_0+t}) + g(t_0, \varphi)$$
$$\left. + K(t_0 + t, \hat{\varphi}_{t_0+t}, \hat{\varphi}_{t_0+t})z(t)] \right|$$
$$= \left| K^{-1}(t_0 + t, \hat{\varphi}_{t_0+t}, \hat{\varphi}_{t_0+t})[K(t_0 + t, \hat{\varphi}_{t_0+t}, \hat{\varphi}_{t_0+t}) \right.$$
$$- K(t_0 + t, \hat{\varphi}_{t_0+t} + z_t, \hat{\varphi}_{t_0+t})]z(t) + K^{-1}(t_0 + t, \hat{\varphi}_{t_0+t}, \hat{\varphi}_{t_0+t})$$
$$\left. \times [-L(t_0 + t, \hat{\varphi}_{t_0+t} + z_t, \hat{\varphi}_{t_0+t}) - g(t_0 + t, \hat{\varphi}_{t_0+t}) + g(t_0, \varphi)] \right|$$
$$= \left| K^{-1}(t_0 + t, \hat{\varphi}_{t_0+t}, \hat{\varphi}_{t_0+t})K(t_0, \varphi, \varphi) \right.$$
$$\times [K^{-1}(t_0, \varphi, \varphi)K(t_0 + t, \hat{\varphi}_{t_0+t}, \hat{\varphi}_{t_0+t}) - I]z(t)$$
$$- K^{-1}(t_0 + t, \hat{\varphi}_{t_0+t}, \hat{\varphi}_{t_0+t})K(t_0, \varphi, \varphi)$$
$$\times [K^{-1}(t_0, \varphi, \varphi)K(t_0 + t, \hat{\varphi}_{t_0+t} + z_t, \hat{\varphi}_{t_0+t}) - I]z(t)$$
$$+ K^{-1}(t_0 + t, \hat{\varphi}_{t_0+t}, \hat{\varphi}_{t_0+t})[-L(t_0 + t, \hat{\varphi}_{t_0+t} + z_t, \hat{\varphi}_{t_0+t})$$
$$\left. - g(t_0 + t, \hat{\varphi}_{t_0+t}) + g(t_0, \varphi)] \right|$$

$$\leq |K^{-1}(t_0 + t, \hat{\varphi}_{t_0+t}, \hat{\varphi}_{t_0+t})K(t_0, \varphi, \varphi)|$$
$$\times \big[ (|K^{-1}(t_0, \varphi, \varphi)K(t_0 + t, \hat{\varphi}_{t_0+t}, \hat{\varphi}_{t_0+t}) - I|$$
$$+ |K^{-1}(t_0, \varphi, \varphi)K(t_0 + t, \hat{\varphi}_{t_0+t} + z_t, \hat{\varphi}_{t_0+t}) - I|)|z(t)|$$
$$+ |K^{-1}(t_0, \varphi, \varphi)L(t_0 + t, \hat{\varphi}_{t_0+t} + z_t, \hat{\varphi}_{t_0+t})| \big]$$
$$+ |K^{-1}(t_0 + t, \hat{\varphi}_{t_0+t}, \hat{\varphi}_{t_0+t})||g(t_0 + t, \hat{\varphi}_{t_0+t}) - g(t_0, \varphi)|.$$

According to (2.45)-(2.48), we have

$$|(Sz)(t)| \leq \alpha(2k_2 + k_1)\gamma + \frac{1 - \alpha(2k_2 + k_1)}{2}\gamma = \frac{1 + \alpha(2k_2 + k_1)}{2}\gamma.$$

Therefore, $|(Sz)(t) + (Uw)(t)| \leq \gamma$ for $t \in [0, \sigma]$. This means that $Sz + Uw \in E(\sigma, \gamma)$ whenever $z, w \in E(\sigma, \gamma)$.

**Claim II.** $S$ is a contraction mapping on $E(\sigma, \gamma)$.

For any $z, w \in E(\sigma, \gamma)$, we obtain

$$|(Sz)(t) - (Sw)(t)|$$
$$\leq |K^{-1}(t_0 + t, \hat{\varphi}_{t_0+t}, \hat{\varphi}_{t_0+t})||K(t_0 + t, \hat{\varphi}_{t_0+t}, \hat{\varphi}_{t_0+t})$$
$$- K(t_0 + t, \hat{\varphi}_{t_0+t} + z_t, \hat{\varphi}_{t_0+t} + w_t)||z(t) - w(t)|$$
$$+ |K^{-1}(t_0 + t, \hat{\varphi}_{t_0+t}, \hat{\varphi}_{t_0+t})L(t_0 + t, \hat{\varphi}_{t_0+t} + z_t, \hat{\varphi}_{t_0+t} + w_t)|$$
$$\leq \big| [I - K^{-1}(t_0 + t, \hat{\varphi}_{t_0+t}, \hat{\varphi}_{t_0+t})K(t_0, \varphi, \varphi)]$$
$$- [K^{-1}(t_0 + t, \hat{\varphi}_{t_0+t}, \hat{\varphi}_{t_0+t})K(t_0, \varphi, \varphi)]$$
$$\times [K^{-1}(t_0, \varphi, \varphi)K(t_0 + t, \hat{\varphi}_{t_0+t} + z_t, \hat{\varphi}_{t_0+t} + w_t) - I] \big| \, |z(t) - w(t)|$$
$$+ |K^{-1}(t_0 + t, \hat{\varphi}_{t_0+t}, \hat{\varphi}_{t_0+t})K(t_0, \varphi, \varphi)K^{-1}(t_0, \varphi, \varphi)$$
$$\times L(t_0 + t, \hat{\varphi}_{t_0+t} + z_t, \hat{\varphi}_{t_0+t} + w_t)|$$
$$\leq (\alpha k_2 + \alpha k_2)|z(t) - w(t)| + \alpha k_1 \sup_{0 \leq s \leq t} |z(s) - w(s)|$$
$$\leq \alpha(2k_2 + k_1) \sup_{0 \leq s \leq t} |z(s) - w(s)|,$$

where $\alpha(2k_2 + k_1) < 1$, and therefore $S$ is a contraction mapping on $E(\sigma, \gamma)$.

**Claim III.** Now we show that $U$ is a completely continuous operator.

For any $z \in E(\sigma, \gamma), 0 \leq \tau < t \leq \sigma$, we get

$$|(Uz)(t) - (Uz)(\tau)|$$
$$= \Big| K^{-1}(t_0 + t, \hat{\varphi}_{t_0+t}, \hat{\varphi}_{t_0+t})\frac{1}{\Gamma(q)}\int_0^t (t - s)^{q-1}f(t_0 + s, \hat{\varphi}_{t_0+s} + z_s)ds$$
$$- K^{-1}(t_0 + \tau, \hat{\varphi}_{t_0+\tau}, \hat{\varphi}_{t_0+\tau})\frac{1}{\Gamma(q)}\int_0^\tau (\tau - s)^{q-1}f(t_0 + s, \hat{\varphi}_{t_0+s} + z_s)ds \Big|$$
$$= \Big| K^{-1}(t_0 + t, \hat{\varphi}_{t_0+t}, \hat{\varphi}_{t_0+t})\frac{1}{\Gamma(q)}\int_\tau^t (t - s)^{q-1}f(t_0 + s, \hat{\varphi}_{t_0+s} + z_s)ds$$
$$+ K^{-1}(t_0 + t, \hat{\varphi}_{t_0+t}, \hat{\varphi}_{t_0+t})\frac{1}{\Gamma(q)}\int_0^\tau (t - s)^{q-1}f(t_0 + s, \hat{\varphi}_{t_0+s} + z_s)ds$$

$$- K^{-1}(t_0+t, \hat{\varphi}_{t_0+t}, \hat{\varphi}_{t_0+t}) \frac{1}{\Gamma(q)} \int_0^\tau (\tau-s)^{q-1} f(t_0+s, \hat{\varphi}_{t_0+s} + z_s) ds$$

$$+ K^{-1}(t_0+t, \hat{\varphi}_{t_0+t}, \hat{\varphi}_{t_0+t}) \frac{1}{\Gamma(q)} \int_0^\tau (\tau-s)^{q-1} f(t_0+s, \hat{\varphi}_{t_0+s} + z_s) ds$$

$$\left. - K^{-1}(t_0+\tau, \hat{\varphi}_{t_0+\tau}, \hat{\varphi}_{t_0+\tau}) \frac{1}{\Gamma(q)} \int_0^\tau (\tau-s)^{q-1} f(t_0+s, \hat{\varphi}_{t_0+s} + z_s) ds \right|$$

$$\leq \frac{|K^{-1}(t_0+t, \hat{\varphi}_{t_0+t}, \hat{\varphi}_{t_0+t})|}{\Gamma(q)} \left| \int_\tau^t (t-s)^{q-1} f(t_0+s, \hat{\varphi}_{t_0+s} + z_s) ds \right|$$

$$+ \frac{|K^{-1}(t_0+t, \hat{\varphi}_{t_0+t}, \hat{\varphi}_{t_0+t})|}{\Gamma(q)}$$

$$\times \left| \int_0^\tau [(t-s)^{q-1} - (\tau-s)^{q-1}] f(t_0+s, \hat{\varphi}_{t_0+s} + z_s) ds \right|$$

$$+ \frac{|K^{-1}(t_0+t, \hat{\varphi}_{t_0+t}, \hat{\varphi}_{t_0+t}) - K^{-1}(t_0+\tau, \hat{\varphi}_{t_0+\tau}, \hat{\varphi}_{t_0+\tau})|}{\Gamma(q)}$$

$$\times \left| \int_0^\tau (\tau-s)^{q-1} f(t_0+s, \hat{\varphi}_{t_0+s} + z_s) ds \right|$$

$$= \frac{|K^{-1}(t_0+t, \hat{\varphi}_{t_0+t}, \hat{\varphi}_{t_0+t})|}{\Gamma(q)} (I_1 + I_2)$$

$$+ \frac{|K^{-1}(t_0+t, \hat{\varphi}_{t_0+t}, \hat{\varphi}_{t_0+t}) - K^{-1}(t_0+\tau, \hat{\varphi}_{t_0+\tau}, \hat{\varphi}_{t_0+\tau})|}{\Gamma(q)} I_3,$$

where

$$I_1 = \left| \int_\tau^t (t-s)^{q-1} f(t_0+s, \hat{\varphi}_{t_0+s} + z_s) ds \right|,$$

$$I_2 = \left| \int_0^\tau \left( (t-s)^{q-1} - (\tau-s)^{q-1} \right) f(t_0+s, \hat{\varphi}_{t_0+s} + z_s) ds \right|,$$

$$I_3 = \left| \int_0^\tau (\tau-s)^{q-1} f(t_0+s, \hat{\varphi}_{t_0+s} + z_s) ds \right|.$$

By using an analogous argument presented in (2.51), we can conclude that

$$I_1 \leq \frac{M^{q_1}}{(1+\beta)^{1-q_1}} \left( (t-\tau)^{1+\beta} \right)^{1-q_1},$$

$$I_3 \leq \frac{M^{q_1}}{(1+\beta)^{1-q_1}} \left( \tau^{1+\beta} \right)^{1-q_1},$$

and

$$I_2 \leq \left( \int_0^\tau \left| (t-s)^{q-1} - (\tau-s)^{q-1} \right|^{\frac{1}{1-q_1}} ds \right)^{1-q_1} \left( \int_{t_0}^{t_0+\tau} |f(s, x_s)|^{\frac{1}{q_1}} ds \right)^{q_1}$$

$$\leq M^{q_1} \left( \int_0^\tau (\tau-s)^\beta - (t-s)^\beta ds \right)^{1-q_1}$$

$$\leq \frac{M^{q_1}}{(1+\beta)^{1-q_1}} \left( \tau^{1+\beta} - t^{1+\beta} + (t-\tau)^{1+\beta} \right)^{1-q_1}$$

$$\leq \frac{M^{q_1}}{(1+\beta)^{1-q_1}} \left( (t-\tau)^{1+\beta} \right)^{1-q_1},$$

where $\beta = \frac{q-1}{1-q_1} \in (-1, 0)$. Therefore

$$|(Uz)(t) - (Uz)(\tau)| \leq \frac{|K^{-1}(t_0 + t, \hat{\varphi}_{t_0+t}, \hat{\varphi}_{t_0+t})|}{\Gamma(q)} \frac{2M^{q_1}}{(1+\beta)^{1-q_1}} \left((t-\tau)^{1+\beta}\right)^{1-q_1}$$

$$+ \frac{|K^{-1}(t_0 + t, \hat{\varphi}_{t_0+t}, \hat{\varphi}_{t_0+t}) - K^{-1}(t_0 + \tau, \hat{\varphi}_{t_0+\tau}, \hat{\varphi}_{t_0+\tau})|}{\Gamma(q)} \frac{M^{q_1}}{(1+\beta)^{1-q_1}} \left(\tau^{1+\beta}\right)^{1-q_1}.$$

Since the property of the matrix function $K(t_0 + t, \hat{\varphi}_{t_0+t}, \hat{\varphi}_{t_0+t})$ which is nonsingular and continuous in $t \in [0, \sigma]$ implies that its inverse matrix $K^{-1}(t_0 + t, \hat{\varphi}_{t_0+t}, \hat{\varphi}_{t_0+t})$ exists and is continuous in $t \in [0, \sigma]$, then $\{Uz : z \in E(\sigma, \gamma)\}$ is equicontinuous. On the other hand, $U$ is continuous from condition (H3) and $\{Uz : z \in E(\sigma, \gamma)\}$ is uniformly bounded from (2.51), thus $U$ is a completely continuous operator by Arzela-Ascoli theorem.

Therefore, Krasnoselskii fixed point theorem shows that $S + U$ has a fixed point on $E(\sigma, \gamma)$, and hence fractional IVP (2.40)-(2.41) has a solution $x(t) = \varphi(0) + z(t - t_0)$ for all $t \in [t_0, t_0 + \sigma]$. $\qquad\square$

**Remark 2.1.** If we replace condition (H1) by

**(H1)′** $f(t, \phi)$ is measurable with respect to $t$ on $[t_0, t_0 + \sigma_1]$,

then we can also conclude that the result of Theorem 2.7 holds.

In fact, for any $x \in A(t_0, \varphi, \sigma_1, \gamma_1)$, suppose $x_{t_0+t} = \hat{\varphi}_{t_0+t} + z_t$, $t \in [0, \sigma_1]$, then, according to the definition of $\hat{\varphi}_{t_0+t}$ and $z_t$, we know that $x_{t_0+t}$ is a measurable function. It follows that from (H1)′ and (H3), $f(t, x_t)$ is measurable in $t$, where $x \in A(t_0, \varphi, \sigma_1, \gamma_1)$ and satisfies $x_{t_0+t} = \hat{\varphi}_{t_0+t} + z_t$, $t \in [0, \sigma_1]$.

**Remark 2.2.** If we replace condition (H3) by a weaker condition

**(H3)′** for any $x, y \in A(t_0, \varphi, \sigma, \gamma)$ with $\sup_{t_0 \leq s \leq t_0+\sigma} |x(s) - y(s)| \to 0$,

$$\left| \int_{t_0}^{t} (t-s)^{q-1} (f(s, x_s) - f(s, y_s)) ds \right| \to 0, \quad t \in [t_0, t_0 + \sigma],$$

where $\sigma$ satisfy (2.48), then we can also conclude that the result of Theorem 2.7 holds.

The following existence and uniqueness result for fractional IVP (2.40)-(2.41) is based on Banach contraction mapping principle.

**Theorem 2.8.** *Assume that $g$ is generalized atomic on $\Omega$, and that for any $(t_0, \varphi) \in [0, \infty) \times \Omega$, there exist constants $\sigma_1, \gamma_1 \in (0, \infty), q_1 \in (0, q)$ and a real-valued function $m(t) \in L^{\frac{1}{q_1}}[t_0, t_0 + \sigma_1]$ such that conditions (H1)-(H2) of Theorem 2.7 hold. Further assume that:*

**(H4)** *there exists a nonnegative function $\ell : [0, \sigma_1] \to [0, \infty)$ continuous at $t = 0$ and $\ell(0) = 0$ such that for any $x, y \in A(t_0, \varphi, \sigma_1, \gamma_1)$ we have*

$$\left| \int_{t_0}^{t} (t-s)^{q-1} [f(s, x_s) - f(s, y_s)] ds \right| \leq \ell(t - t_0) \sup_{t_0 \leq s \leq t} |x(s) - y(s)|, \quad t \in [t_0, t_0 + \sigma_1],$$

*then fractional IVP (2.40)-(2.41) has a unique solution.*

**Proof.** According to the argument of Theorem 2.7, it suffices to prove that $S + U$ has a unique fixed point on $E(\sigma, \gamma)$, where $\sigma, \gamma > 0$ are sufficiently small. Now, choose $\sigma \in (0, \sigma_1)$, $\gamma \in (0, \gamma_1]$, such that (2.49) holds and that

$$c = \alpha(2k_2 + k_1) + \sup_{0 \le s \le \sigma} \frac{|K^{-1}(t_0 + s, \hat{\varphi}_{t_0+s}, \hat{\varphi}_{t_0+s})||\ell(s)|}{\Gamma(q)} < 1.$$

Obviously, $S + U$ is a mapping from $E(\sigma, \gamma)$ into itself. Using the same argument as that of Theorem 2.7, for any $z, w \in E(\sigma, \gamma)$, we get

$$|(Sz)(t) - (Sw)(t)| \le \alpha(2k_2 + k_1) \sup_{0 \le s \le \sigma} |z(s) - w(s)|,$$

and

$$
\begin{aligned}
&|(Uz)(t) - (Uw)(t)| \\
&\le \frac{|K^{-1}(t_0 + t, \hat{\varphi}_{t_0+t}, \hat{\varphi}_{t_0+t})|}{\Gamma(q)} \left| \int_0^t (t - s)^{q-1} f(t_0 + s, \hat{\varphi}_{t_0+s} + z_s) ds \right. \\
&\qquad \left. - \int_0^t (t - s)^{q-1} f(t_0 + s, \hat{\varphi}_{t_0+s} + w_s) ds \right| \\
&\le \frac{|K^{-1}(t_0 + t, \hat{\varphi}_{t_0+t}, \hat{\varphi}_{t_0+t})|}{\Gamma(q)} |\ell(t)| \sup_{0 \le s \le t} |z(s) - w(s)| \\
&\le \frac{\sup_{0 \le s \le \sigma} |K^{-1}(t_0 + s, \hat{\varphi}_{t_0+s}, \hat{\varphi}_{t_0+s})||\ell(s)|}{\Gamma(q)} \sup_{0 \le s \le \sigma} |z(s) - w(s)|.
\end{aligned}
$$

Therefore

$$
\begin{aligned}
&|(S + U)z(t) - (S + U)w(t)| \\
&\le \left[ \alpha(2k_2 + k_1) + \sup_{0 \le s \le \sigma} \frac{|K^{-1}(t_0 + s, \hat{\varphi}_{t_0+s}, \hat{\varphi}_{t_0+s})| \, |\ell(s)|}{\Gamma(q)} \right] \sup_{0 \le s \le \sigma} |z(s) - w(s)| \\
&= c \sup_{0 \le s \le \sigma} |z(s) - w(s)|.
\end{aligned}
$$

Hence, we have

$$\|(S + U)z - (S + U)w\| \le c\|z - w\|,$$

where $c < 1$. By applying Banach contraction mapping principle, we know that $S + U$ has a unique fixed point on $E(\sigma, \gamma)$. $\qquad \square$

**Corollary 2.6.** *If the condition (H4) of Theorem 2.8 is replaced by the following condition:*

**(H4)′** *there exist $q_2 \in (0, q)$ and a function $\ell_1 \in L^{\frac{1}{q_2}}[t_0, t_0 + \sigma_1]$, such that for any $x, y \in A(t_0, \varphi, \sigma_1, \gamma_1)$ we have*

$$|f(t, x_t) - f(t, y_t)| \le \ell_1(t) \sup_{t_0 \le s \le t} |x(s) - y(s)|, \quad t \in [t_0, t_0 + \sigma_1],$$

*then the result of Theorem 2.8 holds.*

**Proof.** It suffices to prove that the condition (H4) of Theorem 2.8 holds. Note that $\ell_1 \in L^{\frac{1}{q_2}}[t_0, t_0 + \sigma_1]$, hence, there must exist a positive number $N$, such that $N = \|\ell_1\|_{L^{\frac{1}{q_2}}[t_0, t_0+\sigma_1]}$. Then for any $x, y \in A(t_0, \varphi, \sigma_1, \gamma_1)$ we have

$$\left| \int_{t_0}^t (t-s)^{q-1} (f(s, x_s) - f(s, y_s)) ds \right|$$

$$\leq \int_{t_0}^t (t-s)^{q-1} |f(s, x_s) - f(s, y_s)| ds$$

$$\leq \int_{t_0}^t (t-s)^{q-1} \ell_1(s) \, ds \sup_{t_0 \leq s \leq t} |x(s) - y(s)|$$

$$\leq \frac{N}{(1+\beta')^{1-q_2}} (t-t_0)^{(1+\beta')(1-q_2)} \sup_{t_0 \leq s \leq t} |x(s) - y(s)|,$$

where $\beta' = \frac{q-1}{1-q_2} \in (-1, 0)$. Let

$$\ell(t - t_0) = \frac{N}{(1+\beta')^{1-q_2}} (t-t_0)^{(1+\beta')(1-q_2)}, \ t \in [t_0, t_0 + \sigma_1].$$

Obviously, $\ell : [0, \sigma_1] \to [0, \infty)$ continuous at $t = 0$ and $\ell(0) = 0$. Then the condition (H4) of Theorem 2.8 holds. □

### 2.4.3 *Continuation of Solutions*

For any $t_0, \varphi \in [0, \infty) \times \Omega$, $\omega \subset \Omega$ and positive constants $\sigma, \gamma > 0$, define $B_\omega(t_0, \varphi, \sigma, \gamma)$ as the set of all maps $x : (-\infty, t_0 + \sigma) \to \mathbb{R}^n$ such that $x_{t_0} = \varphi$, $x : [t_0, t_0 + \sigma) \to \mathbb{R}^n$ is continuous with $|x(t) - \varphi(0)| \leq \gamma$ and $x_t \in \omega$ for all $t \in [t_0, t_0 + \sigma)$. In the following theorem, $W$ is a set of all subsets of $\Omega$ such that for any $(t_0, \varphi) \in [0, \infty) \times \Omega$, constants $\sigma, \gamma > 0$ and a set $\omega \in W$, if $x \in B_\omega(t_0, \varphi, \sigma, \gamma)$ and if $x(t_0 + \sigma) = \lim_{t \to (t_0+\sigma)^-} x(t)$ exists, then $x_{t_0+\sigma} \in \Omega$.

**Theorem 2.9.** *Let all conditions of Theorem 2.7 hold. Besides, suppose that $\sigma \in (0, \sigma_1], \gamma \in (0, \gamma_1]$ and for any $x \in B_\omega(t_0, \varphi, \sigma, \gamma)$,*

**(H5)** *there exist constants $q_\omega \in (0, q)$ and a real-valued function $m_\omega(t) \in L^{\frac{1}{q_\omega}}[t_0, t_0 + \sigma]$ such that $f(t, x_t)$ is measurable and $|f(t, x_t)| \leq m_\omega(t)$ for $t \in [t_0, t_0 + \sigma)$;*

**(H6)** $\lim_{\tau \to 0^+} [g(t, x_{t-\tau}) - g(t - \tau, x_{t-\tau})] = 0$ *uniformly for $t \in [t_0 + \tau, t_0 + \sigma)$;*

**(H7)** $K(t, x_t, x_t) - K(t, x_t, x_{t-\tau}) \to 0$ *uniformly for $t \in [t_0 + \tau, t_0 + \sigma)$ as $\tau \to 0^+$ and as $\sup_{t_0+\tau \leq s \leq t} |x(s) - x(s - \tau)| \to 0$;*

**(H8)** *there exists a constant $H$ such that $|K^{-1}(t, x_t, x_t)| \leq H$ for all $t \in [t_0, t_0+\sigma)$;*

**(H9)** *there exists a continuous function $\ell_\omega : [0, \infty) \to [0, \infty)$ with $\ell_\omega(0) = 0$ such that*

$$|L(t, x_t, x_{t-\tau}) - L_b^*(t, x_t, x_{t-\tau})| \leq \ell_\omega(b) \sup_{-b \leq \theta \leq 0} |x(t + \theta) - x(t - \tau + \theta)|$$

*where for a given $b > 0$, $\lim_{\tau \to 0^+} L_b^*(t, x_t, x_{t-\tau}) = 0$ uniformly for $t \in [t_0 + \tau, t_0 + \sigma)$.*

*Then for any $\omega \in W$ and any $\gamma > 0$, if $x(t)$ is a noncontinuable solution of fractional IVP (2.40)-(2.41) defined on $[t_0, t_0 + \sigma)$, there exists a $t^* \in [t_0, t_0 + \sigma)$ such that $|x(t^*) - \varphi(0)| > \gamma$ or $x_{t^*} \notin \omega$.*

**Proof.** By way of contradiction, if there exists a noncontinuable solution $x(t)$ of fractional IVP (2.40)-(2.41) on $[t_0, t_0 + \sigma)$ such that $|x(t) - \varphi(0)| \le \gamma$ and $x_t \in \omega$ for all $t \in [t_0, t_0 + \sigma)$, that is, $x \in B_\omega(t_0, \varphi, \sigma, \gamma)$, then first, $x(t)$ is not uniformly continuous on $[t_0, t_0 + \sigma)$. Otherwise, $x(t_0 + \sigma) = \lim_{t \to (t_0+\sigma)^-} x(t)$ exists and thus $x_{t_0+\sigma} \in \Omega$. By Theorem 2.7, $x(t)$ can be continued beyond $t_0 + \sigma$.

Therefore, there exist a sufficiently small constant $\varepsilon > 0$, and sequences $\{t_k\} \subseteq [t_0, t_0 + \sigma)$, $\{\Delta_k\}$ with $\Delta_k \to 0^+$ as $k \to \infty$, such that

$$|x(t_k) - x(t_k - \Delta_k)| \ge \varepsilon, \quad \text{for all } k = 1, 2, \dots.$$

Now choose a constant $H > 0$ so that

$$|K^{-1}(t, x_t, x_t)| \le H, \quad \text{for all } t \in [t_0, t_0 + \sigma).$$

For given $H$ and $\varepsilon > 0$, by (H5)-(H7) and (H9), we can find positive constants $b$ and $\sigma_0$ so that

$$\frac{2HM_\omega}{\Gamma(q)(1 + \beta_\omega)^{1-q_\omega}}(\sigma_0^{1+\beta_\omega})^{1-q_\omega} < \frac{\varepsilon}{5},$$

where $\beta_\omega = \frac{q-1}{1-q_\omega} \in (-1, 0)$, $M_\omega = (\int_{t_0}^{t_0+\sigma}(m_\omega(s))^{\frac{1}{q_\omega}} ds)^{q_\omega}$,

$$H|K(t, x_t, x_t) - K(t, x_t, x_{t-\tau})| < \frac{1}{5}, \quad \text{as} \quad \sup_{t_0+\tau \le s \le t} |x(s) - x(s - \tau)| \le \varepsilon,$$

and

$$H|g(t, x_{t-\tau}) - g(t - \tau, x_{t-\tau})| < \frac{\varepsilon}{5},$$

$$H\ell_\omega(b) < \frac{1}{5}, \quad b < \frac{\delta - \sigma_0}{2},$$

$$H|L_b^*(t, x_t, x_{t-\tau})| < \frac{\varepsilon}{5},$$

for all $t \in [t_0 + \tau, t_0 + \sigma)$ and $0 < \tau < \sigma_0$.

Since $x(t)$ is uniformly continuous on $[t_0, t_0 + \sigma - b]$, we can find a constant $H_1 > 0$ so that for all $k \ge H_1$, we have $\Delta_k < \sigma_0$ and $|x(t) - x(t - \Delta_k)| < \varepsilon$ for all $t \in [t_0 + \sigma_k, t_0 + \sigma - b]$. Now for all $k \ge H_1$, define a sequence $\{s_k\}$ in the following pattern

$$s_k = \inf\{t \in (t_0 + \sigma - b, t_0 + \sigma) : |x(t) - x(t - \Delta_k)| \ge \varepsilon\}.$$

Then

$$|x(s_k) - x(s_k - \Delta_k)| = \varepsilon.$$

Thus we get

$$\frac{2HM_\omega}{\Gamma(q)(1 + \beta_\omega)^{1-q_\omega}}(\Delta_k^{1+\beta_\omega})^{1-q_\omega} < \frac{\varepsilon}{5},$$

$$H|K(s_k, x_{s_k}, x_{s_k}) - K(s_k, x_{s_k}, x_{s_k - \Delta_k})| < \frac{1}{5},$$

$$H|g(s_k, x_{s_k - \Delta_k}) - g(s_k - \Delta_k, x_{s_k - \Delta_k})| < \frac{\varepsilon}{5}$$

and

$$H|L(s_k, x_{s_k}, x_{s_k - \Delta_k}) - L_b^*(s_k, x_{s_k}, x_{s_k - \Delta_k})|$$

$$\leq \frac{1}{5} \sup_{-b \leq \theta \leq 0} |x(s_k + \theta) - x(s_k - \Delta_k + \theta)| \leq \frac{\varepsilon}{5}.$$

On the other hand, we see that

$$g(s_k, x_{s_k}) - g(s_k - \Delta_k, x_{s_k - \Delta_k})$$
$$= g(s_k, x_{s_k}) - g(s_k, x_{s_k - \Delta_k}) + g(s_k, x_{s_k - \Delta_k}) - g(s_k - \Delta_k, x_{s_k - \Delta_k})$$
$$= (K(s_k, x_{s_k}, x_{s_k - \Delta_k}) - K(s_k, x_{s_k}, x_{s_k}))\ (x(s_k) - x(s_k - \Delta_k))$$
$$+ K(s_k, x_{s_k}, x_{s_k})\ (x(s_k) - x(s_k - \Delta_k)) + L(s_k, x_{s_k}, x_{s_k - \Delta_k})$$
$$- L_b^*(s_k, x_{s_k}, x_{s_k - \Delta_k}) + L_b^*(s_k, x_{s_k}, x_{s_k - \Delta_k})$$
$$+ g(s_k, x_{s_k - \Delta_k}) - g(s_k - \Delta_k, x_{s_k - \Delta_k}).$$

By using the same argument as that of Claim III in Theorem 2.7, we have

$$|g(s_k, x_{s_k}) - g(s_k - \Delta_k, x_{s_k - \Delta_k})|$$

$$\leq \frac{1}{\Gamma(q)} \left| \int_{s_k - \Delta_k}^{s_k} (s_k - s)^{q-1} f(s, x_s) ds \right|$$

$$+ \frac{1}{\Gamma(q)} \left| \int_{t_0}^{s_k - \Delta_k} [(s_k - s)^{q-1} - (s_k - \Delta_k - s)^{q-1}] f(s, x_s) ds \right|$$

$$\leq \frac{2M_\omega}{\Gamma(q)(1 + \beta_\omega)^{1 - q_\omega}} (\Delta_k^{1 + \beta_\omega})^{1 - q_\omega}.$$

Therefore

$$|x(s_k) - x(s_k - \Delta_k)|$$
$$\leq |K^{-1}(s_k, x_{s_k}, x_{s_k})|[|g(s_k, x_{s_k}) - g(s_k - \Delta_k, x_{s_k - \Delta_k})|$$
$$+ |K(s_k, x_{s_k}, x_{s_k - \Delta_k}) - K(s_k, x_{s_k}, x_{s_k})||x(s_k) - x(s_k - \Delta_k)|$$
$$+ |L(s_k, x_{s_k}, x_{s_k - \Delta_k}) - L_b^*(s_k, x_{s_k}, x_{s_k - \Delta_k})|$$
$$+ |L_b^*(s_k, x_{s_k}, x_{s_k - \Delta_k})| + |g(s_k, x_{s_k - \Delta_k}) - g(s_k - \Delta_k, x_{s_k - \Delta_k})|]$$
$$< \varepsilon.$$

This is contrary to $|x(s_k) - x(s_k - \Delta_k)| = \varepsilon$. The proof is completed. $\square$

**Remark 2.3.** If we replace conditions of Theorem 2.7 by conditions of Remark 2.1, the result of Theorem 2.9 holds.

**Remark 2.4.** If we replace the condition (H3) of Theorem 2.7 by a weaker condition:

**(H3)″** for any $x, y \in A(t_0, \varphi, \sigma, \gamma)$ with $\sup_{t_0 \le s \le t_0+\sigma} |x(s) - y(s)| \to 0$,

$$\left| \int_{t_0}^{t} (t-s)^{q-1}[f(s, x_s) - f(s, y_s)]ds \right| \to 0, \quad t \in [t_0, t_0 + \sigma],$$

where $\sigma \in (0, \sigma_1], \gamma \in (0, \gamma_1]$.

Then we can also conclude that the result of Theorem 2.9 holds.

In the following, for any $(t_0, \varphi) \in [0, \infty) \times \Omega$ and any constants $\varepsilon, \sigma, \gamma > 0$, $C_\varepsilon(t_0, \varphi, \delta, \gamma)$ denotes the set of all functions $x : (-\infty, t_0 + \sigma] \to \mathbb{R}^n$ so that $\|x_{t_0} - \varphi\|_B < \varepsilon$, $x : [t_0, t_0 + \sigma] \to \mathbb{R}^n$ is continuous and $|x(t) - \varphi(0)| \le \gamma$.

**Theorem 2.10.** *Suppose that for any $(t_0, \varphi) \in [0, \infty) \times \Omega$, the solution of fractional IVP (2.40)-(2.41) is unique. Besides, suppose that $\sigma \in (0, \sigma_1], \gamma \in (0, \gamma_1]$ and for any $x \in C_\varepsilon(t_0, \varphi, \sigma, \gamma)$, (H5)-(H9) hold and*

**(H10)** *for any $x, y \in C_\varepsilon(t_0, \varphi, \sigma, \gamma)$, if $\|x_{t_0} - y_{t_0}\|_B \to 0$ and $\sup_{t_0 \le s \le t_0+\sigma} |x(s) - y(s)| \to 0$, then $g(t, x_t) \to g(t, y_t)$ and $|\int_{t_0}^{t} (t-s)^{q-1}[f(s, x_s) - f(s, y_s)]ds| \to 0$ for $t \in [t_0, t_0 + \sigma]$.*

*If $x$ is a noncontinuable solution of fractional IVP(2.40)-(2.41) defined on $[t_0, t_0 + \sigma_1)$, then for any $\varepsilon > 0$ and $\sigma \in (0, \sigma_1)$, we can find a $\sigma > 0$ so that if $\|\varphi - \psi\|_B < \sigma$, then $|x(t) - y(t)| < \varepsilon$ for $t \in [t_0, t_0 + \sigma]$, where $y(t)$ is a solution of (2.40) through $(t_0, \psi)$.*

**Proof.** By way of contradiction, if the conclusion above is not true, then there exist $\varepsilon > 0$, sequences $\{t_k\} \subseteq [t_0, t_0 + \sigma]$ and $\{\varphi^k\} \subseteq \Omega$ such that

$$\|\varphi^k - \varphi\|_B < \frac{1}{k},$$

$$|y^k(t_k) - x(t_k)| = \varepsilon$$

and

$$|y^k(t) - x(t)| < \varepsilon, \quad \text{for } t \in [t_0, t_k),$$

where $y^k(t)$ is a solution of following fractional IVP

$$^C_{t_0}D_t^\alpha g(t, y_t) = f(t, y_t), \quad y_{t_0} = \varphi^k. \tag{2.52}$$

Without loss of generality, we may assume $t_k \to \bar{t} \in [t_0, t_0 + \sigma]$. Now define a sequence of functions $\{z^k\}$ as follows:

$$z^k(t) = \begin{cases} y^k(t), & \text{for } t \in [t_0, t_k], \\ y^k(t_k), & \text{for } t \in [t_k, \bar{t}\,], \text{ if } t_k < \bar{t}. \end{cases}$$

Using the same argument as that of Theorem 2.7, we can assume that $\{z^k\}$ is equicontinuous in $t \in [t_0, \bar{t}\,]$. By Arzela-Ascoli theorem, without loss of generality, we can find a function $y : (-\infty, \bar{t}\,] \to \mathbb{R}^n$ such that $\lim_{k \to \infty} \sup_{t_0 \le s \le \bar{t}} |z^k(s) - y(s)| = 0$ and $y(s) = \varphi(s)$ for $s \le t_0$.

Now considering equation (2.52), we get

$$g(t, y_t^k) - g(t_0, \varphi^k) = \frac{1}{\Gamma(q)} \int_{t_0}^t (t - s)^{q-1} f(s, y_s^k) ds, \quad \text{for } t \in [t_0, \bar{t}\,].$$

By (H10) and Lebesgue dominated convergence theorem and let $k \to \infty$, we obtain

$$g(t, y_t) - g(t_0, \varphi) = \frac{1}{\Gamma(q)} \int_{t_0}^t (t - s)^{q-1} f(s, y_s) ds, \quad \text{for } t \in [t_0, \bar{t}\,].$$

This means that $y(t) = x(t)$ for $t \in [t_0, \bar{t}\,]$ by the uniqueness assumption of the solutions of fractional IVP(2.40)-(2.41). This is contrary to

$$|y^k(t_k) - x(t_k)| = \varepsilon$$

and

$$\lim_{k \to \infty} \sup_{t_0 \le s \le \bar{t}} |z^k(s) - y(s)| = 0.$$

The proof is therefore complete. $\qquad\square$

## 2.5 Iterative Functional Differential Equations

### 2.5.1 *Introduction*

In Section 2.5, we consider the following fractional iterative functional differential equations with parameter

$$\begin{cases} {}_a^C D_t^q x(t) = f(t, x(t), x(x^v(t))) + \lambda, & t \in [a, b], \ v \in \mathbb{R} \setminus \{0\}, \\ & q \in (0, 1), \ \lambda \in \mathbb{R}, \\ x(t) = \varphi(t), & t \in [a_1, a], \\ x(t) = \psi(t), & t \in [b, b_1], \end{cases} \tag{2.53}$$

where ${}_a^C D_t^q$ is Caputo fractional derivative of order $q$ and

**(C1)** $a_1 \le a < b \le b_1$, $a_1 \le a_1^v$ and $b_1^v \le b_1$;
**(C2)** $f \in C([a, b] \times [a_1, b_1]^2, \mathbb{R})$;
**(C3)** $\varphi \in C([a_1, a], [a_1, b_1])$ and $\psi \in C([b, b_1], [a_1, b_1])$.

**Definition 2.12.** A function $x \in C([a_1, b_1], [a_1, b_1])$ is said to be a solution of the problem (2.53) if $x$ satisfies the equation ${}_a^C D_t^q x(t) = f(t, x(t), x(x^v(t))) + \lambda$ on $[a, b]$, and the conditions $x(t) = \varphi(t), t \in [a_1, a]$, $x(t) = \psi(t), t \in [b, b_1]$.

The purpose of this section is to determine the pair $(x, \lambda)$, $x \in C([a_1, b_1], [a_1, b_1])$ (or $C_L^q([a_1, b_1], [a_1, b_1])$), $\lambda \in \mathbb{R}$, which satisfies the problem (2.53). In Subsection 2.5.2, by using Schauder fixed point theorem, we establish existence theorems in $C([a_1, b_1], [a_1, b_1])$ and $C_L^q([a_1, b_1], [a_1, b_1])$ respectively. Unfortunately, uniqueness results cannot be obtained since the solution operator is not Lipschitz continuous but only Hölder continuous. Meanwhile, data dependence results of solutions and parameters provide possible way to describe the error estimates between explicit and approximative solutions for such problems. In Subsection 2,5.4, We make some examples to illustrate our results and conclude some possible extensions to general parametrized fractional iterative functional differential equations.

### 2.5.2    *Existence*

We first give existence result in $C([a_1, b_1], [a_1, b_1])$. Let $(x, \lambda)$ be a solution of the problem (2.53). Then this problem is equivalent to the following fixed point equation

$$x(t) = \begin{cases} \varphi(t), & \text{for } t \in [a_1, a], \\ \varphi(a) + \dfrac{1}{\Gamma(q)} \displaystyle\int_a^t (t-s)^{q-1} f(s, x(s), x(x^v(s))) ds \\ \quad + \dfrac{\lambda}{\Gamma(q+1)} (t-a)^q, & \text{for } t \in [a, b], \\ \psi(t), & \text{for } t \in [b, b_1]. \end{cases} \tag{2.54}$$

From the condition of continuity of $x$ in $t = b$, we have that

$$\lambda = \frac{\Gamma(q+1)(\psi(b) - \varphi(a))}{(b-a)^q} - \frac{q}{(b-a)^q} \int_a^b (b-s)^{q-1} f(s, x(s), x(x^v(s))) ds.$$

Now we consider the operator

$$A : C([a_1, b_1], [a_1, b_1]) \to C([a_1, b_1], \mathbb{R}),$$

where

$$(Ax)(t) := \begin{cases} \varphi(t), & \text{for } t \in [a_1, a], \\ \varphi(a) + \dfrac{(t-a)^q}{(b-a)^q}(\psi(b) - \varphi(a)) - \dfrac{(t-a)^q}{\Gamma(q)(b-a)^q} \\ \quad \times \displaystyle\int_a^b (b-s)^{q-1} f(s, x(s), x(x^v(s))) ds \\ \quad + \dfrac{1}{\Gamma(q)} \displaystyle\int_a^t (t-s)^{q-1} f(s, x(s), x(x^v(s))) ds, & \text{for } t \in [a, b], \\ \psi(t), & \text{for } t \in [b, b_1]. \end{cases} \tag{2.55}$$

It is clear that $(x, \lambda)$ is a solution of the problem (2.53) if and only if $x$ is a fixed point of the operator $A$ and $\lambda$ is given by (2.54). So, the problem is to study the fixed point equation

$$x = A(x).$$

Now, we are ready to state our first result in this section.

**Theorem 2.11.** *We suppose that*

**(i)** *conditions (C1)-(C3) are satisfied;*

**(ii)** *there are $m_f$, $M_f \in \mathbb{R}$ such that*

$$m_f \le f(t, u, w) \le M_f, \ \forall \, t \in [a, b], u, w \in [a_1, b_1],$$

*along with*

$$a_1 \le \min(\varphi(a), \psi(b)) - \max\left\{0, \frac{M_f(b-a)^q}{\Gamma(q+1)}\right\} + \min\left\{0, \frac{m_f(b-a)^q}{\Gamma(q+1)}\right\},$$

*and*

$$\max(\varphi(a), \psi(b)) - \min\left\{0, \frac{m_f(b-a)^q}{\Gamma(q+1)}\right\} + \max\left\{0, \frac{M_f(b-a)^q}{\Gamma(q+1)}\right\} \le b_1.$$

*Then problem (2.53) has a solution in $C([a_1, b_1], [a_1, b_1])$.*

**Proof.** In what follow we consider on $C([a_1, b_1], \mathbb{R})$ with the Chebyshev norm $\|\cdot\|_C$.

Condition (ii) assures that the set $C([a_1, b_1], [a_1, b_1])$ is an invariant subset for the operator $A$, that is, we have

$$A(C([a_1, b_1], [a_1, b_1])) \subset C([a_1, b_1], [a_1, b_1]).$$

Indeed, for $t \in [a_1, a] \cup [b, b_1]$, we have $A(x)(t) \in [a_1, b_1]$. Furthermore, we obtain

$$a_1 \le A(x)(t) \le b_1, \ \forall \, t \in [a, b],$$

if and only if

$$a_1 \le \min_{t \in [a,b]} A(x)(t) \tag{2.56}$$

and

$$\max_{t \in [a,b]} A(x)(t) \le b_1 \tag{2.57}$$

hold.

Since

$$\min_{t \in [a,b]} A(x)(t) \ge \min\{\varphi(a), \psi(b)\} - \max\left\{0, \frac{M_f(b-a)^q}{\Gamma(q+1)}\right\} + \min\left\{0, \frac{m_f(b-a)^q}{\Gamma(q+1)}\right\},$$

and

$$\max_{t \in [a,b]} A(x)(t) \le \max\{\varphi(a), \psi(b)\} - \min\left\{0, \frac{m_f(b-a)^q}{\Gamma(q+1)}\right\} + \max\left\{0, \frac{M_f(b-a)^q}{\Gamma(q+1)}\right\},$$

respectively, the requirements (2.56) and (2.57) are equivalent with the conditions appearing in (ii).

So, in the above conditions we have a self-mapping operator

$$A : C([a_1, b_1], [a_1, b_1]) \to C([a_1, b_1], [a_1, b_1]).$$

Further, we check $A$ is a completely continuous operator.

Let $\{x_n\}$ be a sequence such that $x_n \to x$ in $C([a_1, b_1], [a_1, b_1])$. Then for each $t \in [a_1, b_1]$, we have that

$$|(Ax_n)(t) - (Ax)(t)| \le \begin{cases} 0, & \text{for } t \in [a_1, a], \\ \dfrac{2(b-a)^q}{\Gamma(q+1)} \|f(\cdot, x_n(x_n^v(\cdot))) - f(\cdot, x(x^v(\cdot)))\|_C, & \text{for } t \in [a, b], \\ 0, & \text{for } t \in [b, b_1]. \end{cases}$$

Since $f \in C([a,b] \times [a_1,b_1]^2, \mathbb{R})$, we have that

$$\|Ax_n - Ax\|_C \to 0 \quad \text{as } n \to \infty.$$

Now, consider $a_1 \le t_1 < t_2 \le a$. Then,

$$|(Ax)(t_2) - (Ax)(t_1)| = |\varphi(t_2) - \varphi(t_1)|.$$

Similarly, for $b \le t_1 < t_2 \le b_1$,

$$|(Ax)(t_2) - (Ax)(t_1)| = |\psi(t_2) - \psi(t_1)|.$$

On the other hand, for $a \le t_1 < t_2 \le b$,

$$|(Ax)(t_2) - (Ax)(t_1)| \le \frac{(t_2 - t_1)^q}{(b-a)^q}|\psi(b) - \varphi(a)|$$
$$+ \frac{4(t_2 - t_1)^q \max\{|m_f|, |M_f|\}}{\Gamma(q+1)}. \tag{2.58}$$

Together with Arzela-Ascoli theorem and $A$ is a continuous operator, we can conclude that $A$ is a completely continuous operator.

It is obvious that the set $C([a_1,b_1],[a_1,b_1]) \subseteq C([a_1,b_1],\mathbb{R})$ is a bounded convex closed subset of the Banach space $C([a_1,b_1],\mathbb{R})$. Thus, the operator $A$ has a fixed point due to Schauder fixed point theorem. This completes the proof. $\square$

In the following, we present the existence and estimate results in $C_L^q([a_1,b_1],[a_1,b_1])$. Let $L > 0$ and $I \subset \mathbb{R}$ be a compact interval, and introduce the following notation:

$$C_L^q(I, \mathbb{R}) = \{x \in C(I,\mathbb{R}) | \ |x(t_1) - x(t_2)| \le L|t_1 - t_2|^q\}$$

for all $t_1, t_2 \in I$. Remark that $C_L^q(I,\mathbb{R}) \subseteq C(I,\mathbb{R})$ is a complete metric space. Then (2.58) implies that under assumptions of Theorem 2.11 any solution of problem (2.53) belongs to $C_{L_*}^q([a,b],\mathbb{R})$ for

$$L_* = \frac{|\psi(b) - \varphi(a)|}{(b-a)^q} + \frac{4\max\{|m_f|, |M_f|\}}{\Gamma(q+1)}. \tag{2.59}$$

Now we present our second result in this section.

**Theorem 2.12.** *We suppose that*

**(i)** *conditions of Theorem 2.11 hold and* $\varphi \in C_{L_\varphi}^q([a_1,a],[a_1,b_1])$, $\psi \in C_{L_\psi}^q([b,b_1],[a_1,b_1])$ *for some* $L_\varphi, L_\psi \ge 0$.

*Then problem (2.53) has a solution in* $X = C_L^q([a_1,b_1],[a_1,b_1])$ *and all its solution belongs to* $X$ *for*

$$L = (\sqrt[1-q]{L_\varphi} + \sqrt[1-q]{L_\psi} + \sqrt[1-q]{L_*})^{1-q},$$

*where* $L_*$ *is defined by (2.59).*

*Assume in addition*

**(ii)** *there exist $L_u > 0$ and $L_w > 0$ such that*

$$|f(t, u_1, w_1) - f(t, u_2, w_2)| \leq L_u |u_1 - u_2| + L_w |w_1 - w_2|,$$

*for all $t \in [a, b]$, $u_i, w_i \in [a_1, b_1]$, $i = 1, 2$.*

*Then two solutions $x_1$ and $x_2$ of problem (2.53) satisfy*

$$\|x_1 - x_2\|_C \leq L_A^{\frac{1}{1 - q\min\{1, v\}}} \tag{2.60}$$

*for*

$$L_A := \frac{2(b - a)^q}{\Gamma(q + 1)}((L_u + L_w)b_1^{1 - q\min\{1, v\}} + \max\{1, v^q\}b_1^{q\max\{v - 1, 0\}}L_wL). \tag{2.61}$$

*If in addition*

$$\Gamma(q + 1) > 2(b - a)^q L_u, \tag{2.62}$$

*then*

$$\|x_1 - x_2\|_C$$
$$\leq \left(\frac{2(b - a)^q L_w(b_1^{1 - q\min\{1, v\}} + \max\{1, v^q\}b_1^{q\max\{v - 1, 0\}}L)}{\Gamma(q + 1) - 2(b - a)^q L_u}\right)^{\frac{1}{1 - q\min\{1, v\}}}. \tag{2.63}$$

**Proof.** Consider the operator $A$ given by (2.55). From Theorem 2.11, we have

$$A : C([a_1, b_1], [a_1, b_1]) \to C([a_1, b_1], [a_1, b_1])$$

and $A$ has a fixed point in $C([a_1, b_1], [a_1, b_1])$.

Now, consider $a_1 \leq t_1 < t_2 \leq a$. Then,

$$|(Ax)(t_2) - (Ax)(t_1)| = |\varphi(t_2) - \varphi(t_1)| \leq L_\varphi |t_1 - t_2|^q \leq L_* |t_1 - t_2|^q$$

as $\varphi \in C_{L_\varphi}^q([a_1, a], [a_1, b_1])$, due to (i).

Similarly, for $b \leq t_1 < t_2 \leq b_1$,

$$|(Ax)(t_2) - (Ax)(t_1)| = |\psi(t_2) - \psi(t_1)| \leq L_\psi |t_1 - t_2|^q \leq L_* |t_1 - t_2|^q$$

that follows from (i), too.

On the other hand, for $a \leq t_1 < t_2 \leq b$, we already know (see (2.58))

$$|(Ax)(t_2) - (Ax)(t_1)| \leq L_* |t_1 - t_2|^q.$$

Next, if $a_1 \leq t_1 \leq a \leq t_2 \leq b$, then by Hölder inequality with $q' = \frac{1}{q}$ and $p' = \frac{1}{1 - q}$ (note $q', p' > 1$)

$$|(Ax)(t_2) - (Ax)(t_1)| \leq |(Ax)(a) - (Ax)(t_1)| + |(Ax)(t_2) - (Ax)(a)|$$
$$\leq L_\varphi(a - t_1)^q + L_*(t_2 - a)^q$$
$$\leq \sqrt[p']{L_\varphi^{p'} + L_*^{p'}} \sqrt[q']{(a - t_1)^{qq'} + (t_2 - a)^{qq'}}$$
$$\leq L|t_1 - t_2|^q.$$

Furthermore, if $a_1 \leq t_1 \leq a < b \leq t_2 \leq b_1$, then again by the Hölder inequality with $q' = \frac{1}{q}$ and $p' = \frac{1}{1-q}$

$$|(Ax)(t_2) - (Ax)(t_1)|$$
$$\leq |(Ax)(a) - (Ax)(t_1)| + |(Ax)(b) - (Ax)(a)| + |(Ax)(t_2) - (Ax)(b)|$$
$$\leq L_\varphi(a - t_1)^q + L_*(b - a)^q + L_\psi(t_2 - b)^q$$
$$\leq \sqrt[p']{L_\varphi^{p'} + L_*^{p'} + L_\psi^{p'}} \sqrt[q']{(a - t_1)^{qq'} + (b - a)^{qq'} + (t_2 - b)^{qq'}}$$
$$= L|t_1 - t_2|^q.$$

Therefore, the function $A(x)(t)$ belongs to $X$. This proves the first statement. Take $x_1, x_2 \in X$. Then for all $t \in [a_1, a] \cup [b, b_1]$, we have

$$|A(x_1)(t) - A(x_2)(t)| = 0.$$

Moreover, for $t \in [a, b]$, from our conditions, we get

$$|A(x_1)(t) - A(x_2)(t)|$$
$$\leq \frac{(t - a)^q}{\Gamma(q)(b - a)^q} \int_a^b (b - s)^{q-1} |f(s, x_1(s), x_1(x_1^v(s))) - f(s, x_2(s), x_2(x_2^v(s)))| \, ds$$
$$+ \frac{1}{\Gamma(q)} \int_a^t (t - s)^{q-1} |f(s, x_1(s), x_1(x_1^v(s))) - f(s, x_2(s), x_2(x_2^v(s)))| \, ds$$
$$\leq \frac{1}{\Gamma(q)} \int_a^b (b - s)^{q-1} \left( L_u |x_1(s) - x_2(s)| + L_w |x_1(x_1^v(s)) - x_2(x_2^v(s))| \right) ds$$
$$+ \frac{1}{\Gamma(q)} \int_a^t (t - s)^{q-1} \left( L_u |x_1(s) - x_2(s)| + L_w |x_1(x_1^v(s)) - x_2(x_2^v(s))| \right) ds$$
$$\leq \frac{1}{\Gamma(q)} \int_a^b (b - s)^{q-1} (L_u |x_1(s) - x_2(s)| + L_w |x_1(x_1^v(s)) - x_1(x_2^v(s))|$$
$$+ L_w |x_1(x_2^v(s)) - x_2(x_2^v(s))|) ds$$
$$+ \frac{1}{\Gamma(q)} \int_a^t (t - s)^{q-1} (L_u |x_1(s) - x_2(s)| + L_w |x_1(x_1^v(s)) - x_1(x_2^v(s))|$$
$$+ L_w |x_1(x_2^v(s)) - x_2(x_2^v(s))|) ds$$
$$\leq \frac{1}{\Gamma(q)} \int_a^b (b - s)^{q-1} ((L_u + L_w) \|x_1 - x_2\|_C + L_w L |x_1^v(s) - x_2^v(s)|^q) ds$$
$$+ \frac{1}{\Gamma(q)} \int_a^t (t - s)^{q-1} ((L_u + L_w) \|x_1 - x_2\|_C + L_w L |x_1^v(s) - x_2^v(s)|^q) ds$$
$$\leq \frac{1}{\Gamma(q)} \int_a^b (b - s)^{q-1} ((L_u + L_w) \|x_1 - x_2\|_C$$
$$+ \max\{1, v^q\} b_1^{q \max\{v-1, 0\}} L_w L \|x_1 - x_2\|_C^{q \min\{1, v\}}) ds$$
$$+ \frac{1}{\Gamma(q)} \int_a^t (t - s)^{q-1} ((L_u + L_w) \|x_1 - x_2\|_C$$

$$+ \max\{1, v^q\} b_1^{q\max\{v-1,0\}} L_w L \|x_1 - x_2\|_C^{q\min\{1,v\}}) ds$$

$$\leq \frac{2(b-a)^q}{\Gamma(q+1)} ((L_u + L_w)\|x_1 - x_2\|_C$$

$$+ \max\{1, v^q\} b_1^{q\max\{v-1,0\}} L_w L \|x_1 - x_2\|_C^{q\min\{1,v\}}),$$

where we use the inequality

$$r^v - s^v \leq \max\{1, v\} r^{\max\{v-1,0\}} (r - s)^{\min\{1,v\}}$$

for any $r \geq s \geq 0$ and $v > 0$. From $\|x_1 - x_2\|_C \leq b_1$ we get

$$\|A(x_1) - A(x_2)\|_C \leq \frac{2(b-a)^q}{\Gamma(q+1)} ((L_u + L_w) b_1^{1-q\min\{1,v\}}$$

$$+ \max\{1, v^q\} b_1^{q\max\{v-1,0\}} L_w L) \|x_1 - x_2\|_C^{q\min\{1,v\}}$$

$$= L_A \|x_1 - x_2\|_C^{q\min\{1,v\}}. \tag{2.64}$$

So $A$ is Hölder continuous but never Lipschitz continuous, since $q\min\{1,v\} \leq q < 1$. If $x_1$ and $x_2$ are fixed points of $A$ then

$$\|x_1 - x_2\|_C = \|A(x_1) - A(x_2)\|_C \leq L_A \|x_1 - x_2\|_C^{q\min\{1,v\}}$$

which implies (2.60). In general, we have

$$\|x_1 - x_2\|_C \leq \frac{2(b-a)^q}{\Gamma(q+1)} L_u \|x_1 - x_2\|_C$$

$$+ \frac{2(b-a)^q L_w}{\Gamma(q+1)} (b_1^{1-q\min\{1,v\}}$$

$$+ \max\{1, v^q\} b_1^{q\max\{v-1,0\}} L) \|x_1 - x_2\|_C^{q\min\{1,v\}},$$

which implies (2.63) under (2.62). The proof is completed. □

We do not know about uniqueness. But this is not so surprising, since $A$ is not Lipschitzian in general. So we cannot apply metric fixed point theorems, only topological one. This can be simply illustrated on a simpler problem

$$_0^C D_t^{\frac{1}{2}} x(t) = x(\sqrt{x(t)}), \quad x(0) = 0, \quad t \in [0, 1]. \tag{2.65}$$

Rewriting (2.65) as

$$x(t) = B(x)(t) = \frac{1}{\Gamma(\frac{1}{2})} \int_0^t \frac{x(\sqrt{x(s)})}{\sqrt{t-s}} ds,$$

it follows that $B : C_{\frac{1}{2}}([0,1], [0,1]) \to C_{\frac{1}{2}}([0,1], \mathbb{R})$ satisfies

$$\|B(x_1) - B(x_2)\|_C \leq \frac{2}{\sqrt{\pi}} \left( \|x_1 - x_2\|_C + \frac{1}{2}\sqrt{\|x_1 - x_2\|_C} \right),$$

so it is not Lipschitzian. Hence (2.65) should have a nonzero solution, and it does have $x(t) = \frac{4}{\pi} t$.

### 2.5.3 Data Dependence

Consider the following two problems

$$\begin{cases} {}^{C}_{a}D^q_t x(t) = f_i(t, x(t), x(x^v(t))) + \lambda_i, & t \in [a,b], \ v \in (0,1], \ q \in (0,1), \\ x(t) = \varphi_i(t), & t \in [a_1, a], \\ x(t) = \psi_i(t), & t \in [b, b_1], \end{cases} \quad (2.66)$$

where $f_i$, $\lambda_i$, $\varphi_i$ and $\psi_i$, $i = 1, 2$ be as in Theorem 2.12.

Consider the operators

$$A_i : C([a_1, b_1], [a_1, b_1]) \to C([a_1, b_1], [a_1, b_1])$$

given by (2.55) when $\varphi$, $\psi$, $f$ and $\lambda$ are replaced by $\varphi_i$, $\psi_i$, $f_i$ and $\lambda_i$, respectively. We are ready to state the third result in this section.

**Theorem 2.13.** *Suppose conditions of Theorem 2.12, and, moreover*

**(i)** *there exists $\eta_1 > 0$ such that*

$$|\varphi_1(t) - \varphi_2(t)| \leq \eta_1, \ \forall \ t \in [a_1, a],$$

*and*

$$|\psi_1(t) - \psi_2(t)| \leq \eta_1, \ \forall \ t \in [b, b_1];$$

**(ii)** *there exists $\eta_2 > 0$ such that*

$$|f_1(t, u, w) - f_2(t, u, w)| \leq \eta_2, \ \forall \ t \in [a, b], \ u, w \in [a_1, b_1].$$

*Let $r_*$ be a positive root of equation*

$$r_* = L^* r_*^{q \min\{1, v\}} + 3\eta_1 + \frac{2(b-a)^q}{\Gamma(q+1)} \eta_2, \quad (2.67)$$

*where $L^* = \min\{L_{A_1}, L_{A_2}\}$ (see (2.61)). Then*

$$\|x_1^* - x_2^*\|_C \leq r_*, \quad (2.68)$$

*and*

$$|\lambda_1^* - \lambda_2^*| \leq \frac{\Gamma(q+1)}{(b-a)^q} \left( 2\eta_1 + \frac{L^*}{2} r_*^{q \min\{1, v\}} \right) + \eta_2, \quad (2.69)$$

*where $(x_i^*, \lambda_i^*)$, $i = 1, 2$ are solutions of the corresponding problems (2.66). Note $r_*$ is uniquely defined.*

**Proof.** Using the condition (i), it is easy to see that for $x \in C([a_1, b_1], [a_1, b_1])$ and $t \in [a_1, a] \cup [b, b_1]$, we have

$$\|A_1(x) - A_2(x)\|_C \leq \eta_1.$$

On the other hand, for $t \in [a, b]$, using the condition (ii), we obtain

$$|A_1(x)(t) - A_2(x)(t)|$$

$$\leq |\varphi_1(a) - \varphi_2(a)| + \frac{(t-a)^q}{(b-a)^q}(|\psi_1(b) - \psi_2(b)| + |\varphi_1(a) - \varphi_2(a)|)$$

$$+ \frac{(t-a)^q}{\Gamma(q)(b-a)^q} \int_a^b (b-s)^{q-1} |f_1(s,x(s),x(x^v(s))) - f_2(s,x(s),x(x^v(s)))| \, ds$$

$$+ \frac{1}{\Gamma(q)} \int_a^t (t-s)^{q-1} |f_1(s,x(s),x(x^v(s))) - f_2(s,x(s),x(x^v(s)))| \, ds$$

$$\leq 3\eta_1 + \frac{2(b-a)^q}{\Gamma(q+1)}\eta_2.$$

So, we have

$$\|A_1(x) - A_2(x)\|_C \leq 3\eta_1 + \frac{2(b-a)^q}{\Gamma(q+1)}\eta_2.$$

Next, (2.64) holds for both $A_i$ with $L_{f_i}$. Without loss of generality, we may suppose that $L^* = L_{A_1} = \min\{L_{A_1}, L_{A_2}\}$. Consequently, we obtain

$$\|x_1^* - x_2^*\|_C = \|A_1(x_1^*) - A_2(x_2^*)\|_C$$

$$\leq \|A_1(x_1^*) - A_1(x_2^*)\|_C + \|A_1(x_2^*) - A_2(x_2^*)\|_C$$

$$\leq L^* \|x_1^* - x_2^*\|_C^{q\min\{1,v\}} + 3\eta_1 + \frac{2(b-a)^q}{\Gamma(q+1)}\eta_2,$$

which implies (2.68). Moreover, we get

$$|\lambda_1^* - \lambda_2^*|$$

$$\leq \frac{\Gamma(q+1)(|\psi_1(b) - \psi_2(b)| + |\varphi_1(a) - \varphi_2(a)|)}{(b-a)^q}$$

$$+ \frac{q}{(b-a)^q} \int_a^b (b-s)^{q-1} |f_1(s,x_1^*(s),x_1^*(x_1^{*v}(s))) - f_1(s,x_2^*(s),x_2^*(x_2^{*v}(s)))| \, ds$$

$$+ \frac{q}{(b-a)^q} \int_a^b (b-s)^{q-1} |f_1(s,x_2^*(s),x_2^*(x_2^{*v}(s))) - f_2(s,x_2^*(s),x_2^*(x_2^{*v}(s)))| \, ds$$

$$\leq \frac{\Gamma(q+1)}{(b-a)^q}\left(2\eta_1 + \frac{L^*}{2}r_*^{q\min\{1,v\}}\right) + \eta_2.$$

The proof is completed. $\qquad\square$

### 2.5.4 *Examples and General Cases*

**Example 2.2.** Consider the following problem:

$$\begin{cases} {}^C_0D_t^{\frac{1}{2}}x(t) = \mu x(x(t)) + \lambda, & t \in [0,1], \ \mu > 0, \ \lambda \in \mathbb{R}, \\ x(t) = 0, & t \in [-h,0], \ h > 0, \\ x(t) = 1, & t \in [1,1+h], \end{cases} \qquad (2.70)$$

where $x \in C([-h,1+h],[-h,1+h])$.

**Proposition 2.1.** *Suppose that*

$$\mu \le \frac{\Gamma(\frac{3}{2})h}{1 + 2h}.$$

*Then the problem* (2.70) *has a solution in* $C([-h, 1 + h], [-h, 1 + h])$.

**Proof.** First of all notice that accordingly to Theorem 2.11 we have $v = 1$, $q = \frac{1}{2}$, $a = 0, b = 1$, $\psi(b) = 1, \varphi(a) = 0$ and $f(t, u_1, u_2) = \mu u_2$. Moreover, $a_1 = -h$ and $b_1 = 1 + h$ can be taken. Therefore, from the relation

$$m_f \le f(t, u_1, u_2) \le M_f, \ \forall \, t \in [0, 1], \ u_1, u_2 \in [-h, 1 + h],$$

we can choose $m_f = -h\mu$ and $M_f = (1 + h)\mu$. For these data it can be easily verified that the conditions (ii) from Theorem 2.11 are equivalent to the relation

$$\mu \le \frac{\Gamma(\frac{3}{2})h}{1 + 2h},$$

consequently we complete the proof. $\square$

**Example 2.3.** Consider the following problem:

$$\begin{cases} {}_{2h}^{C}D_t^{\frac{1}{2}} x(t) = \mu x^2(x(t)) + \lambda, & t \in [2h, 3h], \ \mu > 0, \ \lambda \in \mathbb{R}, \\ x(t) = \dfrac{1}{2}, & t \in [h, 2h], \\ x(t) = \dfrac{1}{2}, & t \in [3h, 4h], \ h \in \left[\dfrac{1}{8}, \dfrac{1}{2}\right], \end{cases} \quad (2.71)$$

where $x \in C([h, 4h], [h, 4h])$. Note $\frac{1}{2} \in [h, 4h]$ for $h \in [\frac{1}{8}, \frac{1}{2}]$.

**Proposition 2.2.** *We suppose that*

$$0 < \mu \le \frac{(-1 + 8h)\sqrt{\pi}}{64h^{5/2}}, \quad \text{for } h \in \left(\frac{1}{8}, \frac{1}{5}\right),$$

$$0 < \mu \le \frac{(1 - 2h)\sqrt{\pi}}{64h^{5/2}}, \quad \text{for } h \in \left[\frac{1}{5}, \frac{1}{2}\right).$$

*Then the problem* (2.71) *has a solution in* $C_L^{\frac{1}{2}}([h, 4h], [h, 4h])$ *with* $L = \frac{128\mu h^2}{\sqrt{\pi}}$. *Note* $0 < \mu \le \frac{15\sqrt{5\pi}}{64} \doteq 0.928905$. *Furthermore, any two solutions* $x_1, x_2 \in C_L^{\frac{1}{2}}([h, 4h], [h, 4h])$ *of* (2.71) *satisfy*

$$\|x_1 - x_2\|_C \le \frac{4096h^4\mu^2(64h^{\frac{3}{2}}\mu + \sqrt{\pi})^2}{\pi^2}. \quad (2.72)$$

**Proof.** First of all notice that accordingly to Theorem 2.12 we have $v = 1$, $q = \frac{1}{2}$, $a = 2h, b = 3h$, $\psi(b) = \frac{1}{2}, \varphi(a) = \frac{1}{2}$, $a_1 = h$, $b_1 = 4h$. Observe that $|f(t, u_1, u_2) - f(t, w_1, w_2)| = \mu|u_2 + w_2||u_2 - w_2| \le 8h\mu|u_2 - w_2|$, $u_2, w_2 \in [h, 4h]$. So $L_u = 0$ and $L_w = 8h\mu$. Next, we choose $m_f = \mu h^2$ and $M_f = 16\mu h^2$. By a common check in the conditions of Theorem 2.12 we can make sure that

$$a_1 \le \min(\varphi(a), \psi(b)) - \max\left(0, \frac{M_f(b - a)^q}{\Gamma(q + 1)}\right) + \min\left(0, \frac{m_f(b - a)^q}{\Gamma(q + 1)}\right)$$

$$\Longleftrightarrow h + \frac{16\mu h^{\frac{5}{2}}}{\Gamma(\frac{3}{2})} \le \frac{1}{2},$$

$$\max(\varphi(a), \psi(b)) - \min\left(0, \frac{m_f(b-a)^q}{\Gamma(q+1)}\right) + \max\left(0, \frac{M_f(b-a)^q}{\Gamma(q+1)}\right) \le b_1$$

$$\Longleftrightarrow \frac{1}{2} \le 4h - \frac{16\mu h^{\frac{5}{2}}}{\Gamma(\frac{3}{2})}.$$

These inequalities are equivalent to

$$0 < \mu \le \min\left\{\frac{(1-2h)\sqrt{\pi}}{64h^{\frac{5}{2}}}, \frac{(-1+8h)\sqrt{\pi}}{64h^{\frac{5}{2}}}\right\}.$$

The function $\kappa(h) = \min\{-\frac{(-1+2h)\sqrt{\pi}}{64h^{\frac{5}{2}}}, \frac{(-1+8h)\sqrt{\pi}}{64h^{\frac{5}{2}}}\}$ is increasing from 0 to $\frac{15\sqrt{5\pi}}{64} \doteq 0.928905$ on $[\frac{1}{8}, \frac{1}{5}]$ and then it is decreasing to 0 on $[\frac{1}{5}, \frac{1}{2}]$. Next, we derive $L_\varphi = L_\psi = 0$ and

$$L_* = \frac{|\psi(b) - \varphi(a)|}{(b-a)^q} + \frac{4\max\{|m_f|, |M_f|\}}{\Gamma(q+1)} = \frac{64\mu h^2}{\Gamma(\frac{3}{2})} = \frac{128\mu h^2}{\sqrt{\pi}},$$

so $L = L_*$. By (2.61) we derive

$$L_A = \frac{2\sqrt{h}}{\Gamma(\frac{3}{2})}\left(8h\mu\sqrt{4h} + 8h\mu\frac{128\mu h^2}{\sqrt{\pi}}\right) = \frac{64h^2\mu(64h^{\frac{3}{2}}\mu + \sqrt{\pi})}{\pi}.$$

This gives (2.72) by (2.60). Therefore, by Theorem 2.12 the proof is completed. $\quad\square$

**Example 2.4.** Now take the following problems

$$\begin{cases} {}^C_{2h}D_t^{\frac{1}{2}}x(t) = \mu x^2(x(t)) + \lambda_i, & t \in [2h, 3h], \ \mu_i = \mu, \ \lambda_i \in R, \\ x(t) = \varphi_i, & t \in [h, 2h], \ h > 0, \\ x(t) = \psi_i, & t \in [3h, 4h] \end{cases} \qquad (2.73)$$

for $i = 1, 2$. Suppose the following assumptions.

**(H1)** $\varphi_i \in C_{L_*}^{\frac{1}{2}}([h, 2h], [h, 4h])$, $\psi_i \in C_{L_*}^{\frac{1}{2}}([3h, 4h], [h, 4h])$ such that $\varphi_i(2h) = \frac{1}{2}$, $\psi_i(3h) = \frac{1}{2}$, $i = 1, 2$ and $L_* = \frac{128\mu h^2}{\sqrt{\pi}}$;
**(H2)** we are in the conditions of Proposition 2.2 for both of the problems (2.73).

Let $(x_i^*, \lambda_i^*)$ be solutions of the problems (2.73). We are looking for an estimation for $\|x_1^* - x_2^*\|_C$ and $|\lambda_1^* - \lambda_2^*|$.

Then, build upon Theorem 2.13, by a common substitution one can make sure that we have

**Proposition 2.3.** *Consider the problems (2.73) and suppose the requirements (H1)-(H2) hold. Additionally, there exists $\eta_1 > 0$ such that*

$$|\varphi_1(t) - \varphi_2(t)| \le \eta_1, \ \forall\, t \in [h, 2h],$$

*and*

$$|\psi_1(t) - \psi_2(t)| \le \eta_1, \ \forall \ t \in [3h, 4h].$$

*Then*

$$\|x_1^* - x_2^*\|_C \le r_*,$$

*and*

$$|\lambda_1^* - \lambda_2^*| \le \frac{2}{\sqrt{\pi h}} \left( 2\eta_1 + \frac{L^*}{2} \sqrt{r_*} \right),$$

*where $L^*$ and $r_*$ are given by (2.74) and (2.75), respectively.*

**Proof.** Results follow from Theorem 2.13 as follows. By Proposition 2.2, we have $L = \sqrt{3}L_* = \frac{128\mu h^2 \sqrt{3}}{\sqrt{\pi}}$ and then (see (2.61))

$$L^* = L_{A_1} = L_{A_2} = \frac{64h^2\mu(64\sqrt{3}h^{3/2}\mu + \sqrt{\pi}))}{\pi}. \tag{2.74}$$

Realizing that now $\eta_2 = 0$, equation (2.67) has the form

$$r_* = L^*\sqrt{r_*} + 3\eta_1,$$

which has the positive solution

$$r_* = \frac{1}{\pi^2} \Big( 25165824h^7\mu^4 + 64h^2\mu\pi^{\frac{3}{2}} + \eta_1\pi^2.$$

$$+ 8\sqrt{2}\sqrt{4947802324992h^{14}\mu^8 + 25165824h^9\mu^5\pi^{\frac{3}{2}} + 393216h^7\mu^4\eta_1\pi^2} \Big).$$
$$\tag{2.75}$$

The estimate for $|\lambda_1^* - \lambda_2^*|$ follows directly from (2.69). The proof is finished. $\quad\square$

We conclude this section by considering a general fractional order iterative functional differential equations with parameter given by

$$\begin{cases} {}^C_a D^q_t x(t) = f(t, x(t), x(x^v(t)), \lambda), & t \in [a, b], \ v \in (0, 1], \ q \in (0, 1), \ \lambda \in J, \\ x(t) = \varphi(t), & t \in [a_1, a], \\ x(t) = \psi(t), & t \in [b, b_1], \end{cases}$$
$$\tag{2.76}$$

when $J \subset R$ is an open interval, conditions (C1), (C3) are supposed and (C2) is extended to

**(C4)** $f \in C([a, b] \times [a_1, b_1]^2 \times J, \mathbb{R})$.

Then by (2.55) we have an operator $A(\lambda, x)$. It is easy to see that $A(\lambda, x) = A(x)$ for the problem (2.53). Supposing the assumptions of Theorem 2.11 for the problem (2.76) uniformly with respect to $\lambda \in J$, we can find its fixed point $x^*(\lambda, \cdot) \in C([a_1, b_1], [a_1, b_1])$. In order to get a solution of the problem (2.76), we need to solve

$$\Upsilon(\lambda) = \Gamma(q)(\psi(b) - \varphi(a)) - \int_a^b (b - s)^{q-1} f(s, x^*(\lambda, s), x^*(\lambda, x^*(\lambda, s)^v))ds = 0.$$
$$\tag{2.77}$$

If there is an $\lambda_0 \in J$ solving (2.77) then $x^*(\lambda_0, t)$ is a solution of the problem (2.76). Since $x^*(\lambda, \cdot)$ is not unique in general, function $\Upsilon(\lambda)$ is multivalued. Consequently this way is not very useful. We propose another approach. The problem (2.76) is equivalent to the following fixed point equation

$$x(t) = \begin{cases} \varphi(t), & \text{for } t \in [a_1, a], \\ \varphi(a) + \dfrac{1}{\Gamma(q)} \displaystyle\int_a^t (t-s)^{q-1} f(s, x(s), x(x^v(s)), \lambda) ds, & \text{for } t \in [a, b], \\ \psi(t), & \text{for } t \in [b, b_1]. \end{cases}$$

From the condition of continuity of $x$ in $t = b$, we have that

$$\psi(b) = \varphi(a) + \frac{1}{\Gamma(q)} \int_a^b (b-s)^{q-1} f(s, x(s), x(x^v(s)), \lambda) ds.$$

Now we consider the operator

$$A : C^b([a_1, b_1], [a_1, b_1]) \times J \to C^b([a_1, b_1], \mathbb{R})$$

where

$$C^b([a_1, b_1], [a_1, b_1]) = \left\{ x \in C([a_1, b], [a_1, b_1]) \cap C^b((b, b_1], [a_1, b_1]) : \exists \lim_{s \to b_+} x(s) \right\},$$

$$C^b([a_1, b_1], \mathbb{R}) = \left\{ x \in C([a_1, b], \mathbb{R}) \cap C^b((b, b_1], \mathbb{R}) : \exists \lim_{s \to b_+} x(s) \right\}$$

and

$$A(x, \lambda)(t) := \begin{cases} \varphi(t), & \text{for } t \in [a_1, a], \\ \varphi(a) + \dfrac{1}{\Gamma(q)} \displaystyle\int_a^t (t-s)^{q-1} f(s, x(s), x(x^v(s)), \lambda) ds, & \text{for } t \in [a, b], \\ \psi(t), & \text{for } t \in (b, b_1]. \end{cases}$$

Now, we are ready to state the following result.

**Theorem 2.14.** *Suppose that*

**(i)** *conditions (C1), (C3) and (C4) are satisfied;*
**(ii)** *there are $m_f, M_f \in R$ such that*

$$m_f \le f(t, u, w, \lambda) \le M_f, \ \forall \, t \in [a, b], u, w \in [a_1, b_1], \lambda \in J$$

*along with*

$$a_1 \le \varphi(a) + \min\left(0, \frac{m_f(b-a)^q}{\Gamma(q+1)}\right),$$

*and*

$$\varphi(a) + \max\left(0, \frac{M_f(b-a)^q}{\Gamma(q+1)}\right) \le b_1.$$

*Then operator $A(x, \lambda)$ has a fixed point in $C^b([a_1, b_1], [a_1, b_1])$ for any $\lambda \in J$.*

**Proof.** Like in the proof of Theorem 2.11, condition (ii) assures that the set $C^b([a_1, b_1], [a_1, b_1])$ is an invariant subset for the operator $A$, that is, we have

$$A(C^b([a_1, b_1], [a_1, b_1]) \times J) \subset C^b([a_1, b_1], [a_1, b_1]). \qquad (2.78)$$

Similarly, $A$ is a completely continuous operator. It is obvious that the set $C^b([a_1, b_1], [a_1, b_1]) \subseteq C^b([a_1, b_1], \mathbb{R})$ is a bounded convex closed subset of the Banach space $C^b([a_1, b_1], \mathbb{R})$. Thus, the operator $A(x, \lambda)$ has a fixed point due to Schauder fixed point theorem. This completes the proof. $\qquad \square$

We still do not have uniqueness result. For this purpose, we suppose

**(C5)** $f$ is nonnegative and nondecreasing, i.e. $m_f \geq 0$ and $0 \leq f(s_1, u_1, v_1, \lambda) \leq f(s_2, u_2, v_2, \lambda)$ for any $s_1 \leq s_2 \in [a, b]$, $u_1 \leq u_2, v_1 \leq v_2 \in [a_1, b_1]$ and $\lambda \in J$.

We introduce the Banach space

$$C_m^b([a_1, b_1], [a_1, b_1]) = \left\{ x \in C^b([a_1, b_1], [a_1, b_1]) \mid x \text{ is nondecreasing on } [a_1, b_1] \right\}.$$

Now, we have the next result.

**Theorem 2.15.** *We suppose conditions (i), (ii) of Theorem 2.14, (C5) as well and $\varphi(t)$, $\psi(t)$ are nondecreasing with $\varphi(a) \leq \psi(b)$. Then operator $A(x, \lambda)$ is monotone nondecreasing in $x$ on $C_m^b([a_1, b_1], [a_1, b_1])$ for any $\lambda \in J$. Consequently it has a unique smallest and largest fixed points $x_m(\lambda), x_M(\lambda)$ in $C_m^b([a_1, b_1], [a_1, b_1])$. Moreover, a nondecreasing sequence $\{A^k(a_1, \lambda)(t)\}_{k \geq 1}$ and a nonincreasing sequence $\{A^k(b_1, \lambda)(t)\}_{k \geq 1}$ satisfy*

$$a_1 \leq A^k(a_1, \lambda)(t) \leq x_m(\lambda)(t) \leq x_M(\lambda)(t) \leq A^k(b_1, \lambda)(t) \leq b_1 \quad t \in J$$

*for any $k \geq 1$ and $\lim_{k \to \infty} A^k(a_1, \lambda)(t) = x_m(\lambda)(t)$ and $\lim_{k \to \infty} A^k(b_1, \lambda)(t) = x_M(\lambda)(t)$ uniformly on $[a_1, b_1]$.*

**Proof.** We already know (2.78). Let $x \in C_m([a_1, b_1], [a_1, b_1])$ then clearly $A(x, \lambda)(t_1) \leq A(x, \lambda)(t_2)$ for $t_1 \leq t_2 \in [a_1, a]$ and $t_1 \leq t_2 \in (b, b_1]$. Next for $s_1 \leq s_2 \in [a, b]$ we have $x(s_1) \leq x(s_2)$, $x^v(s_1) \leq x^v(s_2)$ and $x(x^v(s_1)) \leq x(x^v(s_2))$, which imply

$$f(s_1, x(s_1), x(x^v(s_1)), \lambda) \leq f(s_2, x(s_2), x(x^v(s_2)), \lambda). \qquad (2.79)$$

Furthermore, for $t_1 \leq t_2 \in [a, b]$, following El-Sayed, 1995 and Darwish, 2008, and using (2.79), we derive

$$A(x, \lambda)(t_2) - A(x, \lambda)(t_1)$$

$$= \frac{1}{\Gamma(q)} \int_a^{t_2} (t_2 - s)^{q-1} f(s, x(s), x(x^v(s)), \lambda) ds$$

$$- \frac{1}{\Gamma(q)} \int_a^{t_1} (t_1 - s)^{q-1} f(s, x(s), x(x^v(s)), \lambda) ds$$

$$= \frac{1}{\Gamma(q)} \int_a^{t_1} \left( (t_2 - s)^{q-1} - (t_1 - s)^{q-1} \right) f(s, x(s), x(x^v(s)), \lambda) ds$$

$$+ \frac{1}{\Gamma(q)} \int_{t_1}^{t_2} (t_2 - s)^{q-1} f(s, x(s), x(x^v(s)), \lambda) ds$$

$$\geq \frac{1}{\Gamma(q)} f(t_1, x(t_1), x(x^v(t_1)), \lambda)$$

$$\times \left( \int_a^{t_1} (t_2 - s)^{q-1} - (t_1 - s)^{q-1} ds + \int_{t_1}^{t_2} (t_2 - s)^{q-1} ds \right)$$

$$= \frac{1}{\Gamma(q+1)} f(t_1, x(t_1), x(x^v(t_1)), \lambda) \left( (t_2 - a)^q - (t_1 - a)^q \right)$$

$$\geq 0.$$

Consequently, we obtain

$$A(C_m^b([a_1, b_1], [a_1, b_1]), \lambda) \subset C_m^b([a_1, b_1], [a_1, b_1])$$

for any $\lambda \in J$.

Next, if $x_1, x_2 \in C_m^b([a_1, b_1], [a_1, b_1])$ with $x_1(t) \leq x_2(t)$, $t \in [a_1, b_1]$ then clearly we have $A(x_1, \lambda)(t) \leq A(x_2, \lambda)(t)$ for $t \in [a_1, a] \cup (b, b_1]$. For $s \in [a, b]$, we have $x_1(s) \leq x_2(s)$, $x_1^v(s) \leq x_2^v(s)$ and $x_1(x_1^v(s)) \leq x_2(x_2^v(s))$, which imply

$$f(s, x_1(s), x_1(x_1^v(s)), \lambda) \leq f(s, x_2(s), x_2(x_2^v(s)), \lambda).$$

Then for $t \in [a, b]$ we have

$$A(x_2, \lambda)(t) - A(x_1, \lambda)(t) = \frac{1}{\Gamma(q)} \int_a^t (t - s)^{q-1} (f(s, x_2(s), x_2(x_2^v(s)), \lambda)$$

$$- f(s, x_1(s), x_1(x_1^v(s)), \lambda)) ds \geq 0.$$

This means that operator $A(x, \lambda)$ is monotone nondecreasing in $x$ on $C_m^b([a_1, b_1], [a_1, b_1])$ for any $\lambda \in J$. We also know that $A$ is a completely continuous operator. Then results follow from the general theory of nondecreasing compact operators in Banach spaces (see, e.g., Deimling, 1985). The proof is completed. $\square$

To get continuous solution, we need to solve either

$$\Upsilon_m(\lambda) = \psi(b) - \varphi(a) - \frac{1}{\Gamma(q)} \int_a^b (b - s)^{q-1} f(s, x_m(\lambda)(s), x_m(\lambda)(x_m^v(\lambda)(s)), \lambda) ds = 0$$

or

$$\Upsilon_M(\lambda) = \psi(b) - \varphi(a) - \frac{1}{\Gamma(q)} \int_a^b (b - s)^{q-1} f(s, x_M(\lambda)(s), x_M(\lambda)(x_M^v(\lambda)(s)), \lambda) ds = 0.$$

We can use to handle these equations also an analytical-numerical method like in Ronto, 2009. This means that first successive approximation is used

$$x_{n+1}(\lambda, t) = A(x_n(\lambda, t), \lambda)$$

for up to some order $j$ with either $x_0(\lambda, t) = a_1$ or $x_0(\lambda, t) = b_1$. Then approximations

$$\Upsilon_j(\lambda) = \psi(b) - \varphi(a) - \frac{1}{\Gamma(q)} \int_a^b (b - s)^{q-1} f(s, x_j(\lambda, s), x_j(\lambda, x_j^v(\lambda, s)), \lambda) ds$$

of $\Upsilon_m$ and $\Upsilon_M$ are numerically drawn to check if they change the sign over $J$.

## 2.6 Notes and Remarks

The results in Section 2.2 are taken from Agarwal, Zhou and He, 2010. The main results in Section 2.3 are adopted from Zhou, Jiao and Li, 2009a. The material in Section 2.4 due to Zhou, Jiao and Li, 2009b. The results in Section 2.5 are taken from Wang, Fečkan and Zhou, 2013c.

# Chapter 3

# Fractional Ordinary Differential Equations in Banach Spaces

## 3.1 Introduction

In Chapter 3, we discuss the Cauchy problem of fractional ordinary differential equations in Banach spaces under hypotheses based on Carathéodory condition. The tools used include some classical and modern nonlinear analysis methods such as fixed point theory, measure of noncompactness method, topological degree method and Picard operators technique, etc.

Firstly, we give an example which show that the criteria on existence of solutions for the initial value problem of fractional differential equations in finite-dimensional spaces may not be true in infinite-dimensional cases. It is well known that Peano theorem of integer order ordinary differential equations is not true in infinite-dimensional Banach spaces. The first result in this direction was obtained by Dieudonne, 1950. Dieudonne, 1950, produced an example which showed that Peano theorem is not true in the space $c_0$ of sequences which converge to zero. In fact, Peano theorem of fractional differential equations is also not true in infinite-dimensional Banach spaces. In the following, we shall show that the existence result of nonlocal Cauchy problem for fractional abstract differential equations which has been obtained by N'Guerekata in 2009 is not true in the space $c_0$.

**Example 3.1.** Let $E = c_0 = \{z = (z_1, z_2, z_3, \ldots) : z_n \to 0 \text{ as } n \to \infty\}$ with the norm $\|z\| = \sup_{n \geq 1} |z_n|$ and $f(z) = 2(\sqrt{|z_1|}, \sqrt{|z_2|}, \sqrt{|z_3|}, \ldots)$ with $z = (z_1, z_2, z_3, \ldots) \in c_0$. Consider the nonlocal Cauchy problem for fractional differential equations given by

$$_0^C D_t^q x(t) = f(x(t)), \quad x(0) = \xi, \ t \in (0, t_0] \tag{3.1}$$

where $_0^C D_t^q$ is Caputo fractional derivative of order $0 < q < 1$, $\xi = (1, 1/2^2, 1/3^2, \ldots) \in c_0$, $t_0 < (\frac{\Gamma(1+q)}{2})^{\frac{1}{q}}$.

It is obvious that $f : c_0 \to c_0$ is continuous. According to N'Guerekata, 2009, there exists a constant $k^* = \frac{\Gamma(1+q)}{\Gamma(1+q) - 2t_0^q}$, such that IVP (3.1) possesses at least one continuous solution $x \in C([0, t_0], c_0)$ and $x(t) = (x_1(t), x_2(t), x_3(t), \ldots) \in c_0$ on $[0, t_0]$ with $\sup_{t \in [0, t_0]} \|x(t)\| \leq k^*$. According to the definition of the norm of $c_0$, we can conclude that

$$\,_0^C D_t^q x_n(t) = 2\sqrt{|x_n(t)|}, \quad x_n(0) = \frac{1}{n^2}, \quad t \in (0, t_0], \quad n = 1, 2, 3, \ldots, \tag{3.2}$$

where $x_n$ satisfies that $x_n \in C([0, t_0], \mathbb{R})$ with $\sup_{t \in [0, t_0]} |x_n(t)| \leq k^*$.

Let us consider equation (3.2) which can be written as the following equivalent form

$$x_n(t) = \frac{1}{n^2} + \,_0 D_t^{-q} \sqrt{|x_n(t)|} = \frac{1}{n^2} + \frac{2}{\Gamma(q)} \int_0^t (t-s)^{q-1} \sqrt{|x_n(s)|} ds, \quad t \in [0, t_0]. \tag{3.3}$$

Since $(t-s)^{q-1} > 1$ with $s \in [0, t)$ for $t \in (0, t_0]$, we have by (3.3)

$$x_n(t) \geq \frac{1}{n^2} + \frac{2}{\Gamma(q)} \int_0^t \sqrt{|x_n(s)|} ds, \quad t \in [0, t_0], \quad n = 1, 2, 3, \ldots . \tag{3.4}$$

Assume that $y_n \in C([0, t_0], \mathbb{R})$ is a solution of the following integral equation

$$y_n(t) = \frac{1}{4n^2} + \frac{2}{\Gamma(q)} \int_0^t \sqrt{|y_n(s)|} ds, \quad t \in [0, t_0], \quad n = 1, 2, 3, \ldots . \tag{3.5}$$

Then, we get

$$x_n(t) \geq y_n(t), \quad t \in [0, t_0], \quad n = 1, 2, 3, \ldots . \tag{3.6}$$

In fact, suppose (for contraction) that the conclusion (3.6) is not true. Then, because of the continuity of $x$ and $y$, and that $x_n(0) > y_n(0)$, it follows that there exists a $t_1 \in (0, t_0]$ such that

$$x_n(t_1) = y_n(t_1), \quad x_n(t) > y_n(t) \ t \in [0, t_1), \quad n = 1, 2, 3, \ldots . \tag{3.7}$$

Then using (3.4) and (3.7), we get

$$y_n(t_1) = \frac{1}{4n^2} + \frac{2}{\Gamma(q)} \int_0^{t_1} \sqrt{|y_n(s)|} ds$$

$$< \frac{1}{n^2} + \frac{2}{\Gamma(q)} \int_0^{t_1} \sqrt{|x_n(s)|} ds$$

$$\leq x_n(t_1), \quad n = 1, 2, 3, \ldots,$$

which is a contraction in view of (3.7). Hence the conclusion (3.6) is valid.

Since the integral (3.5) is equivalent to the following fractional IVP

$$y_n'(t) = \frac{2}{\Gamma(q)} \sqrt{|y_n(t)|}, \quad y_n(0) = \frac{1}{4n^2}, \quad t \in [0, t_0], \quad n = 1, 2, 3, \ldots, \tag{3.8}$$

and noting $y_n(t) > 0$, $t \in [0, t_0]$, we can conclude that fractional IVP (3.8) has a continuous solution

$$y_n(t) = \left( \frac{t}{\Gamma(q)} + \frac{1}{2n} \right)^2, \quad t \in [0, t_0], \quad n = 1, 2, 3, \ldots,$$

which means that

$$x_n(t) \geq y_n(t) = \left( \frac{t}{\Gamma(q)} + \frac{1}{2n} \right)^2, \quad t \in [0, t_0], \quad n = 1, 2, 3, \ldots . \tag{3.9}$$

Therefore, for $t \in (0, t_0]$, $\lim_{n \to \infty} x_n(t) \neq 0$ by (3.9), contracting $x(t) \in c_0$. Hence fractional IVP (3.1) has no nonlocal solution in $c_0$.

## 3.2 Cauchy Problems via Measure of Noncompactness Method

### 3.2.1 *Introduction*

In Section 3.2, we assume that $X$ is a Banach space with the norm $|\cdot|$. Let $J \subset \mathbb{R}$. Denote $C(J, X)$ be the Banach space of continuous functions from $J$ into $X$.

Let $r > 0$ and $\mathcal{C} = C([-r, 0], X)$ be the space of continuous functions from $[-r, 0]$ into $X$. For any element $z \in \mathcal{C}$, define the norm $\|z\|_* = \sup_{\theta \in [-r,0]} |z(\theta)|$.

Consider the initial value problem (fractional IVP) for fractional functional differential equation given by

$$\begin{cases} {}_0^C D_t^q x(t) = f(t, x_t), & t \in (0, a), \\ x_0 = \varphi \in \mathcal{C}, \end{cases} \tag{3.10}$$

where ${}_0^C D_t^q$ is Caputo fractional derivative of order $0 < q < 1$, $f \colon [0, a) \times \mathcal{C} \to X$ is a given function satisfying some assumptions and define $x_t$ by $x_t(\theta) = x(t + \theta)$, for $\theta \in [-r, 0]$.

In this section, we shall discuss the existence of the solutions for fractional IVP (3.10) under assumptions that $f$ satisfies Carathéodory condition and the condition on measure of noncompactness. Then, we give an example to illustrate the application of our abstract results.

**Definition 3.1.** A function $x \in C([-r, T], X)$ is a solution for fractional IVP (3.10) on $[-r, T]$ for $T \in (0, a)$ if

**(i)** the function $x(t)$ is absolutely continuous on $[0, T]$;
**(ii)** $x_0 = \varphi$;
**(iii)** $x$ satisfies the equation in (3.10).

### 3.2.2 *Existence*

We are now ready to prove the existence of the solutionsfor fractional IVP (3.10) under the following hypotheses:

**(H1)** for almost all $t \in [0, a)$, the function $f(t, \cdot) \colon \mathcal{C} \to X$ is continuous and for each $z \in \mathcal{C}$, the function $f(\cdot, z) \colon [0, a) \to X$ is strongly measurable;
**(H2)** for each $\tau > 0$, there exist a constant $q_1 \in [0, q)$ and $m_1 \in L^{\frac{1}{q_1}}([0, a), \mathbb{R}^+)$ such that $|f(t, z)| \leq m_1(t)$ for all $z \in \mathcal{C}$ with $\|z\|_* \leq \tau$ and almost all $t \in [0, a)$;
**(H3)** there exist a constant $q_2 \in (0, q)$ and $m_2 \in L^{\frac{1}{q_2}}([0, a), \mathbb{R}^+)$ such that $\alpha(f(t, B)) \leq m_2(t)\alpha(B)$ for almost all $t \in [0, a)$ and $B$ a bounded set in $\mathcal{C}$.

In order to prove the existence theorem, we need the following lemma.

**Lemma 3.1.** *Assume that the hypotheses (H1) and (H2) hold. $x \in C([-r, T], X)$ is a solution for fractional IVP (3.10) on $[-r, T]$ for $T \in (0, a)$ if and only if $x$*

*satisfies the following relation*

$$\begin{cases} x(\theta) = \varphi(\theta), & \text{for } \theta \in [-r, 0], \\ x(t) = \varphi(0) + \dfrac{1}{\Gamma(q)} \displaystyle\int_0^t (t-s)^{q-1} f(s, x_s) ds, & \text{for } t \in [0, T]. \end{cases} \qquad (3.11)$$

**Proof.** Since $x_t$ is continuous in $t \in [0, a)$, according to (H1), $f(t, x_t)$ is a measurable function in $[0, a)$. Direct calculation gives that $(t-s)^{q-1} \in L^{\frac{1}{1-q_1}}[0, t]$ for $t \in (0, a)$ and $q_1 \in [0, q)$. Let

$$b_1 = \frac{q-1}{1-q_1} \in (-1, 0), \quad M = \|m_1\|_{L^{\frac{1}{q_1}}[0,a)}.$$

By using Hölder inequality and (H2), for $t \in (0, a)$, we obtain that

$$\int_0^t |(t-s)^{q-1} f(s, x_s)| ds \le \left( \int_0^t (t-s)^{\frac{q-1}{1-q_1}} ds \right)^{1-q_1} \|m_1\|_{L^{\frac{1}{q_1}}[0,t]}$$

$$\le \frac{M}{(1+b_1)^{1-q_1}} a^{(1+b_1)(1-q_1)}.$$

Thus, $|(t-s)^{q-1} f(s, x_s)|$ is Lebesgue integrable with respect to $s \in [0, t)$ for all $t \in (0, a)$. From Lemma 1.5 (Bochner theorem), it follows that $(t-s)^{q-1} f(s, x_s)$ is Bochner integrable with respect to $s \in [0, t)$ for all $t \in (0, a)$.

Let $L(\tau, s) = (t-\tau)^{-q} |\tau - s|^{q-1} m_1(s)$. Since $L(\tau, s)$ is a nonnegative, measurable function on $D = [0, t] \times [0, t]$, then we have

$$\int_0^t \left( \int_0^t L(\tau, s) ds \right) d\tau = \int_D L(\tau, s) ds d\tau = \int_0^t \left( \int_0^t L(\tau, s) d\tau \right) ds$$

and

$$\int_D L(\tau, s) ds d\tau = \int_0^t \left( \int_0^t L(\tau, s) ds \right) d\tau$$

$$= \int_0^t (t-\tau)^{-q} \left( \int_0^t |\tau - s|^{q-1} m_1(s) ds \right) d\tau$$

$$= \int_0^t (t-\tau)^{-q} \left( \int_0^\tau (\tau - s)^{q-1} m_1(s) ds \right) d\tau$$

$$\quad + \int_0^t (t-\tau)^{-q} \left( \int_\tau^t (s-\tau)^{q-1} m_1(s) ds \right) d\tau$$

$$\le \frac{2M}{(1+b_1)^{1-q_1}} a^{(1+b_1)(1-q_1)} \int_0^t (t-\tau)^{-q} d\tau$$

$$\le \frac{2M}{(1-q)(1+b_1)^{1-q_1}} a^{(1+b_1)(1-q_1)+1-q}.$$

Therefore, $L_1(\tau, s) = (t-\tau)^{-q}(\tau - s)^{q-1} f(s, x_s)$ is a Bochner integrable function on $D = [0, t] \times [0, t]$, then we have

$$\int_0^t d\tau \int_0^\tau L_1(\tau, s) ds = \int_0^t ds \int_s^t L_1(\tau, s) d\tau.$$

We now prove that

$$_0D_t^q(_0D_t^{-q}f(t,x_t)) = f(t,x_t), \quad \text{for } t \in (0,T],$$

where $_0D_t^q$ is Riemann-Liouville fractional derivative.

Indeed, we have

$$_0D_t^q(_0D_t^{-q}f(t,x_t)) = \frac{1}{\Gamma(1-q)\Gamma(q)} \frac{d}{dt} \int_0^t (t-\tau)^{-q} \left( \int_0^\tau (\tau-s)^{q-1} f(s,x_s) ds \right) d\tau$$

$$= \frac{1}{\Gamma(1-q)\Gamma(q)} \frac{d}{dt} \int_0^t d\tau \int_0^\tau L_1(\tau,s) ds$$

$$= \frac{1}{\Gamma(1-q)\Gamma(q)} \frac{d}{dt} \int_0^t ds \int_s^t L_1(\tau,s) d\tau$$

$$= \frac{1}{\Gamma(1-q)\Gamma(q)} \frac{d}{dt} \int_0^t f(s,x_s) ds \int_s^t (t-\tau)^{-q}(\tau-s)^{q-1} d\tau$$

$$= \frac{d}{dt} \int_0^t f(s,x_s) ds$$

$$= f(t,x_t) \quad \text{for } t \in (0,T].$$

If $x$ satisfies the relation (3.11), then we can get that $x(t)$ is absolutely continuous on $[0,T]$. In fact, for any disjoint family of open intervals $\{(c_i,d_i)\}_{1\le i \le n}$ in $[0,T]$ with $\sum_{i=1}^n (d_i - c_i) \to 0$, we have

$$\sum_{i=1}^n |x(d_i) - x(c_i)|$$

$$= \sum_{i=1}^n \frac{1}{\Gamma(q)} \left| \int_0^{d_i} (d_i - s)^{q-1} f(s,x_s) ds - \int_0^{c_i} (c_i - s)^{q-1} f(s,x_s) ds \right|$$

$$\le \sum_{i=1}^n \frac{1}{\Gamma(q)} \left| \int_{c_i}^{d_i} (d_i - s)^{q-1} f(s,x_s) ds \right|$$

$$+ \sum_{i=1}^n \frac{1}{\Gamma(q)} \left| \int_0^{c_i} (d_i - s)^{q-1} f(s,x_s) ds - \int_0^{c_i} (c_i - s)^{q-1} f(s,x_s) ds \right|$$

$$\le \sum_{i=1}^n \frac{1}{\Gamma(q)} \int_{c_i}^{d_i} (d_i - s)^{q-1} m_1(s) ds$$

$$+ \sum_{i=1}^n \frac{1}{\Gamma(q)} \int_0^{c_i} \left( (c_i - s)^{q-1} - (d_i - s)^{q-1} \right) m_1(s) ds$$

$$\le \sum_{i=1}^n \frac{1}{\Gamma(q)} \left( \int_{c_i}^{d_i} (d_i - s)^{\frac{q-1}{1-q_1}} ds \right)^{1-q_1} \|m_1\|_{L^{\frac{1}{q_1}}[0,T]}$$

$$+ \sum_{i=1}^n \frac{1}{\Gamma(q)} \left( \int_0^{c_i} (c_i - s)^{\frac{q-1}{1-q_1}} - (d_i - s)^{\frac{q-1}{1-q_1}} ds \right)^{1-q_1} \|m_1\|_{L^{\frac{1}{q_1}}[0,T]}$$

$$= \sum_{i=1}^n \frac{(d_i - c_i)^{(1+b_1)(1-q_1)}}{\Gamma(q)(1+b_1)^{1-q_1}} \|m_1\|_{L^{\frac{1}{q_1}}[0,T]}$$

$$+ \sum_{i=1}^{n} \frac{(c_i^{1+b_1} - d_i^{1+b_1} + (d_i - c_i)^{1+b_1})^{1-q_1}}{\Gamma(q)(1+b_1)^{1-q_1}} \|m_1\|_{L^{\frac{1}{q_1}}[0,T]}$$

$$\leq 2 \sum_{i=1}^{n} \frac{(d_i - c_i)^{(1+b_1)(1-q_1)}}{\Gamma(q)(1+b_1)^{1-q_1}} \|m_1\|_{L^{\frac{1}{q_1}}[0,T]} \to 0.$$

Therefore, $x(t)$ is absolutely continuous on $[0,T]$, which implies that $x(t)$ is differentiable a.e. on $[0,T]$. According to the argument above and Definition 1.3, for $t \in (0,T]$, we have

$$\begin{aligned}
{}_0^C D_t^q x(t) &= {}_0^C D_t^q \left( \varphi(0) + \frac{1}{\Gamma(q)} \int_0^t (t-s)^{q-1} f(s,x_s) ds \right) \\
&= {}_0^C D_t^q \left( \frac{1}{\Gamma(q)} \int_0^t (t-s)^{q-1} f(s,x_s) ds \right) \\
&= {}_0^C D_t^q \left( {}_0 D_t^{-q} f(t,x_t) \right) \\
&= {}_0 D_t^q \left( {}_0 D_t^{-q} f(t,x_t) \right) - [{}_0 D_t^{-q} f(t,x_t)]_{t=0} \frac{t^{-q}}{\Gamma(1-q)} \\
&= f(t,x_t) - [{}_0 D_t^{-q} f(t,x_t)]_{t=0} \frac{t^{-q}}{\Gamma(1-q)}.
\end{aligned}$$

Since $(t-s)^{q-1} f(s,x_s)$ is Lebesgue integrable with respect to $s \in [0,t)$ for all $t \in (0,T]$, we know that $[{}_0 D_t^{-q} f(t,x_t)]_{t=0} = 0$, which means that ${}_0^C D_t^q x(t) = f(t,x_t)$, for $t \in (0,T]$. Hence, $x \in C([-r,T],X)$ is a solution of fractional IVP (3.10). On the other hand, it is obvious that if $x \in C([-r,T],X)$ is a solution of fractional IVP (3.10), then $x$ satisfies the relation (3.11), and this completes the proof. □

**Theorem 3.1.** *Assume that hypotheses (H1)-(H3) hold. Then, for every $\varphi \in \mathcal{C}$, there exists a solution $x \in C([-r,T],X)$ for fractional IVP (3.10) with some $T \in (0,a)$.*

**Proof.** Let $k > 0$ be any number and we can choose $T \in (0,a)$ such that

$$\frac{T^{(1+b_1)(1-q_1)}}{\Gamma(q)(1+b_1)^{1-q_1}} \|m_1\|_{L^{\frac{1}{q_1}}[0,T]} \leq k \qquad (3.12)$$

and

$$\frac{T^{(1+b_2)(1-q_2)}}{\Gamma(q)(1+b_2)^{1-q_2}} \|m_2\|_{L^{\frac{1}{q_2}}[0,T]} < 1, \qquad (3.13)$$

where $b_i = \frac{q-1}{1-q_i} \in (-1,0)$, $i = 1,2$.

Consider the set $B_k$ defined as follows

$$B_k = \left\{ x \in C([-r,T],X) \,\middle|\, x_0 = \varphi, \ \sup_{s \in [0,T]} |x(s) - \varphi(0)| \leq k \right\}.$$

Define the operator $F$ on $B_k$ as follows

$$\begin{cases} Fx(\theta) = \varphi(\theta), & \text{for } \theta \in [-r, 0], \\ Fx(t) = \varphi(0) + \dfrac{1}{\Gamma(q)} \displaystyle\int_0^t (t-s)^{q-1} f(t, x_s) ds, & \text{for } t \in [0, T], \end{cases}$$

where $x \in B_k$. We prove that the operator equation $x = Fx$ has a solution $x \in B_k$, which means that $x$ is a solution of fractional IVP (3.10).

Firstly, we observe that for every $y \in B_k$, $(Fy)(t)$ is continuous on $t \in [-r, T]$ and for $t \in [0, T]$, by (3.12) and Hölder inequality, we have

$$\begin{aligned} |(Fy)(t) - \varphi(0)| &\leq \frac{1}{\Gamma(q)} \int_0^t |(t-s)^{q-1} f(s, y_s)| ds \\ &\leq \frac{1}{\Gamma(q)} \left( \int_0^t (t-s)^{\frac{q-1}{1-q_1}} ds \right)^{1-q_1} \|m_1\|_{L^{\frac{1}{q_1}}[0,T]} \\ &\leq \frac{T^{(1+b_1)(1-q_1)}}{\Gamma(q)(1+b_1)^{1-q_1}} \|m_1\|_{L^{\frac{1}{q_1}}[0,T]} \\ &\leq k, \end{aligned} \tag{3.14}$$

where $b_1 = \frac{q-1}{1-q_1} \in (-1, 0)$. Thus, $\sup_{t \in [0,T]} |(Fy)(t) - \varphi(0)| \leq k$, which implies that $F : B_k \to B_k$.

Further, we prove that $F$ is a continuous operator on $B_k$. Let $\{y^n\} \subseteq B_k$ with $y^n \to y$ on $B_k$. Then by (H1) and the fact that $y_t^n \to y_t$, $t \in [0, T]$, we have

$$f(s, y_s^n) \to f(s, y_s), \quad \text{a.e. } s \in [0, T], \quad \text{as } n \to \infty.$$

Noting that $(t-s)^{q-1}|f(s, y_s^n) - f(s, y_s)| \leq (t-s)^{q-1} 2m_1(s)$, by Lebesgue dominated convergence theorem, as $n \to \infty$, we have

$$|(Fy^n)(t) - (Fy)(t)| \leq \frac{1}{\Gamma(q)} \int_0^t (t-s)^{q-1} |f(s, y_s^n) - f(s, y_s)| ds \to 0.$$

Therefore $Fy^n \to Fy$ as $n \to \infty$ which implies that $F$ is continuous.

For each $n \geq 1$, we define a sequence $\{x^n : n \geq 1\}$ in the following way

$$x^n(t) = \begin{cases} \varphi^0(t), & \text{for } t \in \left[-r, \dfrac{T}{n}\right], \\ \varphi(0) + \dfrac{1}{\Gamma(q)} \displaystyle\int_0^{t-\frac{T}{n}} (t-s)^{q-1} f(t, x_s^n) ds, & \text{for } t \in \left[\dfrac{T}{n}, T\right], \end{cases}$$

where $\varphi^0 \in C([-r, a), X)$ denotes the function defined by

$$\varphi^0(t) = \begin{cases} \varphi(t), & \text{for } t \in [-r, 0], \\ \varphi(0), & \text{for } t \in [0, a). \end{cases}$$

Using the similar method as we did in (3.14), we get that $x^n \in B_k$ for all $n \geq 1$.

Let $A = \{x^n : n \geq 1\}$. It follows that the set $A$ is uniformly bounded. Further, we show that the set $A$ is equicontinuous on $[-r, T]$.

If $-r \le t_1 < t_2 \le \frac{T}{n}$, then for each $x^n \in A$, we have $\lim_{t_1 \to t_2} |x^n(t_2) - x^n(t_1)| = \lim_{t_1 \to t_2} |\varphi^0(t_2) - \varphi^0(t_1)| = 0$ independently of $x^n \in A$. Next, if $-r \le t_1 \le \frac{T}{n} < t_2 \le T$, then for each $x^n \in A$, by using Hölder inequality, we have

$$
\begin{aligned}
&|x^n(t_2) - x^n(t_1)| \\
&\le |\varphi(0) - \varphi^0(t_1)| + \left| \frac{1}{\Gamma(q)} \int_0^{t_2 - \frac{T}{n}} (t_2 - s)^{q-1} f(s, x_s^n) ds \right| \\
&\le |\varphi(0) - \varphi^0(t_1)| + \frac{1}{\Gamma(q)} \int_0^{t_2 - \frac{T}{n}} (t_2 - s)^{q-1} m_1(s) ds \\
&\le |\varphi(0) - \varphi^0(t_1)| + \frac{1}{\Gamma(q)} \left( \int_0^{t_2 - \frac{T}{n}} (t_2 - s)^{\frac{q-1}{1-q_1}} ds \right)^{1-q_1} \|m_1\|_{L^{\frac{1}{q_1}}[0,T]} \\
&= |\varphi(0) - \varphi^0(t_1)| + \frac{(t_2^{1+b_1} - (\frac{T}{n})^{1+b_1})^{1-q_1}}{\Gamma(q)(1+b_1)^{1-q_1}} \|m_1\|_{L^{\frac{1}{q_1}}[0,T]}.
\end{aligned}
$$

According to the definition of $\varphi^0$, and using the last inequality, we obtain that $|x^n(t_2) - x^n(t_1)| \to 0$ independently of $x^n \in A$, as $t_1 \to t_2$.

Finally, if $\frac{T}{n} \le t_1 < t_2 \le T$, then for each $x^n \in A$, by using Hölder inequality, we have

$$
\begin{aligned}
&|x^n(t_2) - x^n(t_1)| \\
&= \left| \frac{1}{\Gamma(q)} \int_0^{t_2 - \frac{T}{n}} (t_2 - s)^{q-1} f(s, x_s^n) ds - \frac{1}{\Gamma(q)} \int_0^{t_1 - \frac{T}{n}} (t_1 - s)^{q-1} f(s, x_s^n) ds \right| \\
&\le \left| \frac{1}{\Gamma(q)} \int_{t_1 - \frac{T}{n}}^{t_2 - \frac{T}{n}} (t_2 - s)^{q-1} f(s, x_s^n) ds \right| + \left| \frac{1}{\Gamma(q)} \int_0^{t_1 - \frac{T}{n}} (t_2 - s)^{q-1} f(s, x_s^n) ds \right. \\
&\qquad \left. - \frac{1}{\Gamma(q)} \int_0^{t_1 - \frac{T}{n}} (t_1 - s)^{q-1} f(s, x_s^n) ds \right| \\
&\le \frac{1}{\Gamma(q)} \int_{t_1 - \frac{T}{n}}^{t_2 - \frac{T}{n}} (t_2 - s)^{q-1} m_1(s) ds \\
&\quad + \frac{1}{\Gamma(q)} \int_0^{t_1 - \frac{T}{n}} \left( (t_1 - s)^{q-1} - (t_2 - s)^{q-1} \right) m_1(s) ds \\
&\le \frac{1}{\Gamma(q)} \left( \int_{t_1 - \frac{T}{n}}^{t_2 - \frac{T}{n}} (t_2 - s)^{\frac{q-1}{1-q_1}} ds \right)^{1-q_1} \|m_1\|_{L^{\frac{1}{q_1}}[0,T]} \\
&\quad + \frac{1}{\Gamma(q)} \left( \int_0^{t_1 - \frac{T}{n}} (t_1 - s)^{\frac{q-1}{1-q_1}} - (t_2 - s)^{\frac{q-1}{1-q_1}} ds \right)^{1-q_1} \|m_1\|_{L^{\frac{1}{q_1}}[0,T]} \\
&= \frac{((t_2 - t_1 + \frac{T}{n})^{1+b_1} - (\frac{T}{n})^{1+b_1})^{1-q_1}}{\Gamma(q)(1+b_1)^{1-q_1}} \|m_1\|_{L^{\frac{1}{q_1}}[0,T]} \\
&\quad + \frac{(t_1^{1+b_1} - (\frac{T}{n})^{1+b_1} - t_2^{1+b_1} + (t_2 - t_1 + \frac{T}{n})^{1+b_1})^{1-q_1}}{\Gamma(q)(1+b_1)^{1-q_1}} \|m_1\|_{L^{\frac{1}{q_1}}[0,T]}
\end{aligned}
$$

$$\leq 2\frac{((t_2 - t_1 + \frac{T}{n})^{1+b_1} - (\frac{T}{n})^{1+b_1})^{1-q_1}}{\Gamma(q)(1 + b_1)^{1-q_1}}\|m_1\|_{L^{\frac{1}{q_1}}[0,T]}.$$

It is easy to see that the last inequality tends to zero independently of $x^n \in A$, as $t_1 \to t_2$, which means that the set $A$ is equicontinuous.

Set $A(t) = \{x^n(t) : n \geq 1\}$ and $A_t = \{x_t^n : n \geq 1\}$ for any $t \in [0,T]$. By the properties (iv) and (vi) of the measure of noncompactness, for any fixed $t \in (0,T]$ and $\delta \in (0,t)$, we have

$$\alpha(A(t)) \leq \alpha\left(\left\{\frac{1}{\Gamma(q)}\int_0^t (t-s)^{q-1}f(s,x_s^n)ds : n \geq 1\right\}\right)$$

$$+ \alpha\left(\left\{\frac{1}{\Gamma(q)}\int_{t-\frac{T}{n}}^t (t-s)^{q-1}f(s,x_s^n)ds : n \geq 1\right\}\right),$$

for all $\epsilon > 0$, we can find $\delta$ sufficiently small such that

$$\frac{\delta^{(1+b_1)(1-q_1)}}{\Gamma(q)(1+b_1)^{1-q_1}}\|m_1\|_{L^{\frac{1}{q_1}}[0,T]} < \frac{\epsilon}{2}.$$

Therefore, for each $t \in (0,T]$, we can choose $N_\delta \geq 1$ such that $\frac{T}{n} \leq \delta$ for $n \geq N_\delta$. Then we obtain that

$$\alpha\left(\left\{\frac{1}{\Gamma(q)}\int_{t-\frac{T}{n}}^t (t-s)^{q-1}f(s,x_s^n)ds : n \geq N_\delta\right\}\right)$$

$$\leq \frac{2}{\Gamma(q)}\sup_{n\geq N_\delta}\int_{t-\frac{T}{n}}^t (t-s)^{q-1}m_1(s)ds$$

$$< \epsilon,$$

for each $t \in (0,T]$. Hence, by the properties (iii) and (v) of the measure of noncompactness, it follows that

$$\alpha\left(\left\{\frac{1}{\Gamma(q)}\int_{t-\frac{T}{n}}^t (t-s)^{q-1}f(s,x_s^n)ds : n \geq 1\right\}\right) < \epsilon.$$

Then, we obtain that

$$\alpha(A(t)) \leq \alpha\left(\left\{\frac{1}{\Gamma(q)}\int_0^t (t-s)^{q-1}f(s,x_s^n)ds : n \geq 1\right\}\right) + \epsilon,$$

for $t \in (0,T]$. By Proposition 1.16 and (H3), we have that

$$\alpha(A(t)) \leq \frac{2}{\Gamma(q)}\int_0^t (t-s)^{q-1}\alpha(f(s,A_s))ds + \epsilon$$

$$\leq \frac{2}{\Gamma(q)}\int_0^t (t-s)^{q-1}m_2(s)\alpha(A_s)ds + \epsilon,$$

where $t \in (0,T]$. Since $x^n(\theta) = \varphi(\theta)$, $\theta \in [-r,0]$, we have $\alpha(\{x^n(\theta) : n \geq 1\}) = 0$ for $\theta \in [-r,0]$. Moreover, by Proposition 1.15, for $s \in [0,t]$ with $t \in (0,T]$, we deduce that

$$\alpha(A_s) = \max_{\theta\in[-r,0]} \alpha(\{x_s^n(\theta) : n \geq 1\}) \leq \sup_{s\in[0,t]} \alpha(\{x^n(s) : n \geq 1\}) = \sup_{s\in[0,t]} \alpha(A(s)).$$

Since $\epsilon$ is arbitrary, we have that

$$\alpha(A(t)) \leq \frac{2T^{(1+b_2)(1-q_2)}}{\Gamma(q)(1+b_2)^{1-q_2}} \|m_2\|_{L^{\frac{1}{q_2}}[0,T]} \sup_{s\in[0,t]} \alpha(A(s)),$$

where $t \in (0, T]$ and $b_2 = \frac{q-1}{1-q_2} \in (-1, 0)$.

Since (3.13) and $x_0^n = \varphi$, we must have that $\alpha(A(t)) = 0$ for every $t \in [-r, T]$. Then, by Proposition 1.15, we have that $\alpha(A) = \sup_{t\in[-r,T]} \alpha(A(t)) = 0$. Therefore, $A$ is a relatively compact subset of $B_k$. Then, there exists a subsequence if necessary, we may assume that the sequence $\{x^n\}_{n\geq 1}$ converges uniformly on $[-r, T]$ to a continuous function $x \in B_k$ with $x(\theta) = \varphi(\theta)$, $\theta \in [-r, 0]$.

Moreover, for $t \in [0, \frac{T}{n}]$, we have

$$|(Fx^n)(t) - x^n(t)| \leq \frac{1}{\Gamma(q)} \int_0^{\frac{T}{n}} (t-s)^{q-1}|f(t, x_s^n)|ds \leq \frac{1}{\Gamma(q)} \int_0^{\frac{T}{n}} (t-s)^{q-1}m_1(s)ds$$

and for $t \in [\frac{T}{n}, T]$, we have

$$\begin{aligned}
|(Fx^n)(t) - x^n(t)| &= \frac{1}{\Gamma(q)}\left| \int_0^t (t-s)^{q-1}f(t, x_s^n)ds - \int_0^{t-\frac{T}{n}} (t-s)^{q-1}f(t, x_s^n)ds \right| \\
&= \frac{1}{\Gamma(q)}\left| \int_{t-\frac{T}{n}}^t (t-s)^{q-1}f(t, x_s^n)ds \right| \\
&\leq \frac{1}{\Gamma(q)} \int_{t-\frac{T}{n}}^t (t-s)^{q-1}m_1(s)ds.
\end{aligned}$$

Therefore, it follows that

$$\sup_{t\in[0,T]} |(Fx^n)(t) - x^n(t)| \to 0 \quad \text{as} \quad n \to \infty. \tag{3.15}$$

Since

$$\begin{aligned}
\sup_{t\in[0,T]} |(Fx)(t) - x(t)| &\leq \sup_{t\in[0,T]} |(Fx)(t) - (Fx^n)(t)| \\
&\quad + \sup_{t\in[0,T]} |(Fx^n)(t) - x^n(t)| + \sup_{t\in[0,T]} |x^n(t) - x(t)|,
\end{aligned}$$

then, by (3.15) and the fact that $F$ is a continuous operator, we obtain that $\sup_{t\in[0,T]} |(Fx)(t) - x(t)| = 0$. It follows that $x(t) = (Fx)(t)$ for every $t \in [0, T]$. Hence

$$x(t) = \begin{cases} \varphi(t), & \text{for } t \in [-r, 0], \\ \varphi(0) + \frac{1}{\Gamma(q)} \int_0^t (t-s)^{q-1}f(t, x_s)ds, & \text{for } t \in [0, T] \end{cases}$$

solve fractional IVP (3.10), and this completes the proof. $\qquad\square$

**Corollary 3.1.** *Assume that hypotheses (H1)-(H3) hold. Then, for every $\varphi \in \mathcal{C}$, there exist $T \in (0, a)$ and a sequence of continuous function $x^n : [-r, T] \to X$, such that*

**(i)** $x^n(t)$ *are absolutely continuous on* $[0,T]$;

**(ii)** $x_0^n = \varphi$, *for every* $n \geq 1$, *and*

**(iii)** *extracting a subsequence which is labeled in the same way such that* $x^n(t) \to$ $x(t)$ *uniformly on* $[-r,T]$ *and* $x : [-r,T] \to X$ *is a solution for fractional IVP* (3.10).

We now give an example to illustrate the application of our abstract results.

**Example 3.2.** Consider the infinite system of fractional functional differential equations

$$
\begin{cases}
{}_0^C D_t^{\frac{1}{2}} x_n(t) = \dfrac{1}{nt^{1/3}} x_n^2(t-r), & \text{for } t \in (0,a), \\[2mm]
x_n(\theta) = \varphi(\theta) = \dfrac{\theta}{n}, & \text{for } \theta \in [-r,0], \ n = 1,2,3,\ldots.
\end{cases}
\tag{3.16}
$$

Let $E = c_0 = \{x = (x_1, x_2, x_3, \ldots) : x_n \to 0\}$ with norm $|x| = \sup_{n \geq 1} |x_n|$. Then the infinite system (3.16) can be regarded as a fractional IVP of form (3.10) in $E$. In this situation, $q = \frac{1}{2}$, $x = (x_1, \ldots, x_n, \ldots)$, $x_t = x(t-r) = (x_1(t-r), \ldots, x_n(t-r), \ldots)$, $\varphi(\theta) = (\theta, \frac{\theta}{2}, \ldots, \frac{\theta}{n}, \ldots)$ for $\theta \in [-r,0]$ and $f = (f_1, \ldots, f_n, \ldots)$, in which

$$
f_n(t, x_t) = \frac{1}{nt^{1/3}} x_n^2(t-r).
\tag{3.17}
$$

It is obvious that conditions (H1) and (H2) are satisfied. Now, we check the condition (H3) and the argument is similar to Section 2.4. Let $t \in (0,a)$, $R > 0$ be given and $\{w^{(m)}\}$ be any sequence in $f(t,B)$, where $w^{(m)} = (w_1^{(m)}, \ldots, w_n^{(m)}, \ldots)$ and $B = \{z \in \mathcal{C} : \|z\|_* \leq R\}$ is a bounded set in $\mathcal{C}$. By (3.17), we have

$$
0 \leq w_n^{(m)} \leq \frac{R^2}{nt^{1/3}}, \quad n,m = 1,2,3,\ldots.
\tag{3.18}
$$

So, $\{w_n^{(m)}\}$ is bounded and, by the diagonal method, we can choose a subsequence $\{m_i\} \subset \{m\}$ such that

$$
w_n^{(m_i)} \to w_n \text{ as } i \to \infty, \quad n = 1,2,3,\ldots,
\tag{3.19}
$$

which implies by virtue of (3.18) that

$$
0 \leq w_n \leq \frac{R^2}{nt^{1/3}}, \quad n = 1,2,3,\ldots.
\tag{3.20}
$$

Hence $w = (w_1, \ldots, w_n, \ldots) \in c_0$. It is easy to see from (3.18)-(3.20) that

$$
|w^{(m_i)} - w| = \sup_n |w_n^{(m_i)} - w_n| \to 0 \text{ as } i \to \infty.
$$

Thus, we have proved that $f(t,B)$ is relatively compact in $c_0$ for $t \in (0,a)$, which means that $f(t,B) = 0$ for almost all $t \in [0,a)$ and $B$ a bounded set in $\mathcal{C}$. Hence, the condition (H3) is satisfied. Finally, from Theorem 3.1, we can conclude that the infinite system (3.16) has a continuous solution.

## 3.3    Cauchy Problems via Topological Degree Method

### 3.3.1    *Introduction*

It is well known that the topological methods is proved to be a powerful tool in the study of various problems which appear in nonlinear analysis. Particularly, a priori estimate method has been often used together with the Brouwer degree, the Leray-Schauder degree or the coincidence degree in order to prove the existence of solutionsfor some boundary value problems and bifurcation problems for nonlinear differential equations or nonlinear partial differential equations. See, for example, Fečkan, 2008; Mawhin, 1979.

In Section 3.3, we consider the following nonlocal problem via a coincidence degree for condensing mapping in a Banach space $X$

$$\begin{cases} {}^{C}_{0}D^{q}_{t}u(t) = f(t, u(t)), & t \in J := [0, T], \\ u(0) + g(u) = u_0, \end{cases} \tag{3.21}$$

where ${}^{C}_{0}D^{q}_{t}$ is Caputo fractional derivative of order $q \in (0, 1)$, $u_0$ is an element of $X$, $f : J \times X \to X$ is continuous. The nonlocal term $g : C(J, X) \to X$ is a given function, here $C(J, X)$ is the Banach space of all continuous functions from $J$ into $X$ with the norm $\|u\| := \sup_{t \in J} |u(t)|$ for $u \in C(J, X)$.

### 3.3.2    *Qualitative Analysis*

This subsection deals with existence of solutions for the nonlocal problem (3.21).

**Definition 3.2.** A function $u \in C^1(J, X)$ is said to be a solution of the nonlocal problem (3.21) if $u$ satisfies the equation ${}^{C}_{0}D^{q}_{t}u(t) = f(t, u(t))$ a.e. on $J$, and the condition $u(0) + g(u) = u_0$.

**Lemma 3.2.** *A function $u \in C(J, X)$ is a solution of the fractional integral equation*

$$u(t) = u_0 - g(u) + \frac{1}{\Gamma(q)} \int_0^t (t - s)^{q-1} f(s, u(s))ds, \tag{3.22}$$

*if and only if $u$ is a solution of the nonlocal problem (3.21).*

We make some following assumptions:

**(H1)** for arbitrary $u, v \in C(J, X)$, there exists a constant $K_g \in [0, 1)$ such that

$$|g(u) - g(v)| \le K_g \|u - v\|;$$

**(H2)** for arbitrary $u \in C(J, X)$, there exist $C_g, M_g > 0$, $q_1 \in [0, 1)$ such that

$$|g(u)| \le C_g \|u\|^{q_1} + M_g;$$

**(H3)** for arbitrary $(t, u) \in J \times X$, there exist $C_f, M_f > 0$, $q_2 \in [0, 1)$ such that

$$|f(t, u)| \le C_f |u|^{q_2} + M_f;$$

**(H4)** for any $r > 0$, there exists a constant $\beta_r > 0$ such that

$$\alpha(f(s, \mathcal{M})) \leq \beta_r \alpha(\mathcal{M}),$$

for all $t \in J$, $\mathcal{M} \subset \mathcal{B}_r := \{\|u\| \leq r : u \in C(J, X)\}$ and

$$\frac{2T^q \beta_r}{\Gamma(q+1)} < 1.$$

Under the assumptions (H1)-(H4), we show that fractional integral equation (3.22) has at least one solution $u \in C(J, X)$.

Define operators

$$F : C(J, X) \to C(J, X), \quad (Fu)(t) = u_0 - g(u), \qquad t \in J,$$

$$G : C(J, X) \to C(J, X), \quad (Gu)(t) = \frac{1}{\Gamma(q)} \int_0^t (t - s)^{q-1} f(s, u(s)) ds, \quad t \in J,$$

$$\mathbb{T} : C(J, X) \to C(J, X), \quad \mathbb{T}u = Fu + Gu.$$

It is obvious that $\mathbb{T}$ is well defined. Then, fractional integral equation (3.22) can be written as the following operator equation

$$u = \mathbb{T}u = Fu + Gu.$$

Thus, the existence of a solution for the nonlocal problem (3.21) is equivalent to the existence of a fixed point for operator $\mathbb{T}$.

**Lemma 3.3.** *The operator $F : C(J, X) \to C(J, X)$ is Lipschitz with constant $K_g$. Consequently $F$ is $\alpha$-Lipschitz with the same constant $K_g$. Moreover, $F$ satisfies the following growth condition:*

$$\|Fu\| \leq |u_0| + C_g \|u\|^{q_1} + M_g, \tag{3.23}$$

*for every $u \in C(J, X)$.*

**Proof.** Using (H1), we have $\|Fu - Fv\| \leq |g(u) - g(v)| \leq K_g \|u - v\|$, for every $u, v \in C(J, X)$. By Proposition 1.20, $F$ is $\alpha$-Lipschitz with constant $K_g$. Relation (3.23) is a simple consequence of (H2). $\qquad\square$

**Lemma 3.4.** *The operator $G : C(J, X) \to C(J, X)$ is continuous. Moreover, $G$ satisfies the following growth condition:*

$$\|Gu\| \leq \frac{T^q (C_f \|u\|^{q_2} + M_f)}{\Gamma(q+1)}, \tag{3.24}$$

*for every $u \in C(J, X)$.*

**Proof.** For that, let $\{u_n\}$ be a sequence of a bounded set $\mathcal{B}_K \subseteq C(J, X)$ such that $u_n \to u$ in $\mathcal{B}_K (K > 0)$. We have to show that $\|Gu_n - Gu\| \to 0$.

It is easy to see that $f(s, u_n(s)) \to f(s, u(s))$ as $n \to \infty$ due to the continuity of $f$. On the one hand, using (H3), we get for each $t \in J$, $(t - s)^{q-1}|f(s, u_n(s)) - f(s, u(s))| \leq (t - s)^{q-1} 2(C_f K^{q_2} + M_f)$. On the other hand, using the fact that the function $s \to (t-s)^{q-1}2(C_fK^{q_2} + M_f)$ is integrable for $s \in [0,t]$, $t \in J$, Lebesgue dominated convergence theorem yields $\int_0^t (t-s)^{q-1}|f(s,u_n(s)) - f(s,u(s))|ds \to 0$ as $n \to \infty$. Then, for all $t \in J$,

$$|(Gu_n)(t) - (Gu)(t)| \le \frac{1}{\Gamma(q)} \int_0^t (t-s)^{q-1}|f(s,u_n(s)) - f(s,u(s))|ds \to 0.$$

Therefore, $Gu_n \to Gu$ as $n \to \infty$ which implies that $G$ is continuous. Relation (3.24) is a simple consequence of (H3). □

**Lemma 3.5.** *The operator* $G : C(J,X) \to C(J,X)$ *is compact. Consequently* $G$ *is* $\alpha$*-Lipschitz with zero constant.*

**Proof.** In order to prove the compactness of $G$, we consider a bounded set $\mathcal{M} \subseteq C(J,X)$ and the key step is to show that $G(\mathcal{M})$ is relatively compact in $C(J,X)$.

Let $\{u_n\}$ be a sequence on $\mathcal{M} \subset \mathfrak{B}_K$, for every $u_n \in \mathcal{M}$. By relation (3.24), we have

$$\|Gu_n\| \le \frac{T^q(C_fK^{q_2} + M_f)}{\Gamma(q+1)} =: r,$$

for every $u_n \in \mathcal{M}$, so $G(\mathcal{M})$ is bounded in $\mathfrak{B}_r$.

Now we prove that $\{Gu_n\}$ is equicontinuous. For $0 \le t_1 < t_2 \le T$, we get

$$|(Gu_n)(t_1) - (Gu_n)(t_2)|$$

$$\le \frac{1}{\Gamma(q)} \int_0^{t_1} \left((t_1 - s)^{q-1} - (t_2 - s)^{q-1}\right) |f(s,u_n(s))|ds$$

$$+ \frac{1}{\Gamma(q)} \int_{t_1}^{t_2} (t_2 - s)^{q-1}|f(s,u_n(s))|ds$$

$$\le \frac{1}{\Gamma(q)} \int_0^{t_1} \left((t_1 - s)^{q-1} - (t_2 - s)^{q-1}\right) (C_f|u_n(s)|^{q_2} + M_f)ds$$

$$+ \frac{1}{\Gamma(q)} \int_{t_1}^{t_2} (t_2 - s)^{q-1} (C_f|u_n(s)|^{q_2} + M_f)\, ds$$

$$\le \frac{(C_fK^{q_2} + M_f)}{\Gamma(q)} \left(\frac{t_1^q}{q} - \frac{t_2^q}{q} + \frac{(t_2 - t_1)^q}{q} + \frac{(t_2 - t_1)^q}{q}\right)$$

$$\le \frac{2(C_fK^{q_2} + M_f)(t_2 - t_1)^q}{\Gamma(q+1)}.$$

As $t_2 \to t_1$, the right-hand side of the above inequality tends to zero. Therefore $\{Gu_n\}$ is equicontinuous.

Consider a bounded set

$$\mathcal{M}(t) := \left\{v_n(t) : v_n(t) = \frac{1}{\Gamma(q)} \int_0^t (t-s)^{q-1}f(s,v_n(s))ds\right\} \subset \mathfrak{B}_r.$$

Applying Proposition 1.15, we know that the function $t \to \alpha(\mathcal{M}(t))$ is continuous on $J$. Moreover,

$$(t-s)^{q-1}|f(s,v_n(s))| \le (t-s)^{q-1}(C_fr^{q_2} + M_f) \in L^1(J,\mathbb{R}_+), \text{ for } s \in [0,t], \ t \in J.$$

Using (H4) and Proposition 1.16, we have

$$\alpha(\mathcal{M}(t)) \leq \alpha \left( \left\{ \frac{1}{\Gamma(q)} \int_0^t (t-s)^{q-1} f(s, \mathcal{M}(s)) ds \right\} \right)$$

$$\leq \frac{2}{\Gamma(q)} \int_0^t (t-s)^{q-1} \alpha \left( f(s, \mathcal{M}(s)) \right) ds$$

$$\leq \frac{2\beta_r}{\Gamma(q)} \int_0^t (t-s)^{q-1} \alpha(\mathcal{M}(s)) ds,$$

which implies that

$$\alpha(\mathcal{M}) \leq \left( \frac{2\beta_r}{\Gamma(q)} \int_0^t (t-s)^{q-1} ds \right) \alpha(\mathcal{M})$$

$$\leq \frac{2T^q \beta_r}{\Gamma(q+1)} \alpha(\mathcal{M})$$

$$< \alpha(\mathcal{M}),$$

due to the condition

$$\frac{2T^q \beta_r}{\Gamma(q+1)} < 1.$$

Then we can deduce that $\alpha(\mathcal{M}) = 0$. Therefore, $G(\mathcal{M})$ is a relatively compact subset of $C(J, X)$, and so, there exists a subsequence $v_n$ which converge uniformly on $J$ to some $v_* \in C(J, X)$ together with Arzela-Ascoli theorem. By Proposition 1.19, $G$ is $\alpha$-Lipschitz with zero constant. $\qquad \square$

**Theorem 3.2.** *Assume that (H1)-(H4) hold, then the nonlocal problem (3.21) has at least one solution $u \in C(J, X)$ and the set of the solutions of system (3.21) is bounded in $C(J, X)$.*

**Proof.** Let $F, G, \mathbb{T} : C(J, X) \to C(J, X)$ be the operators defined in the beginning of this section. They are continuous and bounded. Moreover, $F$ is $\alpha$-Lipschitz with constant $K_g \in [0, 1)$ and $G$ is $\alpha$-Lipschitz with zero constant (see Lemmas 3.3-3.5). Proposition 1.18 shows that $\mathbb{T}$ is a strict $\alpha$-contraction with constant $K_g$.

Set

$$S_0 = \{ u \in C(J, X) : \exists \lambda \in [0, 1] \text{ such that } u = \lambda \mathbb{T}u \}.$$

Next, we prove that $S_0$ is bounded in $C(J, X)$. Consider $u \in S_0$ and $\lambda \in [0, 1]$ such that $u = \lambda \mathbb{T}u$. It follows from (3.23) and (3.24) that

$$\|u\| = \lambda \|\mathbb{T}u\| \leq \lambda(\|Fu\| + \|Gu\|)$$

$$\leq |u_0| + C_g \|u\|^{q_1} + M_g + \frac{T^q(C_f \|u\|^{q_2} + M_f)}{\Gamma(q+1)}. \qquad (3.25)$$

This inequality (3.25), together with $q_1 < 1$ and $q_2 < 1$, shows that $S_0$ is bounded in $C(J, X)$. If not, we suppose by contradiction, $\rho := \|u\| \to \infty$. Dividing both sides of (3.25) by $\rho$, and taking $\rho \to \infty$, we have

$$1 \leq \lim_{\rho \to \infty} \rho^{-1} \left( |u_0| + C_g \rho^{q_1} + M_g + \frac{T^q(C_f \rho^{q_2} + M_f)}{\Gamma(q+1)} \right) = 0.$$

This is a contradiction. Consequently, by Theorem 1.2 we deduce that $\mathbb{T}$ has at least one fixed point and the set of the fixed points of $\mathbb{T}$ is bounded in $C(J, X)$. $\quad \square$

## Remark 3.1.

(i) If the growth condition (H2) is formulated for $q_1 = 1$, then the conclusions of Theorem 3.2 remain valid provided that $C_g < 1$.

(ii) If the growth condition (H3) is formulated for $q_2 = 1$, then the conclusions of Theorem 3.2 remain valid provided that $\frac{T^q C_f}{\Gamma(q+1)} < 1$.

(iii) If the growth conditions (H2) and (H3) are formulated for $q_1 = 1$ and $q_2 = 1$, then the conclusions of Theorem 3.2 remain valid provided that $C_g + \frac{T^q C_f}{\Gamma(q+1)} < 1$.

## 3.4 Cauchy Problems via Picard Operators Technique

### 3.4.1 *Introduction*

Assume that $(X, |\cdot|)$ is a Banach space, and $J := [0, T]$, $T > 0$. Let $C(J, X)$ be the Banach space of all continuous functions from $J$ into $X$ with the norm $\|x\| := \sup\{|x(t)| : t \in J\}$ for $x \in C(J, X)$.

Consider the following Cauchy problem of fractional differential equation

$$\begin{cases} {}^{C}_{0}D^q_t x(t) = f(t, x(t)), \quad \text{a.e. } t \in J, \\ x(0) = x_0 \in X, \end{cases} \tag{3.26}$$

where ${}^{C}_{0}D^q_t$ is Caputo fractional derivative of order $q \in (0, 1)$, the function $f : J \times X \to X$ satisfies some assumptions that will be specified later.

To our knowledge, Picard operators and weak Picard operators technique due to Rus 1979, 1987, 1993, 2003; Rus and Muresan 2000 have been used to study the existence for the solutions of some integer differential equations (see, Mureşan, 2004; Şerban, Rus and Petruşel, 2010). In the present section we consider suitable Bielecki norms in some convenient spaces and obtain existence, uniqueness and data dependence results for the solutions of the fractional Cauchy problem (3.26) via Picard operators and weak Picard operators technique.

**Definition 3.3.** A function $x \in C^1(J, X)$ is said to be a solution of the fractional Cauchy problem (3.26) if $x$ satisfies the equation ${}^{C}_{0}D^q_t x(t) = f(t, x(t))$ a.e. on $J$, and the condition $x(0) = x_0$.

In Subsection 3.4.2, we give the existence, uniqueness and data dependence results for the solutions of (3.26) via Picard operator by the successive approximation method. In Subsection 3.4.3, we obtain the existence results for the solutions of (3.26) via weak Picard operator.

### 3.4.2 *Results via Picard Operators*

Consider a Banach space $(X, |\cdot|)$, let $\|\cdot\|_B$ and $\|\cdot\|_C$ be the Bielecki and Chebyshev norms on $C(J, X)$ defined by

$$\|x\|_B = \max_{t \in J} |x(t)| e^{-\tau t} (\tau > 0) \quad \text{and} \quad \|x\|_C = \max_{t \in J} |x(t)|$$

and denote by $d_B$ and $d_C$ their corresponding metrics. We consider the set

$$C_L^{q-q^*}(J, X) = \{x \in C(J, X) : |x(t_1) - x(t_2)| \leq L|t_1 - t_2|^{q-q^*} \text{ for all } t_1, t_2 \in J\}$$

where $L > 0$, $q^* \in (0, q)$, and

$$C_{\bar{L}}^q(J, X) = \{x \in C(J, X) : |x(t_1) - x(t_2)| \leq \bar{L}|t_1 - t_2|^q \text{ for all } t_1, t_2 \in J\}$$

where $\bar{L} > 0$, and

$$C_{\bar{L}}^q(J, B_R) = \{x \in C(J, B_R) : |x(t_1) - x(t_2)| \leq \bar{L}|t_1 - t_2|^q \text{ for all } t_1, t_2 \in J\}$$

where $B_R = \{x \in X : |x| \leq R\}$ with $R > 0$.

If $d \in \{d_C, d_B\}$, then $(C(J, X), d)$, $(C_L^{q-q^*}(J, X), d)$, $(C_{\bar{L}}^q(J, X), d)$ and $(C_{\bar{L}}^q(J, B_R), d)$ are complete metric spaces.

Let $q_i \in (0, q)$, $i = 1, 2, 3$ and the functions $m(t) \in L^{\frac{1}{q_1}}(J, \mathbb{R}_+)$, $\eta(t) \in L^{\frac{1}{q_2}}(J, \mathbb{R}_+)$, $\mu(t) \in L^{\frac{1}{q_3}}(J, \mathbb{R}_+)$ and $l(t) \in C(J, \mathbb{R}_+)$.

For brevity, let

$$M = \|m\|_{L^{\frac{1}{q_1}}J}, \quad N = \|\eta\|_{L^{\frac{1}{q_2}}J}, \quad V = \|\mu\|_{L^{\frac{1}{q_2}}J}, \quad L_0 = \max_{t \in J}\{l(t)\},$$

$$\beta = \frac{q-1}{1-q_1} \in (-1, 0), \quad \gamma = \frac{q-1}{1-q_2} \in (-1, 0), \quad \nu = \frac{q-1}{1-q_3} \in (-1, 0).$$

**Lemma 3.6.** *A function $x \in C(J, X)$ is a solution of the fractional integral equation*

$$x(t) = x_0 + \frac{1}{\Gamma(q)} \int_0^t (t-s)^{q-1} f(s, x(s)) ds, \quad (3.27)$$

*if and only if $x$ is a solution of the fractional Cauchy problem (3.26).*

**Theorem 3.3.** *Suppose the following conditions hold:*

**(C1)** $f \in C(J \times X, X)$;

**(C2)** *there exist a constant $q_1 \in (0, q)$ and function $m(\cdot) \in L^{\frac{1}{q_1}}(J, \mathbb{R}_+)$ such that*

$$|f(t, x)| \leq m(t)$$

*for all $x \in X$ and all $t \in J$;*

**(C3)** *there exists a constant $L > 0$ such that*

$$L \geq \frac{2M}{\Gamma(q)(1+\beta)^{1-q_1}};$$

**(C4)** *there exists a function $l(\cdot) \in C(J, \mathbb{R}_+)$ such that*

$$|f(t, u_1) - f(t, u_2)| \leq l(t)|u_1 - u_2|$$

*for all $u_i \in X$ $(i = 1, 2)$ and all $t \in J$;*

**(C5)** *there exist constants $q_1$ and $\tau$ such that*

$$\frac{L_0}{\Gamma(q)} \frac{T^{(1+\beta)(1-q_1)}}{(1+\beta)^{1-q_1}} \left(\frac{q_1}{\tau}\right)^{q_1} < 1.$$

*Then the fractional Cauchy problem (3.26) has a unique solution $x^*$ in $C_L^{q-q_1}(J,X)$, and this solution can be obtained by the successive approximation method, starting from any element of $C_L^{q-q_1}(J,X)$.*

**Proof.** Consider the operator

$$A : (C_L^{q-q_1}(J,X), \|\cdot\|_B) \to (C_L^{q-q_1}(J,X), \|\cdot\|_B)$$

defined by

$$Ax(t) = x_0 + \frac{1}{\Gamma(q)} \int_0^t (t-s)^{q-1} f(s, x(s))\, ds.$$

It is easy to see the operator $A$ is well defined due to (C1).

Firstly, we check that $Ax \in C(J,X)$ for every $x \in C_L^{q-q_1}(J,X)$.

For any $\delta > 0$, every $x \in C_L^{q-q_1}(J,X)$, by (C2) and Hölder inequality,

$$|(Ax)(t+\delta) - (Ax)(t)|$$

$$\leq \frac{1}{\Gamma(q)} \int_0^t \left((t-s)^{q-1} - (t+\delta-s)^{q-1}\right) |f(s,x(s))| ds$$

$$+ \frac{1}{\Gamma(q)} \int_t^{t+\delta} (t+\delta-s)^{q-1} |f(s,x(s))| ds$$

$$\leq \frac{1}{\Gamma(q)} \int_0^t \left((t-s)^{q-1} - (t+\delta-s)^{q-1}\right) m(s) ds + \frac{1}{\Gamma(q)} \int_t^{t+\delta} (t+\delta-s)^{q-1} m(s) ds$$

$$\leq \frac{1}{\Gamma(q)} \left(\int_0^t \left((t-s)^{q-1} - (t+\delta-s)^{q-1}\right)^{\frac{1}{1-q_1}} ds\right)^{1-q_1} \left(\int_0^t (m(s))^{\frac{1}{q_1}} ds\right)^{q_1}$$

$$+ \frac{1}{\Gamma(q)} \left(\int_t^{t+\delta} \left((t+\delta-s)^{q-1}\right)^{\frac{1}{1-q_1}} ds\right)^{1-q_1} \left(\int_t^{t+\delta} (m(s))^{\frac{1}{q_1}} ds\right)^{q_1}$$

$$\leq \frac{M}{\Gamma(q)} \left(\int_0^t \left((t-s)^{\beta} - (t+\delta-s)^{\beta}\right) ds\right)^{1-q_1} + \frac{M}{\Gamma(q)} \left(\int_t^{t+\delta} (t+\delta-s)^{\beta} ds\right)^{1-q_1}$$

$$\leq \frac{M}{\Gamma(q)(1+\beta)^{1-q_1}} \left(|t^{1+\beta} - (t+\delta)^{1+\beta}| + \delta^{1+\beta}\right)^{1-q_1} + \frac{M}{\Gamma(q)(1+\beta)^{1-q_1}} \delta^{(1+\beta)(1-q_1)}$$

$$\leq \frac{2M}{\Gamma(q)(1+\beta)^{1-q_1}} \delta^{(1+\beta)(1-q_1)} + \frac{M}{\Gamma(q)(1+\beta)^{1-q_1}} \delta^{(1+\beta)(1-q_1)}$$

$$\leq \frac{3M}{\Gamma(q)(1+\beta)^{1-q_1}} \delta^{(1+\beta)(1-q_1)}.$$

It is easy to see that the right-hand side of the above inequality tends to zero as $\delta \to 0$. Therefore $Ax \in C(J,X)$.

Secondly, we show that $Ax \in C_L^{q-q_1}(J,X)$.

Without loss of generality, for any $t_1 < t_2$, $t_1, t_2 \in J$, applying (C2) and Hölder inequality, we have

$$|(Ax)(t_2) - (Ax)(t_1)|$$

$$\leq \frac{1}{\Gamma(q)} \left| \int_0^{t_1} [(t_2 - s)^{q-1} - (t_1 - s)^{q-1}] f(s, x(s)) ds + \int_{t_1}^{t_2} (t_2 - s)^{q-1} f(s, x(s)) ds \right|$$

$$\leq \frac{1}{\Gamma(q)} \int_0^{t_1} \left( (t_1 - s)^{q-1} - (t_2 - s)^{q-1} \right) |f(s, x(s))| ds$$

$$+ \frac{1}{\Gamma(q)} \int_{t_1}^{t_2} (t_2 - s)^{q-1} |f(s, x(s))| ds$$

$$\leq \frac{1}{\Gamma(q)} \int_0^{t_1} \left( (t_1 - s)^{q-1} - (t_2 - s)^{q-1} \right) m(s) ds + \frac{1}{\Gamma(q)} \int_{t_1}^{t_2} (t_2 - s)^{q-1} m(s) ds$$

$$\leq \frac{1}{\Gamma(q)} \left( \int_0^{t_1} \left( (t_1 - s)^{q-1} - (t_2 - s)^{q-1} \right)^{\frac{1}{1-q_1}} ds \right)^{1-q_1} \left( \int_0^{t_1} (m(s))^{\frac{1}{q_1}} ds \right)^{q_1}$$

$$+ \frac{1}{\Gamma(q)} \left( \int_{t_1}^{t_2} \left( (t_2 - s)^{q-1} \right)^{\frac{1}{1-q_1}} ds \right)^{1-q_1} \left( \int_{t_1}^{t_2} (m(s))^{\frac{1}{q_1}} ds \right)^{q_1}$$

$$\leq \frac{M}{\Gamma(q)} \left( \int_0^{t_1} (t_1 - s)^{\beta} - (t_2 - s)^{\beta} ds \right)^{1-q_1} + \frac{M}{\Gamma(q)} \left( \int_{t_1}^{t_2} (t_2 - s)^{\beta} ds \right)^{1-q_1}$$

$$\leq \frac{M}{\Gamma(q)(1 + \beta)^{1-q_1}} \left( t_1^{1+\beta} - t_2^{1+\beta} + (t_2 - t_1)^{1+\beta} \right)^{1-q_1}$$

$$+ \frac{M}{\Gamma(q)(1 + \beta)^{1-q_1}} (t_2 - t_1)^{(1+\beta)(1-q_1)}$$

$$\leq \frac{2M}{\Gamma(q)(1 + \beta)^{1-q_1}} |t_1 - t_2|^{(1+\beta)(1-q_1)}$$

$$\leq \frac{2M}{\Gamma(q)(1 + \beta)^{1-q_1}} |t_1 - t_2|^{q-q_1}.$$

Similarly, for any $t_1 > t_2$, $t_1, t_2 \in J$, we also have the above inequality. This implies that $Ax$ is belong to $C_L^{q-q_1}(J, X)$ due to (C3).

Thirdly, $A$ is continuous.

For that, let $\{x_n\}$ be a sequence of $B_R$ such that $x_n \to x$ in $B_R$. Then, $f(s, x_n(s)) \to f(s, x(s))$ as $n \to \infty$ due to (C1). On the one hand, by using (C2), we get for each $s \in [0, t]$, $|f(s, x_n(s)) - f(s, x(s))| \leq 2m(s)$. On the other hand, using the fact that the function $s \to 2(t - s)^{q-1} m(s)$ is integrable on $[0, t]$, Lebesgue dominated convergence theorem yields

$$\int_0^t (t - s)^{q-1} |f(s, x_n(s)) - f(s, x(s))| ds \to 0.$$

For all $t \in J$, we have

$$|(Ax_n)(t) - (Ax)(t)| \leq \frac{1}{\Gamma(q)} \int_0^t (t - s)^{q-1} |f(s, x_n(s)) - f(s, x(s))| ds.$$

Thus, $Ax_n \to Ax$ as $n \to \infty$ which implies that $A$ is continuous.

Moreover, for all $x, z \in C_L^{q-q_1}(J, X)$, using (C4) and Hölder inequality we have

$$|(Ax)(t) - (Az)(t)| \leq \frac{1}{\Gamma(q)} \int_0^t (t - s)^{q-1} |f(s, x(s)) - f(s, z(s))| ds$$

$$\leq \frac{1}{\Gamma(q)} \int_0^t (t-s)^{q-1} l(s) |x(s) - z(s)| ds$$

$$\leq \frac{1}{\Gamma(q)} \int_0^t (t-s)^{q-1} \max_{s \in [0,t]} \{l(s)\} \left( |x(s) - z(s)| e^{-\tau s} \right) e^{\tau s} ds$$

$$\leq \frac{L_0}{\Gamma(q)} \|x - z\|_B \int_0^t (t-s)^{q-1} e^{\tau s} ds$$

$$\leq \frac{L_0}{\Gamma(q)} \|x - z\|_B \left( \int_0^t (t-s)^{\beta} ds \right)^{1-q_1} \left( \int_0^t e^{\frac{\tau s}{q_1}} ds \right)^{q_1}$$

$$\leq \frac{L_0}{\Gamma(q)} \frac{T^{(1+\beta)(1-q_1)}}{(1+\beta)^{1-q_1}} \left( \frac{q_1}{\tau} \right)^{q_1} e^{\tau t} \|x - z\|_B.$$

It follows that

$$|(Ax)(t) - (Az)(t)| e^{-\tau t} \leq \frac{L_0}{\Gamma(q)} \frac{T^{(1+\beta)(1-q_1)}}{(1+\beta)^{1-q_1}} \left( \frac{q_1}{\tau} \right)^{q_1} \|x - z\|_B$$

for all $t \in J$. So we have

$$\|Ax - Az\|_B \leq \frac{L_0}{\Gamma(q)} \frac{T^{(1+\beta)(1-q_1)}}{(1+\beta)^{1-q_1}} \left( \frac{q_1}{\tau} \right)^{q_1} \|x - z\|_B$$

for all $x, z \in C_L^{q-q_1}(J, X)$. The operator $A$ is of Lipschitz type with constant

$$L_A = \frac{L_0}{\Gamma(q)} \frac{T^{(1+\beta)(1-q_1)}}{(1+\beta)^{1-q_1}} \left( \frac{q_1}{\tau} \right)^{q_1} \tag{3.28}$$

and $0 < L_A < 1$ due to (C5). By applying Banach contraction mapping principle to this operator we obtain that $A$ is a Picard operator. This completes the proof. □

**Example 3.3.** Consider the fractional Cauchy problem

$$\begin{cases} {}^C_0 D_t^q x(t) = x(t), & q = \frac{1}{2}, \\ x(0) = 0 \in X, \end{cases}$$

on $[0, 1]$. Set $L_0 = 1$, $T = 1$, $q_1 = \frac{1}{3}$, then $\beta = -\frac{3}{4}$. Indeed

$$\frac{L_0}{\Gamma(q)} \frac{T^{(1+\beta)(1-q_1)}}{(1+\beta)^{1-q_1}} \left( \frac{q_1}{\tau} \right)^{q_1} < 1 \iff \frac{qL_0}{\Gamma(q+1)} \frac{T^{(1+\beta)(1-q_1)}}{(1+\beta)^{1-q_1}} \left( \frac{q_1}{\tau} \right)^{q_1} < 1,$$

which implies that we must choose a suitable $\tau_0 > 0$ such that $\frac{\frac{1}{2}}{\Gamma(\frac{3}{2})} \frac{1}{(\frac{1}{4})^{\frac{2}{3}}} \left( \frac{1}{\tau_0} \right)^{\frac{1}{3}} < 1$.

Noting that $\Gamma(\frac{3}{2}) = \frac{\sqrt{\pi}}{2}$, for $\tau_0 = \frac{16}{9} > \frac{16}{3\sqrt{\pi^3}}$ we have the condition (C5) in Theorem 3.3.

**Theorem 3.4.** *Suppose the following conditions hold:*

**(C1)** $f \in C(J \times X, X)$;

**(C2)'** *there exists a constant $\bar{M} > 0$ such that $|f(t, x)| \leq \bar{M}$ for all $x \in X$ and all $t \in J$;*

**(C3)**′ *there exists a constant $\bar{L} > 0$ such that $\bar{L} \geq \frac{2\bar{M}}{\Gamma(q+1)}$;*

**(C4)**′ *there exists a constant $\bar{L}_0 > 0$ such that $|f(t, u_1) - f(t, u_2)| \leq \bar{L}_0|u_1 - u_2|$ for all $u_i \in X$ $(i = 1, 2)$ and all $t \in J$;*

**(C5)**′ *there exist constants $q_1$ and $\tau$ such that $\bar{L}_{\bar{A}} = \frac{\bar{L}_0}{\Gamma(q)} \frac{T^{(1+\beta)(1-q_1)}}{(1+\beta)^{1-q_1}} \left(\frac{q_1}{\tau}\right)^{q_1} < 1$.*

*Then the fractional Cauchy problem (3.26) has a unique solution $x^*$ in $C_{\bar{L}}^q(J, X)$, and this solution can be obtained by the successive approximation method, starting from any element of $C_{\bar{L}}^q(J, X)$.*

**Proof.** Consider the following continuous operator

$$\bar{A} : (C_{\bar{L}}^q(J, X), \|\cdot\|_B) \to (C_{\bar{L}}^q(J, X), \|\cdot\|_B)$$

defined by

$$\bar{A}x(t) = x_0 + \frac{1}{\Gamma(q)} \int_0^t (t - s)^{q-1} f(s, x(s))\, ds.$$

As the proof in Theorem 3.3, applying the given conditions one can verify that

$$\|\bar{A}(x) - \bar{A}(z)\|_B \leq \frac{\bar{L}_0}{\Gamma(q)} \frac{T^{(1+\beta)(1-q_1)}}{(1+\beta)^{1-q_1}} \left(\frac{q_1}{\tau}\right)^{q_1} \|x - z\|_B$$

for all $x, z \in C_{\bar{L}}^q(J, X)$. So, the operator $\bar{A}$ is a Picard operator. $\qquad\square$

Similarly, we can prove the following theorem.

**Theorem 3.5.** *Suppose the following conditions hold:*

**(C1)**′ *$f \in C(J \times B_R, X)$;*

**(C2)**″ *there exists a constant $\bar{M}(R) > 0$ such that $|f(t, x)| \leq \bar{M}(R)$ for all $x \in B_R$ and all $t \in J$ with $R \geq |x_0| + \frac{\bar{M}(R)T^q}{\Gamma(q+1)}$;*

**(C3)**″ *there exists a constant $\bar{L} > 0$ such that $\bar{L} \geq \frac{2\bar{M}(R)}{\Gamma(q+1)}$;*

**(C4)**″ *there exists a constant $\bar{L}_0 > 0$ such that $|f(t, u_1) - f(t, u_2)| \leq \bar{L}_0|u_1 - u_2|$ for all $u_i \in B_R$ $(i = 1, 2)$ and all $t \in J$;*

**(C5)**′ *there exist constants $q_1$ and $\tau$ such that $\bar{L}_{\bar{A}} = \frac{\bar{L}_0}{\Gamma(q)} \frac{T^{(1+\beta)(1-q_1)}}{(1+\beta)^{1-q_1}} \left(\frac{q_1}{\tau}\right)^{q_1} < 1$.*

*Then the fractional Cauchy problem (3.26) has a unique solution $x^*$ in $C_{\bar{L}}^q(J, B_R)$, and this solution can be obtained by the successive approximation method, starting from any element of $C_{\bar{L}}^q(J, B_R)$.*

Consider the following fractional Cauchy problem

$$\begin{cases} {}_0^C D_t^q x(t) = g(t, x(t)), & t \in J, \\ x(0) = y_0 \in X, \end{cases} \tag{3.29}$$

where $g \in C(J \times X, X)$. By Lemma 3.6, a function $x \in C(J, X)$ is a solution of the fractional integral equation

$$x(t) = y_0 + \frac{1}{\Gamma(q)} \int_0^t (t - s)^{q-1} g(s, x(s))\, ds, \tag{3.30}$$

if and only if $x$ is a solution of the fractional Cauchy problem (3.29).

Now, we consider both fractional integral equation (3.27) and (3.30).

**Theorem 3.6.** *Suppose the following:*

**(D1)** *all conditions in Theorem 3.3 are satisfied and $x^* \in C_L^{q-q_1}(J, X)$ is the unique solution of the fractional integral equation (3.27);*

**(D2)** *with the same function $m(\cdot)$ as in Theorem 3.3, $|g(t,x)| \le m(t)$ for all $x \in X$ and all $t \in J$;*

**(D3)** *with the same function $l(\cdot)$ as in Theorem 3.3, $|g(t,u_1) - g(t,u_2)| \le l(t)|u_1 - u_2|$ for all $u_i \in X$ $(i=1,2)$ and all $t \in J$;*

**(D4)** $L \ge \frac{2M}{\Gamma(q)(1+\beta)^{1-q_1}}$;

**(D5)** *there exists a constant $q_2 \in (0,q)$ and function $\eta(\cdot) \in L^{\frac{1}{q_2}}(J, \mathbb{R}_+)$ such that $|f(t,u) - g(t,u)| \le \eta(t)$ for all $u \in X$ and all $t \in J$.*

*If $y^*$ is the solution of the fractional integral equation (3.30), then*

$$\|x^* - y^*\|_B \le \frac{|x_0 - y_0| + \frac{NT^{(1+\gamma)(1-q_2)}}{\Gamma(q)(1+\gamma)^{1-q_2}}}{1 - L_A}, \tag{3.31}$$

*where $L_A$ is given by (3.28) with $\tau = \tau_0 > 0$ such that $0 < L_A < 1$.*

**Proof.** Consider the following two operators

$$A, \, B : (C_L^{q-q_1}(J, X), \|\cdot\|_B) \to (C_L^{q-q_1}(J, X), \|\cdot\|_B)$$

defined by

$$Ax(t) = x_0 + \frac{1}{\Gamma(q)} \int_0^t (t-s)^{q-1} f(s, x(s)) \, ds,$$

$$Bx(t) = y_0 + \frac{1}{\Gamma(q)} \int_0^t (t-s)^{q-1} g(s, x(s)) \, ds,$$

on $J$. We have

$$|Ax(t) - Bx(t)| \le |x_0 - y_0| + \frac{1}{\Gamma(q)} \int_0^t (t-s)^{q-1} |f(s, x(s)) - g(s, x(s))| \, ds$$

$$\le |x_0 - y_0| + \frac{1}{\Gamma(q)} \int_0^t (t-s)^{q-1} \eta(s) \, ds$$

$$\le |x_0 - y_0| + \frac{NT^{(1+\gamma)(1-q_2)}}{\Gamma(q)(1+\gamma)^{1-q_2}},$$

for $t \in J$. It follows that

$$\|Ax - Bx\|_B \le |x_0 - y_0| + \frac{NT^{(1+\gamma)(1-q_2)}}{\Gamma(q)(1+\gamma)^{1-q_2}}.$$

So we can apply Theorem 1.13 to obtain (3.31) which completes the proof. $\square$

**Remark 3.2.** All the results obtained in Theorem 3.3 hold even if the condition (C2) is replaced by the following:

(C2-E) there exist a constant $q_1 \in [0, q)$ and function $m(\cdot) \in L^{\frac{1}{q_1}}(J, \mathbb{R}_+)$ such that $|f(t, x)| \le m(t)$ for all $x \in X$ and all $t \in J$.

In fact, we only need extend the space $L^p(J, \mathbb{R}_+)$ $(1 < p < \infty)$ to $L^p(J, \mathbb{R}_+)$ $(1 \le p \le \infty)$ where $L^p(J, \mathbb{R}_+)$ $(1 \le p \le \infty)$ be the Banach space of all Lebesgue measurable functions $\phi : J \to \mathbb{R}_+$ with $\|\phi\|_{L^p J} < \infty$.

### 3.4.3 Results via Weakly Picard Operators

Now, we consider another fractional integral equation

$$x(t) = x(0) + \frac{1}{\Gamma(q)} \int_0^t (t - s)^{q-1} f(s, x(s))\, ds \qquad (3.32)$$

on $J$, where $f \in C(J \times X, X)$ is as in the fractional Cauchy problem (3.26).

**Theorem 3.7.** *Suppose that for the fractional integral equation (3.32) the same conditions as in Theorem 3.3 are satisfied. Then this equation has solutions in $C_L^{q-q_1}(J, X)$. If $S \subset C_L^{q-q_1}(J, X)$ is its solutions set, then* card $S =$ card $X$.

**Proof.** Consider the operator

$$A_* : (C_L^{q-q_1}(J, X), \|\cdot\|_B) \to (C_L^{q-q_1}(J, X), \|\cdot\|_B)$$

defined by

$$A_* x(t) = x(0) + \frac{1}{\Gamma(q)} \int_0^t (t - s)^{q-1} f(s, x(s))\, ds.$$

This is a continuous operator, but not a Lipschitz one. We can write

$$C_L^{q-q_1}(J, X) = \bigcup_{\alpha \in X} X_\alpha, \quad X_\alpha = \left\{ x \in C_L^{q-q_1}(J, X) : x(0) = \alpha \right\}.$$

We have that $X_\alpha$ is an invariant set of $A_*$ and we apply Theorem 3.3 to $A_*|_{X_\alpha}$. By using Theorem 1.3 we obtain that $A_*$ is a weak Picard operator.

Consider the operator

$$A_*^\infty : C_L^{q-q_1}(J, X) \to C_L^{q-q_1}(J, X), \quad A_*^\infty x = \lim_{n \to \infty} A_*^n x.$$

From $A_*^{n+1}(x) = A_*(A_*^n(x))$ and the continuity of $A_*$, $A_*^\infty(x) \in F_{A_*}$. Then

$$A_*^\infty(C_L^{q-q_1}(J, X)) = F_{A_*} = S \text{ and } S \ne \varnothing.$$

So, card $S =$ card $X$. $\qquad\square$

**Theorem 3.8.** *Suppose that for the fractional integral equation (3.32) the same conditions as in Theorem 3.4 are satisfied. Then this equation has solutions in $C_{\bar{L}}^q(J, X)$. If $S \subset C_{\bar{L}}^q(J, X)$ is its solutions set, then* card $S =$ card $X$.

**Proof.** As the proof in Theorem 3.7, we need to consider the continuous operator (but not a Lipschitz one)

$$\bar{A}_* : (C_{\bar{L}}^q(J, X), \|\cdot\|_B) \to (C_{\bar{L}}^q(J, X), \|\cdot\|_B)$$

defined by

$$\bar{A}_* x(t) = x(0) + \frac{1}{\Gamma(q)} \int_0^t (t-s)^{q-1} f(s, x(s))\, ds.$$

We can write $C_{\bar{L}}^q(J, X) = \bigcup_{\alpha \in X} \bar{X}_\alpha$, $\bar{X}_\alpha = \left\{ x \in C_{\bar{L}}^q(J, X) : x(0) = \alpha \right\}$. We have that $\bar{X}_\alpha$ is an invariant set of $\bar{A}_*$ and we apply Theorem 3.4 to $\bar{A}_*|_{X_\alpha}$. By using Theorem 1.3 we obtain that $\bar{A}_*$ is a weak Picard operator. Consider the operator $\bar{A}_*^\infty : C_{\bar{L}}^q(J, X) \to C_{\bar{L}}^q(J, X)$, $\bar{A}_*^\infty(x) = \lim_{n \to \infty} \bar{A}_*^n(x)$. From $\bar{A}_*^{n+1}(x) = \bar{A}_*(\bar{A}_*^n(x))$ and the continuity of $\bar{A}_*$, $\bar{A}_*^\infty(x) \in F_{\bar{A}_*}$. Then $\bar{A}_*^\infty(C_{\bar{L}}^q(J, X)) = F_{\bar{A}_*} = \mathcal{S}$ and $\mathcal{S} \neq \varnothing$. So, *card* $\mathcal{S} = $ *card* $X$.  □

Similarly as above, we can prove the following theorem.

**Theorem 3.9.** *Suppose that for the fractional integral equation (3.32) the same conditions as in Theorem 3.5 are satisfied. Then this equation has solutions in $C_{\bar{L}}^q(J, B_R)$. If $\mathcal{S} \subset C_{\bar{L}}^q(J, B_R)$ is its solutions set, then card $\mathcal{S} = $ card $B_R$.*

In order to study data dependence for the solutions set of the fractional integral equation (3.32), we consider both (3.32) and the following fractional integral equation

$$x(t) = x(0) + \frac{1}{\Gamma(q)} \int_0^t (t-s)^{q-1} g(s, x(s))\, ds$$

on $J$ where $g \in C(J \times X, X)$. Let $\mathcal{S}_1$ be the solutions set of this equation.

**Theorem 3.10.** *Suppose the following conditions:*

**(E1)** *there exists a function $l(t) \in C(J, \mathbb{R}_+)$ such that*

$$|f(t, u_1) - f(t, u_2)| \le l(t)|u_1 - u_2| \quad and \quad |g(t, u_1) - g(t, u_2)| \le l(t)|u_1 - u_2|$$

*for all $u_i \in X$ $(i = 1, 2)$ and all $t \in J$;*

**(E2)** *there exist $q_1, q_3 \in (0, q)$ and functions $m(t) \in L^{\frac{1}{q_1}}(J, \mathbb{R}_+)$, $\mu(t) \in L^{\frac{1}{q_3}}(J, \mathbb{R}_+)$ such that*

$$|f(t, x)| \le m(t) \quad and \quad |g(t, x)| \le \mu(t)$$

*for all $x \in X$ and all $t \in J$;*

**(E3)** *there exists a constant $L > 0$ such that*

$$L \ge \frac{2 \max\{M, V\}}{\Gamma(q) \min\left\{(1 + \beta)^{1 - q_1}, (1 + \nu)^{1 - q_3}\right\}};$$

**(E4)** *there exist a constant $q_2 \in (0, q)$ and function $\eta \in L^{\frac{1}{q_2}}(J, \mathbb{R}_+)$*

$$|f(t, u) - g(t, u)| \le \eta(t)$$

*for all $u \in X$ and all $t \in J$;*

**(E5)** $\frac{L_0 T^q}{\Gamma(q+1)} < 1$.

*Then*

$$H_{\|\cdot\|_C}(\mathcal{S}, \mathcal{S}_1) \leq \frac{q N T^{(1+\gamma)(1-q_2)}}{(\Gamma(q+1) - L_0 T^q)(1+\gamma)^{1-q_2}}$$

*where by $H_{\|\cdot\|_C}$ we denote the Pompeiu-Hausdorff functional with respect to $\|\cdot\|_C$ on $C_L^{q-q_1}(J, X)$.*

**Proof.** Consider the operator

$$B_* : (C_L^{q-q_1}(J, X), \|\cdot\|_B) \to (C_L^{q-q_1}(J, X), \|\cdot\|_B)$$

defined by

$$B_* x(t) = x(0) + \frac{1}{\Gamma(q)} \int_0^t (t-s)^{q-1} g(s, x(s)) \, ds, \text{ for } t \in J.$$

Because of (E1)-(E3), $A_*, B_* : (C_L^{q-q_1}(J, X), \|\cdot\|_B) \to (C_L^{q-q_1}(J, X), \|\cdot\|_B)$ are two orbitally continuous operators. Moreover, we have

$$|A_*^2 x(t) - A_* x(t)| \leq \frac{L_0}{\Gamma(q)} \int_0^t (t-s)^{q-1} |A_* x(s) - x(s)| ds$$

$$\leq \frac{L_0 T^q}{\Gamma(q+1)} \|A_* x - x\|_C,$$

for all $x \in C_L^{q-q_1}(J, X)$. Similarly,

$$|B_*^2 x(t) - B_* x(t)| \leq \frac{L_0 T^q}{\Gamma(q+1)} \|B_* x - x\|_C$$

for all $x \in C_L^{q-q_1}(J, X)$. It follows that

$$\|A_*^2 x - A_* x\|_C \leq \frac{L_0 T^q}{\Gamma(q+1)} \|A_* x - x\|_C,$$

$$\|B_*^2 x - B_* x\|_C \leq \frac{L_0 T^q}{\Gamma(q+1)} \|B_* x - x\|_C.$$

Because of (E4),

$$\|A_* x - B_* x\|_C \leq \frac{1}{\Gamma(q)} \int_0^t (t-s)^{q-1} \eta(s) ds$$

$$\leq \frac{N T^{(1+\gamma)(1-q_2)}}{\Gamma(q)(1+\gamma)^{1-q_2}},$$

for all $x \in C_L^{q-q_1}(J, X)$.

By (E5) and applying Theorem 1.14, we obtain

$$H_{\|\cdot\|_C}(F_{A_*}, F_{B_*}) \leq \frac{q N T^{(1+\gamma)(1-q_2)}}{(\Gamma(q+1) - L_0 T^q)(1+\gamma)^{1-q_2}}$$

and the theorem is proved. $\qquad\square$

**Theorem 3.11.** *Suppose the following conditions:*

**(E1)**′ *there exists a constant $L_* > 0$ such that*

$$|f(t, u_1) - f(t, u_2)| \leq L_*|u_1 - u_2| \quad and \quad |g(t, u_1) - g(t, u_2)| \leq L_*|u_1 - u_2|$$

*for all $u_i \in X$ ($i = 1, 2$) and all $t \in J$;*

**(E2)**′ *there exists a constant $M_* > 0$ such that*

$$|f(t, x)| \leq M_* \quad and \quad |g(t, x)| \leq M_*$$

*for all $x \in X$ and all $t \in J$;*

**(E3)**′ *there exists a constant $\bar{L} > 0$ such that*

$$\bar{L} \geq \frac{2M_*}{\Gamma(q + 1)};$$

**(E4)**′ *there exists a constant $\eta_* > 0$ such that*

$$|f(t, u) - g(t, u)| \leq \eta_*$$

*for all $u \in X$ and all $t \in J$;*

**(E5)**′ $\frac{L_* T^q}{\Gamma(q+1)} < 1$.

*Then we have*

$$\bar{H}_{\|\cdot\|_C}(\mathcal{S}, \mathcal{S}_1) \leq \frac{\eta_* T^q}{\Gamma(q + 1) - L_* T^q}$$

*where by $\bar{H}_{\|\cdot\|_C}$ we denote the Pompeiu-Hausdorff functional with respect to $\|\cdot\|_C$ on $C_{\bar{L}}^q(J, X)$.*

**Proof.** Consider the operator

$$\bar{B}_* : (C_{\bar{L}}^q(J, X), \|\cdot\|_B) \to (C_{\bar{L}}^q(J, X), \|\cdot\|_B)$$

defined by

$$\bar{B}_* x(t) = x(0) + \frac{1}{\Gamma(q)} \int_0^t (t - s)^{q-1} g(s, x(s)) \, ds, \text{ for } t \in J.$$

Applying (E1)′-(E3)′, $\bar{A}_*, \bar{B}_* : (C_{\bar{L}}^{q-q_1}(J, X), \|\cdot\|_B) \to (C_{\bar{L}}^{q-q_1}(J, X), \|\cdot\|_B)$ are two orbitally continuous operators. Moreover, we have

$$|\bar{A}_*^2 x(t) - \bar{A}_* x(t)| \leq \frac{L_* T^q}{\Gamma(q + 1)} \|\bar{A}_*(x) - x\|_C,$$

$$|\bar{B}_*^2 x(t) - \bar{B}_* x(t)| \leq \frac{L_* T^q}{\Gamma(q + 1)} \|\bar{B}_*(x) - x\|_C,$$

for all $x \in C_{\bar{L}}^q(J, X)$. It follows that

$$\|\bar{A}_*^2(x) - \bar{A}_*(x)\|_C \leq \frac{L_* T^q}{\Gamma(q + 1)} \|\bar{A}_*(x) - x\|_C$$

$$\|\bar{B}_*^2(x) - \bar{B}_*(x)\|_C \leq \frac{L_* T^q}{\Gamma(q + 1)} \|\bar{B}_*(x) - x\|_C.$$

Because of (E4)′, we obtain

$$\|\bar{A}_*(x) - \bar{B}_*(x)\|_C \leq \frac{1}{\Gamma(q)} \int_0^t (t-s)^{q-1} \eta_* ds \ \leq\ \frac{\eta_* T^q}{\Gamma(q+1)},$$

for all $x \in C_{\bar{L}}^q(J, X)$.

By (E5)′ and applying Theorem 1.14, we obtain the result and the theorem is proved. □

Similarly, we can prove the following theorem.

**Theorem 3.12.** *Suppose the following:*

**(E1)″** *there exists a constant $L_* > 0$ such that*

$$|f(t, u_1) - f(t, u_2)| \leq L_*|u_1 - u_2| \quad and \quad |g(t, u_1) - g(t, u_2)| \leq L_*|u_1 - u_2|$$

*for all $u_i \in B_R$ $(i = 1, 2)$ and all $t \in J$;*
**(E2)″** *there exists a constant $M_*(R) > 0$ such that*

$$|f(t, x)| \leq M_*(R) \quad and \quad |g(t, x)| \leq M_*(R)$$

*for all $x \in B_R$ and all $t \in J$ with*

$$R \geq |x(0)| + \frac{M_*(R)T^q}{\Gamma(q+1)};$$

**(E3)″** *there exists a constant $\bar{L} > 0$ such that*

$$\bar{L} \geq \frac{2M_*(R)}{\Gamma(q+1)};$$

**(E4)″** *there exists a constant $\eta_* > 0$ such that*

$$|f(t, u) - g(t, u)| \leq \eta_*$$

*for all $u \in B_R$ and all $t \in J$;*
**(E5)″** $\frac{L_* T^q}{\Gamma(q+1)} < 1$.

*Then*

$$\bar{H}_{\|\cdot\|_C}(\mathcal{S}, \mathcal{S}_1) \leq \frac{\eta_* T^q}{\Gamma(q+1) - L_* T^q}$$

*where by $\bar{H}_{\|\cdot\|_C}$ we denote the Pompeiu-Hausdorff functional with respect to $\|\cdot\|_C$ on $C_{\bar{L}}^q(J, B_R)$.*

## 3.5   Notes and Remarks

The results in Sections 3.1 and 3.2 are taken from Zhou, Jiao and Pečarić, 2013. The material in Section 3.3 due to Wang, Zhou and Medved, 2012. The main results in Section 3.4 are adopted from Wang, Zhou and Wei, 2013.

# Chapter 4

# Fractional Abstract Evolution Equations

## 4.1 Introduction

The existence of mild solutions for the Cauchy problem of fractional evolution equations has been considered in several recent papers (see, e.g., Agarwal and Shmad, 2011; Belmekki and Benchohra, 2010; Chang, Kavitha and Mallika, 2009; Darwish, Henderson and Ntouyas, 2009; Hernandez, O'Regan and Balachandran, 2010; Hu, Ren and Sakthivel, 2009; Kumar and Sukavanam, 2012; Li, Peng and Jia, 2012; Shu, Lai and Chen, 2011; Wang, Chen and Xiao, 2012; Wang and Zhou, 2011; Wang, Fečkan and Zhou, 2011; Zhou and Jiao, 2010), much less is known about the fractional evolution equations with Riemann-Liouville fractional derivative. In most of the existing articles, Schauder fixed point theorem, Krasnoselskii fixed point theorem or Darbo fixed point theorem, Kuratowski measure of noncompactness are employed to obtain the fixed points of the solution operator of the Cauchy problems under some restrictive conditions. In order to show that the solution operator is compact, a very common approach is to use Arzela-Ascoli theorem. However, it is difficult to check the relative compactness of the solution operator and the equicontinuity of certain family of functions which is given by the solution operator.

In this chapter, we discuss the existence of mild solutions of fractional abstract evolution equations. The suitable mild solutions of fractional evolution equations with Riemann-Liouville fractional derivative and Caputo fractional derivative are introduced respectively. In Sections 4.2 and 4.3, by using the theory of Hausdorff measure of noncompactness, we investigate the existence of mild solutions for the Cauchy problems in the cases $C_0$-semigroup is compact and noncompact, respectively. In Section 4.4, the existence results of mild solutions of nonlocal problem of fractional evolution equations are presented. Section 4.5 is devoted to study of the evolution equations with almost sectorial operator.

## 4.2   Evolution Equations with Riemann-Liouville Derivative

### 4.2.1   *Introduction*

Assume that $X$ is a Banach space with the norm $|\cdot|$. Let $a \in \mathbb{R}^+$, $J = [0, a]$ and $J' = (0, a]$. Denote $C(J, X)$ as the Banach space of continuous functions from $J$ into $X$ with the norm

$$\|x\| = \sup_{t \in [0, a]} |x(t)|,$$

where $x \in C(J, X)$, and $B(X)$ be the space of all bounded linear operators from $X$ to $X$ with the norm $\|Q\|_{B(X)} = \sup\{|Q(x)| \mid |x| = 1\}$, where $Q \in B(X)$ and $x \in X$.

Consider the following nonlocal Cauchy problem of fractional evolution equation with Riemann-Liouville fractional derivative

$$\begin{cases} {}_0D_t^q x(t) - Ax(t) + f(t, x(t)), & \text{a.e. } t \in [0, a], \\ {}_0D_t^{q-1} x(0) + g(x) = x_0, \end{cases} \tag{4.1}$$

where ${}_0D_t^q$ is Riemann-Liouville fractional derivative of order $q$, ${}_0D_t^{q-1}$ is Riemann-Liouville fractional integral of order $1 - q$, $0 < q < 1$, $A$ is the infinitesimal generator of a $C_0$-semigroup of bounded linear operators $\{Q(t)\}_{t \geq 0}$ in Banach space $X$, $f : J \times X \to X$ is a given function, $g : C(J, X) \to L(J, X)$ is a given operator satisfying some assumptions and $x_0$ is an element of the Banach space $X$.

A strong motivation for investigating the nonlocal Cauchy problem (4.1) comes from physics. For example, fractional diffusion equations are abstract partial differential equations that involve fractional derivatives in space and time. They are useful to model anomalous diffusion, where a plume of particles spreads in a different manner than the classical diffusion equation predicts. The time fractional diffusion equation is obtained from the standard diffusion equation by replacing the first-order time derivative with a fractional derivative of order $q \in (0, 1)$, namely

$$\partial_t^q u(z, t) = Au(z, t), \quad t \geq 0, \quad z \in \mathbb{R}.$$

We can take $A = \partial_z^{\beta_1}$, for $\beta_1 \in (0, 1]$, or $A = \partial_z + \partial_z^{\beta_2}$ for $\beta_2 \in (1, 2]$, where $\partial_t^q, \partial_z^{\beta_1}, \partial_z^{\beta_2}$ are the fractional derivatives of order $q, \beta_1, \beta_2$ respectively. We refer the interested reader to Eidelman and Kochubei, 2004; Hanyga, 2002; Hayashi, Kaikina and Naumkin, 2005; Meerschaert, Benson, Scheffler *et al.*, 2002; Schneider and Wayes, 1989; Zaslavsky, 1994 and the references therein for more details.

The nonlocal conditions ${}_0D_t^{q-1} x(0) + g(x) = x_0$ and $x(0) + g(x) = x_0$ can be applied in physics with better effect than the classical initial conditions ${}_0D_t^{q-1} x(0) = x_0$ and $x(0) = x_0$ respectively. For example, $g(x)$ may be given by

$$g(x) = \sum_{i=1}^{m} c_i x(t_i),$$

where $c_i$ $(i = 1, 2, \ldots, n)$ are given constants and $0 < t_1 < t_2 < \cdots < t_n \leq a$. Nonlocal conditions were initiated by Byszewski, 1991, which he proved the existence and uniqueness of mild solutions and classical solutions for nonlocal Cauchy problems. As remarked by Byszewski and Lakshmikantham, 1991, the nonlocal condition can be more useful than the standard initial condition to describe some physical phenomena.

In this section, we study the nonlocal Cauchy problems of fractional evolution equations with Riemann-Liouville fractional derivative by considering an integral equation which is given in terms of probability density. By using the theory of Hausdorff measure of noncompactness, we establish various existence theorems of mild solutions for the Cauchy problem (4.1) in the cases $C_0$-semigroup is compact and noncompact, respectively. Subsection 4.2.2 is devoted to obtain the appropriate definition on the mild solutions of the problem (4.1) by considering an integral equation which is given in terms of probability density. In Subsection 4.2.3, we give some preliminary lemmas. Subsection 4.2.4 provides various existence theorems of mild solutions for the Cauchy problem (4.1) in the case $C_0$-semigroup is compact. In Subsection 4.2.5, we establish various existence theorems of mild solutions for the Cauchy problem (4.1) in the case $C_0$-semigroup is noncompact.

### 4.2.2 Definition of Mild Solutions

The following lemma is the special case of Proposition 1.3.

**Lemma 4.1.** *(Zain and Tazali, 1982)*

**(i)** *Let $\xi, \eta \in \mathbb{R}$ such that $\eta > -1$. If $t > 0$, then*

$$_0D_t^{-\xi}\frac{t^\eta}{\Gamma(\eta+1)} = \begin{cases} \dfrac{t^{\xi+\eta}}{\Gamma(\xi+\eta+1)}, & \text{if } \xi+\eta \neq -n \\ 0, & \text{if } \xi+\eta = -n \end{cases} \quad (n \in \mathbb{N}).$$

**(ii)** *Let $\xi > 0$ and $\varphi \in L((0,a), X)$. Define*

$$G_\xi(t) = {}_0D_t^{-\xi}\varphi, \quad \text{for } t \in (0,a),$$

*then*

$$_0D_t^{-\eta}G_\xi(t) = {}_0D_t^{-(\xi+\eta)}\varphi(t), \quad \eta > 0, \quad \text{almost all } t \in [0,a].$$

**Lemma 4.2.** *The nonlocal Cauchy problem (4.1) is equivalent to the integral equation*

$$x(t) = \frac{t^{q-1}}{\Gamma(q)}(x_0 - g(x)) + \frac{1}{\Gamma(q)}\int_0^t (t-s)^{q-1}[Ax(s) + f(s,x(s))]ds, \quad \text{for } t \in (0,a],$$
(4.2)

*provided that the integral in (4.2) exists.*

**Proof.** Suppose (4.2) is true, then

$$_0D_t^{q-1}x(t) = {}_0D_t^{q-1}\left(\frac{t^{q-1}}{\Gamma(q)}(x_0 - g(x)) + \frac{1}{\Gamma(q)}\int_0^t (t-\tau)^{q-1}(Ax(\tau) + f(\tau, x(\tau)))d\tau\right),$$

applying Lemma 4.1 we obtain that

$$_0D_t^{q-1}x(t) = x_0 - g(x) + \int_0^t (Ax(s) + f(s, x(s)))]ds, \quad \text{almost all } t \in [0, a],$$

and this proves that $_0D_t^{q-1}x(t)$ is absolutely continuous on $[0, a]$. Then we have

$$_0D_t^q x(t) = \frac{d}{dt} {}_0D_t^{q-1}x(t) = Ax(t) + f(t, x(t)), \quad \text{almost all } t \in [0, a]$$

and

$$_0D_t^{q-1}x(0) + g(x) = x_0.$$

The proof of the converse is given as follows.

Suppose (4.1) is true, then

$$_0D_t^{-q}({}_0D_t^q x(t)) = {}_0D_t^{-q}(Ax(t) + f(t, x(t))).$$

Since

$$_0D_t^{-q}({}_0D_t^q x(t)) = x(t) - \frac{t^{q-1}}{\Gamma(q)} {}_0D_t^{q-1}x(0)$$

$$= x(t) - \frac{t^{q-1}}{\Gamma(q)}(x_0 - g(x)), \quad \text{for } t \in (0, a],$$

then we have

$$x(t) = \frac{t^{q-1}}{\Gamma(q)}(x_0 - g(x)) + {}_0D_t^{-q}(Ax(t) + f(t, x(t)))$$

$$= \frac{t^{q-1}}{\Gamma(q)}(x_0 - g(x)) + \frac{1}{\Gamma(q)}\int_0^t (t - s)^{q-1}(Ax(s) + f(s, x(s)))ds, \quad \text{for } t \in (0, a].$$

The proof is completed.                                                                    □

Before giving the definition of mild solution of (4.1), we firstly prove the following lemma.

**Lemma 4.3.** *If*

$$x(t) = \frac{t^{q-1}}{\Gamma(q)}(x_0 - g(x)) + \frac{1}{\Gamma(q)}\int_0^t (t - s)^{q-1}[Ax(s) + f(s, x(s))]ds, \quad \text{for } t > 0 \quad (4.3)$$

*holds, then we have*

$$x(t) = t^{q-1}P_q(t)(x_0 - g(x)) + \int_0^t (t - s)^{q-1}P_q(t - s)f(s, x(s))ds, \quad \text{for } t > 0,$$

*where*

$$P_q(t) = \int_0^\infty q\theta M_q(\theta)Q(t^q\theta)d\theta.$$

**Proof.** Let $\lambda > 0$. Applying the Laplace transform

$$\nu(\lambda) = \int_0^\infty e^{-\lambda s} x(s) ds \quad \text{and} \quad \omega(\lambda) = \int_0^\infty e^{-\lambda s} f(s, x(s)) ds, \quad \text{for } \lambda > 0$$

to (4.3), we have

$$\nu(\lambda) = \frac{1}{\lambda^q}(x_0 - g(x)) + \frac{1}{\lambda^q} A\nu(\lambda) + \frac{1}{\lambda^q}\,\omega(\lambda)$$

$$= (\lambda^q I - A)^{-1}(x_0 - g(x)) + (\lambda^q I - A)^{-1}\omega(\lambda)$$

$$= \int_0^\infty e^{-\lambda^q s} Q(s)(x_0 - g(x)) ds + \int_0^\infty e^{-\lambda^q s} Q(s)\omega(\lambda) ds, \qquad (4.4)$$

provided that the integrals in (4.4) exist, where $I$ is the identity operator defined on $X$.

Set

$$\psi_q(\theta) = \frac{q}{\theta^{q+1}} M_q(\theta^{-q}),$$

whose Laplace transform is given by

$$\int_0^\infty e^{-\lambda\theta} \psi_q(\theta) d\theta = e^{-\lambda^q}, \quad \text{where } q \in (0,1). \qquad (4.5)$$

Using (4.5), we get

$$\int_0^\infty e^{-\lambda^q s} Q(s)(x_0 - g(x)) ds = \int_0^\infty q t^{q-1} e^{-(\lambda t)^q} Q(t^q)(x_0 - g(x)) dt$$

$$= \int_0^\infty \int_0^\infty q\psi_q(\theta) e^{-(\lambda t\theta)} Q(t^q) t^{q-1}(x_0 - g(x)) d\theta dt$$

$$= \int_0^\infty \int_0^\infty q\psi_q(\theta) e^{-\lambda t} Q\left(\frac{t^q}{\theta^q}\right) \frac{t^{q-1}}{\theta^q}(x_0 - g(x)) d\theta dt$$

$$= \int_0^\infty e^{-\lambda t}\left[q \int_0^\infty \psi_q(\theta) Q\left(\frac{t^q}{\theta^q}\right) \frac{t^{q-1}}{\theta^q}(x_0 - g(x)) d\theta\right] dt, \qquad (4.6)$$

$$\int_0^\infty e^{-\lambda^q s} Q(s)\omega(\lambda) ds$$

$$= \int_0^\infty \int_0^\infty q t^{q-1} e^{-(\lambda t)^q} Q(t^q) e^{-\lambda s} f(s, x(s)) ds dt$$

$$= \int_0^\infty \int_0^\infty \int_0^\infty q\psi_q(\theta) e^{-(\lambda t\theta)} Q(t^q) e^{-\lambda s} t^{q-1} f(s, x(s)) d\theta ds dt$$

$$= \int_0^\infty \int_0^\infty \int_0^\infty q\psi_q(\theta) e^{-\lambda(t+s)} Q\left(\frac{t^q}{\theta^q}\right) \frac{t^{q-1}}{\theta^q} f(s, x(s)) d\theta ds dt$$

$$= \int_0^\infty e^{-\lambda t}\left[q \int_0^t \int_0^\infty \psi_q(\theta) Q\left(\frac{(t-s)^q}{\theta^q}\right) \frac{(t-s)^{q-1}}{\theta^q} f(s, x(s)) d\theta ds\right] dt. \qquad (4.7)$$

According to (4.6) and (4.7), we have

$$\nu(\lambda) = \int_0^\infty e^{-\lambda t}\left[q \int_0^\infty \psi_q(\theta) Q\left(\frac{t^q}{\theta^q}\right) \frac{t^{q-1}}{\theta^q}(x_0 - g(x)) d\theta\right.$$

$$\left. + q \int_0^t \int_0^\infty \psi_q(\theta) Q\left(\frac{(t-s)^q}{\theta^q}\right) \frac{(t-s)^{q-1}}{\theta^q} f(s, x(s)) d\theta ds\right] dt.$$

Now we can invert the last Laplace transform to get

$$x(t) = q \int_0^\infty \theta t^{q-1} M_q(\theta) Q(t^q \theta)(x_0 - g(x)) d\theta$$

$$+ q \int_0^t \int_0^\infty \theta(t-s)^{q-1} M_q(\theta) Q((t-s)^q \theta) f(s, x(s)) d\theta ds$$

$$= t^{q-1} P_q(t)(x_0 - g(x)) + \int_0^t (t-s)^{q-1} P_q(t-s) f(s, x(s)) ds.$$

The proof is completed.                                                    □

Due to Lemma 4.3, we give the following definition of the mild solution of (4.1).

**Definition 4.1.** By the mild solution of the nonlocal Cauchy problem (4.1), we mean that the function $x \in C(J', X)$ which satisfies

$$x(t) = t^{q-1} P_q(t)(x_0 - g(x)) + \int_0^t (t-s)^{q-1} P_q(t-s) f(s, x(s)) ds, \quad \text{for } t \in (0, a].$$

Suppose that $A$ is the infinitesimal generator of a $C_0$-semigroup $\{Q(t)\}_{t\geq 0}$ of uniformly bounded linear operators on Banach space $X$. This means that there exists $M > 1$ such that $M = \sup_{t \in [0,\infty)} \|Q(t)\|_{B(X)} < \infty$.

**Proposition 4.1.** *(Zhou and Jiao, 2010a) For any fixed $t > 0$, $P_q(t)$ is linear and bounded operator, i.e., for any $x \in X$*

$$|P_q(t)x| \leq \frac{M}{\Gamma(q)}|x|.$$

**Proposition 4.2.** *(Zhou and Jiao, 2010a) Operator $\{P_q(t)\}_{t>0}$ is strongly continuous, which means that, for all $x \in X$ and $0 < t' < t'' \leq a$, we have*

$$|P_q(t'')x - P_q(t')x| \to 0 \quad \text{as } t'' \to t'.$$

**Proposition 4.3.** *(Zhou and Jiao, 2010a) Assume that $\{Q(t)\}_{t>0}$ is compact operator. Then $\{P_q(t)\}_{t>0}$ is also compact operator.*

**Proposition 4.4.** *(Pazy, 1983) Assume that $\{Q(t)\}_{t>0}$ is compact operator. Then $\{Q(t)\}_{t>0}$ is equicontinuous.*

### 4.2.3  *Preliminary Lemmas*

Define

$$X^{(q)}(J') = \left\{ x \in C(J', X) : \lim_{t \to 0+} t^{1-q} x(t) \text{ exists and is finite} \right\}.$$

For any $x \in X^{(q)}(J')$, let the norm $\| \cdot \|_q$ defined by

$$\|x\|_q = \sup_{t \in (0,a]} \{t^{1-q}|x(t)|\}.$$

Then $(X^{(q)}(J'), \| \cdot \|_q)$ is a Banach space.

For $r > 0$, define a closed subset $B_r^{(q)}(J') \subset X^{(q)}(J')$ as follows

$$B_r^{(q)}(J') = \{x \in X^{(q)}(J') : \|x\|_q \le r\}.$$

Thus, $B_r^{(q)}(J')$ is a bounded closed and convex subset of $X^{(q)}(J')$.

Let $B(J)$ be the closed ball of the space $C(J, X)$ with radius $r$ and center at $0$, that is

$$B(J) = \{y \in C(J, X) : \|y\| \le r\}.$$

Thus $B(J)$ is a bounded closed and convex subset of $C(J, X)$.

We introduce the following hypotheses:

**(H0)** $Q(t)(t > 0)$ is equicontinuous, i.e., $Q(t)$ is continuous in the uniform operator topology for $t > 0$;

**(H1)** for each $t \in J'$, the function $f(t, \cdot) : X \to X$ is continuous and for each $x \in X$, the function $f(\cdot, x) : J' \to X$ is strongly measurable;

**(H2)** there exists a function $m \in L(J', \mathbb{R}^+)$ such that

$$_0D_t^{-q}m \in C(J', \mathbb{R}^+), \quad \lim_{t \to 0+} t^{1-q} {}_0D_t^{-q}m(t) = 0,$$

and

$$|f(t, x)| \le m(t) \text{ for all } x \in B_r^{(q)}(J') \text{ and almost all } t \in [0, a];$$

**(H3)** there exists a constant $L \in (0, \frac{\Gamma(q)}{M})$ such that the operator $g : C(J', X) \to L(J', X)$ satisfies

$$|g(x_1) - g(x_2)| \le L\|x_1 - x_2\|_q, \text{ for } x_1, x_2 \in B_r^{(q)}(J');$$

**(H4)** there exists a constant $r > 0$ such that

$$\frac{M}{\Gamma(q) - ML}\left(|x_0| + |g(0)| + \sup_{t \in (0,a]}\left\{t^{1-q}\int_0^t (t-s)^{q-1}m(s)ds\right\}\right) \le r;$$

**(H3)$'$** the operator $g : C(J', X) \to L(J', X)$ is a continuous and compact map, and there exist positive constants $L_1, L_2$ such that $L_1 \in (0, \frac{\Gamma(q)}{M})$ and $|g(x)| \le L_1\|x\|_q + L_2$ for all $x \in B_r^{(q)}(J')$;

**(H4)$'$** there exists a constant $r > 0$ such that

$$\frac{M}{\Gamma(q) - ML_1}\left(|x_0| + L_2 + \sup_{t \in (0,a]}\left\{t^{1-q}\int_0^t (t-s)^{q-1}m(s)ds\right\}\right) \le r.$$

For any $x \in B_r^{(q)}(J')$, define an operator $T$ as follows

$$(Tx)(t) = (T_1x)(t) + (T_2x)(t),$$

where

$$(T_1x)(t) = t^{q-1}P_q(t)(x_0 - g(x)), \qquad\qquad \text{for } t \in (0, a],$$

$$(T_2x)(t) = \int_0^t (t-s)^{q-1}P_q(t-s)f(s, x(s))ds, \quad \text{for } t \in (0, a].$$

It is easy to see that $\lim_{t\to 0+} t^{1-q}(Tx)(t) = \frac{x_0 - g(x)}{\Gamma(q)}$. For any $y \in B(J)$, set

$$x(t) = t^{q-1}y(t), \quad \text{for } t \in (0, a].$$

Then, $x \in B_r^{(q)}(J')$. Define $\mathscr{T}$ as follows

$$(\mathscr{T}y)(t) = (\mathscr{T}_1 y)(t) + (\mathscr{T}_2 y)(t),$$

where

$$(\mathscr{T}_1 y)(t) = \begin{cases} t^{1-q}(T_1 x)(t), & \text{for } t \in (0, a], \\ \dfrac{x_0 - g(x)}{\Gamma(q)}, & \text{for } t = 0, \end{cases}$$

$$(\mathscr{T}_2 y)(t) = \begin{cases} t^{1-q}(T_2 x)(t), & \text{for } t \in (0, a], \\ 0, & \text{for } t = 0. \end{cases}$$

Obviously, $x$ is a mild solution of (4.1) in $B_r^{(q)}(J')$ if and only if the operator equation $x = Tx$ has a solution $x \in B_r^{(q)}(J')$. Before giving the main results, we firstly prove the following lemmas.

**Lemma 4.4.** *Assume that (H0)-(H4) hold, then $\{\mathscr{T}y : y \in B(J)\}$ is equicontinuous.*

**Proof. Claim I.** $\{\mathscr{T}_1 y : y \in B(J)\}$ is equicontinuous.

For any $y \in B(J)$, let $x(t) = t^{q-1}y(t)$, $t \in (0, a]$. Then $x \in B_r^{(q)}(J')$. For $t_1 = 0$, $0 < t_2 \le a$, we get

$$\begin{aligned}
|(\mathscr{T}_1 y)(t_2) - (\mathscr{T}_1 y)(0)| &\le \left| P_q(t_2)(x_0 - g(x)) - \frac{x_0 - g(x)}{\Gamma(q)} \right| \\
&\le \left| \left( P_q(t_2) - \frac{1}{\Gamma(q)} \right)(x_0 - g(x)) \right| \\
&\le \left| \left( P_q(t_2) - \frac{1}{\Gamma(q)} \right) \right| (|x_0| + L\|x\|_q + |g(0)|) \\
&\le \left| \left( P_q(t_2) - \frac{1}{\Gamma(q)} \right) \right| (|x_0| + Lr + |g(0)|) \\
&\to 0, \quad \text{as } t_2 \to 0.
\end{aligned}$$

For $0 < t_1 < t_2 \le a$, we get

$$\begin{aligned}
|(\mathscr{T}_1 y)(t_2) - (\mathscr{T}_1 y)(t_1)| &\le |P_q(t_2)(x_0 - g(x)) - P_q(t_1)(x_0 - g(x))| \\
&\le |(P_q(t_2) - P_q(t_1))(x_0 - g(x))| \\
&\le |(P_q(t_2) - P_q(t_1))|(|x_0| + L\|x\|_q + |g(0)|) \\
&\le |(P_q(t_2) - P_q(t_1))|(|x_0| + Lr + |g(0)|) \\
&\to 0, \quad \text{as } t_2 \to t_1.
\end{aligned}$$

Hence, $\{\mathscr{T}_1 y : y \in B(J)\}$ is equicontinuous.

**Claim II.** $\{\mathscr{T}_2 y : y \in B(J)\}$ is equicontinuous.

For any $y \in B(J)$, let $x(t) = t^{q-1}y(t)$, $t \in (0, a]$. Then $x \in B_r^{(q)}(J')$. For $t_1 = 0$, $0 < t_2 \leq a$, we get

$$|(\mathscr{T}_2 y)(t_2) - (\mathscr{T}_2 y)(0)| = \left| t_2^{1-q} \int_0^{t_2} (t_2 - s)^{q-1} P_q(t_2 - s) f(s, x(s)) ds \right|$$

$$\leq \frac{M}{\Gamma(q)} t_2^{1-q} \int_0^{t_2} (t_2 - s)^{q-1} m(s) ds$$

$$\to 0, \quad \text{as } t_2 \to 0.$$

For $0 < t_1 < t_2 \leq a$, we have

$$|(\mathscr{T}_2 y)(t_2) - (\mathscr{T}_2 y)(t_1)|$$

$$\leq \left| \int_{t_1}^{t_2} t_2^{1-q}(t_2 - s)^{q-1} P_q(t_2 - s) f(s, x(s)) ds \right|$$

$$+ \left| \int_0^{t_1} t_2^{1-q}(t_2 - s)^{q-1} P_q(t_2 - s) f(s, x(s)) ds \right.$$

$$- \int_0^{t_1} t_1^{1-q}(t_1 - s)^{q-1} P_q(t_2 - s) f(s, x(s)) ds \Big|$$

$$+ \left| \int_0^{t_1} t_1^{1-q}(t_1 - s)^{q-1} P_q(t_2 - s) f(s, x(s)) ds \right.$$

$$- \int_0^{t_1} t_1^{1-q}(t_1 - s)^{q-1} P_q(t_1 - s) f(s, x(s)) ds \Big|$$

$$\leq \frac{M}{\Gamma(q)} \left| \int_{t_1}^{t_2} t_2^{1-q}(t_2 - s)^{q-1} m(s) ds \right|$$

$$+ \frac{M}{\Gamma(q)} \int_0^{t_1} \left( t_1^{1-q}(t_1 - s)^{q-1} - t_2^{1-q}(t_2 - s)^{q-1} \right) m(s) ds$$

$$+ \left| \int_0^{t_1} t_1^{1-q}(t_1 - s)^{q-1} \left( P_q(t_2 - s) f(s, x(s)) - P_q(t_1 - s) f(s, x(s)) \right) ds \right|$$

$$\leq I_1 + I_2 + I_3,$$

where

$$I_1 = \frac{M}{\Gamma(q)} \left| \int_0^{t_2} t_2^{1-q}(t_2 - s)^{q-1} m(s) ds - \int_0^{t_1} t_1^{1-q}(t_1 - s)^{q-1} m(s) ds \right|,$$

$$I_2 = \frac{2M}{\Gamma(q)} \int_0^{t_1} \left( t_1^{1-q}(t_1 - s)^{q-1} - t_2^{1-q}(t_2 - s)^{q-1} \right) m(s) ds,$$

$$I_3 = \left| \int_0^{t_1} t_1^{1-q}(t_1 - s)^{q-1} \left( P_q(t_2 - s) - P_q(t_1 - s) \right) f(s, x(s)) ds \right|.$$

One can reduce that $\lim_{t_2 \to t_1} I_1 = 0$, since $_0 D_t^{-q} m \in C(J', \mathbb{R}^+)$. Noting that

$$(t_1^{1-q}(t_1 - s)^{q-1} - t_2^{1-q}(t_2 - s)^{q-1}) m(s) \leq t_1^{1-q}(t_1 - s)^{q-1} m(s),$$

and $\int_0^{t_1} t_1^{1-q}(t_1-s)^{q-1}m(s)ds$ exists ($s \in [0,t_1]$), then by Lebesgue dominated convergence theorem, we have

$$\int_0^{t_1}(t_1^{1-q}(t_1-s)^{q-1}-t_2^{1-q}(t_2-s)^{q-1})m(s)ds \to 0, \quad \text{as } t_2 \to t_1,$$

then one can deduce that $\lim_{t_2 \to t_1} I_2 = 0$.

For $\varepsilon > 0$ be enough small, we have

$$I_3 \leq \int_0^{t_1-\varepsilon} t_1^{1-q}(t_1-s)^{q-1}\|P_q(t_2-s)-P_q(t_1-s)\|_{B(X)} \ |f(s,x(s))|ds$$

$$+ \int_{t_1-\varepsilon}^{t_1} t_1^{1-q}(t_1-s)^{q-1}\|P_q(t_2-s)-P_q(t_1-s)\|_{B(X)} \ |f(s,x(s))|ds$$

$$\leq t_1^{1-q}\int_0^{t_1}(t_1-s)^{q-1}m(s)ds \sup_{s\in[0,t_1-\varepsilon]}\|P_q(t_2-s)-P_q(t_1-s)\|_{B(X)}$$

$$+ \frac{2M}{\Gamma(q)}\int_{t_1-\varepsilon}^{t_1} t_1^{1-q}(t_1-s)^{q-1}m(s)ds$$

$$\leq I_{31}+I_{32}+I_{33},$$

where

$$I_{31} = \frac{r\Gamma(q)}{M}\sup_{s\in[0,t_1-\varepsilon]}\|P_q(t_2-s)-P_q(t_1-s)\|_{B(X)},$$

$$I_{32} = \frac{2M}{\Gamma(q)}\left|\int_0^{t_1} t_1^{1-q}(t_1-s)^{q-1}m(s)ds - \int_0^{t_1-\varepsilon}(t_1-\varepsilon)^{1-q}(t_1-\varepsilon-s)^{q-1}m(s)ds\right|,$$

$$I_{33} = \frac{2M}{\Gamma(q)}\int_0^{t_1-\varepsilon}((t_1-\varepsilon)^{1-q}(t_1-\varepsilon-s)^{q-1}-t_1^{1-q}(t_1-s)^{q-1})m(s)ds.$$

By (H0), it is easy to see that $I_{31} \to 0$ as $t_2 \to t_1$. Similar to the proof that $I_1, I_2$ tend to zero, we get $I_{32} \to 0$ and $I_{33} \to 0$ as $\varepsilon \to 0$. Thus, $I_3$ tends to zero independently of $y \in B(J)$ as $t_2 \to t_1$, $\varepsilon \to 0$. Therefore, $|(\mathscr{T}_2y)(t_2)-(\mathscr{T}_2y)(t_1)|$ tends to zero independently of $y \in B(J)$ as $t_2 \to t_1$, which means that $\{\mathscr{T}_2y : y \in B(J)\}$ is equicontinuous.

Therefore, $\{\mathscr{T}y : y \in B(J)\}$ is equicontinuous. $\qquad\square$

**Lemma 4.5.** *Assume that (H1)-(H4) hold. Then $\mathscr{T}$ maps $B(J)$ into $B(J)$, and $\mathscr{T}$ is continuous in $B(J)$.*

**Proof. Claim I.** $\mathscr{T}$ maps $B(J)$ into $B(J)$. For any $y \in B(J)$, let $x(t) = t^{q-1}y(t)$. Then $x \in B_r^{(q)}(J')$.

For $t \in [0,a]$, by (H1)-(H4), we have

$$|(\mathscr{T}y)(t)| \leq |P_q(t)(x_0-g(x))| + t^{1-q}\left|\int_0^t(t-s)^{q-1}P_q(t-s)f(s,x(s))ds\right|$$

$$\leq \frac{M}{\Gamma(q)}(|x_0|+L\|x\|_q+|g(0)|) + \frac{Mt^{1-q}}{\Gamma(q)}\int_0^t(t-s)^{q-1}|f(s,x(s))|ds$$

$$\leq \frac{M}{\Gamma(q)}\left(|x_0| + Lr + |g(0)| + \sup_{t\in[0,a]}\left\{t^{1-q}\int_0^t (t-s)^{q-1}m(s)ds\right\}\right)$$

$$\leq r.$$

Hence, $\|\mathscr{T}y\| \leq r$, for any $y \in B(J)$.

**Claim II.** $\mathscr{T}$ is continuous in $B(J)$. For any $y_m, y \in B(J)$, $m = 1, 2, \ldots$, with $\lim_{m\to\infty} y_m = y$, we have

$$\lim_{m\to\infty} y_m(t) = y(t) \quad \text{and} \quad \lim_{m\to\infty} t^{q-1}y_m(t) = t^{q-1}y(t), \quad \text{for } t \in (0, a].$$

Then by (H1), we have

$$f(t, x_m(t)) = f(t, t^{q-1}y_m(t)) \to f(t, t^{q-1}y(t)) = f(t, x(t)), \quad \text{as } m \to \infty,$$

where $x_m(t) = t^{q-1}y_m(t)$ and $x(t) = t^{q-1}y(t)$.

On the one hand, using (H2), we get for each $t \in J'$,

$$(t-s)^{q-1}|f(s, x_m(s)) - f(s, x(s))| \leq (t-s)^{q-1}2m(s), \quad \text{a.e. in } [0, t].$$

On the other hand, the function $s \to (t-s)^{q-1}2m(s)$ is integrable for $s \in [0, t]$ and $t \in J$. By Lebesgue dominated convergence theorem, we get

$$\int_0^t (t-s)^{q-1}|f(s, x_m(s)) - f(s, x(s))|ds \to 0, \quad \text{as } m \to 0.$$

For $t \in [0, a]$

$$|(\mathscr{T}y_m)(t) - (\mathscr{T}y)(t)| = |t^{1-q}(Tx_m(t) - Tx(t))|$$

$$\leq |P_q(t)(g(x_m) - g(x))| + t^{1-q}\left|\int_0^t (t-s)^{q-1}P_q(t-s)(f(s, x_m(s)) - f(s, x(s)))ds\right|$$

$$\leq \frac{ML}{\Gamma(q)}\|x_m - x\|_q + \frac{Mt^{1-q}}{\Gamma(q)}\int_0^t (t-s)^{q-1}|f(s, x_m(s)) - f(s, x(s))|ds$$

$$\leq \frac{ML}{\Gamma(q)}\|y_m - y\| + \frac{Mt^{1-q}}{\Gamma(q)}\int_0^t (t-s)^{q-1}|f(s, x_m(s)) - f(s, x(s))|ds.$$

Therefore, $\mathscr{T}y_m \to \mathscr{T}y$ pointwise on $J$ as $m \to \infty$, by which Lemma 4.4 implies that $\mathscr{T}y_m \to \mathscr{T}y$ uniformly on $J$ as $m \to \infty$ and so $\mathscr{T}$ is continuous. $\square$

**Lemma 4.6.** *Assume that (H0)-(H2), (H3)′ and (H4)′ hold. Then $\{\mathscr{T}y : y \in B(J)\}$ is equicontinuous.*

**Proof.** For any $y \in B(J)$, for $t_1 = 0$, $0 < t_2 \leq a$, then, we get

$$|(\mathscr{T}y)(t_2) - (\mathscr{T}y)(0)|$$

$$\leq \left|P_q(t_2)(x_0 - g(x)) - \frac{x_0 - g(x)}{\Gamma(q)}\right| + \left|t_2^{1-q}\int_0^{t_2} (t_2-s)^{q-1}P_q(t_2-s)f(s, x(s))ds\right|$$

$$\leq \left|P_q(t_2)(x_0 - g(x)) - \frac{x_0 - g(x)}{\Gamma(q)}\right| + \frac{M}{\Gamma(q)}t_2^{1-q}\int_0^{t_2} (t_2-s)^{q-1}m(s)ds$$

$$\to 0, \quad \text{as } t_2 \to 0.$$

For any $y \in B(J)$ and $0 < t_1 < t_2 \leq a$, we get

$$|(\mathscr{T}y)(t_2) - (\mathscr{T}y)(t_1)| \leq |(\mathscr{T}_1 y)(t_2) - (\mathscr{T}_1 y)(t_1)| + |(\mathscr{T}_2 y)(t_2) - (\mathscr{T}_2 y)(t_1)|$$
$$\leq |(P_q(t_2) - P_q(t_1))(x_0 - g(x))| + I_1 + I_2 + I_3,$$

where $I_1, I_2$ and $I_3$ are defined as in the proof of Lemma 4.4. According to Proposition 4.2, we know that $|(\mathscr{T}y)(t_2) - (\mathscr{T}y)(t_1)|$ tends to zero independently of $y \in B(J)$ as $t_2 \to t_1$, which means that $\{\mathscr{T}y : y \in B(J)\}$ is equicontinuous. $\qquad \square$

**Lemma 4.7.** *Assume that (H1), (H2), (H3)' and (H4)' hold. Then $\mathscr{T}$ maps $B(J)$ into $B(J)$, and $\mathscr{T}$ is continuous in $B(J)$.*

**Proof.** For any $y \in B(J)$, we have

$$|(\mathscr{T}y)(t)| \leq \frac{|x_0| + L_1 r + L_2}{\Gamma(q)} \leq r, \quad \text{for } t = 0$$

and

$$|(\mathscr{T}y)(t)| = t^{1-q}|(Tx)(t)| \leq r, \quad \text{for } t \in (0, a].$$

Hence, $\|\mathscr{T}y\|_B \leq r$, for any $y \in B(J)$. Using the similar argument as that we did in the proof of Lemma 4.5, we know that $\mathscr{T}$ is continuous in $B(J)$. $\qquad \square$

### 4.2.4  Compact Semigroup Case

In the following, we suppose that the operator $A$ generates a compact $C_0$-semigroup $\{Q(t)\}_{t \geq 0}$ on $X$, that is, for any $t > 0$, the operator $Q(t)$ is compact.

**Theorem 4.1.** *Assume that $Q(t)(t > 0)$ is compact. Furthermore assume that (H1)-(H4) hold. Then the nonlocal Cauchy problem (4.1) has at least one mild solution in $B_r^{(q)}(J')$.*

**Proof.** Obviously, $x$ is a mild solution of (4.1) in $B_r^{(q)}(J')$ if and only if $y$ is a fixed point of $y = \mathscr{T}y$ in $B(J)$, where $x(t) = t^{q-1}y(t)$. So, it is enough to prove that $y = \mathscr{T}y$ has a fixed point in $B(J)$.

For any $y_1, y_2 \in B(J)$, according to (H3), we have

$$|\mathscr{T}_1 y_1(t) - \mathscr{T}_1 y_2(t)| = t^{1-q}|(T_1 x_1)(t) - (T_1 x_2)(t)|$$
$$\leq \frac{M}{\Gamma(q)}|g(x_1) - g(x_2)|$$
$$\leq \frac{ML}{\Gamma(q)}\|x_1 - x_2\|_q$$
$$= \frac{ML}{\Gamma(q)}\|y_1 - y_2\|$$

which implies that $\|\mathscr{T}_1 y_1 - \mathscr{T}_1 y_2\| \leq \frac{ML}{\Gamma(q)}\|y_1 - y_2\|$. Thus, we obtain that

$$\alpha(\mathscr{T}_1(B(J))) \leq \frac{ML}{\Gamma(q)}\alpha(B(J)). \qquad (4.8)$$

Next, we show that for any $t \in [0,a]$, $V(t) = \{(\mathscr{T}_2 y)(t), y \in B(J)\}$ is relatively compact in $X$. Obviously, $V(0)$ is relatively compact in $X$. Let $t \in (0,a]$ be fixed. For every $\varepsilon \in (0,t)$ and $\delta > 0$, define an operator $\mathscr{T}_{\varepsilon,\delta}$ on $B(J)$ by the formula

$$(\mathscr{T}_{\varepsilon,\delta} y)(t) = qt^{1-q} \int_0^{t-\varepsilon} \int_\delta^\infty \theta(t-s)^{q-1} M_q(\theta) Q((t-s)^q \theta) f(s, x(s)) d\theta ds$$

$$= qt^{1-q} Q(\varepsilon^q \delta) \int_0^{t-\varepsilon} \int_\delta^\infty \theta(t-s)^{q-1} M_q(\theta) Q((t-s)^q \theta - \varepsilon^q \delta) f(s, x(s)) d\theta ds,$$

where $x \in B_r^{(q)}(J')$. Then from the compactness of $Q(\varepsilon^q \delta)(\varepsilon^q \delta > 0)$, we obtain that the set $V_{\varepsilon,\delta}(t) = \{(\mathscr{T}_{\varepsilon,\delta} y)(t), y \in B(J)\}$ is relatively compact in $X$ for every $\varepsilon \in (0,t)$ and $\delta > 0$. Moreover, for every $y \in B(J)$, we have

$$|(\mathscr{T}_2 y)(t) - (\mathscr{T}_{\varepsilon,\delta} y)(t)|$$

$$\leq \left| qt^{1-q} \int_0^t \int_0^\delta \theta(t-s)^{q-1} M_q(\theta) Q((t-s)^q \theta) f(s, x(s)) d\theta ds \right|$$

$$+ \left| qt^{1-q} \int_{t-\varepsilon}^t \int_\delta^\infty \theta(t-s)^{q-1} M_q(\theta) Q((t-s)^q \theta) f(s, x(s)) d\theta ds \right|$$

$$\leq qMt^{1-q} \int_0^t (t-s)^{q-1} m(s) ds \int_0^\delta \theta M_q(\theta) d\theta$$

$$+ qMt^{1-q} \int_{t-\varepsilon}^t (t-s)^{q-1} m(s) ds \int_0^\infty \theta M_q(\theta) d\theta$$

$$\leq qMt^{1-q} \int_0^t (t-s)^{q-1} m(s) ds \int_0^\delta \theta M_q(\theta) d\theta + \frac{M}{\Gamma(q)} t^{1-q} \int_{t-\varepsilon}^t (t-s)^{q-1} m(s) ds$$

$$\to 0, \quad \text{as } \varepsilon \to 0, \ \delta \to 0.$$

Therefore, there are relatively compact sets arbitrarily close to the set $V(t)$, $t > 0$. Hence the set $V(t)$, $t > 0$ is also relatively compact in $X$. Therefore, $\{(\mathscr{T}_2 y)(t), y \in B(J)\}$ is relatively compact by Arzela-Ascoli theorem. Thus, we have $\alpha(T_2(B_r^{(q)}(J'))) = 0$. By (4.8), we have

$$\alpha(\mathscr{T}(B(J))) \leq \alpha(\mathscr{T}_1(B(J))) + \alpha(\mathscr{T}_2(B(J)))$$

$$\leq \frac{ML}{\Gamma(q)} \alpha(B(J)).$$

Thus, the operator $\mathscr{T}$ is an $\alpha$-contraction in $B(J)$. By Lemma 4.5, we know that $\mathscr{T}$ is continuous. Hence, Theorem 1.10 shows that $\mathscr{T}$ has a fixed point $y^* \in B(J)$. Let $x^*(t) = t^{q-1} y^*(t)$. Then $x^*$ is a mild solution of (4.1). $\qquad\square$

**Theorem 4.2.** *Assume that $Q(t)(t > 0)$ is compact. Furthermore assume that (H1), (H2), (H3)′ and (H4)′ hold. Then the nonlocal Cauchy problem (4.1) has at least one mild solution in $B_r^{(q)}(J')$.*

**Proof.** Since Proposition 4.4, $Q(t)(t > 0)$ is equicontinuous, which implies (H0) is satisfied. Then, by Lemmas 4.4-4.5, we know that $\mathscr{T} : B(J) \to B(J)$ is bounded, continuous and $\{\mathscr{T} y : y \in B(J)\}$ is equicontinuous.

According to the argument of Theorem 4.1, we only need prove that for any $t \in J$, the set $V_1(t) = \{(\mathscr{T}_1 y)(t) : y \in B(J)\}$ is relatively compact in $X$. Obviously, $V_1(0)$ is relatively compact in $X$. Let $0 < t \le a$ be fixed. For every $\delta > 0$, define an operator $\mathscr{T}_1^\delta$ on $B(J)$ by the formula

$$(\mathscr{T}_1^\delta y)(t) = q \int_\delta^\infty \theta M_q(\theta) Q(t^q \theta)(x_0 - g(x)) d\theta$$

$$= qQ(t^q \delta) \int_\delta^\infty \theta M_q(\theta) Q(t^q \theta - t^q \delta)(x_0 - g(x)) d\theta,$$

where $x(t) = t^{q-1} y(t)$, $t \in (0, a]$. From the compactness of $Q(t^q \delta)$ $(t^q \delta > 0)$, we obtain that the set $V_1^\delta(t) = \{(\mathscr{T}_1^\delta y)(t), \ y \in B(J)\}$ is relatively compact in $X$ for every $\delta > 0$. Moreover, for any $y \in B(J)$, we have

$$|(\mathscr{T}_1 y)(t) - (\mathscr{T}_1^\delta y)(t)| = \left| q \int_0^\delta \theta M_q(\theta) Q(t^q \theta)(x_0 - g(x)) d\theta \right|$$

$$\le qM(|x_0| + L_1 r + L_2) \int_0^\delta \theta M_q(\theta) d\theta.$$

Therefore, there are relatively compact sets arbitrarily close to the set $V_1(t)$, $t > 0$. Hence the set $V_1(t)$, $t > 0$ is also relatively compact in $X$. Moreover, $\{\mathscr{T} y : y \in B(J)\}$ is uniformly bounded by Lemma 4.7. Therefore, $\{(\mathscr{T} y)(t), \ y \in B(J)\}$ is relatively compact by Arzela-Ascoli theorem. Hence, Theorem 1.10 shows that $\mathscr{T}$ has a fixed point $y^* \in B(J)$. Let $x^*(t) = t^{q-1} y^*(t)$. Then $x^*$ is a mild solution of (4.1). $\qquad \square$

**Remark 4.1.** If $g$ is not a compact map, we use another method given in Zhu and Li, 2008 to consider the following integral equations

$$x(t) = t^{q-1} P_q \left( t + \frac{1}{n} \right)(x_0 - g(x)) + \int_0^t (t - s)^{q-1} P_q(t - s) f(s, x(s)) ds, \quad t \in (0, a].$$
$$(4.9)$$

For any $n \in \mathbb{N}$, noticing that the operator $Q(\frac{1}{n})$ is compact, one can easily derive the relative compactness of $V(0)$ and $V(t)(t > 0)$. Then, (4.9) has a solution in $B_r^{(q)}(J')$. By passing the limit, as $n \to \infty$, one obtains a mild solution of the nonlocal Cauchy problem (4.1). However, because $Q(t)$ is replaced by $Q(\frac{1}{n})$, one needs a more restrictive condition than (H4)$'$, such as

**(H4)$''$** there exists a constant $r > 0$ such that

$$\frac{M_\varepsilon}{\Gamma(q)} \left( |x_0| + L_1 r + L_2 + \sup_{t \in (0, a]} \left\{ t^{1-q} \int_0^t (t - s)^{q-1} m(s) ds \right\} \right) \le r,$$

where $M_\varepsilon = \sup_{t \in [0, a+\varepsilon]} \|Q(t)\|_{B(X)}$, $\varepsilon$ is a small constant.

**Remark 4.2.** The condition (H2) of Theorems 4.1-4.2 can be replaced by the following condition.

**(H2)′** There exist a constant $q_1 \in (0, q)$ and $m \in L^{\frac{1}{q_1}}(J, \mathbb{R}^+)$ such that

$$|f(t, x)| \leq m(t) \quad \text{for all } x \in B_r^{(q)}(J') \text{ and almost all } t \in [0, a].$$

In fact, if (H2)′ holds, by using the Hölder inequality, for any $t_1, t_2 \in J'$ and $t_1 < t_2$, we obtain

$$|_0 D_t^{-q} m(t_2) - {}_0 D_t^{-q} m(t_1)|$$

$$= \frac{1}{\Gamma(q)} \left| \int_0^{t_1} ((t_2 - s)^{q-1} - (t_1 - s)^{q-1}) m(s) ds + \int_{t_1}^{t_2} (t_2 - s)^{q-1} m(s) ds \right|$$

$$\leq \frac{1}{\Gamma(q)} \left( \int_0^{t_1} ((t_1 - s)^{q-1} - (t_2 - s)^{q-1})^{\frac{1}{1-q_1}} ds \right)^{1-q_1} \left( \int_0^{t_1} (m(s))^{\frac{1}{q_1}} ds \right)^{q_1}$$

$$+ \frac{1}{\Gamma(q)} \left( \int_{t_1}^{t_2} ((t_2 - s)^{q-1})^{\frac{1}{1-q_1}} ds \right)^{1-q_1} \left( \int_{t_1}^{t_2} (m(s))^{\frac{1}{q_1}} ds \right)^{q_1}$$

$$\leq \frac{1}{\Gamma(q)} \left( \int_0^{t_1} ((t_1 - s)^{\frac{q-1}{1-q_1}} - (t_2 - s)^{\frac{q-1}{1-q_1}}) ds \right)^{1-q_1} \|m\|_{L^{\frac{1}{q_1}}}$$

$$+ \frac{1}{\Gamma(q)} \left( \int_{t_1}^{t_2} (t_2 - s)^{\frac{q-1}{1-q_1}} ds \right)^{1-q_1} \|m\|_{L^{\frac{1}{q_1}}}$$

$$\leq \frac{\|m\|_{L^{\frac{1}{q_1}}}}{\Gamma(q)} \left( \frac{1-q_1}{q-q_1} \right)^{1-q_1} \left( t_1^{\frac{q-q_1}{1-q_1}} + (t_2 - t_1)^{\frac{q-q_1}{1-q_1}} - t_2^{\frac{q-q_1}{1-q_1}} \right)^{1-q_1}$$

$$+ \frac{\|m\|_{L^{\frac{1}{q_1}}}}{\Gamma(q)} \left( \frac{1-q_1}{q-q_1} \right)^{1-q_1} \left( (t_2 - t_1)^{\frac{q-q_1}{1-q_1}} \right)^{1-q_1}$$

$$\leq \frac{2\|m\|_{L^{\frac{1}{q_1}}}}{\Gamma(q)} \left( \frac{1-q_1}{q-q_1} \right)^{1-q_1} (t_2 - t_1)^{q-q_1} \to 0, \quad \text{as } t_2 \to t_1, \tag{4.10}$$

where

$$\|m\|_{L^{\frac{1}{q_1}}} = \left( \int_0^a (m(t))^{\frac{1}{q_1}} dt \right)^{q_1}.$$

Furthermore,

$$t^{1-q} \int_0^t (t - s)^{q-1} m(s) ds$$

$$\leq t^{1-q} \left( \int_0^t (t - s)^{\frac{q-1}{1-q_1}} ds \right)^{1-q_1} \left( \int_0^t (m(s))^{\frac{1}{q_1}} ds \right)^{q_1}$$

$$\leq \left( \frac{1-q_1}{q-q_1} \right)^{1-q_1} t^{1-q_1} \|m\|_{L^{\frac{1}{q_1}}}$$

$$\to 0, \quad \text{as } t \to 0. \tag{4.11}$$

Thus, (4.10) and (4.11) mean that $_0 D_t^{-q} m \in C(J', \mathbb{R}^+)$, and $\lim_{t \to 0^+} t^{1-q} {}_0 D_t^{-q} m(t) = 0$. Hence, (H2) holds.

**Example 4.1.** Let $X = L^2([0, \pi], \mathbb{R})$. Consider the following fractional partial differential equations.

$$\begin{cases} \partial_t^q u(t, z) = \partial_z^2 u(t, z) + \partial_z G(t, u(t, z)), & z \in [0, \pi], \ t \in (0, a], \\ u(t, 0) = u(t, \pi) = 0, & t \in (0, a], \\ u(0, z) + \sum_{i=0}^{n} \int_0^\pi k(z, y) u(t_i, y) dy = u_0(z), & z \in [0, \pi], \end{cases} \tag{4.12}$$

where $\partial_t^q$ is Riemann-Liouville fractional partial derivative of order $0 < q < 1$, $a > 0$, $G$ is a given function, $n$ is a positive integer, $0 < t_0 < t_1 < \ldots < t_n \leq a$, $u_0(z) \in X = L^2([0, \pi], \mathbb{R})$, $k(z, y) \in L^2([0, \pi] \times [0, \pi], \mathbb{R}^+)$.

We define an operator $A$ by $Av = v''$ with the domain

$$D(A) = \{v(\cdot) \in X : v, v' \text{ absolutely continuous}, v'' \in X, v(0) = v(\pi) = 0\}.$$

Then $A$ generates a $C_0$-semigroup $\{Q(t)\}_{t \geq 0}$ which is compact, analytic and self-adjoint. Clearly the nonlocal Cauchy problem (4.2) and (H1) are satisfied.

The system (4.12) can be reformulated as the following nonlocal Cauchy problem in $X$

$$\begin{cases} {}_0D_t^q x(t) = Ax(t) + f(t, x(t)), & \text{almost all } t \in [0, a], \\ {}_0D_t^{q-1} x(0) + g(x) = x_0, \end{cases}$$

where $x(t) = u(t, \cdot)$, that is $x(t)(z) = u(t, z)$, $t \in (0, a]$, $z \in [0, \pi]$. The function $f : J' \times X \to X$ is given by

$$f(t, x(t))(z) = \partial_z G(t, u(t, z)),$$

and the operator $g : C(J', X) \to L(J', X)$ is given by

$$g(x)(z) = \sum_{i=0}^{n} K_g x(t_i)(z),$$

where $K_g v(z) = \int_0^\pi k(z, y) v(y) dy$, for $v \in X = L^2([0, \pi], \mathbb{R})$, $z \in [0, \pi]$.

We can take $q = 1/3$ and $f(t, x(t)) = t^{-1/4} \sin x(t)$, and choose

$$m(t) = t^{-1/4}, \quad L = (n+1) \left( \int_0^\pi \int_0^\pi k^2(z, y) dy dz \right)^{\frac{1}{2}}$$

and

$$r = \frac{M}{\Gamma(\frac{1}{3}) - ML} \left( |x_0| + g(0) + \frac{\Gamma(\frac{1}{3})\Gamma(\frac{3}{4})}{\Gamma(\frac{12}{13})} a^{\frac{3}{4}} \right).$$

Then, (H1)-(H4) are satisfied (noting that $K_g : X \to X$ is completely continuous). According to Theorem 4.1, system (4.12) has a mild solution in $B_r^{(1/3)}((0, a])$ provided that $\frac{ML}{\Gamma(1/3)} < 1$.

### 4.2.5 Noncompact Semigroup Case

If $Q(t)$ is noncompact, we give an assumption as follows.

(H5) There exists a constant $\ell > 0$ such that for any bounded $D \subset X$,

$$\alpha(f(t, D)) \leq \ell\alpha(D).$$

**Theorem 4.3.** *Assume that (H0)-(H5) hold. Then the nonlocal Cauchy problem (4.1) has at least one mild solution in $B_r^{(q)}(J')$.*

**Proof.** By Lemmas 4.5-4.6, we know that $\mathscr{T}_2 : B(J) \to B(J)$ is bounded, continuous and $\{\mathscr{T}_2 y : y \in B(J)\}$ is equicontinuous. Next, we show that $\mathscr{T}_2$ is compact in a subset of $B(J)$.

For each bounded subset $B_0 \subset B(J)$, set

$$\mathscr{T}^1(B_0) = \mathscr{T}_2(B_0), \ \mathscr{T}^n(B_0) = \mathscr{T}_2\left(\overline{co}(\mathscr{T}^{n-1}(B_0))\right), \ n = 2, 3, \ldots.$$

Then, from Propositions 1.16-1.17, for any $\varepsilon > 0$, there is a sequence $\{y_n^{(1)}\}_{n=1}^\infty \subset B_0$ such that

$$\alpha(\mathscr{T}^1(B_0(t))) = \alpha(\mathscr{T}_2(B_0(t)))$$

$$\leq 2\alpha\left(t^{1-q}\int_0^t (t-s)^{q-1} P_q(t-s) f(s, \{s^{q-1}y_n^{(1)}(s)\}_{n=1}^\infty) ds\right) + \varepsilon$$

$$\leq \frac{4M}{\Gamma(q)} t^{1-q}\int_0^t (t-s)^{q-1}\alpha\left(f(s, \{s^{q-1}y_n^{(1)}(s)\}_{n=1}^\infty)\right) ds + \varepsilon$$

$$\leq \frac{4M\ell}{\Gamma(q)} t^{1-q}\int_0^t (t-s)^{q-1}\alpha(\{s^{q-1}y_n^{(1)}(s)\}_{n=1}^\infty) ds + \varepsilon$$

$$\leq \frac{4M\ell\alpha(B_0)}{\Gamma(q)} t^{1-q}\int_0^t (t-s)^{q-1} s^{q-1} ds + \varepsilon$$

$$\leq \frac{4M\ell\Gamma(q) t^q \alpha(B_0)}{\Gamma(2q)} + \varepsilon.$$

Since $\varepsilon > 0$ is arbitrary, we have

$$\alpha(\mathscr{T}^1(B_0(t))) \leq \frac{4M\ell\Gamma(q) t^q}{\Gamma(2q)} \alpha(B_0).$$

From Propositions 1.16-1.17, for any $\varepsilon > 0$, there is a sequence $\{y_n^{(2)}\}_{n=1}^\infty \subset \overline{co}(\mathscr{T}^1(B_0))$ such that

$$\alpha(\mathscr{T}^2(B_0(t))) = \alpha(\mathscr{T}_2(\overline{co}(\mathscr{T}^1(B_0(t)))))$$

$$\leq 2\alpha\left(t^{1-q}\int_0^t (t-s)^{q-1} P_q(t-s) f(s, \{s^{q-1}y_n^{(2)}(s)\}_{n=1}^\infty) ds\right) + \varepsilon$$

$$\leq \frac{4Mt^{1-q}}{\Gamma(q)}\int_0^t (t-s)^{q-1}\alpha(f(s, \{s^{q-1}y_n^{(2)}(s)\}_{n=1}^\infty)) ds + \varepsilon$$

$$\leq \frac{4M\ell t^{1-q}}{\Gamma(q)} \int_0^t (t-s)^{q-1} \alpha(\{s^{q-1}y_n^{(2)}(s)\}_{n=1}^\infty) ds + \varepsilon$$

$$\leq \frac{4M\ell t^{1-q}}{\Gamma(q)} \int_0^t (t-s)^{q-1} s^{q-1} \alpha(\{y_n^{(2)}(s)\}_{n=1}^\infty) ds + \varepsilon$$

$$\leq \frac{(4M\ell)^2 t^{1-q}}{\Gamma(2q)} \int_0^t (t-s)^{q-1} s^{2q-1} ds + \varepsilon$$

$$= \frac{(4M\ell)^2 \Gamma(q)}{\Gamma(3q)} t^{2q} \alpha(B_0) + \varepsilon.$$

It can be shown, by mathematical induction, that for every $\bar{n} \in \mathbb{N}$,

$$\alpha(\mathscr{T}^{\bar{n}}(B_0(t))) \leq \frac{(4M\ell)^{\bar{n}} \Gamma(q)}{\Gamma((\bar{n}+1)q)} t^{\bar{n}q} \alpha(B_0).$$

Since

$$\lim_{\bar{n}\to\infty} \frac{(4M\ell a^q)^{\bar{n}} \Gamma(q)}{\Gamma((\bar{n}+1)q)} = 0,$$

there exists a positive integer $\hat{n}$ such that

$$\frac{(4M\ell)^{\hat{n}} \Gamma(q)}{\Gamma((\hat{n}+1)q)} t^{\hat{n}q} \leq \frac{(4M\ell a^q)^{\hat{n}} \Gamma(q)}{\Gamma((\hat{n}+1)q)} = k < 1.$$

Then

$$\alpha(\mathscr{T}^{\hat{n}}(B_0(t))) \leq k\alpha(B_0).$$

We know from Proposition 1.14, $\mathscr{T}^{\hat{n}}(B_0(t))$ is bounded and equicontinuous. Then, from Proposition 1.15, we have

$$\alpha(\mathscr{T}^{\hat{n}}(B_0)) = \max_{t\in[0,a]} \alpha(\mathscr{T}^{\hat{n}}(B_0(t))).$$

Hence

$$\alpha(\mathscr{T}^{\hat{n}}(B_0)) \leq k\alpha(B_0).$$

Let

$$D_0 = B(J), \quad D_1 = \overline{co}(\mathscr{T}^{\hat{n}}(D)), \ldots, \quad D_n = \overline{co}(\mathscr{T}^{\hat{n}}(D_{n-1})), \quad n = 2, 3, \ldots.$$

Then, we can get

**(i)** $D_0 \supset D_1 \supset D_2 \supset \cdots \supset D_{n-1} \supset D_n \supset \cdots$;
**(ii)** $\lim_{n\to\infty} \alpha(D_n) = 0$.

Then $\hat{D} = \bigcap_{n=0}^\infty D_n$ is a nonempty, compact and convex subset in $B(J)$.
We prove $\mathscr{T}_2(\hat{D}) \subset \hat{D}$. Firstly, we show

$$\mathscr{T}_2(D_n) \subset D_n, \quad n = 0, 1, 2, \ldots. \tag{4.13}$$

From $\mathscr{T}^1(D_0) = \mathscr{T}_2(D_0) \subset D_0$, we know $\overline{co}(\mathscr{T}^1(D_0)) \subset D_0$. Therefore

$$\mathscr{T}^2(D_0) = \mathscr{T}_2(\overline{co}(\mathscr{T}^1(D_0))) \subset \mathscr{T}_2(D_0) = \mathscr{T}^1(D_0),$$

$$\mathscr{T}^3(D_0) = \mathscr{T}_2(\overline{co}(\mathscr{T}^2(D_0))) \subset \mathscr{T}_2(\overline{co}(\mathscr{T}^1(D_0))) = \mathscr{T}^2(D_0),$$

$$\vdots$$

$$\mathscr{T}^{\hat{n}}(D_0) = \mathscr{T}_2(\overline{co}(\mathscr{T}^{\hat{n}-1}(D_0))) \subset \mathscr{T}_2(\overline{co}(\mathscr{T}^{\hat{n}-2}(D_0))) = \mathscr{T}^{\hat{n}-1}(D_0).$$

Hence, $D_1 = \overline{co}(\mathscr{T}^{\hat{n}}(D_0)) \subset \overline{co}(\mathscr{T}^{\hat{n}-1}(D_0))$, so $\mathscr{T}(D_1) \subset \mathscr{T}(\overline{co}(\mathscr{T}^{\hat{n}-1}(D_0))) = \mathscr{T}^{\hat{n}}(D_0) \subset \overline{co}(\mathscr{T}^{\hat{n}}(D_0)) = D_1$. Employing the same method, we can prove $\mathscr{T}_2(D_n) \subset D_n (n = 0, 1, 2, \ldots)$. By (4.13), we get $\mathscr{T}_2(\hat{D}) \subset \bigcap_{n=0}^{\infty} \mathscr{T}_2(D_n) \subset \bigcap_{n=0}^{\infty} D_n = \hat{D}$. Then $\mathscr{T}_2(\hat{D})$ is compact. Hence, $\alpha(\mathscr{T}_2(\hat{D})) = 0$.

On the other hand, for any $y_1, y_2 \in \hat{D}$ and $t \in (0, a]$, according to (H3), we have

$$
\begin{aligned}
|\mathscr{T}_1 y_1(t) - \mathscr{T}_1 y_2(t)| &= t^{1-q}|(T_1 x_1)(t) - (T_1 x_2)(t)| \\
&\leq \frac{M}{\Gamma(q)}|g(x_1) - g(x_2)| \\
&\leq \frac{ML}{\Gamma(q)}\|x_1 - x_2\|_q \\
&= \frac{ML}{\Gamma(q)}\|y_1 - y_2\|,
\end{aligned}
$$

which implies that $\|\mathscr{T}_1 y_1 - \mathscr{T}_1 y_2\| \leq \frac{ML}{\Gamma(q)}\|y_1 - y_2\|$. Thus, we obtain that

$$\alpha(\mathscr{T}_1(\hat{D})) \leq \frac{ML}{\Gamma(q)}\alpha(\hat{D}). \tag{4.14}$$

By (4.14), we have

$$
\begin{aligned}
\alpha(\mathscr{T}(\hat{D})) &\leq \alpha(\mathscr{T}_1(\hat{D})) + \alpha(\mathscr{T}_2(\hat{D})) \\
&\leq \frac{ML}{\Gamma(q)}\alpha(\hat{D}).
\end{aligned}
$$

Thus, the operator $\mathscr{T}$ is an $\alpha$-contraction in $\hat{D}$. By Lemma 4.5, we know that $\mathscr{T}$ is continuous. Hence, Theorem 1.10 shows that $\mathscr{T}$ has a fixed point $y^* \in B(J)$. Let $x^*(t) = t^{q-1}y^*(t)$. Then $x^*$ is a mild solution of (4.1). $\qquad\square$

**Theorem 4.4.** *Assume that (H0)-(H2), (H3)$'$, (H4)$'$ and (H5) hold, then the non-local Cauchy problem (4.1) has at least one mild solution in $B_r^{(q)}(J')$.*

**Proof.** Since $g(x)$ is compact and $P_q(t)$ is bounded, for every $t > 0$, $\{(\mathscr{T}_1 y)(t), y \in B(J)\}$ is relatively compact. Thus, we have $\alpha(\mathscr{T}_1(B(J))) = 0$.

By the proof of Theorem 4.3, we know that there exists a $\hat{D} \subset B(J)$ such that $\mathscr{T}_2(\hat{D})$ is relatively compact, i.e., $\alpha(\mathscr{T}_2(\hat{D})) = 0$. Hence, we have

$$\alpha(\mathscr{T}(\hat{D})) \leq \alpha(\mathscr{T}_1(\hat{D})) + \alpha(\mathscr{T}_2(\hat{D})) = 0.$$

Hence, Theorem 1.10 shows that $\mathscr{T}$ has a fixed point $y^* \in B(J)$. Let $x^*(t) = t^{q-1}y^*(t)$. Then $x^*$ is a mild solution of (4.1). $\qquad\square$

## 4.3 Evolution Equations with Caputo Derivative

### 4.3.1 Introduction

Consider the following nonlocal Cauchy problems of fractional evolution equation with Caputo fractional derivative

$$\begin{cases} {}_0^C D_t^q x(t) = Ax(t) + f(t, x(t)), & \text{a.e. } t \in [0, a], \\ x(0) + g(x) = x_0, \end{cases} \tag{4.15}$$

where ${}_0^C D_t^q$ is Caputo fractional derivative of order $q$, $0 < q < 1$, $A$ is the infinitesimal generator of a $C_0$-semigroup of bounded linear operators $\{Q(t)\}_{t \geq 0}$ in Banach space $X$, $f : J \times X \to X$ , $g : C(J, X) \to L(J, X)$ are given operators satisfying some assumptions and $x_0$ is an element of the Banach space $X$.

In this section, by using the theory of Hausdorff measure of noncompactness and fixed point theorems, we study the nonlocal Cauchy problem (4.15) in the cases $Q(t)$ is compact and noncompact, respectively. Subsection 4.3.2 is devoted to obtaining the appropriate definition on the mild solutions of the problem (4.15) by considering a integral equation which is given in terms of probability density. In Subsection 4.3.3, we give some preliminary lemmas. Subsection 4.3.4 provides various existence theorems of mild solutions for the Cauchy problem (4.15) in the case $Q(t)$ is compact. In Subsection 4.3.5, we establish various existence theorems of mild solutions for the Cauchy problem (4.15) in the case $Q(t)$ is noncompact.

### 4.3.2 Definition of Mild Solutions

**Lemma 4.8.** *If*

$$x(t) = x_0 - g(x) + \frac{1}{\Gamma(q)} \int_0^t (t-s)^{q-1} [Ax(s) + f(s, x(s))] ds, \quad \text{for } t \geq 0 \tag{4.16}$$

*holds, then we have*

$$x(t) = S_q(t)(x_0 - g(x)) + \int_0^t (t-s)^{q-1} P_q(t-s) f(s, x(s)) ds, \quad \text{for } t \geq 0,$$

*where*

$$S_q(t) = \int_0^\infty M_q(\theta) Q(t^q \theta) d\theta, \quad P_q(t) = \int_0^\infty q\theta M_q(\theta) Q(t^q \theta) d\theta.$$

**Proof.** Let $\lambda > 0$. Applying the Laplace transform

$$\nu(\lambda) = \int_0^\infty e^{-\lambda s} x(s) ds \text{ and } \omega(\lambda) = \int_0^\infty e^{-\lambda s} f(s, x(s)) ds, \quad \lambda > 0$$

to (4.16), we have

$$\begin{aligned} \nu(\lambda) &= \frac{1}{\lambda}(x_0 - g(x)) + \frac{1}{\lambda^q} A\nu(\lambda) + \frac{1}{\lambda^q} \omega(\lambda) \\ &= \lambda^{q-1}(\lambda^q I - A)^{-1}(x_0 - g(x)) + (\lambda^q I - A)^{-1} \omega(\lambda) \\ &= \lambda^{q-1} \int_0^\infty e^{-\lambda^q s} Q(s)(x_0 - g(x)) ds + \int_0^\infty e^{-\lambda^q s} Q(s) \omega(\lambda) ds, \quad (4.17) \end{aligned}$$

provided that the integrals in (4.17) exist, where $I$ is the identity operator defined on $X$.

Using (4.5) and (4.17), we get

$$
\lambda^{q-1}\int_0^\infty e^{-\lambda^q s}Q(s)(x_0-g(x))ds
$$
$$
=\int_0^\infty q(\lambda t)^{q-1}e^{-(\lambda t)^q}Q(t^q)(x_0-g(x))dt
$$
$$
=\int_0^\infty -\frac{1}{\lambda}\frac{d}{dt}\left(e^{-(\lambda t)^q}\right)Q(t^q)(x_0-g(x))dt
$$
$$
=\int_0^\infty\int_0^\infty \theta\psi_q(\theta)e^{-\lambda t\theta}Q(t^q)(x_0-g(x))d\theta dt
$$
$$
=\int_0^\infty e^{-\lambda t}\left(\int_0^\infty \psi_q(\theta)Q\left(\frac{t^q}{\theta^q}\right)(x_0-g(x))d\theta\right)dt. \tag{4.18}
$$

According to (4.7), (4.17) and (4.18), we have

$$
\nu(\lambda)=\int_0^\infty e^{-\lambda t}\left(\int_0^\infty \psi_q(\theta)Q\left(\frac{t^q}{\theta^q}\right)(x_0-g(x))d\theta\right.
$$
$$
\left.+q\int_0^t\int_0^\infty \psi_q(\theta)Q\left(\frac{(t-s)^q}{\theta^q}\right)f(s,x(s))\frac{(t-s)^{q-1}}{\theta^q}d\theta ds\right)dt.
$$

Now we can invert the last Laplace transform to get

$$
x(t)=\int_0^\infty \psi_q(\theta)Q\left(\frac{t^q}{\theta^q}\right)(x_0-g(x))d\theta
$$
$$
+q\int_0^t\int_0^\infty \psi_q(\theta)Q\left(\frac{(t-s)^q}{\theta^q}\right)f(s,x(s))\frac{(t-s)^{q-1}}{\theta^q}d\theta ds
$$
$$
=\int_0^\infty M_q(\theta)Q(t^q\theta)(x_0-g(x))d\theta
$$
$$
+q\int_0^t\int_0^\infty \theta(t-s)^{q-1}M_q(\theta)Q((t-s)^q\theta)f(s,x(s))d\theta ds
$$
$$
=S_q(t)(x_0-g(x))+\int_0^t (t-s)^{q-1}P_q(t-s)f(s,x(s))ds.
$$

The proof is completed. $\square$

Due to Lemma 4.8, we give the following definition of the mild solution of (4.15).

**Definition 4.2.** By the mild solution of the nonlocal Cauchy problem (4.15), we mean that the function $x\in C(J,X)$ which satisfies

$$
x(t)=S_q(t)(x_0-g(x))+\int_0^t (t-s)^{q-1}P_q(t-s)f(s,x(s))ds,\quad\text{for }t\in[0,a].
$$

Suppose that $A$ is the infinitesimal generator of a $C_0$-semigroup $\{Q(t)\}_{t\geq 0}$ of uniformly bounded linear operators on Banach space $X$. This means that there exists $M > 1$ such that $M = \sup_{t \in [0,\infty)} \|Q(t)\|_{B(X)} < \infty$.

**Proposition 4.5.** *(Zhou and Jiao, 2010a) For any fixed $t > 0$, $\{S_q(t)\}_{t>0}$ and $\{P_q(t)\}_{t>0}$ are linear and bounded operators, i.e., for any $x \in X$*

$$|S_q(t)x| \leq M|x|, \quad |P_q(t)x| \leq \frac{M}{\Gamma(q)}|x|.$$

**Proposition 4.6.** *(Zhou and Jiao, 2010a) Operators $\{S_q(t)\}_{t>0}$ and $\{P_q(t)\}_{t>0}$ are strongly continuous, which means that, for all $x \in X$ and $0 < t' < t'' \leq a$, we have*

$$|S_q(t'')x - S_q(t')x| \to 0, \quad |P_q(t'')x - P_q(t')x| \to 0, \quad \text{as } t'' \to t'.$$

**Proposition 4.7.** *(Zhou and Jiao, 2010a) Assume that $\{Q(t)\}_{t>0}$ is compact operator. Then $\{S_q(t)\}_{t>0}$ and $\{P_q(t)\}_{t>0}$ are also compact operators.*

**Remark 4.3.** Since $S_q(\cdot)$ and $P_q(\cdot)$ are associated with the $q$, there are no analogue of the semigroup property, i.e., $S_q(t + s) \neq S_q(t)S_q(s)$, $P_q(t + s) \neq P_q(t)P_q(s)$ for $t, s > 0$.

### 4.3.3 Preliminary Lemmas

For $r > 0$, let $B_r(J)$ be the closed ball of the space $C(J, X)$ with radius $r$ and center at 0, that is,

$$B_r(J) = \{x \in C(J, X) : \|x\| \leq r\},$$

where $\|x\| = \sup_{t \in [0,a]} |x(t)|$.

We introduce the following hypotheses:

**(H0)** $Q(t)(t > 0)$ is equicontinuous, i.e., $Q(t)$ is continuous in the uniform operator topology for $t > 0$;

**(H1)** for each $t \in J$, the function $f(t, \cdot) : X \to X$ is continuous and for each $x \in X$, the function $f(\cdot, x) : J \to X$ is strongly measurable;

**(H2)** there exists a function $m$ such that

$$_0D_t^{-q}m \in C(J, \mathbb{R}^+), \quad \lim_{t \to 0+} {_0D_t^{-q}m(t)} = 0$$

and

$$|f(t, x)| \leq m(t) \quad \text{for all } x \in B_r(J) \text{ and almost all } t \in [0, a];$$

**(H3)** there exists a constant $L \in (0, \frac{1}{M})$, the operator $g : C(J, X) \to L(J, X)$ satisfies

$$|g(x_1) - g(x_2)| \leq L\|x_1 - x_2\|, \quad \text{for } x_1, x_2 \in B_r(J);$$

**(H4)** there exists a constant $r > 0$ such that

$$\frac{M}{1 - ML}\left(|x_0| + |g(0)| + \sup_{t \in [0,a]}\left\{\frac{1}{\Gamma(q)}\int_0^t (t-s)^{q-1}m(s)ds\right\}\right) \le r;$$

**(H3)′** the operator $g : C(J, X) \to L(J, X)$ is a continuous and compact map, and there exist positive constants $L_1, L_2$ such that $|g(x)| \le L_1\|x\| + L_2$ for all $x \in B_r(J)$;

**(H4)′** there exists a constant $r > 0$ such that

$$\frac{M}{1 - ML_1}\left(|x_0| + L_2 + \sup_{t \in [0,a]}\left\{\frac{1}{\Gamma(q)}\int_0^t (t-s)^{q-1}m(s)ds\right\}\right) \le r.$$

For any $x \in B_r(J)$, we define an operator $T$ as follows

$$(Tx)(t) = (T_1x)(t) + (T_2x)(t),$$

where

$$(T_1x)(t) = S_q(t)(x_0 - g(x)), \qquad \text{for } t \in [0, a],$$

$$(T_2x)(t) = \int_0^t (t-s)^{q-1}P_q(t-s)f(s, x(s))ds, \quad \text{for } t \in [0, a].$$

Obviously, $x$ is a mild solution of (4.15) in $B_r(J)$ if and only if the operator equation $x = Tx$ has a solution $x \in B_r(J)$.

**Lemma 4.9.** *Assume that (H0)-(H3) hold. Then $\{T_2x : x \in B_r(J)\}$ is equicontinuous.*

**Proof.** For any $x \in B_r(J)$, for $t_1 = 0$, $0 < t_2 \le a$, by (H2), we get

$$|(T_2x)(t_2) - (T_2x)(0)| = \left|\int_0^{t_2} (t_2-s)^{q-1}P_q(t_2-s)f(s, x(s))ds\right|$$

$$\le \frac{M}{\Gamma(q)}\int_0^{t_2} (t_2-s)^{q-1}m(s)ds \to 0 \text{ as } t_2 \to 0.$$

For $0 < t_1 < t_2 \le a$, we have

$$|(T_2x)(t_2) - (T_2x)(t_1)|$$

$$= \left|\int_0^{t_2} (t_2-s)^{q-1}P_q(t_2-s)f(s, x(s))ds - \int_0^{t_1} (t_1-s)^{q-1}P_q(t_1-s)f(s, x(s))ds\right|$$

$$= \left|\int_{t_1}^{t_2} (t_2-s)^{q-1}P_q(t_2-s)f(s, x(s))ds\right|$$

$$+ \left|\int_0^{t_1} (t_2-s)^{q-1}P_q(t_2-s)f(s, x(s))ds - \int_0^{t_1} (t_1-s)^{q-1}P_q(t_2-s)f(s, x(s))ds\right|$$

$$+ \left|\int_0^{t_1} (t_1-s)^{q-1}P_q(t_2-s)f(s, x(s))ds - \int_0^{t_1} (t_1-s)^{q-1}P_q(t_1-s)f(s, x(s))ds\right|$$

$$\le \frac{M}{\Gamma(q)}\int_{t_1}^{t_2} (t_2-s)^{q-1}m(s)ds + \frac{M}{\Gamma(q)}\int_0^{t_1} \left((t_1-s)^{q-1} - (t_2-s)^{q-1}\right)m(s)ds$$

$$+ \int_0^{t_1} (t_1 - s)^{q-1} |P_q(t_2 - s)f(s, x(s)) - P_q(t_1 - s)f(s, x(s))| ds$$

$$\leq \frac{M}{\Gamma(q)} \left| \int_0^{t_2} (t_2 - s)^{q-1} m(s) ds - \int_0^{t_1} (t_1 - s)^{q-1} m(s) ds \right|$$

$$+ \frac{2M}{\Gamma(q)} \int_0^{t_1} \left( (t_1 - s)^{q-1} - (t_2 - s)^{q-1} \right) m(s) ds$$

$$+ \int_0^{t_1} (t_1 - s)^{q-1} \| P_q(t_2 - s) - P_q(t_1 - s) \|_{B(X)} m(s) ds$$

$$=: I_1 + I_2 + I_3,$$

Since $_0D_t^{-q} m \in C(J, \mathbb{R}^+)$, thus $I_1 \to 0$ as $t_2 \to t_1$.
For $t_1 < t_2$,

$$I_2 \leq \frac{2M}{\Gamma(q)} \int_0^{t_1} (t_1 - s)^{q-1} m(s) ds,$$

then by Lebesgue dominated convergence theorem, we have that $I_2 \to 0$ as $t_2 \to t_1$.
For $\varepsilon > 0$ be small enough, we have

$$I_3 \leq \int_0^{t_1 - \varepsilon} (t_1 - s)^{q-1} \| P_q(t_2 - s) - P_q(t_1 - s) \|_{B(X)} m(s) ds$$

$$+ \int_{t_1 - \varepsilon}^{t_1} (t_1 - s)^{q-1} \| P_q(t_2 - s) - P_q(t_1 - s) \|_{B(X)} m(s) ds$$

$$\leq \int_0^{t_1} (t_1 - s)^{q-1} m(s) ds \sup_{s \in [0, t_1 - \varepsilon]} \| P_q(t_2 - s) - P_q(t_1 - s) \|_{B(X)}$$

$$+ \frac{2M}{\Gamma(q)} \int_{t_1 - \varepsilon}^{t_1} (t_1 - s)^{q-1} m(s) ds$$

$$\leq \int_0^{t_1} (t_1 - s)^{q-1} m(s) ds \sup_{s \in [0, t_1 - \varepsilon]} \| P_q(t_2 - s) - P_q(t_1 - s) \|_{B(X)}$$

$$+ \frac{2M}{\Gamma(q)} \left| \int_0^{t_1} (t_1 - s)^{q-1} m(s) ds - \int_0^{t_1 - \varepsilon} (t_1 - \varepsilon - s)^{q-1} m(s) ds \right|$$

$$+ \frac{2M}{\Gamma(q)} \int_0^{t_1 - \varepsilon} [(t_1 - \varepsilon - s)^{q-1} - (t_1 - s)^{q-1}] m(s) ds.$$

$$=: I_{31} + I_{32} + I_{33}.$$

By (H0), it is easy to see that $I_{31} \to 0$ as $t_2 \to t_1$. Similar to the proof that $I_1, I_2$ tend to zero, we get $I_{32} \to 0$ and $I_{33} \to 0$ as $\varepsilon \to 0$. Thus, $I_3$ tends to zero independently of $x \in B_r(J)$ as $t_2 \to t_1$, $\varepsilon \to 0$. Therefore, $|(T_2 x)(t_1) - (T_2 x)(t_2)|$ tends to zero independently of $x \in B_r(J)$ as $t_2 \to t_1$, which means that $\{T_2 x : x \in B_r(J)\}$ is equicontinuous. $\qquad \square$

**Lemma 4.10.** *Assume that (H1)-(H4) hold. Then $T$ maps $B_r(J)$ into $B_r(J)$, and $T$ is continuous in $B_r(J)$.*

**Proof. Claim I.** $T$ maps $B_r(J)$ into $B_r(J)$.

For any $x \in B_r(J)$ and $t \in J$, by using (H1)-(H4), we have

$$|(Tx)(t)| = |(T_1x)(t) + (T_2x)(t)|$$

$$\leq |S_q(t)(x_0 - g(x))| + \left| \int_0^t (t-s)^{q-1} P_q(t-s) f(s, x(s)) ds \right|$$

$$\leq M(|x_0| + L\|x - 0\| + |g(0)|) + \frac{M}{\Gamma(q)} \int_0^t (t-s)^{q-1} |f(s, x(s))| ds$$

$$\leq M\left( |x_0| + Lr + |g(0)| + \sup_{t \in [0,a]} \left\{ \frac{1}{\Gamma(q)} \int_0^t (t-s)^{q-1} m(s) ds \right\} \right)$$

$$\leq r.$$

Hence, $\|Tx\| \leq r$ for any $x \in B_r(J)$.

**Claim II.** $T$ is continuous in $B_r(J)$.

For any $\{x_m\}_{m=1}^{\infty} \subseteq B_r(J), x \in B_r(J)$ with $\lim_{n \to \infty} \|x_m - x\| = 0$, by the condition (H1), we have

$$\lim_{m \to \infty} f(s, x_m(s)) = f(s, x(s)).$$

On the one hand, using (H2), we get for each $t \in J$, $(t-s)^{q-1} |f(s, x_m(s)) - f(s, x(s))| \leq (t-s)^{q-1} 2m(s)$. On the other hand, the function $s \to (t-s)^{q-1} 2m(s)$ is integrable for $s \in [0, t]$ and $t \in J$. By Lebesgue dominated convergence theorem, we get

$$\int_0^t (t-s)^{q-1} |f(s, x_m(s)) - f(s, x(s))| ds \to 0, \quad \text{as } m \to \infty.$$

Then, for $t \in J$,

$$|(Tx_m)(t) - (Tx)(t)|$$

$$\leq |S_q(t)(g(x_m) - g(x))| + \left| \int_0^t (t-s)^{q-1} P_q(t-s) [f(s, x_m(s)) - f(s, x(s))] ds \right|$$

$$\leq ML\|x_m - x\| + \frac{M}{\Gamma(q)} \int_0^t (t-s)^{q-1} |f(s, x_m(s)) - f(s, x(s))| ds.$$

Therefore, $Tx_m \to Tx$ pointwise on $J$ as $m \to \infty$, by which Lemma 4.9 implies that $Tx_m \to Tx$ uniformly on $J$ as $m \to \infty$ and so $T$ is continuous. $\square$

**Lemma 4.11.** *Assume that there exists a constant $r > 0$ such that the conditions (H0)-(H2) and (H3)$'$ are satisfied. Then $\{Tx : x \in B_r(J)\}$ is equicontinuous.*

**Proof.** For any $x \in B_r(J)$ and $0 \leq t_1 < t_2 \leq a$, we get

$$|(Tx)(t_2) - (Tx)(t_1)| \leq |(S_q(t_2) - S_q(t_1))(x_0 - g(x))| + I_1 + I_2 + I_3,$$

where $I_1, I_2$ and $I_3$ are defined as in the proof of Lemma 4.9. According to Proposition 4.6, we know that $|(Tx)(t_2) - (Tx)(t_1)|$ tends to zero independently of $x \in B_r(J)$ as $t_2 \to t_1$, which means that $\{Tx, x \in B_r(J)\}$ is equicontinuous. $\square$

**Lemma 4.12.** *Assume that there exists a constant $r > 0$ such that the conditions (H1), (H2), (H3)' and (H4)' are satisfied. Then $T$ maps $B_r(J)$ into $B_r(J)$, and $T$ is continuous in $B_r(J)$.*

**Proof.** For any $x \in B_r(J)$ and $t \in J$, by using (H1), (H2), (H3)' and (H4)', we have

$$|(Tx)(t)| = |(T_1x)(t) + (T_2x)(t)|$$

$$\leq |S_q(t)(x_0 - g(x))| + \left| \int_0^t (t-s)^{q-1} P_q(t-s) f(s, x(s)) ds \right|$$

$$\leq M(|x_0| + L_1 r + L_2) + \frac{M}{\Gamma(q)} \int_0^t (t-s)^{q-1} |f(s, x(s))| ds$$

$$\leq M \left( |x_0| + L_1 r + L_2 + \sup_{t \in [0,a]} \left\{ \frac{1}{\Gamma(q)} \int_0^t (t-s)^{q-1} m(s) ds \right\} \right)$$

$$\leq r.$$

Hence, $\|Tx\| \leq r$ for any $x \in B_r(J)$. Using the similar argument as that we did in the proof of Lemma 4.10, we know that $T$ is continuous in $B_r(J)$. □

### 4.3.4 Compact Semigroup Case

In the following, we suppose that the operator $A$ generates a compact $C_0$-semigroup $\{Q(t)\}_{t \geq 0}$ on $X$, that is, for any $t > 0$, the operator $Q(t)$ is compact.

**Theorem 4.5.** *Assume that $Q(t)(t > 0)$ is compact. Furthermore assume that there exists a constant $r > 0$ such that the conditions (H1)-(H4) are satisfied. Then the nonlocal Cauchy problem (4.15) has at least one mild solution in $B_r(J)$.*

**Proof.** Since Proposition 4.4, $Q(t)(t > 0)$ is equicontinuous, which implies (H0) is satisfied.

For any $x_1, x_2 \in B_r(J)$ and $t \in J$, according to (H3), we have

$$|(T_1x_1)(t) - (T_1x_2)(t)| \leq M|g(x_1) - g(x_2)|$$
$$\leq ML\|x_1 - x_2\|,$$

which implies that $\|T_1x_1 - T_1x_2\| \leq ML\|x_1 - x_2\|$. Thus, we obtain that

$$\alpha(T_1 B_r(J)) \leq ML\alpha(B_r(J)). \tag{4.19}$$

Next, we show that $\{T_2x, x \in B_r(J)\}$ is relatively compact, i.e., $\alpha(T_2(B_r(J))) = 0$. It suffices to show that the family of functions $\{T_2x : x \in B_r(J)\}$ is uniformly bounded and equicontinuous, and for any $t \in J$, $\{(T_2x)(t) : x \in B_r(J)\}$ is relatively compact in $X$.

By Lemma 4.10, we have $\|T_2x\| \leq r$, for any $x \in B_r(J)$, which means that $\{T_2x : x \in B_r(J)\}$ is uniformly bounded. By Lemma 4.9, $\{T_2x : x \in B_r(J)\}$ is equicontinuous.

It remains to prove that for any $t \in J$, $V(t) = \{(T_2 x)(t) : x \in B_r(J)\}$ is relatively compact in $X$.

Obviously, $V(0)$ is relatively compact in $X$. Let $0 < t \leq a$ be fixed. For every $\varepsilon \in (0, t)$ and $\delta > 0$, define an operator $T_\varepsilon^\delta$ on $B_r(J)$ by the formula

$$(T_\varepsilon^\delta x)(t) = q \int_0^{t-\varepsilon} \int_\delta^\infty \theta(t-s)^{q-1} M_q(\theta) Q((t-s)^q \theta) f(s, x(s)) d\theta ds$$

$$= q\, Q(\varepsilon^q \delta) \int_0^{t-\varepsilon} \int_\delta^\infty \theta(t-s)^{q-1} M_q(\theta) Q((t-s)^q \theta - \varepsilon^q \delta) f(s, x(s)) d\theta ds,$$

where $x \in B_r(J)$. Then from the compactness of $Q(\varepsilon^q \delta)(\varepsilon^q \delta > 0)$, we obtain that the set $V_\varepsilon^\delta(t) = \{(T_\varepsilon^\delta x)(t) : x \in B_r(J)\}$ is relatively compact in $X$ for every $\varepsilon \in (0, t)$ and $\forall \delta > 0$. Moreover, for any $x \in B_r(J)$, we have

$$|(T_2 x)(t) - (T_\varepsilon^\delta x)(t)| \leq q \left| \int_0^t \int_0^\delta \theta(t-s)^{q-1} M_q(\theta) Q((t-s)^q \theta) f(s, x(s)) d\theta ds \right|$$

$$+ q \left| \int_{t-\varepsilon}^t \int_\delta^\infty \theta(t-s)^{q-1} M_q(\theta) Q((t-s)^q \theta) f(s, x(s)) d\theta ds \right|$$

$$\leq q M \int_0^t (t-s)^{q-1} m(s) ds \int_0^\delta \theta M_q(\theta) d\theta$$

$$+ \frac{M}{\Gamma(q)} \int_{t-\varepsilon}^t (t-s)^{q-1} m(s) ds \to 0, \quad \text{as } \varepsilon \to 0, \ \delta \to 0.$$

Therefore, there are relatively compact sets arbitrarily close to the set $V(t)$, $t > 0$. Hence the set $V(t)$, $t > 0$ is also relatively compact in $X$.

Therefore, $\{(T_2 x)(t) : x \in B_r(J)\}$ is relatively compact by Arzela-Ascoli theorem. Thus, we have $\alpha(T_2(B_r(J))) = 0$. By (4.19), we have

$$\alpha(T(B_r(J))) \leq \alpha(T_1(B_r(J))) + \alpha(T_2(B_r(J))) \leq ML\alpha(B_r(J)).$$

Thus, the operator $T$ is an $\alpha$-contraction in $B_r(J)$. By Lemma 4.10, we know that $T$ is continuous. Hence, Theorem 1.10 shows that $T$ has a fixed point in $B_r(J)$. Therefore, the nonlocal Cauchy problem (4.15) has a mild solution in $B_r(J)$. $\square$

**Theorem 4.6.** *Assume that $Q(t)(t > 0)$ is compact. Furthermore assume that (H1), (H2), (H3)′ and (H4)′ hold. Then the nonlocal Cauchy problem (4.15) has at least a mild solution in $B_r(J)$.*

**Proof.** Since Proposition 4.4, $Q(t)(t > 0)$ is equicontinuous, which implies (H0) is satisfied. By Lemma 4.12, we have $\|Tx\| \leq r$, for any $x \in B_r(J)$, which means that $\{Tx : x \in B_r(J)\}$ is uniformly bounded. By Lemmas 4.11-4.12, we know that $T$ is continuous, $\{Tx : x \in B_r(J)\}$ is equicontinuous. It remains to prove that for $t \in J$, the set $\{(Tx)(t) : x \in B_r(J)\}$ is relatively compact in $X$.

According to the argument of Theorem 4.5, we only need to prove that for any $t \in J$, the set $V_1(t) = \{(T_1 x)(t) : x \in B_r(J)\}$ is relatively compact in $X$.

Obviously, $V_1(0)$ is relatively compact in $X$. Let $0 < t \le a$ be fixed. For every $\delta > 0$, define an operator $T_1^\delta$ on $B_r(J)$ by the formula

$$(T_1^\delta x)(t) = \int_\delta^\infty M_q(\theta)Q(t^q\theta)(x_0 - g(x))d\theta$$

$$= Q(t^q\delta)\int_\delta^\infty M_q(\theta)Q(t^q\theta - t^q\delta)(x_0 - g(x))d\theta$$

where $x \in B_r(J)$. From the compactness of $Q(t^q\delta)(t^q\delta > 0)$, we obtain that the set $V_1^\delta(t) = \{(T_1^\delta x)(t) : x \in B_r(J)\}$ is relatively compact in $X$ for every $\delta > 0$. Moreover, for every $x \in B_r(J)$, we have

$$|(T_1 x)(t) - (T_1^\delta x)(t)|$$

$$= \left| \int_0^\infty M_q(\theta)Q(t^q\theta)(x_0 - g(x))d\theta - \int_\delta^\infty M_q(\theta)Q(t^q\theta)(x_0 - g(x))d\theta \right|$$

$$= \left| \int_0^\delta M_q(\theta)Q(t^q\theta)(x_0 - g(x))d\theta \right|$$

$$\le M(|x_0| + L_1 r + L_2)\int_0^\delta M_q(\theta)d\theta.$$

Therefore, there are relatively compact sets arbitrarily close to the set $V_1(t)$, $t > 0$. Hence the set $V_1(t)$, $t > 0$ is also relatively compact in $X$. Moreover, $\{Tx : x \in B_r(J)\}$ is uniformly bounded by Lemma 4.10. Therefore, $\{Tx : x \in B_r(J)\}$ is relatively compact by Arzela-Ascoli theorem. Therefore, $\alpha(T(B_r(J))) = 0$. Hence, Theorem 1.10 shows that $T$ has a fixed point in $B_r(J)$, which means that the nonlocal Cauchy problem (4.15) has a mild solution.                                      □

**Remark 4.4.** If $g$ is not a compact mapping, we consider the following integral equations

$$x(t) = t^{q-1}P_q\left(t + \frac{1}{n}\right)(x_0 - g(x)) + \int_0^t (t - s)^{q-1}P_q(t - s)f(s, x(s))ds, \quad t \in (0, a].$$
$$(4.20)$$

For any $n \in \mathbb{N}$, noticing that the operator $Q(\frac{1}{n})$ is compact, one can easily derive the relative compactness of $V(0)$ and $V(t)(t > 0)$. Then, (4.20) has a solution in $B_r^{(q)}(J')$. By passing the limit, as $n \to \infty$, one obtains a mild solution of the nonlocal Cauchy problem (4.15). However, because $Q(t)$ is replaced by $Q(\frac{1}{n})$, one needs a more restrictive condition than (H4)$'$, such as

**(H4)$''$** there exists a constant $r > 0$ such that

$$M_\varepsilon\left(|x_0| + L_1 r + L_2 + \sup_{t \in [0,a]}\left\{\frac{1}{\Gamma(q)}\int_0^t (t - s)^{q-1}m(s)ds\right\}\right) \le r,$$

where $M_\varepsilon = \sup_{t \in [0, a+\varepsilon]}\|Q(t)\|_{B(X)}$, $\varepsilon$ is a small constant.

**Remark 4.5.** The condition (H2) of Theorems 4.5-4.6 can be replaced by the following condition.

**(H2)′** There exist a constant $q_1 \in (0, q)$ and $m \in L^{\frac{1}{q_1}}(J, \mathbb{R}^+)$ such that

$$|f(t, x)| \leq m(t) \text{ for all } x \in B_r(J) \text{ and almost all } t \in [0, a].$$

We emphasize that (H2) is weaker than the condition (H2)′.

### 4.3.5 Noncompact Semigroup Case

If $Q(t)$ is noncompact, we give an assumption as follows.

**(H5)** There exists $\ell > 0$ such that for any bounded $D \subset X$,

$$\alpha(f(t, D)) \leq \ell\alpha(D).$$

**Theorem 4.7.** *Assume that (H0)-(H5) hold. Then the nonlocal Cauchy problem (4.15) has at least one mild solution in $B_r(J)$.*

**Proof.** By Lemmas 4.9-4.10, we know that $T : B_r(J) \to B_r(J)$ is bounded, continuous and $\{T_2 x : x \in B_r(J)\}$ is equicontinuous. For each bounded subset $B_0 \subset B_r(J)$, set

$$T^1(B_0) = T_2(B_0), \quad T^n(B_0) = T_2\left(\overline{co}(T^{n-1}(B_0))\right), \quad n = 2, 3, \ldots.$$

Then for any $\varepsilon > 0$, there is a sequence $\{x_n^{(1)}\}_{n=1}^{\infty}$ such that

$$\alpha(T^1(B_0(t))) = \alpha(T_2(B_0(t)))$$

$$\leq 2\alpha\left(\int_0^t (t-s)^{q-1} P_q(t-s) f(s, \{x_n^{(1)}(s)\}_{n=1}^{\infty}) ds\right) + \varepsilon$$

$$\leq \frac{4M}{\Gamma(q)} \int_0^t (t-s)^{q-1} \alpha(f(s, \{x_n^{(1)}(s)\}_{n=1}^{\infty})) ds + \varepsilon$$

$$\leq \frac{4M\ell\alpha(B_0)}{\Gamma(q)} \int_0^t (t-s)^{q-1} ds + \varepsilon$$

$$= \frac{4M\ell t^q \alpha(B_0)}{\Gamma(q+1)} + \varepsilon.$$

Since $\varepsilon > 0$ is arbitrary, we have

$$\alpha(T^1(B_0(t))) \leq \frac{4M\ell t^q}{\Gamma(q+1)} \alpha(B_0).$$

From Propositions 1.16-1.17, for any $\varepsilon > 0$, there is a sequence $\{x_n^{(2)}\}_{n=1}^{\infty} \subset \overline{co}(T^1(B_0))$ such that

$$\alpha(T^2(B_0(t))) = \alpha(T_2(\overline{co}(T^1(B_0(t)))))$$

$$\leq 2\alpha\left(\int_0^t (t-s)^{q-1} P_q(t-s) f(s, \{x_n^{(2)}(s)\}_{n=1}^{\infty}) ds\right) + \varepsilon$$

$$\leq \frac{4M}{\Gamma(q)} \int_0^t (t-s)^{q-1} \alpha(f(s, \{x_n^{(2)}(s)\}_{n=1}^{\infty})) ds + \varepsilon$$

$$\leq \frac{4M\ell}{\Gamma(q)} \int_0^t (t-s)^{q-1} \alpha(\{x_n^{(2)}(s)\}_{n=1}^\infty) ds + \varepsilon$$

$$\leq \frac{(4M\ell)^2 \alpha(B_0)}{\Gamma(q)\Gamma(q+1)} \int_0^t (t-s)^{q-1} s^q ds + \varepsilon$$

$$= \frac{(4M\ell)^2 t^{2q} \alpha(B_0)}{\Gamma(2q+1)} + \varepsilon.$$

It can be shown, by mathematical induction, that for every $\bar{n} \in \mathbb{N}$,

$$\alpha(T^{\bar{n}}(B_0(t))) \leq \frac{(4M\ell)^{\bar{n}} t^{\bar{n}q} \alpha(B_0)}{\Gamma(\bar{n}q+1)}.$$

Since

$$\lim_{\bar{n}\to\infty} \frac{(4M\ell a^q)^{\bar{n}}}{\Gamma(\bar{n}q+1)} = 0,$$

there exists a positive integer $\hat{n}$ such that

$$\frac{(4M\ell)^{\hat{n}} t^{\hat{n}q}}{\Gamma(\hat{n}q+1)} \leq \frac{(4M\ell a^q)^{\hat{n}}}{\Gamma(\hat{n}q+1)} = k < 1.$$

Then

$$\alpha(T^{\hat{n}}(B_0(t))) \leq k\alpha(B_0).$$

We know from Proposition 1.14, $T^{\hat{n}}(B_0(t))$ is bounded and equicontinuous, Then, from Proposition 1.15, we have

$$\alpha(T^{\hat{n}}(B_0)) = \max_{t\in[0,a]} \alpha(T^{\hat{n}}(B_0(t))).$$

Hence

$$\alpha(T^{\hat{n}}(B_0)) \leq k\alpha(B_0).$$

By using the similar method as in the proof of Theorem 4.3, we can prove that there exists a $D \subset B_r(J)$ such that

$$\alpha(T_2(D)) = 0. \tag{4.21}$$

On the other hand, for any $x_1, x_2 \in D$ and $t \in J$, according to (H3), we have

$$|(T_1x_1)(t) - (T_1x_2)(t)| \leq M|g(x_1) - g(x_2)|$$

$$\leq ML\|x_1 - x_2\|,$$

which implies that $\|T_1x_1 - T_1x_2\| \leq ML\|x_1 - x_2\|$. Thus, we obtain that

$$\alpha(T_1D) \leq ML\alpha(D). \tag{4.22}$$

By (4.21) and (4.22), we have

$$\alpha(T(D)) \leq \alpha(T_1(D)) + \alpha(T_2(D)) \leq ML\alpha(T(D)).$$

Thus, the operator $T$ is an $\alpha$-contraction in $D$. By Lemma 4.10, we know that $T$ is continuous. Hence, Theorem 1.10 shows that $T$ has a fixed point in $D \subset B_r(J)$. Therefore, the nonlocal Cauchy problem (4.15) has a mild solution in $B_r(J)$. $\qquad\square$

**Theorem 4.8.** *Assume that (H0)-(H2), (H3)′, (H4)′ and (H5) hold, then the non-local Cauchy problem (4.15) has at least a mild solution in $B_r(J)$.*

**Proof.** By the proof of Theorem 4.7, we know that there exists a $D \subset B_r(J)$ such that $T_2(D)$ is relatively compact, i.e., $\alpha(T_2(D)) = 0$. Clearly, $\alpha(T_1(D)) = 0$, since $g(x)$ is compact and $S_q(t)$ is bounded. Hence, we have

$$\alpha(T(D)) \leq \alpha(T_1(D)) + \alpha(T_2(D)) = 0.$$

Therefore, Theorem 1.10 shows that $T$ has a fixed point in $D \subset B_r(J)$. Therefore, the nonlocal Cauchy problem (4.15) has a mild solution in $B_r(J)$. $\square$

## 4.4 Nonlocal Problems for Evolution Equations

### 4.4.1 *Introduction*

The nonlocal condition has a better effect on the solution and is more precise for physical measurements than the classical condition alone. In this section, we discuss the existence of mild solutions of Cauchy problem for fractional evolution equations with nonlocal conditions

$$\begin{cases} {}^C_0D^\alpha_t x(t) = Ax(t) + f\left(t, x(t)\right), & \alpha \in (0,1), \ t \in J = [0,1], \\ x(0) = \sum_{k=1}^m a_k x(t_k), & k = 1, 2, \ldots, m, \end{cases} \tag{4.23}$$

where $A : D(A) \subset X \to X$ is the generator of a $C_0$-semigroup $\{Q(t)\}_{t \geq 0}$ on a Banach space $X$, $f : J \times X \to X$ is a given function and $a_k$ $(k = 1, 2, \ldots, m)$ are real numbers with $\Sigma_{k=1}^m a_k \neq 1$ and $t_k$, $k = 1, 2, \ldots, m$ are given points satisfying $0 \leq t_1 \leq t_2 \leq \ldots \leq t_m < 1$.

In Subsection 4.4.2, a suitable definition of mild solution of equation (4.23) is introduced by defining a bounded operator $B$. Meanwhile, two sufficient conditions are given to guarantee such $B$ exists. In Subsection 4.4.3, we state two existence result, the first one relies on a growth condition on $J$ and the other one relies on a growth condition involving two parts, one for $[0, t_m]$, and the other for $[t_m, 1]$.

### 4.4.2 *Definition of mild solutions*

Assume that $P_\alpha, S_\alpha$ are defined as in Subsection 4.3.2. Suppose that there exists a bounded operator $B : X \to X$ given by

$$B = \left(I - \sum_{k=1}^m a_k S_\alpha(t_k)\right)^{-1}. \tag{4.24}$$

We can give two sufficient conditions to guarantee such $B$ exists and is bounded.

**Lemma 4.13.** *The operator $B$ defined in (4.24) exists and is bounded, if one of the following two conditions holds:*

**(C1)** *there exist some real numbers $a_k$ such that*

$$M \sum_{k=1}^{m} |a_k| < 1, \tag{4.25}$$

*where $M = \sup_{t \in (0,\infty)} \|Q(t)\|_{B(X)} < \infty;$*

**(C2)** *$Q(t)$ is compact for each $t > 0$ and homogeneous linear nonlocal problems*

$$\begin{cases} {}^{C}_{0}D_t^{\alpha} x(t) = Ax(t), & \alpha \in (0,1), \ t \in J, \\ x(0) = \sum_{k=1}^{m} a_k x(t_k), \end{cases} \tag{4.26}$$

*has no non-trivial mild solutions.*

**Proof.** For (C1), it is easy to see

$$\left\| \sum_{k=1}^{m} a_k S_\alpha(t_k) \right\|_{B(X)} \leq M \sum_{k=1}^{m} |a_k| < 1,$$

where Proposition 4.5 and (4.25) are used. Thus by Neumann theorem, $B$ defined by (4.24) exists and it is bounded.

For (C2), it is obvious that the mild solutions of the problem (4.26) is given by

$$x(t) = S_\alpha(t)x(0),$$

which implies that

$$x(0) = \sum_{k=1}^{m} a_k x(t_k) = \sum_{k=1}^{m} a_k S_\alpha(t_k)x(0).$$

By Proposition 4.7, $S_\alpha(t_k)$ is compact for each $t_k > 0$, $k = 1, 2, \ldots, m$. Then $\sum_{k=1}^{m} a_k S_\alpha(t_k)$ is also compact. Since the problem (4.26) has no non-trivial mild solutions, one can obtain the desired result via Fredholm alternative theorem. □

Now we introduce the following definition of mild solutions of equation (4.23).

**Definition 4.3.** By a mild solution of equation (4.23), we mean a function $x \in C(J, X)$ satisfying

$$x(t) = S_\alpha(t) \sum_{k=1}^{m} a_k B(g(t_k)) + g(t), \ t \in J, \tag{4.27}$$

where

$$g(t_k) = \int_0^{t_k} (t_k - s)^{\alpha-1} P_\alpha(t_k - s)f(s, x(s))ds, \tag{4.28}$$

and

$$g(t) = \int_0^t (t - s)^{\alpha-1} P_\alpha(t - s)f(s, x(s))ds, \ t \in J. \tag{4.29}$$

**Remark 4.6.** To explain the formula (4.27), we note that a mild solution of the fractional evolution equation (4.23) with the initial condition is just $x(t) = S_\alpha(t)x(0) + g(t)$, so taking into account also the nonlocal condition, we get

$$x(0) = \sum_{k=1}^{m} a_k S_\alpha(t_k)x(0) + \sum_{k=1}^{m} a_k g(t_k).$$

So $x(0) = \sum_{k=1}^{m} a_k B(g(t_k))$ and hence $x(t) = S_\alpha(t) \sum_{k=1}^{m} a_k B(g(t_k)) + g(t)$ which is just (4.27).

### 4.4.3  *Existence*

Our first existence result is based on the well-known Schaefer fixed point theorem.

In this subsection, we study our problem under the following assumptions:

**(H1)** $f : J \times X \to X$ satisfies the Carathéodory conditions;

**(H2)** there exists a function $h$ such that $_0D_t^{-\alpha}h(t)$ exists for all $t \in J$ and $_0D_t^{-\alpha}h(\cdot) \in C((0,1], \mathbb{R}^+)$ with $\lim_{t\to 0+} {_0D_t^{-\alpha}}h(t) = 0$ and a nondecreasing continuous function $\Omega : \mathbb{R}^+ \to \mathbb{R}^+$ such that

$$|f(t,x)| \le h(t)\Omega(|x|)$$

for all $x \in X$ and all $t \in J$;

**(H3)** the inequality

$$\limsup_{\rho\to\infty} \rho \left( M^2 B\Omega(\rho) \sum_{k=1}^{m} |a_k|_0 D_t^{-\alpha}h(t_k) + M\Omega(\rho) \sup_{t\in J} {_0D_t^{-\alpha}}h(t) \right)^{-1} > 1$$

hold;

**(H4)** $Q(t)$ is compact for each $t > 0$.

We begin to consider the following problem

$$\begin{cases} {_0^C}D_t^\alpha x(t) = Ax(t) + \lambda f\left(t, x(t)\right), & \alpha \in (0,1], \ \lambda, t \in J, \\ x(0) = \sum_{k=1}^{m} a_k x(t_k). \end{cases} \tag{4.30}$$

Define an operator $F : C(J, X) \to C(J, X)$ as follows

$$(Fx)(t) = (F_1 x)(t) + (F_2 x)(t), \ t \in J,$$

where $F_i : C(J, X) \to C(J, X)$, $i = 1, 2$ are given by the formulas

$$(F_1 x)(t) = S_\alpha(t) \sum_{k=1}^{m} a_k B(g(t_k)), \ (F_2 x)(t) = g(t),$$

where $B$ is the operator defined in (4.24), $g(t_k)$ is defined in (4.28) and $g(t)$ is defined in (4.29).

Obviously, a mild solution of equation (4.30) is a solution of the operator equation

$$x = \lambda F x \qquad (4.31)$$

and conversely. Thus, we can apply Schaefer fixed point theorem to derive the existence results of solutions of equation (4.23).

**Lemma 4.14.** *Let $x$ be any solution of equation (4.31). Then, there exists $R^* > 0$ such that $\|x\|_C \leq R^*$ which is independent of the parameter $\lambda \in J$.*

**Proof.** Denote $R_0 := \|x\|$. Taking into accounts our conditions and Proposition 4.5 and Proposition 4.6, it follows from (4.27) that

$$|x(t)| \leq |(F_1 x)(t)| + |(F_2 x)(t)|$$

$$\leq M \sum_{k=1}^{m} |a_k| \|B\|_{B(X)} |g(t_k)| + |g(t)|, \ t \in J. \qquad (4.32)$$

Note that

$$|g(t_k)| \leq \int_0^{t_k} (t_k - s)^{\alpha-1} \|P_\alpha(t_k - s)\|_{B(X)} |f(s, x(s))| ds$$

$$\leq \frac{M}{\Gamma(\alpha)} \int_0^{t_k} (t_k - s)^{\alpha-1} h(s) \Omega(\|x\|) ds \qquad (4.33)$$

$$\leq \frac{M\Omega(R_0)}{\Gamma(\alpha)} \int_0^{t_k} (t_k - s)^{\alpha-1} h(s) ds$$

$$= M\Omega(R_0) {}_0 D_{t_k}^{-\alpha} h(t_k), \ k = 1, 2, \ldots, m,$$

and

$$|g(t)| \leq \frac{M\Omega(R_0)}{\Gamma(\alpha)} \int_0^t (t - s)^{\alpha-1} h(s) ds \qquad (4.34)$$

$$= M\Omega(R_0) \sup_{t \in J} {}_0 D_t^{-\alpha} h(t), \ t \in J.$$

In view of (4.32)-(4.34), one can obtain

$$R_0 := \|x\| \leq M^2 \|B\|_{B(X)} \Omega(R_0) \sum_{k=1}^{m} |a_k| {}_0 D_{t_k}^{-\alpha} h(t_k) + M\Omega(R_0) \sup_{t \in J} {}_0 D_t^{-\alpha} h(t), \ t \in J,$$

which implies that

$$R_0 \left( M^2 \|B\|_{B(X)} \Omega(R_0) \sum_{k=1}^{m} |a_k| {}_0 D_{t_k}^{-\alpha} h(t_k) + M\Omega(R_0) \sup_{t \in J} {}_0 D_t^{-\alpha} h(t) \right)^{-1} \leq 1. $$

$$\qquad (4.35)$$

However, it follows (H3) that there exists a $R^* > 0$ such that for all $R > R^*$ we can derive that

$$R \left( M^2 \|B\|_{B(X)} \Omega(R) \sum_{k=1}^{m} |a_k| {}_0 D_{t_k}^{-\alpha} h(t_k) + M\Omega(R) \sup_{t \in J} {}_0 D_t^{-\alpha} h(t) \right)^{-1} > 1. \qquad (4.36)$$

Now, comparing the equalities (4.35) and (4.36), we claim that $R_0 \leq R^*$. As a result, we find that $\|x\| \leq R^*$ which independents the parameter $\lambda$. This completes the proof. $\qquad \square$

Let

$$\mathfrak{B}_{R^*} = \{x \in C(J, X) : \|x\| \leq R^*\}.$$

Then $\mathfrak{B}_{R^*}$ is a bounded closed and convex subset in $C(J, X)$.

By Lemma 4.14, we can derive the following result.

**Lemma 4.15.** *The operator $F$ maps $\mathfrak{B}_{R^*}$ into itself.*

**Lemma 4.16.** *The operator $F : \mathfrak{B}_{R^*} \to \mathfrak{B}_{R^*}$ is completely continuous.*

**Proof.** For our purpose, we only need to check that $F_i : \mathfrak{B}_{R^*} \to \mathfrak{B}_{R^*}$, $i = 1, 2$ is completely continuous. Firstly, by repeating the same producers of Lemma 4.10 and Theorem 4.5, one can obtain $F_2 : \mathfrak{B}_{R^*} \to \mathfrak{B}_{R^*}$ is completely continuous.

Secondly, one can check that $F_1 : \mathfrak{B}_{R^*} \to \mathfrak{B}_{R^*}$ is continuous (by (H1), (H2) and Proposition 4.5) and $F_1 : \mathfrak{B}_{R^*} \to \mathfrak{B}_{R^*}$ is compact in viewing of $S_\alpha(t)$ is compact for each $t > 0$ (by (H4) and Proposition 4.7). The proof is completed. $\qquad \square$

Now, we can state the main result of this section.

**Theorem 4.9.** *Assume that (H1)-(H4) hold and the condition (C1) (or (C2)) is satisfied. Then equation (4.23) has at least one solution $u \in C(J, X)$ and the set of the solutions of equation (4.23) is bounded in $C(J, X)$.*

**Proof.** Obviously, the set $\{x \in C(J, X) : x = \lambda Fx, 0 < \lambda < 1\}$ is bounded due to Lemma 4.15. Now we can apply Theorem 1.6 to derive that $F$ has a fixed point in $\mathfrak{B}_{R^*}$ which is just the mild solution of equation (4.23). This completes the proof. $\qquad \square$

Our second existence result is based on O'Regan fixed point theorem.

In addition to (H1), (H4) and (C1) (or (C2)), motivated by Boucherif and Precup, 2003, 2007, we introduce the following two assumptions:

**(H5)** there exists a function $h$ such that $_0D_t^{-\alpha}h(t)$ exists for all $t \in [0, t_m]$ and $_0D_t^{-\alpha}h(\cdot) \in C((0, t_m], \mathbb{R}^+)$ with $\lim_{t \to 0^+} {_0D_t^{-\alpha}h(t)} = 0$ and a nondecreasing continuous function $\Omega : \mathbb{R}^+ \to \mathbb{R}^+$ such that

$$|f(t, x)| \leq h(t)\Omega(|x|)$$

for all $x \in X$, and for all $t \in [t_m, 1]$ there exists a function $l$ such that $_{t_m}D_t^{-\alpha}l(t)$ exists and $_{t_m}D_t^{-\alpha}l(\cdot) \in C([t_m, 1], \mathbb{R}^+)$ such that

$$|f(t, x)| \leq l(t), \tag{4.37}$$

for all $x \in X$. Moreover, $\Omega$ has the property

$$r > M\Omega(r) \left( \sum_{k=1}^{m} |a_k| \|B\|_{B(X)} + 1 \right) \sup_{t \in [0, t_m]} {_0D_t^{-\alpha}h(t)} \tag{4.38}$$

for all $r > R_1^* > 0$;

**(H6)** there exists a function $q$ such that $_{t_m}D_t^{-\alpha}q(t)$ exists for all $t \in [t_m, 1]$ and $_{t_m}D_t^{-\alpha}q(\cdot) \in C([t_m, 1], \mathbb{R}^+)$ with $M \sup_{t \in [t_m, 1]} {}_0D_t^{-\alpha}q(t) \leq 1$ and a nondecreasing continuous function $\Psi : \mathbb{R}^+ \to \mathbb{R}^+$ with $\Psi(r) < r$ for $r > 0$ such that

$$|f(t, x) - f(t, y)| \leq q(t)\Psi(|x - y|)$$

for all $t \in [t_m, 1]$ and $x, y \in X$.

Consider equation (4.30) again and the equivalent equation

$$x = \lambda T x \tag{4.39}$$

where $T : C(J, X) \to C(J, X)$ is defined by

$$(Tx)(t) = (T_1 x)(t) + (T_2 x)(t), \ t \in J,$$

where $T_i : C(J, X) \to C(J, X)$, $i = 1, 2$ given by

$$(T_1 x)(t) = \begin{cases} S_\alpha(t) \sum_{k=1}^m a_k B(g(t_k)) + g(t), & t \in [0, t_m), \\ S_\alpha(t) \sum_{k=1}^m a_k B(g(t_k)) \\ \qquad + \int_0^{t_m} (t-s)^{\alpha-1} P_\alpha(t-s)f(s, x(s))ds, & t \in [t_m, 1], \end{cases}$$

and

$$(T_2 x)(t) = \begin{cases} 0, & t \in [0, t_m), \\ \int_{t_m}^t (t-s)^{\alpha-1} P_\alpha(t-s)f(s, x(s))ds, & t \in [t_m, 1]. \end{cases}$$

We first prove that any solutions of equation (4.39) have a priori bound.

**Lemma 4.17.** *Let $x$ be any solution of equation (4.39). Then, there exist $R_i^* > 0$ $i = 1, 2$ such that $\|x\|_{C([0,t_m],X)} \leq R_1^*$ and $\|x\|_{C([t_m,1],X)} \leq R_2^*$. In other words, $\|x\| \leq R^* = \max\{R_1^*, R_2^*\}$ which is independent of the parameter $\lambda$.*

**Proof. Case I.** We prove that there exists $R_1^* > 0$ such that $\|x\|_{C([0,t_m],X)} \leq R_1^*$. For $t \in [0, t_m]$ and $\lambda \in J$, denote $R_{[0,t_m]} := \|x\|_{C([0,t_m],X)}$, we have

$$|x(t)| \leq \lambda |(T_1 x)(t)| + |(T_2 x)(t)|$$

$$\leq M \sum_{k=1}^m |a_k| \|B\|_{B(X)} |g(t_k)| + |g(t)|$$

$$\leq M \sum_{k=1}^m |a_k| \|B\|_{B(X)} \frac{M}{\Gamma(\alpha)} \int_0^{t_k} (t_k - s)^{\alpha-1} h(s)\Omega(R_{[0,t_m]})ds$$

$$\quad + \frac{M}{\Gamma(\alpha)} \int_0^t (t-s)^{\alpha-1} h(s)\Omega(R_{[0,t_m]})ds$$

$$\leq M\Omega(R_{[0,t_m]}) \left( \sum_{k=1}^{m} |a_k| \|B\|_{B(X)} + 1 \right) \sup_{t\in[0,t_m]} {_0}D_t^{-\alpha}h(t),$$

which implies that

$$R_{[0,t_m]} \leq M\Omega(R_{[0,t_m]}) \left( \sum_{k=1}^{m} |a_k| \|B\|_{B(X)} + 1 \right) \sup_{t\in[0,t_m]} {_0}D_t^{-\alpha}h(t).$$

From (4.38) we find that there exists $R_1^* \geq R_{[0,t_m]} > 0$ such that

$$\|x\|_{C([0,t_m],X)} \leq R_1^*.$$

**Case II.** We prove that there exists $R_2^* > 0$ such that $\|x\|_{C([t_m,1],X)} \leq R_2^*$. For $t \in [t_m, 1]$ and $\lambda \in J$, keeping in mind our assumptions, we find that

$$|x(t)| \leq M \sum_{k=1}^{m} |a_k| \|B\|_{B(X)} \frac{M}{\Gamma(\alpha)} \int_0^{t_k} (t_k - s)^{\alpha-1} h(s)\Omega(R_1^*) ds$$

$$+ \frac{M}{\Gamma(\alpha)} \int_0^{t_m} (t - s)^{\alpha-1} h(s)\Omega(R_1^*) ds + \frac{M}{\Gamma(\alpha)} \int_{t_m}^{t} (t - s)^{\alpha-1} h(s) ds$$

$$\leq M\Omega(R_1^*) \left( \sum_{k=1}^{m} |a_k| \|B\|_{B(X)} + 1 \right) \sup_{t\in[0,t_m]} {_0}D_t^{-\alpha}h(t)$$

$$+ M \sup_{t\in[t_m,1]} {_{t_m}}D_t^{-\alpha}l(t),$$

which implies that

$$\|x\|_{C([t_m,1],X)} \leq R_2^*,$$

where

$$R_2^* = M \left[ \Omega(R_1^*) \left( \sum_{k=1}^{m} |a_k| \|B\|_{B(X)} + 1 \right) \sup_{t\in[0,t_m]} {_0}D_t^{-\alpha}h(t) + \sup_{t\in[t_m,1]} {_{t_m}}D_t^{-\alpha}l(t) \right].$$

Let $R^* = \max\{R_1^*, R_2^*\}$. Then we find that any possible solutions of equation (4.39) satisfy $\|x\| \leq R^*$ which are independent of the parameter $\lambda$. This completes the proof. $\square$

Denote

$$\mathcal{D} = \{x \in C(J,X) : \|x\| < R^* + 1\}.$$

We can proceed as in the proof of Lemma 4.17 to derive the following result.

**Lemma 4.18.** $T(\overline{\mathcal{D}})$ *is bounded.*

One can proceed as in the proof of Lemma 4.16 to obtain the following result.

**Lemma 4.19.** *The operator $T_1 : \overline{\mathcal{D}} \to C(J,X)$ is completely continuous.*

Next, we show the following result.

**Lemma 4.20.** *The operator $T_2 : \overline{\mathcal{D}} \to C(J,X)$ is nonlinear contraction.*

**Proof.** It follows from the definition of $T_2$ we only need to show $T_2 : \overline{\mathcal{D}} \to C([t_m, 1], X)$ is a nonlinear contraction.

In fact, for any $x, y \in \overline{\mathcal{D}}$ and $t \in [t_m, 1]$, we have

$$|(T_2 x)(t) - (T_2 y)(t)| \leq \int_{t_m}^{t} (t - s)^{\alpha - 1} \|P_\alpha(t - s)[f(s, x(s)) - f(s, y(s))]\| \, ds$$

$$\leq \frac{M}{\Gamma(\alpha)} \int_{t_m}^{t} (t - s)^{\alpha - 1} q(s) \Psi(|x(s) - y(s)|) ds$$

$$\leq \frac{M \Psi(\|x - y\|)}{\Gamma(\alpha)} \int_{t_m}^{t} (t - s)^{\alpha - 1} q(s) ds$$

$$\leq \left( M \sup_{t \in [t_m, 1]} {}_{t_m} D_t^{-\alpha} q(t) \right) \Psi(\|x - y\|),$$

which implies that

$$\|T_2 x - T_2 y\| \leq \Psi(\|x - y\|).$$

This completes the proof.                                                    □

Now, we are ready to present the main result of this section.

**Theorem 4.10.** *Assume that (H1), (H4), (H5) and (H6) hold and the condition (C1) (or (C2)) is satisfied. Then equation (4.23) has at least one solution $u \in C(J, X)$.*

**Proof.** By Lemma 4.17 we see that (ii) of Theorem 1.8 does not hold. Thus, there is no solution of equation (4.39) with $x \in \partial \mathcal{D}$. Therefore, one can apply Theorem 1.8 to derive that $T$ has a fixed point in $\mathcal{D}$ which is just the mild solution of equation (4.23). This completes the proof.                          □

**Remark 4.7.** Theorem 4.10 also holds even if (H5) and (H6) are replaced by the following conditions:

**(H5)′** condition (H5) is assumed without (4.37);
**(H6)′** denoting $\delta := \lim_{r \to +\infty} \inf \frac{\Psi(r)}{r} \leq 1$, condition (H6) is assumed in addition with

$$M \delta \sup_{t \in [t_m, 1]} {}_{t_m} D_t^{-\alpha} q(t) < 1.$$

Indeed, we can modify Case II in the proof of Lemma 4.17 as follows

$$|x(t)| \leq M \sum_{k=1}^{m} |a_k| \|B\|_{B(X)} \frac{M}{\Gamma(\alpha)} \int_{0}^{t_k} (t_k - s)^{\alpha - 1} h(s) \Omega(R_1^*) ds$$

$$+ \frac{M}{\Gamma(\alpha)} \int_{0}^{t_m} (t - s)^{\alpha - 1} h(s) \Omega(R_1^*) ds$$

$$+ \frac{M}{\Gamma(\alpha)} \int_{t_m}^{t} (t - s)^{\alpha - 1} |f(s, 0)| ds + \frac{M}{\Gamma(\alpha)} \int_{t_m}^{t} (t - s)^{\alpha - 1} q(s) \Psi(|x(s)|) ds$$

$$\leq M\Omega(R_1^*) \left( \sum_{k=1}^{m} |a_k| \|B\|_{B(X)} + 1 \right) \sup_{t\in[0,t_m]} t_m D_t^{-\alpha} h(t)$$

$$+ \frac{M \sup_{t\in[t_m,t]} |f(t,0)|(1-t_m)^\alpha}{\Gamma(\alpha+1)}$$

$$+ \frac{M}{\Gamma(\alpha)} \int_{t_m}^{t} (t-s)^{\alpha-1} q(s)(\delta|x(s)| + \delta_1) ds,$$

for some $\delta_1 \geq 0$.

Then we have

$$R_2^* = \frac{1}{1 - M\delta \sup_{t\in[t_m,1]} t_m D_t^{-\alpha} q(t)}$$

$$\times \left[ M\Omega(R_1^*) \left( \sum_{k=1}^{m} |a_k| \|B\|_{B(X)} + 1 \right) \sup_{t\in[0,t_m]} t_m D_t^{-\alpha} h(t) \right.$$

$$\left. + \frac{M \sup_{t\in[t_m,t]} |f(t,0)|(1-t_m)^\alpha}{\Gamma(\alpha+1)} + M\delta_1 \sup_{t\in[t_m,1]} t_m D_t^{-\alpha} q(t) \right].$$

## 4.5 Abstract Cauchy Problems with Almost Sectorial Operators

### 4.5.1 *Introduction*

Let $X$ be a complex Banach space with norm $|\cdot|$. As usual, for a linear operator $A$, we denote by $D(A)$ the domain of $A$, by $\sigma(A)$ its spectrum, while $\rho(A) := \mathbb{C} - \sigma(A)$ is the resolvent set of $A$, and denote by the family $R(z;A) = (zI - A)^{-1}$, $z \in \rho(A)$ of bounded linear operators the resolvent of $A$. Moreover, we denote by $B(X,Y)$ the space of all bounded linear operators from Banach space $X$ to Banach space $Y$ with the usual operator norm $\|\cdot\|_{B(X,Y)}$, and we abbreviate this notation to $B(X)$ when $Y = X$, and write $\|T\|_{B(X)}$ as $\|T\|$ for every $T \in B(X)$.

When dealing with parabolic evolution equations, it is usually assumed that the partial differential operator in the linear part is a sectorial operator, stimulated by the fact that this class of operators appears very often in the applications. For example, one can find from Henry, 1981; Lunardi, 1995 and Pazy, 1983 that many elliptic differential operators equipped homogeneous boundary conditions are sectorial when they are considered in Lebesgue spaces (e.g., $L^p$-spaces) or in the space of continuous functions. We here mention that the operator $A_\varepsilon$ in Example 4.2, which acts on a domain of "dumb-bell with a thin handle", is sectorial on $V_\varepsilon^p$. However, as presented in Example 4.2 and Example 4.3, though the resolvent set of some partial differential operators considered in some special domains such as the limit "domain" of dumb-bell with a thin handle or in some spaces of more regular functions such as the space of Hölder continuous functions, contains a sector, but for which the resolvent operators do not satisfy the required estimate to be a sectorial operator.

**Example 4.2.** In this notation the "dumb-bell with a thin handle" has the form

$$\Omega_\varepsilon = D_1 \cup Q_\varepsilon \cup D_2 \quad (\varepsilon \in (0,1]; \text{ small}),$$

where $D_1$ and $D_2$ are mutually disjoint bounded domains in $\mathbb{R}^N (N \geq 2)$ with smooth boundaries, joined by a thin channel, $Q_\varepsilon$ (which is not required to be cylindrical.), which degenerates to a 1-dim line segment $Q_0$ as $\varepsilon$ approaches zero. This implies that passing to the limit as $\varepsilon \to 0$, the limit "domain" of $\Omega_\varepsilon$ consists of the fixed part $D_1$, $D_2$ and the line segment $Q_0$. Without loss of generality, we may assume that $Q_0 = \{(x,0,\ldots,0); 0 < x < 1\}$. Let $P_0 = (0,0,\ldots,0), P_1 = (1,0,\ldots,0)$ be the points where the line segment touches the boundary of $D_1$ and $D_2$. Put $\Omega = D_1 \cup D_2$.

Firstly, consider the evolution equation of parabolic type equipped with Neumann boundary condition in the form

$$\begin{cases} u_t - \Delta u + u = f(u), & x \in \Omega_\epsilon, \ t > 0, \\ \dfrac{\partial u}{\partial n} = 0, & x \in \partial\Omega_\epsilon, \end{cases} \tag{4.40}$$

where $\Delta$ stands for the Laplacian operator with respect to the spatial variable $x \in \Omega_\epsilon$, $\partial\Omega_\epsilon$ is the boundary of $\Omega_\epsilon$, $\frac{\partial}{\partial n}$ denotes the outward normal derivative on $\partial\Omega_\epsilon$ and $f : \mathbb{R} \to \mathbb{R}$ is a nonlinearity. Let $V_\epsilon^p$ ($1 \leq p < \infty$) denote the family of spaces based on $L^p(\Omega_\epsilon)$, equipped with the norm

$$\|u\|_{V_\epsilon^p} = \left( \int_\Omega |u|^p + \frac{1}{\varepsilon^{N-1}} \int_{Q_\epsilon} |u|^p \right)^{1/p}.$$

Define the linear operator $A_\varepsilon : D(A_\varepsilon) \subset V_\epsilon^p \mapsto V_\epsilon^p$ by

$$D(A_\varepsilon) = \left\{ u \in W^{2,p}(\Omega_\epsilon) : \ \Delta u \in V_\epsilon^p, \ \frac{\partial u}{\partial n}\Big|_{\partial\Omega_\epsilon} = 0 \right\},$$

$$A_\varepsilon u = -\Delta u + u, \quad u \in D(A_\varepsilon).$$

It follows from a standard argument that the operator $A_\varepsilon$ generates an analytic semigroup on $V_\epsilon^p$. Moreover, the following estimate holds

$$\|R(\lambda; -A_\varepsilon)\|_{B(L^p(\Omega_\epsilon))} \leq \frac{C}{|\lambda|}, \quad \text{for } \lambda \in \Sigma_\theta',$$

where $\Sigma_\theta' = \{\lambda \in \mathbb{C} : |\arg(\lambda - 1)| \leq \theta\}$ with $\theta > \frac{\pi}{2}$, and $C$ is a constant that does not depend on $\varepsilon$ (see, e.g., Henry, 1981 and Pazy, 1983).

The limit problem of (4.40) as $\varepsilon \to 0$ is the following problem studied in Carvalho, Dlotko and Neseimento, 2008

$$\begin{cases} w_t - \Delta w + w = f(w), & x \in \Omega, \ t > 0, \\ \dfrac{\partial w}{\partial n} = 0, & x \in \partial\Omega, \\ v_t - \dfrac{1}{g}(gv_x)_x + v = f(v), & x \in Q_0 = (0,1), \\ v(0) = w(P_0), v(1) = w(P_1), \end{cases}$$

where $w$ is a function that lives in $\Omega$ and $v$ lives in the line segment $Q_0$, the function $g : [0, 1] \to (0, \infty)$ is a smooth function related to the geometry of the channel $Q_\varepsilon$, more exactly, on the way the channel $Q_\varepsilon$ collapses to the segment line $Q_0$. Observe that the vector $(w, v)$ is continuous in the junction between $\Omega$ and $Q_0$ and the variable $w$ does not depend on the variable $v$, but $v$ depends on $w$.

We identify $V_0^p$ with $L^p(\Omega) \oplus L_g^p(0, 1)$ ($1 \le p < \infty$) endowed with the norm

$$\|(w, v)\|_{V_0^p} = \left( \int_\Omega |w|^p + \int_0^1 g|v|^p \right)^{1/p}.$$

Consider the operator $A_0 : D(A_0) \subset V_0^p \mapsto V_0^p$ defined by

$$D(A_0) = \{(w, v) \in V_0^P; w \in D(\Delta_\Omega), v \in L_g^p(0, 1), w(P_0) = v(0), w(P_1) = v(1)\},$$

$$A_0(w, v) = \left( -\Delta w + w, -\frac{1}{g}(gv')' + v \right), \quad (w, v) \in V_0^p, \qquad (4.41)$$

where $\Delta_\Omega$ is the Laplace operator with homogeneous Neumann boundary conditions in $L^p(\Omega)$ and

$$D(\Delta_\Omega) = \left\{ u \in W^{2,p}(\Omega) : \frac{\partial u}{\partial n}\bigg|_{\partial\Omega} = 0 \right\}.$$

As pointed out by Arrieta, Carvalho and Lozada-Cruz, 2009a, the operator $A_0$ defined by (4.41) is not a sectorial operator. Its spectrum is all real and, therefore, it is contained in a sector but the resolvent estimate is different from the case of sectorial operator. More precisely, the operator $A_0$ has the following properties (see also Arrieta, Carvalho and Lozada-Cruz, 2006, 2009a):

(a) the domain $D(A_0)$ is dense in $V_0^P$;
(b) if $p > \frac{N}{2}$, then $A_0$ is a closed operator;
(c) $A_0$ has compact resolvent;
(d) for some $\mu \in (0, \frac{\pi}{2})$, $\Sigma_\mu := \{\lambda \in \mathbb{C} \setminus \{0\} : |\arg \lambda| \le \pi - \mu\} \cup \{0\} \subset \rho(-A_0)$, and for $\frac{N}{2} < q \le p$, the following estimate holds:

$$\|R(\lambda; -A_0)\|_{B(V_0^q, V_0^p)} \le \frac{C}{1 + |\lambda|^{\gamma'}}, \quad \lambda \in \Sigma_\mu, \qquad (4.42)$$

for each $0 < \gamma' < 1 - \frac{N}{2q} - \frac{1}{2}(\frac{1}{q} - \frac{1}{p}) < 1$, where $C$ is a positive constant.

**Remark 4.8.** In fact, it is easy to prove that the estimate (4.42) with $p = q > \frac{N}{2}$ is equivalent to

$$\|R(\lambda; -A_0)\|_{B(V_0^p)} \le \frac{\tilde{C}}{|\lambda|^{\gamma'}}, \quad \lambda \in \Sigma_\mu \setminus \{0\},$$

for $0 < \gamma' < 1 - \frac{N}{2p}$, where $\tilde{C}$ is a positive constant.

We refer to Section 2 of Arrieta, Carvalho and Lozada-Cruz, 2006 for a complete and rigorous definition of the dumb-bell domain, and to Arrieta, 1995; Arrieta, Carvalho and Lozada-Cruz, 2006, 2009a,b; Dancer and Daners, 1997; Gadyl'shin, 2005; Jimbo, 1989 for related studies of partial differential equations involving dumb-bell domain.

**Example 4.3.** Assume that $\Omega$ is a bounded domain in $\mathbb{R}^N$ ($N \geq 1$) with boundary $\partial\Omega$ of class $C^{4m}$ ($m \in \mathbb{N}$). Let $C^l(\overline{\Omega})$, $l \in (0,1)$, denote the usual Banach space with norm $|\cdot|_l$. Consider the elliptic differential operator $A' : D(A') \subset C^l(\Omega) \mapsto C^l(\Omega)$ in the form

$$D(A') = \left\{ u \in C^{2m+l}(\overline{\Omega}) : \ D^\beta u|_{\partial\Omega} = 0, |\beta| \leq m-1 \right\},$$

$$A'u = \sum_{|\beta| \leq 2m} a_\beta(x) D^\beta u(x), \quad u \in D(A'),$$

where $\beta$ is a multiindex in $(\mathbb{N} \cup \{0\})^n$,

$$|\beta| = \sum_{j=1}^n \beta_j, \quad D^\beta = \prod_{j=1}^n \left( -i\frac{\partial}{\partial x_j} \right)^{\beta_j}.$$

The coefficients $a_\beta : \overline{\Omega} \mapsto \mathbb{C}$ of $A'$ are assumed to satisfy

**(i)** $a_\beta \in C^l(\overline{\Omega})$ for all $|\beta| \leq 2m$;
**(ii)** $a_\beta(x) \in \mathbb{R}$ for all $x \in \overline{\Omega}$ and $|\beta| = 2m$;
**(iii)** there exists a constant $M > 0$ such that

$$M^{-1}|\xi|^2 \leq \sum_{|\beta|=2m} a_\beta(x)\xi^\beta \leq M|\beta|^2, \quad \text{for all } \xi \in \mathbb{R}^N \text{ and } x \in \overline{\Omega}.$$

Then, the following statements hold.

**(a)** $A'$ is not densely defined in $C^l(\overline{\Omega})$;
**(b)** there exist $\nu$, $\varepsilon > 0$ such that

$$\sigma(A' + \nu) \subset S_{\frac{\pi}{2}-\varepsilon} = \left\{ \lambda \in \mathbb{C} \setminus \{0\} : \ |\arg \lambda| \leq \frac{\pi}{2} - \varepsilon \right\} \cup \{0\},$$

$$\|R(\lambda; A' + \nu)\|_{B(C^l(\overline{\Omega}))} \leq \frac{C}{|\lambda|^{1-\frac{l}{2m}}}, \quad \lambda \in \mathbb{C} \setminus S_{\frac{\pi}{2}-\varepsilon};$$

**(c)** the exponent $\frac{l}{2m} - 1 \in (-1,0)$ is sharp. In particular, the operator $A' + \nu$ is not sectorial.

Notice in particular that the Laplace operator satisfies the conditions (a)-(c) in Example 4.3. For more details we refer to Wahl, 1972.

Observe that from Example 4.2 and Remark 4.8, if $p > \frac{N}{2}$, then $A_0 \in \Theta_\mu^{-\gamma'}(V_0^p)$ for some $\gamma' \in (0, 1 - \frac{N}{2p})$ and $\mu \in (0, \frac{\pi}{2})$, that is, $A_0$ is an almost sectorial operator on $V_0^p$. Also, from Example 4.3 one can find that $(A' + \nu) \in \Theta_{\frac{\pi}{2}-\varepsilon}^{\frac{l}{2m}-1}(C^l(\overline{\Omega}))$, which implies that $A' + \nu$ is an almost sectorial operator on $C^l(\overline{\Omega})$.

In this section, motivated by the above consideration, we are interested in studying the Cauchy problem for the linear evolution equation

$$\begin{cases} {}_0^C D_t^\alpha u(t) + Au(t) = f(t), & t > 0, \\ u(0) = u_0, \end{cases} \tag{4.43}$$

as well as the Cauchy problem for the corresponding semilinear fractional evolution equation

$$\begin{cases} {}_0^C D_t^\alpha u(t) + Au(t) = f(t, u(t)), & t > 0, \\ u(0) = u_0 \end{cases} \tag{4.44}$$

in $X$, where ${}_0^C D_t^\alpha$, $0 < \alpha < 1$, is Caputo fractional derivative of order $\alpha$ and $A$ is an almost sectorial operator, that is, $A \in \Theta_\omega^\gamma(X)$ $(-1 < \gamma < 0, 0 < \omega < \pi/2)$. The main purpose is to study the existence and uniqueness of mild solutions and classical solutions of Cauchy problems (4.43) and (4.44). To do this, we construct two operator families based on the generalized Mittag-Leffler functions and the resolvent operators associated with $A$, present deep anatomy on basic properties for these families consisting on the study of the compactness, and prove that, under natural assumptions, reasonable concept of solutions can be given to problems (4.43) and (4.44), which in turn is used to find solutions to the Cauchy problems.

**Remark 4.9.** We make no assumption on the density of the domain of $A$.

**Remark 4.10.**

(i) M. M. Dzhrbashyan and A. B. Nersessyan in Dzhrbashyan and Nersessyan, 1968 (see also Miller and Ross, 1993) showed that the solution of the Cauchy problem

$$\begin{cases} {}_0^C D_t^\alpha u(t) + \lambda u(t) = 0, & t > 0, \\ u(0) = 1, & 0 < \alpha < 1, \end{cases}$$

has the form $u(t) = E_\alpha(-\lambda t^\alpha)$, where $E_\alpha$ is the known Mittag-Leffler function. This result issues a warning to us that no matter how smooth the data $u_0$ is, it is inappropriate to define the mild solution of problem (4.43) as follows

$$u(t) = T(t)u_0 + \frac{1}{\Gamma(\alpha)} \int_0^t (t-s)^{\alpha-1} T(t-s) f(s) ds,$$

where $T(t)$ is the semigroup generated by $A$ (see Remark 1.6(i)), though this fashion was used in some situations of previous research (see, e.g., Jaradat, Ao-Omari and Momani, 2008).

(ii) Let us point out that in the treatment of problems (4.43) and (4.44), one of the difficult points is to give reasonable concept of solutions (see also the works of Zhou and Jiao, 2010a; Hernandez, O'Regan and Balachandran, 2010). Another is that even though the operator $A$ generates a semigroup $T(t)$ in $X$, it is not continuous at $t = 0$ for nonsmooth initial data $u_0$.

**(iii)** It is worth mentioning that if it is the case when $A$ is a matrix (or even bounded linear operators ) then Kilbas, Srivastava and Trujill, 2006, obtained an explicit representation of mild solution to problem (4.43).

Let us now give a short summary of this section, which is organized in a way close to that given by Carvalho, Dlotko and Nescimento, 2008. In Subsection 4.5.2, we construct a pair of families of operators and present deep anatomy on the properties for these families. Based on the families of operators defined in Subsection 4.5.2, a reasonable concept of solution is given in Subsection 4.5.3 to problems (4.43), which in turn is used to analyze the existence of mild solutions and classical solutions to the Cauchy problem. The corresponding semilinear problem (4.44) is studied in Subsection 4.5.4. We investigate the existence of mild solutions, and then the existence of classical solutions. Finally, based mainly in Carvalho, Dlotko and Nescimento, 2008; Periago and Straub, 2002, we present three examples in Subsection 4.5.5 to illustrate our results.

**Remark 4.11.** Let us note that results in this section can be easily extended to the case of (general) sectorial operators.

### 4.5.2 *Properties of Operators*

Throughout this subsection we let $A$ be an operator in the class $\Theta_\omega^\gamma(X)$ and $-1 < \gamma < 0$, $0 < \omega < \pi/2$. In the sequel, we succeed in defining two families of operators based on the generalized Mittag-Leffler functions and the resolvent operators associated with $A$. They are two families of linear and bounded operators. In order to check the properties of the families, we need a third object, namely the semigroup associated with $A$. We stress that these families are used very frequently throughout the rest of this section. Below the letter $C$ denotes various positive constants.

Define operator families $\{\mathcal{S}_\alpha(t)\}|_{t \in S_{\frac{\pi}{2}-\omega}^0}$, $\{\mathcal{P}_\alpha(t)\}|_{t \in S_{\frac{\pi}{2}-\omega}^0}$ by

$$\mathcal{S}_\alpha(t) := E_\alpha(-zt^\alpha)(A) = \frac{1}{2\pi i} \int_{\Gamma_\theta} E_\alpha(-zt^\alpha) R(z; A) dz,$$

$$\mathcal{P}_\alpha(t) := e_\alpha(-zt^\alpha)(A) = \frac{1}{2\pi i} \int_{\Gamma_\theta} e_\alpha(-zt^\alpha) R(z; A) dz,$$

where the integral contour $\Gamma_\theta := \{\mathbb{R}_+ e^{i\theta}\} \cup \{\mathbb{R}_+ e^{-i\theta}\}$, is oriented counter-clockwise and $\omega < \theta < \mu < \frac{\pi}{2} - |\arg t|$.

We need some basic properties of these families which are used further in this section.

**Theorem 4.11.** *For each fixed $t \in S_{\frac{\pi}{2}-\omega}^0$, $\mathcal{S}_\alpha(t)$ and $\mathcal{P}_\alpha(t)$ are linear and bounded operators on $X$. Moreover, there exists constants $C_s = C(\alpha, \gamma) > 0$, $C_p = C(\alpha, \gamma) > 0$ such that for all $t > 0$,*

$$\|\mathcal{S}_\alpha(t)\|_{B(X)} \le C_s t^{-\alpha(1+\gamma)}, \quad \|\mathcal{P}_\alpha(t)\|_{B(X)} \le C_p t^{-\alpha(1+\gamma)}. \tag{4.45}$$

**Proof.** Note, from the asymptotic expansion of $E_{\alpha,\beta}$ that for each fixed $t \in S^0_{\frac{\pi}{2}-\omega}$,

$$E_\alpha(-zt^\alpha),\ e_\alpha(-zt^\alpha) \in \mathcal{F}^\gamma_0(S^0_\mu).$$

Therefore, by (1.20), the operators families $\{\mathcal{S}_\alpha(t)\}|_{t \in S^0_{\frac{\pi}{2}-\omega}}$, $\{\mathcal{P}_\alpha(t)\}|_{t \in S^0_{\frac{\pi}{2}-\omega}}$ are well-defined, and for each $t \in S^0_{\frac{\pi}{2}-\omega}$, $\mathcal{S}_\alpha(t)$ and $\mathcal{P}_\alpha(t)$ are linear bounded operators on $X$. So, to prove the theorem, it is sufficient to prove that the estimates in (4.45) hold.

Let $T(t)$, $t \in S^0_{\frac{\pi}{2}-\omega}$, be the semigroup defined by (1.21). Then by (W4) and the Fubini Theorem, we get

$$\begin{aligned}
\mathcal{S}_\alpha(t)x &= \frac{1}{2\pi i} \int_{\Gamma_\theta} E_\alpha(-zt^\alpha) R(z;A)x\, dz \\
&= \frac{1}{2\pi i} \int_0^\infty M_\alpha(\lambda) \int_{\Gamma_\theta} e^{-\lambda z t^\alpha} R(z;A)x\, dz\, d\lambda \\
&= \int_0^\infty M_\alpha(s) T(st^\alpha)x\, ds, \quad t \in S^0_{\frac{\pi}{2}-\omega},\ x \in X.
\end{aligned} \tag{4.46}$$

A similar argument shows that

$$\mathcal{P}_\alpha(t)x = \int_0^\infty \alpha s M_\alpha(s) T(st^\alpha)x\, ds, \quad t \in S^0_{\frac{\pi}{2}-\omega},\ x \in X. \tag{4.47}$$

Hence, by (4.46), (4.47), Proposition 4.56 (iii), (W1) and (W3), we have

$$\begin{aligned}
|\mathcal{S}_\alpha(t)x| &\leq C_0 \int_0^\infty M_\alpha(s) s^{-(1+\gamma)} t^{-\alpha(1+\gamma)}|x|\, ds \\
&\leq C_0 \frac{\Gamma(-\gamma)}{\Gamma(1-\alpha(1+\gamma))} t^{-\alpha(1+\gamma)}|x|, \quad t > 0,\ x \in X, \\
|\mathcal{P}_\alpha(t)x| &\leq \alpha C_0 \int_0^\infty M_\alpha(s) s^{-\gamma} t^{-\alpha(1+\gamma)}|x|\, ds \\
&\leq \alpha C_0 \frac{\Gamma(1-\gamma)}{\Gamma(1-\alpha\gamma)} t^{-\alpha(1+\gamma)}|x|, \quad t > 0,\ x \in X.
\end{aligned}$$

Therefore, the estimates in (4.45) hold. This completes the proof. $\qquad\square$

From now on, we frequently use the representations (4.46) and (4.47) for operators $\mathcal{S}_\alpha(t)$ and $\mathcal{P}_\alpha(t)$, respectively.

**Theorem 4.12.** *For $t > 0$, $\mathcal{S}_\alpha(t)$ and $\mathcal{P}_\alpha(t)$ are continuous in the uniform operator topology. Moreover, for every $r > 0$, the continuity is uniform on $[r,\infty)$.*

**Proof.** Let $\epsilon > 0$ be given. For every $r > 0$, it follows from (W3) that we may choose $\delta_1, \delta_2 > 0$ such that

$$\frac{2C_0}{r^{\alpha(1+\gamma)}} \int_0^{\delta_1} \Psi_\alpha(s) s^{-(1+\gamma)} ds \leq \frac{\epsilon}{3}, \quad \frac{2C_0}{r^{\alpha(1+\gamma)}} \int_{\delta_2}^\infty \Psi_\alpha(s) s^{-(1+\gamma)} ds \leq \frac{\epsilon}{3}. \tag{4.48}$$

Then we deduce, by Proposition 1.22(i), that there exists a positive constant $\delta$ such that

$$\int_{\delta_1}^{\delta_2} M_\alpha(s)\|T(t_1^\alpha s) - T(t_2^\alpha s)\|_{B(X)}ds \le \frac{\epsilon}{3}, \tag{4.49}$$

for $t_1, t_2 \ge r$ and $|t_1 - t_2| < \delta$.

On the other hand, using (4.48), (4.49) and Theorem 4.11, we get

$$|\mathcal{S}_\alpha(t_1)x - \mathcal{S}_\alpha(t_2)x| \le \int_0^{\delta_1} M_\alpha(s)\left(\|T(t_1^\alpha s)\|_{B(X)} + \|T(t_2^\alpha s)\|_{B(X)}\right)|x|ds$$

$$+ \int_{\delta_1}^{\delta_2} M_\alpha(s)\|T(t_1^\alpha s) - T(t_2^\alpha s)\|_{B(X)}|x|ds$$

$$+ \int_{\delta_2}^{\infty} M_\alpha(s)\left(\|T(t_1^\alpha s)\|_{B(X)} + \|T(t_2^\alpha s)\|_{B(X)}\right)|x|ds$$

$$\le \frac{2C_0}{r^{\alpha(1+\gamma)}} \int_0^{\delta_1} \Psi_\alpha(s)s^{-(1+\gamma)}|x|ds$$

$$+ \int_{\delta_1}^{\delta_2} M_\alpha(s)\|T(t_1^\alpha s) - T(t_2^\alpha s)\|_{B(X)}|x|ds$$

$$+ \frac{2C_0}{r^{\alpha(1+\gamma)}} \int_{\delta_2}^{\infty} \Psi_\alpha(s)s^{-(1+\gamma)}|x|ds$$

$$\le \epsilon|x|, \quad \text{for any } x \in X,$$

that is,

$$\|\mathcal{S}_\alpha(t_1) - \mathcal{S}_\alpha(t_2)\|_{B(X)} \le \epsilon,$$

which implies that $\mathcal{S}_\alpha(t)$ is uniformly continuous on $[r, \infty)$ in the uniform operator topology and hence, by the arbitrariness of $r > 0$, $\mathcal{S}_\alpha(t)$ is continuous in the uniform operator topology for $t > 0$. A similar argument enables us to give the characterization of continuity on $\mathcal{P}_\alpha(t)$. This completes the proof. $\square$

**Theorem 4.13.** *Let* $0 < \beta < 1 - \gamma$. *Then*

(i) *The range* $R(\mathcal{P}_\alpha(t))$ *of* $\mathcal{P}_\alpha(t)$ *for* $t > 0$, *is contained in* $D(A^\beta)$;

(ii) $\mathcal{S}'_\alpha(t)x = -t^{\alpha-1}A\mathcal{P}_\alpha(t)x$ $(x \in X)$, *and* $\mathcal{S}'_\alpha(t)x$ *for* $x \in D(A)$ *is locally integrable on* $(0, \infty)$;

(iii) *for all* $x \in D(A)$ *and* $t > 0$, $|A\mathcal{S}_\alpha(t)x| \le Ct^{-\alpha(1+\gamma)}|Ax|$, *here* $C$ *is a constant depending on* $\gamma, \alpha$.

**Proof.** It follows from Proposition 1.22 (iv) that for all $x \in X$, $t > 0$, $T(t)x \in D(A^\beta)$ with $\beta > 0$. Therefore, in view of (4.47), Proposition 1.22 (iv) and (W3) we have

$$|A^\beta\mathcal{P}_\alpha(t)x| \le \int_0^\infty \alpha s M_\alpha(s)\|A^\beta T(t^\alpha s)\|_{B(X)}|x|ds$$

$$\le \alpha C' t^{-\alpha(\gamma+\beta+1)} \int_0^\infty M_\alpha(s)s^{-(\beta+\gamma)}ds|x|$$

$$\le \alpha C' \frac{\Gamma(1 - \beta - \gamma)}{\Gamma(1 - \alpha(\beta + \gamma + 1))}t^{-\alpha(1+\beta+\gamma)}|x|,$$

which implies that the assertion (i) holds.

From (i), it is easy to see that for all $x \in X$,

$$\mathcal{S}'_\alpha(t)x = -t^{\alpha-1}A\mathcal{P}_\alpha(t)x.$$

Moreover, for every $x \in D(A)$, one has by Proposition 4.56(iv),

$$|t^{\alpha-1}A\mathcal{P}_\alpha(t)x| \leq t^{\alpha-1}\int_0^\infty \alpha s M_\alpha(s)\|T(t^\alpha s)\|_{B(X)}|Ax|ds$$

$$\leq \alpha C_0 \frac{\Gamma(1-\gamma)}{\Gamma(1-\alpha\gamma)}t^{-\alpha\gamma-1}|Ax|.$$

Since $-\alpha\gamma-1 > -1$, this shows that $\mathcal{S}'_\alpha(t)x$ for each $x \in D(A)$, is locally integrable on $(0,\infty)$, that is , (ii) is true.

Moreover, Proposition 1.22 (iv) and (4.46) imply that

$$|A\mathcal{S}_\alpha(t)x| \leq C_0 t^{-\alpha(1+\gamma)}\int_0^\infty M_\alpha(s)s^{-1-\gamma}ds|Ax|$$

$$\leq C_0 \frac{\Gamma(-\gamma)}{\Gamma(1-\alpha(1+\gamma))}t^{-\alpha(1+\gamma)}|Ax|, \quad x \in D(A).$$

This means that (iii) holds, and completes the proof. $\qquad\square$

**Remark 4.12.** Particularly, from the proof of Theorem 4.13 (i), we can conclude that

$$\|A\mathcal{P}_\alpha(t)\|_{B(X)} \leq Ct^{-\alpha(2+\gamma)}.$$

where $C$ is a constant depending on $\gamma,\alpha$. Moreover, using a similar argument with that in Theorem 4.12, we have that $A\mathcal{P}_\alpha(t)$ for $t > 0$ is continuous in the uniform operator topology.

**Theorem 4.14.** *The following properties hold.*

(i) *Let $\beta > 1+\gamma$. For all $x \in D(A^\beta)$, $\lim_{t\to 0;t>0}\mathcal{S}_\alpha(t)x = x$;*
(ii) *for all $x \in D(A)$, $(\mathcal{S}_\alpha(t) - I)x = \int_0^t -s^{\alpha-1}A\mathcal{P}_\alpha(s)xds$,*
(iii) *for all $x \in D(A)$, $t > 0$, $_0D_t^\alpha \mathcal{S}_\alpha(t)x = -A\mathcal{S}_\alpha(t)x$;*
(iv) *for all $t > 0$, $\mathcal{S}_\alpha(t) = {_0D_t^{\alpha-1}}(t^{\alpha-1}\mathcal{P}_\alpha(t))$.*

**Proof.** For any $x \in X$, note by (4.46) and $(W_3)$ that

$$\mathcal{S}_\alpha(t)x - x = \int_0^\infty \Psi_\alpha(s)(T(t^\alpha s)x - x)ds.$$

On the other hand, by Proposition 1.22(v) it follows that $D(A^\beta) \subset \Sigma_T$ in view of $\beta > 1 + \gamma$. Therefore, we deduce, using Proposition 1.22 (iii), that for any $x \in D(A^\beta)$, there exists a function $\eta(s) \in L^1(0,+\infty)$ depending on $\Psi_\alpha(s)$ such that

$$\|\Psi_\alpha(s)(T(t^\alpha s)x - x)\|_{B(X)} \leq \eta(s).$$

Hence, by means of Lebesgue dominated convergence theorem we obtain that

$$\mathcal{S}_\alpha(t)x - x \to 0, \quad \text{as } t \to 0,$$

that is, the assertion (i) remains true.

From (i) and Theorem 4.13(ii) we get for all $x \in D(A)$,

$$(S_\alpha(t) - I)x = \lim_{s \to 0}(S_\alpha(t)x - S_\alpha(s)x) = \int_0^t -\lambda^{\alpha-1} A \mathcal{P}_\alpha(\lambda)x d\lambda,$$

which implies that the assertion (ii) holds.

To prove (iii), first it is easy to see that $\frac{1}{\varphi_0} \in \mathcal{F}(S_\mu^0)$ and the operator $\varphi_0(A)$ is injective. Taking $x \in D(A)$, by Proposition 4.55 (iii) one has

$$S_\alpha(t)x = E_\alpha(-zt^\alpha)(A)x = (E_\alpha(-zt^\alpha)\varphi_0)(A)\left(\frac{1}{\varphi_0}\right)(A)x.$$

Moreover, by (1.14) we have $\sup_{z \to \infty} |zt^\alpha E_\alpha(-zt^\alpha)| < \infty$, which implies that

$$|zE_\alpha(-zt^\alpha)(1+z)^{-1}| \le C|z|^{-1}t^{-\alpha}, \quad \text{as } z \to \infty,$$

where $C$ is a constant which is independent of $t$. Consequently,

$$- zE_\alpha(-zt^\alpha)(1+z)^{-1} \in \mathcal{F}_0^\gamma(S_\mu^0). \tag{4.50}$$

Notice also that

$$_0^C D_t^\alpha E_\alpha(-zt^\alpha)(1+z)^{-1}R(z;A) = (-z)E_\alpha(-zt^\alpha)(1+z)^{-1}R(z;A).$$

Combining Proposition 1.21(ii) and (4.50), we get

$$_0^C D_t^\alpha((E_\alpha(-zt^\alpha)(1+z^\beta)^{-1})(A)) = \frac{1}{2\pi i}\int_{\Gamma_\theta}(-z)E_\alpha(-zt^\alpha)(1+z)^{-1}R(z;A)dz$$

$$= (-z)(A)\big(E_\alpha(-zt^\alpha)(1+z)^{-1}\big)(A)$$

$$= - A\big(E_\alpha(-zt^\alpha)(1+z)^{-1}\big)(A).$$

Hence, we obtain

$$_0 D_t^\alpha S_\alpha(t)x = - A\big(E_\alpha(-zt^\alpha)(1+z)^{-1}\big)(A)(1+z)(A)x$$

$$= - A(E_\alpha(-zt^\alpha))(A)x$$

$$= - AS_\alpha(t)x.$$

This proves (iii).

For (iv), by a similar argument with (iii), one can prove that $t^{\alpha-1}e_\alpha(-zt^\alpha)$ belongs to $\mathcal{F}_0^\gamma(S_\mu^0)$ for $t > 0$ and hence

$$_0 D_t^{\alpha-1}(t^{\alpha-1}\mathcal{P}_\alpha(t)) =_0 D_t^{\alpha-1}((t^{\alpha-1}e_\alpha(-zt^\alpha)(A)) = (E_\alpha(-zt^\alpha))(A) = S_\alpha(t),$$

in view of

$$_0 D_t^{\alpha-1}(t^{\alpha-1}e_\alpha(-zt^\alpha)) = E_\alpha(-zt^\alpha).$$

This completes the proof.                                              □

Before proceeding with our theory further, we present the following result.

**Lemma 4.21.** *If $R(\lambda; -A)$ is compact for every $\lambda > 0$, then $T(t)$ is compact for every $t > 0$.*

**Proof.** Note first that as a consequence of Theorem 3.13 in Periago and Straub, 2002., for every $\lambda \in \mathbb{C}$ with $Re(\lambda) > 0$, $R(\lambda; -A) = \int_0^\infty e^{-\lambda s} T(s) ds$ defines a bounded linear operator on $X$. Therefore, we obtain

$$\lambda R(\lambda; -A)T(t) - T(t) = \lambda \int_0^\infty e^{-\lambda s}(T(t+s) - T(t)) ds. \qquad (4.51)$$

Let $\epsilon > 0$ be given. For every $\lambda > 0$ and $t > 0$, it follows from Theorem 4.12 that there exists a $\nu > 0$ such that

$$\sup_{s \in [0,\nu]} \|T(s+t) - T(t)\|_{B(X)} \leq \frac{\epsilon}{2}.$$

So

$$\lambda \int_0^\nu e^{-s\lambda} \|T(t+s) - T(t)\|_{B(X)} ds \leq \frac{\epsilon}{2}. \qquad (4.52)$$

On the other hand, by Proposition 1.22(iii) we get

$$\lambda \left\| \int_\nu^\infty e^{-s\lambda}(T(s+t) - T(t)) ds \right\|_{B(X)} \leq \lambda C \int_\nu^\infty e^{-s\lambda}((t+s)^{-1-\gamma} + t^{-\gamma-1}) ds$$
$$\leq 2Ct^{-\gamma-1} e^{-\lambda\nu},$$

which implies that there exists a $\lambda_0 > 0$ such that

$$\lambda \left\| \int_\nu^\infty e^{-s\lambda}(T(s+t) - T(t)) ds \right\|_{B(X)} \leq \frac{\epsilon}{2}, \quad \lambda \geq \lambda_0. \qquad (4.53)$$

Thus, for all $\lambda \geq \lambda_0$, using (4.51), (4.52) and (4.53) we deduce that

$$\|\lambda R(\lambda; -A)T(t) - T(t)\|_{B(X)} \leq \lambda \int_0^\nu e^{-s\lambda} \|T(t+s) - T(t)\|_{B(X)} ds$$
$$+ \lambda \int_\nu^\infty e^{-s\lambda} \|T(s+t) - T(t)\|_{B(X)} ds$$
$$\leq \epsilon.$$

It follows from the arbitrariness of $\nu > 0$ that

$$\lim_{\lambda \to \infty} \|\lambda R(\lambda; -A)T(t) - T(t)\|_{B(X)} = 0.$$

Since $\lambda R(\lambda; -A)T(t)$ is compact for every $\lambda > 0$ and $t > 0$, $T(t)$ is compact for every $t > 0$. $\qquad \square$

With the help of this lemma we now show the following result.

**Theorem 4.15.** *If $R(\lambda; -A)$ is compact for every $\lambda > 0$, then $\mathcal{S}_\alpha(t)$, $\mathcal{P}_\alpha(t)$ are compact for every $t > 0$.*

**Proof.** Let $\epsilon > 0$ be arbitrary. Put

$$\varsigma_\epsilon(t) = \int_\epsilon^\infty \Psi_\alpha(s)T(st^\alpha - \epsilon t^\alpha)ds, \quad \zeta_\epsilon(t) = \int_\epsilon^\infty \Psi_\alpha(s)T(st^\alpha)ds.$$

Then, one has $\zeta_\epsilon(t) = T(\epsilon t^\alpha)\varsigma_\epsilon(t)$, and it is easy to prove that for every $t > 0$, $\varsigma_\epsilon(t)$ is bounded linear operators on $X$. Therefore, from the compactness of $T(t), t > 0$, we see that $\zeta_\epsilon(t)$ is compact for every $t > 0$.

On the other hand, note that

$$\|\zeta_\epsilon(t) - \mathcal{S}_\alpha(t)\|_{B(X)} \leq \left\| \int_0^\epsilon \Psi_\alpha(s)T(st^\alpha)ds \right\|_{B(X)} \leq C_0 t^{-\alpha(1+\gamma)} \int_0^\epsilon \Psi_\alpha(s)s^{-1-\gamma}ds.$$

Hence, it follows from the compactness of $\zeta_\epsilon(t), t > 0$ that $\mathcal{S}_\alpha(t)$ is compact for every $t > 0$. By a similar technique we can conclude that $\mathcal{P}_\alpha(t)$ is compact for every $t > 0$. The proof is completed. $\qquad\square$

### 4.5.3 *Linear Problems*

Let $A \in \Theta_\omega^\gamma(X)$ with $-1 < \gamma < 0$ and $0 < \omega < \pi/2$. We discuss the existence and uniqueness of mild solutions and classical solutions for inhomogeneous linear abstract Cauchy problem (4.43). We assume the following condition.
$(H^*)$ $u \in C([0,T],X)$, $g_{1-\alpha} * u \in C^1((0,T],X)$, $u(t) \in D(A)$ for $t \in (0,T]$, $Au \in L^1((0,T),X)$, and $u$ satisfies (4.43).

Then, by Definitions 1.1 and 1.3, one can rewrite (4.43) as

$$u(t) = u_0 - \frac{1}{\Gamma(\alpha)} \int_0^t (t-s)^{\alpha-1}Au(s)ds + \frac{1}{\Gamma(\alpha)} \int_0^t (t-s)^{\alpha-1}f(s)ds, \qquad (4.54)$$

for $t \in [0,T]$.

Before presenting the definition of mild solution of problem (4.43), we first prove the following lemma.

**Lemma 4.22.** *If $u : [0,T] \to X$ is a function satisfying the assumption $(H^*)$, then $u(t)$ satisfies the following integral equation*

$$u(t) = \mathcal{S}_\alpha(t)u_0 + \int_0^t (t-s)^{\alpha-1}\mathcal{P}_\alpha(t-s)f(s)ds, \quad t \in (0,T].$$

**Proof.** Note that the Laplace transform of an abstract function $f \in L^1(\mathbb{R}^+, X)$ is defined by $\widehat{f}(\lambda) = \int_0^\infty e^{-\lambda t}f(t)dt$, $\lambda > 0$. Applying Laplace transform to (4.54) we get $\widehat{u}(\lambda) = \frac{u_0}{\lambda} - \frac{1}{\lambda^\alpha}A\widehat{u}(\lambda) + \frac{\widehat{f}(\lambda)}{\lambda^\alpha}$, that is,

$$\widehat{u}(\lambda) = \lambda^{\alpha-1}(\lambda^\alpha + A)^{-1}u_0 + (\lambda^\alpha + A)^{-1}\widehat{f}(\lambda).$$

On the other hand, using Proposition 1.23 and (W2) we deduce that

$$\lambda^{\alpha-1}(\lambda^\alpha + A)^{-1}u_0 + (\lambda^\alpha + A)^{-1}\widehat{f}(\lambda)$$
$$= \lambda^{\alpha-1}\int_0^\infty e^{-\lambda^\alpha t}T(t)u_0dt + \int_0^\infty e^{-\lambda^\alpha t}T(t)\widehat{f}(\lambda)dt$$

$$= \int_0^\infty \frac{d}{d\lambda} e^{-(\lambda t)^\alpha} T(t^\alpha) u_0 dt + \int_0^\infty \int_0^\infty \alpha t^{\alpha-1} e^{-(\lambda t)^\alpha t} T(t^\alpha) f(s) e^{-s\lambda} ds dt$$

$$= \int_0^\infty \int_0^\infty \frac{\alpha t}{\tau^\alpha} \Psi_\alpha\left(\frac{1}{\tau^\alpha}\right) e^{-\lambda t\tau} T(t^\alpha) d\tau dt$$

$$+ \int_0^\infty \int_0^\infty \int_0^\infty \frac{\alpha}{\tau^{2\alpha}} t^{\alpha-1} \Psi\left(\frac{1}{\tau^\alpha}\right) e^{-\lambda t} T\left(\frac{t^\alpha}{\tau^\alpha}\right) f(s) e^{-s\lambda} d\tau ds dt$$

$$= \int_0^\infty \int_0^\infty \frac{\alpha}{\tau^{\alpha+1}} \Psi_\alpha\left(\frac{1}{\tau^\alpha}\right) e^{-\lambda t} T\left(\frac{t^\alpha}{\tau^\alpha}\right) d\tau dt$$

$$+ \int_0^\infty \int_0^\infty \int_0^\infty \alpha \tau t^{\alpha-1} \Psi(\tau) T(t^\alpha \tau) f(s) e^{-(s+t)\lambda} d\tau ds dt$$

$$= \int_0^\infty e^{-\lambda t} \int_0^\infty \Psi_\alpha(\tau) T(t^\alpha \tau) d\tau$$

$$+ \int_0^\infty e^{-t\lambda} \int_0^t (t-s)^{\alpha-1} f(s) \left(\int_0^\infty \alpha \tau \Psi(\tau) T((t-s)^\alpha \tau) d\tau\right) ds dt$$

$$= \int_0^\infty e^{-\lambda t} \mathcal{S}_\alpha(t) dt + \int_0^\infty e^{-\lambda t} \int_0^t (t-s)^\alpha \mathcal{P}_\alpha(t-s) f(s) ds dt$$

$$= \int_0^\infty e^{-\lambda t} \left(\mathcal{S}_\alpha(t) u_0 + \int_0^t (t-s)^{\alpha-1} \mathcal{P}_\alpha(t-s) f(s) ds\right) dt.$$

This implies that

$$\widehat{u}(\lambda) = \int_0^\infty e^{-\lambda t} \left(\mathcal{S}_\alpha(t) u_0 + \int_0^t (t-s)^{\alpha-1} \mathcal{P}_\alpha(t-s) f(s) ds\right) dt.$$

Now using the uniqueness of the Laplace transform (cf. Theorem 1.1.6 of Xiao and Liang, 1998), we deduce that the assertion of lemma holds. This completes this proof. □

Motivated by Lemma 4.22, we adopt the following concept of mild solution to problem (4.43).

**Definition 4.4.** By a mild solution of problem (4.43), we mean a function $u \in C((0, T], X)$ satisfying

$$u(t) = \mathcal{S}_\alpha(t) u_0 + \int_0^t (t-s)^{\alpha-1} \mathcal{P}_\alpha(t-s) f(s) ds, \quad t \in (0, T].$$

**Remark 4.13.** It is to be noted that

(a) unlike the case of strongly continuous operator semigroups, we do not require the mild solution of problem (4.43) to be continuous at $t = 0$. Moreover, in general, since the operator $\mathcal{S}_\alpha(t)$ is singular at $t = 0$, solutions to problem (4.43) are assumed to have the same kind of singularity at $t = 0$ as the operator $\mathcal{S}_\alpha(t)$. This is the case, for instance, if $f \equiv 0$ so that we have that $u(t) = \mathcal{S}_\alpha(t) u_0$, which presents a discontinuity at the initial time;

**(b)** when $u_0 \in D(A^\beta)$, $\beta > 1 + \gamma$, it follows from Theorem 4.14 (i) that the mild solution is continuous at $t = 0$.

For $f \in L^1((0, T), X)$, the initial problem (4.43) has a unique mild solution for every $u_0 \in X$. We now be interested in imposing further condition on $f$ and $u_0$ so that the mild solution becomes a classical solution. To this end we first introduce the following definition.

**Definition 4.5.** By a classical solution to problem (4.43), we mean a function $u(t) \in C([0, T], X)$ with ${}^C_0 D^\alpha_t u(t) \in C((0, T], X)$, which, for all $t \in (0, T]$, takes values in $D(A)$ and satisfies (4.43).

We are now ready to state our main result in this subsection.

**Theorem 4.16.** Let $A \in \Theta^\gamma_\omega(X)$ with $0 < \omega < \frac{\pi}{2}$. Suppose that $f(t) \in D(A)$ for all $0 < t \leq T$, $Af(t) \in L^\infty((0, T), X)$, and $f(t)$ is Hölder continuous with an exponent $\theta' > \alpha(1 + \gamma)$, that is,

$$|f(t) - f(s)| \leq K|t - s|^{\theta'}, \quad \text{for all } 0 < t, s \leq T.$$

*Then, for every $u_0 \in D(A)$, there exists a classical solution to problem (4.43) and this solution is unique.*

**Proof.** For $u_0 \in D(A)$, let $u(t) = S_\alpha(t)u_0$ $(t > 0)$. Then it follows from Theorem 4.14 (i, iii) that $u(t)$ is a classical solution of the following problem

$$\begin{cases} {}^C_0 D^\alpha_t u(t) + Au(t) = 0, & 0 < t \leq T, \\ u(0) = u_0. \end{cases} \tag{4.55}$$

Moreover, from Lemma 4.22, it is easy to see that $u(t)$ is the only solution to problem (4.55). Put

$$w(t) = \int_0^t (t-s)^{\alpha-1} \mathcal{P}_\alpha(t-s)f(s)ds, \quad 0 < t \leq T.$$

Then from the assumptions on $f$ and Theorem 4.11 we obtain

$$|Aw(t)| \leq \int_0^t (t-s)^{\alpha-1} \|\mathcal{P}_\alpha(t-s)\|_{B(X)} \|Af(t)\|_{L^\infty(0,T)} ds$$

$$\leq C_p \|Af(t)\|_{L^\infty(0,T)} \frac{1}{-\alpha\gamma} t^{-\gamma\alpha},$$

which implies that $w(t) \in D(A)$ for all $0 < t \leq T$.

Next, we show ${}^C_0 D^\alpha_t w(t) \in C((0, T], X)$. Since $w(0) = 0$ and hence

$${}^C_0 D^\alpha_t w(t) = {}_0 D^1_t {}_0 D^{\alpha-1}_t w(t) = \frac{d}{dt} \int_0^t S_\alpha(t-s)f(s)ds, \tag{4.56}$$

in view of Propositions 1.4, 1.8, 1.9 and Theorem 4.14 (iv). Let

$$v(t) = \int_0^t S_\alpha(t-s)f(s)ds,$$

it remains to prove $v(t) \in C^1((0, T], X)$. Let $h > 0$ and $h \leq T - t$. Then it is easy to verify the identity

$$\frac{v(t + h) - v(t)}{h} = \int_0^t \frac{S_\alpha(t + h - s) - S_\alpha(t - s)}{h} f(s) ds$$

$$+ \frac{1}{h} \int_t^{t+h} S_\alpha(t + h - s) f(s) ds.$$

Again by the assumptions on $f$ and Theorem 4.11, we have, for $t > 0$ fixed,

$$|(t - s)^{\alpha-1} A P_\alpha(t - s) f(s)| \leq C_p (t - s)^{-\alpha\gamma-1} |A f(s)| \in L^1((0, T), X),$$

for all $s \in [0, t)$. Therefore, using Theorem 4.13 (ii) and the Lebesgues dominated convergence theorem we get

$$\lim_{h \to 0} \int_0^t \frac{S_\alpha(t + h - s) - S_\alpha(t - s)}{h} f(s) ds$$

$$= \int_0^t (t - s)^{\alpha-1}(-A) P_\alpha(t - s) f(s) ds \qquad (4.57)$$

$$= -A w(t).$$

Furthermore, note that

$$\frac{1}{h} \int_t^{t+h} S_\alpha(t + h - s) f(s) ds$$

$$= \frac{1}{h} \int_0^h S_\alpha(s) f(t + h - s) ds$$

$$= \frac{1}{h} \int_0^h S_\alpha(s)(f(t + h - s) - f(t - s)) ds$$

$$+ \frac{1}{h} \int_0^h S_\alpha(s)(f(t - s) - f(t)) ds + \frac{1}{h} \int_0^h S_\alpha(s) f(t) ds.$$

From Theorem 4.11 and the Hölder continuity on $f$ we have

$$\frac{1}{h} \left| \int_0^h S_\alpha(s)(f(t + h - s) - f(t - s)) ds \right| \leq \frac{C_s K h^{\theta'-\alpha(1+\gamma)}}{1 - \alpha(1 + \gamma)},$$

and

$$\frac{1}{h} \left| \int_0^h S_\alpha(s)(f(t - s) - f(t)) ds \right| \leq \frac{C_s K h^{\theta'-\alpha(1+\gamma)}}{1 + \theta - \alpha(1 + \gamma)}.$$

And since $f(t) \in D(A)$ $(0 < t \leq T)$, $\lim_{h \to 0} \frac{1}{h} \int_0^h S_\alpha(s) f(t) ds = f(t)$ in view of Theorem 4.14 (i). Hence,

$$\frac{1}{h} \int_t^{t+h} S_\alpha(t + h - s) f(s) ds \to f(t), \quad \text{as } h \to 0^+. \qquad (4.58)$$

Combining (4.57) and (4.58) we deduce that $v$ is differentiable from the right at $t$ and $v'_+(t) = f(t) - A w(t)$, $t \in (0, T]$. By a similar argument with the above one has that $v$ is differentiable from the left at $t$ and $v'_-(t) = f(t) - Aw(t)$, $t \in (0, T]$. Next, we prove $Aw(t) \in C((0, T], X)$. To the end, let $Aw(t) = I_1(t) + I_2(t)$, where

$$I_1(t) = \int_0^t (t-s)^{\alpha-1} A\mathcal{P}_\alpha(t-s)\big(f(s) - f(t)\big)ds,$$

$$I_2(t) = \int_0^t A(t-s)^{\alpha-1}\mathcal{P}_\alpha(t-s)f(t)ds.$$

By Theorem 4.14(ii), we obtain

$$I_2(t) = -(\mathcal{S}_\alpha(t) - I)f(t).$$

So, by the assumption of $f$ and Theorem 4.12 note that $I_2(t)$ is continuous for $0 < t \leq T$. To prove the same conclusion for $I_1(t)$, we let $0 < h \leq T - t$ and write

$I_1(t+h) - I_1(t)$

$$= \int_0^t \Big((t+h-s)^{\alpha-1}A\mathcal{P}_\alpha(t+h-s) - (t-s)^{\alpha-1}A\mathcal{P}_\alpha(t-s)\Big)\big(f(s) - f(t)\big)ds$$

$$+ \int_0^t (t+h-s)^{\alpha-1}A\mathcal{P}_\alpha(t+h-s)(f(t) - f(t+h))ds$$

$$+ \int_t^{t+h} (t+h-s)^{\alpha-1}A\mathcal{P}_\alpha(t+h-s)\big(f(s) - f(t+h)\big)ds$$

$$=: h_1(t) + h_2(t) + h_3(t).$$

For $h_1(t)$, on the one hand, it follows from Theorem 4.12 that

$$\lim_{h\to 0} (t+h-s)^{\alpha-1}A\mathcal{P}_\alpha(t+h-s)(f(s) - f(t))$$

$$= (t-s)^{\alpha-1}A\mathcal{P}_\alpha(t-s)(f(s) - f(t)).$$

On the other hand, for $t \in (0, T]$ fixed, by Remark 4.12 and the assumption on $f$, we get

$$|(t+h-s)^{\alpha-1}A\mathcal{P}_\alpha(t+h-s)\big(f(s) - f(t)\big)|$$

$$\leq C'_p K(t+h-s)^{-\alpha(1+\gamma)-1}(t-s)^{\theta'}$$

$$\leq C'_p K(t-s)^{(\theta'-\alpha-\alpha\gamma)-1} \in L^1((0,t), X).$$

Thus, by mean of the Lebesgues dominated convergence theorem one has

$$\lim_{h\to 0} \int_0^t (t+h-s)^{\alpha-1}A\mathcal{P}_\alpha(t+h-s)\big(f(s) - f(t)\big)ds$$

$$= \int_0^t (t-s)^{\alpha-1}A\mathcal{P}_\alpha(t-s)\big(f(s) - f(t)\big)ds,$$

which implies that $h_1(t) \to 0$ as $h \to 0^+$.

For $h_2(t)$, using Theorem 4.13(i), Remark 4.12,

$$\int_0^t (t+h-s)^{\alpha-1}\|A\mathcal{P}_\alpha(t+h-s)\|_{B(X)}|f(t) - f(t+h)|ds$$

$$\leq \int_0^t C_p' K(t+h-s)^{-\alpha(1+\gamma)-1} h^{\theta'} ds$$

$$= \frac{C_p' K h^{\theta'}}{\alpha(1+\gamma)} (h^{-\alpha(1+\gamma)} - (h+t)^{-\alpha(1+\gamma)}).$$

This yields $h_2(t) \to 0$ as $h \to 0^+$.

Moreover, $h_3(t) \to 0$ as $h \to 0^+$ by the following estimate

$$\left| \int_t^{t+h} (t+h-s)^{\alpha-1} \mathcal{P}_\alpha(t+h-s)(Af(s) - Af(t+h)) ds \right|$$

$$\leq \frac{2C_p}{-\alpha\gamma} \|Af(s)\|_{L^\infty(0,T)} h^{-\alpha\gamma}$$

in view of $Af(s) \in L^\infty((0,T),X)$ and Theorem 4.12.

The same reasoning establishes $I_1(t-h) - I_1(h) \to 0$ as $h \to 0^+$. Consequently, $Aw \in C((0,T],X)$, which implies that $v' \in C((0,T],X)$, provided that $f$ is continuous on $(0,T]$. Then, by (4.56) we have $^C_0 D_t^\alpha w \in C((0,T],X)$. Hence, we prove that $u+w$ is a classical solution to problem (4.43), and Lemma 4.22 implies that it is unique. This completes the proof. $\qquad\square$

### 4.5.4 Nonlinear Problems

In this subsection we apply the theory developed in the previous sections to nonlinear abstract Cauchy problem (4.44).

**Definition 4.6.** By a mild solution to problem (4.44), we mean a function $u \in C((0,T],X)$ satisfying

$$u(t) = \mathcal{S}_\alpha(t)u_0 + \int_0^t (t-s)^{\alpha-1} \mathcal{P}_\alpha(t-s) f(s,u(s)) ds, \quad t \in (0,T].$$

**Theorem 4.17.** *Let* $A \in \Theta_\omega^\gamma(X)$ *with* $-1 < \gamma < -\frac{1}{2}$ *and* $0 < \omega < \frac{\pi}{2}$. *Suppose that the nonlinear mapping* $f : (0,T] \times X \to X$ *is continuous with respect to $t$ and there exist constants* $M, N > 0$ *such that*

$$|f(t,x) - f(t,y)| \leq M \left( 1 + |x|^{\nu-1} + |y|^{\nu-1} \right) |x-y|,$$

$$|f(t,x)| \leq N(1 + |x|^\nu),$$

*for all* $t \in (0,T]$ *and for each* $x, y \in X$, *where $\nu$ is a constant in* $[1, -\frac{\gamma}{1+\gamma})$. *Then, for every* $u_0 \in X$, *there exists a* $T_0 > 0$ *such that the problem (4.44) has a unique mild solution defined on* $(0, T_0]$.

**Proof.** For fixed $r > 0$, we introduce the metric space

$$F_r(T, u_0) = \{u \in C((0,T],X) : \rho_T(u, \mathcal{S}_\alpha(t)u_0) \leq r\},$$

$$\rho_T(u_1, u_2) = \sup_{t \in (0,T]} |u_1(t) - u_2(t)|.$$

It is not difficult to see that, with this metric, $F_r(T, u_0)$ is a complete metric space. Take $L = T^{\alpha(1+\gamma)}r + C_s|u_0|$, then for any $u \in F_r(T, u_0)$, we have

$$|s^{\alpha(1+\gamma)}u(s)| \le s^{\alpha(1+\gamma)}|u - S_\alpha(t)u_0| + s^{\alpha(1+\gamma)}|S_\alpha(t)u_0| \le L.$$

Choose $0 < T_0 \le T$ such that

$$C_p N \frac{T_0^{-\alpha\gamma}}{-\alpha\gamma} + C_p N L^\nu T_0^{-\alpha(\nu(1+\gamma)+\gamma)}B(-\gamma\alpha, 1 - \nu\alpha(1+\gamma)) \le r, \qquad (4.59)$$

$$MC_p \frac{T_0^{-\alpha\gamma}}{-\alpha\gamma} + 2L^{\rho-1}T_0^{-\alpha(\gamma+(1+\gamma)(\nu-1))}B(-\alpha\gamma, 1 - \alpha(1+\gamma)(\nu-1)) \le \frac{1}{2}, \quad (4.60)$$

where $B(\eta_1, \eta_2)$ with $\eta_i > 0$, $i = 1, 2$, denotes the Beta function. Assume that $u_0 \in X$. Consider the mapping $\Gamma^\alpha$ given by

$$(\Gamma^\alpha u)(t) = S_\alpha(t)u_0 + \int_0^t (t - s)^{\alpha-1}P_\alpha(t - s)f(s, u(s))ds, \quad u \in F_r(T_0, u_0).$$

By the assumptions on $f$, Theorem 4.11 and Theorem 4.12, we see that $(\Gamma^\alpha u)(t) \in C((0, T], X)$ and

$$|(\Gamma^\alpha u)(t) - S_\alpha(t)u_0| \le C_p N \int_0^t (t - s)^{-\alpha\gamma-1}(1 + |u(s)|^\nu)ds$$

$$\le C_p N \frac{T_0^{-\alpha\gamma}}{-\alpha\gamma} + \int_0^t C_p N L^\nu (t - s)^{-\alpha\gamma-1}s^{-\nu\alpha(1+\gamma)}ds$$

$$\le C_p N \frac{T_0^{-\alpha\gamma}}{-\alpha\gamma} + C_p N L^\nu T_0^{-\alpha(\nu(1+\gamma)+\gamma)}B(-\gamma\alpha, 1 - \nu\alpha(1+\gamma))$$

$$\le r,$$

in view of (4.59). So, $\Gamma^\alpha$ maps $F_r(T_0, u_0)$ into itself. Next, for any $u, v \in F_r(T_0, u_0)$, by the assumptions on $f$ and Theorem 4.11, we have

$$|(\Gamma^\alpha u)(t) - (\Gamma^\alpha v)(t)|$$

$$= \left| \int_0^t (t - s)^{\alpha-1}P_\alpha(t - s)\big(f(s, u(s)) - f(s, v(s))\big)ds \right|$$

$$\le C_p M \int_0^t (t - s)^{-\alpha\gamma-1}\left(1 + |u(s)|^{\rho-1} + |v(s)|^{\rho-1}\right)|u(s) - v(s)|ds$$

$$\le C_p M \rho_t(u, v) \int_0^t (t - s)^{-\alpha\gamma-1}\left(1 + 2L^{\nu-1}s^{-\alpha(\nu-1)(1+\gamma)}\right)ds$$

$$\le 2L^{\rho-1}T_0^{-\alpha(\gamma+(1+\gamma)(\nu-1))}B(-\alpha\gamma, 1 - \alpha(1 + \gamma)(\nu - 1))\rho_{T_0}(u, v)$$

$$+ MC_p \frac{T_0^{-\alpha\gamma}}{-\alpha\gamma}\rho_{T_0}(u, v).$$

This yields that $\Gamma^\alpha$ is a contraction on $F_r(T_0, u_0)$ due to (4.60). So, $\Gamma_\alpha$ has a unique fixed point $u \in F_r(T_0, u_0)$ in view of Banach contraction mapping principle, this means that $u$ is a mild solution to problem (4.44) defined on $(0, T_0]$. The proof is completed. $\qquad\square$

By a similar argument with the proof of Theorem 4.17 we have:

**Corollary 4.1.** *Assume that $A \in \Theta_\omega^\gamma(X)$ with $-1 < \gamma < -\frac{2}{3}$ and $0 < \omega < \frac{\pi}{2}$. Suppose in addition that the nonlinear mapping $f : (0,T] \times X^\beta \to X$, $\beta \in (1 + \gamma, -1 - 2\gamma)$, is continuous with respect to $t$ and there exist constants $M, N > 0$ such that*

$$|f(t,x) - f(t,y)| \leq M(1 + |x|_\beta^{\nu-1} + |y|_\beta^{\nu-1})|x - y|_\beta,$$

$$|f(t,x)| \leq N\left(1 + |x|_\beta^\nu\right),$$

*for all $t \in (0,T]$ and for each $x, y \in X^\beta$, where $\nu$ is a constant in $[1, -\frac{\gamma+\beta}{1+\gamma})$. Then, for every $u_0 \in X^\beta$, there exists a $T_0 > 0$ such that the problem (4.44) has a unique mild solution $u \in C((0,T_0], X^\beta)$.*

**Remark 4.14.** If $A \in \Theta_\omega^\gamma(X)$ with $-1 < \gamma < 0$ and $0 < \omega < \frac{\pi}{2}$, then we can derive the local existence and uniqueness of mild solutions to problem (4.44), under the conditions:

(i) $u_0 \in X^\beta$ with $\beta > 1 + \gamma$;
(ii) the nonlinear mapping $f : [0,T] \times X \to X$ is continuous with respect to $t$ and there exists a continuous function $L_f(\cdot) : \mathbb{R}^+ \to \mathbb{R}^+$ such that

$$|f(t,x) - f(t,y)| \leq L_f(r)|x - y|,$$

for all $0 \leq t \leq T$ and for each $x, y \in X$ satisfying $|x|, |y| \leq r$.

Indeed, for $r > \frac{C_p T_0^{-\alpha\gamma}}{-\alpha\gamma} \sup_{t \in [0,T]} |f(t, u_0)|$ fixed, we may choose $0 < T_0 \leq T$ such that

$$\sup_{t \in [0,T_0]} |(\mathcal{S}_\alpha(t) - I)u_0| + \frac{C_p T_0^{-\alpha\gamma}}{-\alpha\gamma}\left(L_f(r)r + \sup_{t \in [0,T_0]} |f(t, u_0)|\right) < r \qquad (4.61)$$

in view of Theorem 4.14 (i). Assume that the map $\Gamma^\alpha$ is defined the same as in Theorem 4.17 and the space $F_r(T_0, u_0)$ is replaced by the following Banach space:

$$F_r'(T_0, u_0) = \left\{ u \in C([0,T_0], X) : \ u(0) = u_0 \ \text{and} \ \sup_{t \in [0,T_0]} |u - u_0| \leq r \right\}.$$

Then, it is easy to verify, thanks to the assumptions on $f$ and (4.61), that $\Gamma^\alpha$ maps $F_r'(T_0, u_0)$ into itself and is a contraction on $F_r'(T_0, u_0)$, which implies that the problem (4.44) has a unique mild solution defined on $[0, T_0]$.

Since $1 > 1 + \gamma$ ($-1 < \gamma < -\frac{1}{2}$), $X^1 = D(A)$ is a Banach space endowed with the graph norm $|x|_{X^1} = |Ax|$, for $x \in X^1$.

The following is the existence of $X^1$-smooth solutions.

**Theorem 4.18.** *Let $A \in \Theta_\omega^\gamma(X)$ with $-1 < \gamma < -\frac{1}{2}$ and $0 < \omega < \frac{\pi}{2}$ and $u_0 \in X^1$. Assume that there exists a continuous function $M_f(\cdot) : \mathbb{R}^+ \to \mathbb{R}^+$ and a constant $N_f > 0$ such that the nonlinear mapping $f : (0,T] \times X^1 \to X^1$ satisfies*

$$|f(t,x) - f(t,y)|_{X^1} \leq M_f(r)|x - y|_{X^1},$$

$$|f(t, \mathcal{S}_\alpha(t)u_0)|_{X^1} \leq N_f\big(1 + t^{-\alpha(1+\gamma)}|u_0|_{X^1}\big),$$

for all $0 < t \leq T$ and for each $x, y \in X^1$ satisfying $\sup_{t\in(0,T]} |x(t) - \mathcal{S}_\alpha(t)u_0|_{X^1} \leq r$, $\sup_{t\in(0,T]} |y(t) - \mathcal{S}_\alpha(t)u_0|_{X^1} \leq r$. Then there exists a $T_0 > 0$ such that the problem (4.44) has a unique mild solution defined on $(0, T_0]$.

**Proof.** For $u_0 \in X^1$ and $r > 0$, set

$$F_r''(T, u_0) = \Big\{u \in C((0,T], X^1) : \sup_{t\in(0,T]} |u - \mathcal{S}_\alpha(t)u_0|_{X^1} \leq r\Big\}.$$

For any $u \in F_r''(T, u_0)$, by the assumptions on $f$ and Theorem 4.11 we have

$$|(\Gamma^\alpha u)(t) - \mathcal{S}_\alpha(t)u_0|_{X^1}$$

$$\leq \int_0^t (t-s)^{\alpha-1}\|\mathcal{P}_\alpha(t-s)\|_{B(X)}|f(s, u(s)) - f(s, \mathcal{S}_\alpha(t)u_0)|_{X^1}ds$$

$$+ \int_0^t (t-s)^{\alpha-1}\|\mathcal{P}_\alpha(t-s)\|_{B(X)}|f(s, \mathcal{S}_\alpha(t)u_0))|_{X^1}ds$$

$$\leq C_p \int_0^t (t-s)^{-\alpha\gamma-1}\big(M_f(r)r + N_f + N_f s^{-\alpha(1+\gamma)}|u_0|\big)ds$$

$$\leq C_p(M_f(r)r + N_f)\frac{T^{-\alpha\gamma}}{-\alpha\gamma} + C_p N_f T^{-\alpha(1+2\gamma)}\beta(-\gamma\alpha, 1 - \alpha(1+\gamma))|u_0|.$$

Using this result, it follows from an analogous idea with Theorem 4.17 that the claim of theorem follows. Here we omit the details.          □

Next, we derive mild solutions under the condition of compactness on the resolvent of $A$.

**Theorem 4.19.** *Let $A \in \Theta_\omega^\gamma(X)$ with $-1 < \gamma < 0$ and $0 < \omega < \frac{\pi}{2}$. Let*

**(H1)** *$R(\lambda; -A)$ is compact for every $\lambda > 0$;*
**(H2)** *$f : [0,T] \times X \to X$ is a Carathéodory function and for any $r > 0$, there exists a function $m_r(t) \in L^p((0,T), \mathbb{R}^+)$ with $p > -\frac{1}{\alpha\gamma}$ such that*

$$|f(t,x)| \leq m_r(t), \quad and \quad \liminf_{r\to+\infty} \frac{|m_r(t)|_{L^p(0,T)}}{r} = \sigma < \infty$$

*for a.e $t \in [0,T]$ and all $x \in X$ satisfying $|x| \leq r$.*

*Then for every $u_0 \in D(A^\beta)$ with $\beta > 1 + \gamma$, the problem (4.44) has at least one mild solution, provided that*

$$C_p\sigma\left(\frac{T^{1-(1+\alpha\gamma)q}}{1-(1+\alpha\gamma)q}\right)^{1/q} < 1, \tag{4.62}$$

*where $q = p/(p-1)$.*

**Proof.** Assume that $u_0 \in D(A^\beta)$. On $C([0,T],X)$ define the map

$$(\Gamma^\alpha u)(t) = S_\alpha(t)u_0 + \int_0^t (t-s)^{\alpha-1}\mathcal{P}_\alpha(t-s)f(s,u(s))ds.$$

From our assumptions it is easy to see that $\Gamma_\mu$ is well defined and maps $C([0,T],X)$ into itself. Put

$$\Omega_r = \{u \in C([0,T],X): \|u\| \le r, \text{ for all } 0 \le t \le T\},$$

for $r > 0$ as selected below. We seek for solutions in $\Omega_r$. We claim that there exists an integer $r > 0$ such that $\Gamma^\alpha$ maps $\Omega_r$ into $\Omega_r$. In fact, if this is not the case, then for each $r > 0$, there would exist $u^r \in \Omega_r$ and $t^r \in [0,T]$ such that $\|(\Gamma^\alpha u^r)(t^r)\| > r$. On the other hand, by (H2) and Theorem 4.11 we get

$$r < |(\Gamma^\alpha u^r)(t^r)|$$

$$\le |S_\alpha(t^r)u_0| + \int_0^{t^r} |(t^r-s)^{\alpha-1}\mathcal{P}_\alpha(t^r-s)f(s,u(s))|ds$$

$$\le \sup_{t\in[0,T]} |S_\alpha(t)u_0| + \int_0^{t^r} C_p(t^r-s)^{-1-\alpha\gamma}m_r(s)ds$$

$$\le \sup_{t\in[0,T]} |S_\alpha(t)u_0| + C_p\left(\int_0^{t^r} s^{-(1+\alpha\gamma)q}ds\right)^{\frac{1}{q}}\left(\int_0^{t^r} m_r^p(s)ds\right)^{\frac{1}{p}}$$

$$\le \sup_{t\in[0,T]} |S_\alpha(t)u_0| + C_p\|m_r\|_{L^p(0,T)}\left(\frac{T^{1-(1+\alpha\gamma)q}}{1-(1+\alpha\gamma)q}\right)^{\frac{1}{q}},$$

where $q = p/(p-1)$. Dividing on both sides by $r$ and taking the lower limit as $r \to \infty$, one has

$$1 \le C_p\sigma\left(\frac{T^{1-(1+\alpha\gamma)q}}{1-(1+\alpha\gamma)q}\right)^{1/q},$$

which contradicts (4.62). Hence for some positive integer $r$, $\Gamma^\alpha(\Omega_r) \subset \Omega_r$.

The rest of the proof is divided into three steps.

**Claim I.** $\Gamma^\alpha$ is continuous on $\Omega_r$.

Take $\{u_n\}_{n=1}^\infty \subset \Omega_r$ with $u_n \to u$ in $C([0,T],X)$. Then by the continuity of $f$ with respect to the second argument we deduce that

$$f(s,u_n(s)) \to f(s,u(s)) \quad \text{a.e.} \quad s \in [0,T].$$

Moreover, observe from (H2) and Theorem 4.11, that for fixed $0 < t \le T$,

$$(t-s)^{\alpha-1}|\mathcal{P}_\alpha(t-s)f(s,u_n(s))| \le C_p(t-s)^{-1-\alpha\gamma}m_r(s).$$

Thus, by means of Lebesgue dominated convergence theorem we obtain that

$$\int_0^t (t-s)^{\alpha-1}\|\mathcal{P}_\alpha(t-s)\|_{B(X)}|f(s,u_n(s)) - f(s,u(s))|ds \to 0,$$

which means that $\lim_{n\to\infty}\|\Gamma^\alpha u_n - \Gamma^\alpha u\|_\infty = 0$, that is, $\Gamma^\alpha$ is continuous on $\Omega_r$.

**Claim II.** $P = \{(\Gamma^\alpha u): u \in \Omega_r\}$ is equicontinuous.

For $0 < t_1 < t_2 \leq T$ and $\delta > 0$ small enough, we have

$$|(\Gamma^\alpha u)(t_1) - (\Gamma^\alpha u)(t_2)| \leq I_1 + I_2 + I_3 + I_4 + I_5,$$

where

$$I_1 = |\mathcal{S}_\alpha(t_1)u_0 - \mathcal{S}_\alpha(t_2)u_0|,$$

$$I_2 = \int_{t_1}^{t_2} (t_2 - s)^{\alpha-1} |\mathcal{P}_\alpha(t_2 - s)f(s, u(s))| ds,$$

$$I_3 = \int_0^{t_1-\delta} (t_1 - s)^{\alpha-1} \|\mathcal{P}_\alpha(t_2 - s) - \mathcal{P}_\alpha(t_1 - s)\|_{B(X)} |f(s, u(s))| ds,$$

$$I_4 = \int_{t_1-\delta}^{t_1} (t_1 - s)^{\alpha-1} \|\mathcal{P}_\alpha(t_2 - s) - \mathcal{P}_\alpha(t_1 - s)\|_{B(X)} |f(s, u(s))| ds,$$

$$I_5 = \int_0^{t_1} |(t_2 - s)^{\alpha-1} - (t_1 - s)^{\alpha-1}| \|\mathcal{P}_\alpha(t_2 - s)\|_{B(X)} |f(s, u(s))| ds.$$

From Theorem 4.12 and Theorem 4.14 (i) it is easy to see that $I_1 \to 0$ when $t_1 \to t_2$. Moreover, using (H2) and Theorem 4.11 we get

$$I_2 \leq C_p \left( \frac{(t_2 - t_1)^{1-(1+\alpha\gamma)q}}{1 - (1+\alpha\gamma)q} \right)^{1/q} \|m_r\|_{L^p(0,T)},$$

$$I_3 \leq \sup_{s \in [0, t_1-\delta]} \|\mathcal{P}_\alpha(t_2 - s) - \mathcal{P}_\alpha(t_1 - s)\|_{B(X)} \left( \int_0^{t_1-\delta} (t_1 - s)^{q\alpha-q} q ds \right)^{1/q} \|m_r\|_{L^p(0,T)}$$

$$\leq \sup_{s \in [0, t_1-\delta]} \|\mathcal{P}_\alpha(t_2 - s) - \mathcal{P}_\alpha(t_1 - s)\|_{B(X)} \left( \frac{t_1^{1+q(\alpha-1)} - \delta^{1+q(\alpha-1)}}{1 + q(\alpha-1)} \right)^{1/q} \|m_r\|_{L^p(0,T)},$$

$$I_4 \leq C_p \int_{t_1-\delta}^{t_1} (t_1 - s)^{\alpha-1} \cdot 2(t_1 - s)^{-\alpha(\gamma+1)} m_r(s) ds$$

$$\leq 2C_p \frac{\delta^{1-(1+\alpha\gamma)q}}{1 - (1+\alpha\gamma)q} \|m_r\|_{L^p(0,T)},$$

$$I_5 \leq \int_0^{t_1} C_p \left( (t_1 - s)^{\alpha-1} - (t_2 - s)^{\alpha-1} \right) (t_2 - s)^{-\alpha(1+\gamma)} m_r(s) ds$$

$$\leq \int_0^{t_1} C_p \left( (t_1 - s)^{-\gamma\alpha-1} - (t_2 - s)^{-\alpha\gamma-1} \right) m_r(s) ds$$

$$\leq C_p \left( \int_0^{t_1} (t_1 - s)^{-q(\gamma\alpha+1)} - (t_2 - s)^{-q(\alpha\gamma+1)} ds \right)^{1/q} \|m_r\|_{L^p(0,T)}$$

$$= C_p \left( \frac{(t_2 - t_1)^{1-(1+\alpha\gamma)q}}{1 - (1+\alpha\gamma)q} + \frac{t_1^{1-(1+\alpha\gamma)q} - t_2^{1-(1+\alpha\gamma)q}}{1 - (1+\alpha\gamma)q} \right)^{1/q} \|m_r\|_{L^p(0,T)}.$$

It follows from Theorem 4.12 that $I_i$ $(i = 2, 3, 4, 5)$ tends to zero independent of $u \in \Omega_r$ as $t_2 - t_1 \to 0$, $\delta \to 0$. Hence, we can conclude that

$$|(\Gamma^\alpha u)(t_1) - (\Gamma^\alpha u)(t_2)| \to 0, \quad \text{as } t_2 - t_1 \to 0,$$

and the limit is independent of $u \in \Omega_r$. For the case when $0 = t_1 < t_2 \leq T$, since

$$\int_0^{t_2} (t_2 - s)^{\alpha-1} |P(t_2 - s) f(s, u(s))| ds \leq C_p \left( \frac{t_2^{1-q(\alpha\gamma+1)}}{1 - q(\alpha\gamma + 1)} \right)^{1/q} \|m_r\|_{L^p(0,T)},$$

in view of (H2) and Theorem 4.11, $|(\Gamma^\alpha u)(t_2)|$ can be made small when $t_2$ is small independently of $u \in \Omega_r$. Thus, we prove that the assertion in Claim II holds.

**Claim III.** For each $t \in [0, T]$, $\{(\Gamma^\alpha u)(t); u \in \Omega_r\}$ is precompact in X.

For the case when $t = 0$, it is not difficult to see that $\{(\Gamma^\alpha u)(0) : u \in \Omega_r\} = \{u_0 : u \in \Omega_r\}$ is compact. Let $t \in (0, T]$ be fixed and $\epsilon, \delta > 0$. For $u \in \Omega_r$, define the map $\Gamma^\alpha_{\epsilon,\delta}$ by

$$(\Gamma^\alpha_{\epsilon,\delta} u)(t) = \mathcal{S}_\alpha(t) u_0 + \int_0^{t-\epsilon} \int_\delta^\infty \alpha\tau(t - s)^{\alpha-1} M_\alpha(\tau) T((t - s)^\alpha \tau) f(s, u(s)) d\tau ds$$

Since $A$ has compact resolvent, $\{T(t)\}_{t>0}$ is compact in view of Theorem 4.15. Thus, for each $t \in (0, T]$, $\{(\mathcal{F}_{\epsilon,\delta} u)(t) : u \in \Omega_r, \delta > 0, 0 < \epsilon < t\}$ is precompact in X. On the other hand, using (H2) and Theorem 4.11, a direct calculation yields

$$|(\Gamma^\alpha u)(t) - (\Gamma^\alpha_{\epsilon,\delta} u)(t)|$$

$$\leq \left| \int_0^t \int_0^\delta \alpha\tau(t - s)^{\alpha-1} M_\alpha(\tau) T((t - s)^\alpha \tau) f(s, u(s)) d\tau ds \right|$$

$$+ \left| \int_{t-\epsilon}^t \int_\delta^\infty \alpha\tau(t - s)^{\alpha-1} M_\alpha(\tau) T((t - s)^\alpha \tau) f(s, u(s)) d\tau ds \right|$$

$$\leq \int_0^t C_p(t - s)^{-1-\alpha\gamma} m_r(s) ds \int_0^\delta \tau^{-\gamma} M_\alpha(\tau) d\tau$$

$$+ \int_{t-\epsilon}^t C_p(t - s)^{-1-\alpha\gamma} m_r(s) ds \int_\delta^\infty \tau^{-\gamma} M_\alpha(\tau) d\tau$$

$$\leq C_p \left( \frac{T^{1-(1+\alpha\gamma)q}}{1 - (1 + \alpha\gamma)q} \right)^{1/q} \|m_r\|_{L^p(0,T)} \int_0^\delta \tau^{-\gamma} M_\alpha(\tau) d\tau$$

$$+ C_p \left( \frac{\epsilon^{1-(1+\alpha\gamma)q}}{1 - (1 + \alpha\gamma)q} \right)^{1/q} \|m_r\|_{L^p(0,T)} \frac{\Gamma(1 - \gamma)}{\Gamma(1 - \gamma\alpha)}.$$

Using the total boundedness we have that for each $t \in (0, T]$ $\{(\Gamma^\alpha u)(t) : u \in \Omega_r\}$ is precompact in X. Therefore, for each $t \in [0, T]$, $\{(\Gamma^\alpha u)(t) : u \in \Omega_r\}$ is precompact in X.

Finally, by Claims I-III and Arzela-Ascoli theorem, we conclude that $\Gamma^\alpha$ is a compact operator. Hence, $\Gamma^\alpha$ has a fixed point, which gives rise to a mild solution. This completes the proof. □

**Theorem 4.20.** *Let $A \in \Theta^\gamma_\omega(X)$ with $0 < \omega < \frac{\pi}{2}$ and $-1 < \gamma < -\frac{1}{2}$. Suppose that there exists a continuous function $M'_f(\cdot) : \mathbb{R}^+ \to \mathbb{R}^+$ and a constant $\kappa > \alpha(1 + \gamma)$ such that the nonlinear mapping $f : [0, T] \times X \to X$ satisfies*

$$|f(t, x) - f(s, y)| \leq M'_f(r)(|t - s|^\kappa + |x - y|),$$

*for all $0 \le t \le T$ and $x, y \in X$ satisfying $|x|, |y| \le r$. In addition, let the assumptions of Theorem 4.18 be satisfies and $u$ be a mild solution corresponding to $u_0$, defined on $[0, T_0]$. Then $u$ is in fact the unique classical solution to problem (4.44), existing on $[0, T_0]$, provided that $u_0 \in D(A)$ with $Au_0 \in D(A^\beta)$, $\beta > (1 + \gamma)$.*

**Proof.** In order to prove that $u$ is a classical solution, by Theorem 4.16 and the condition on $f$, we only have to verify that $u$ is Hölder continuous with an exponent $\varsigma > \alpha(1 + \gamma)$ on $(0, T_0]$. For fixed $t \in (0, T_0]$, take $0 < h < 1$ such that $h + t \le T_0$. We estimate the difference

$$|u(t+h) - u(t)| \le |S_\alpha(t+h)u_0 - S_\alpha(t)u_0|$$
$$+ \left| \int_0^h (t+h-s)^{\alpha-1} \mathcal{P}(t+h-s) f(s, u(s)) ds \right|$$
$$+ \left| \int_0^t (t-s)^{\alpha-1} \mathcal{P}(t-s) \big( f(s+h, u(s+h)) - f(s, u(s)) \big) ds \right|$$
$$= I_1 + I_2 + I_3.$$

According to Theorem 4.11, Theorem 4.13(ii) and the assumptions on $f$ we obtain

$$I_1 = \left| \int_0^t -s^{\alpha-1} A \mathcal{P}_\alpha(s) u_0 ds \right| \le \frac{C_p}{-\alpha\gamma} \big( (t+h)^{-\alpha\gamma} - t^{-\alpha\gamma} \big),$$

$$I_3 \le M' C_p \int_0^t (t-s)^{-\alpha\gamma-1} (|h|^\kappa + |u(s+h) - u(s)|) ds$$
$$\le \frac{M' C_p}{-\alpha\gamma} T_0^{-\alpha\gamma} h^\kappa + M' C_p \int_0^t (t-s)^{-\alpha\gamma-1} |u(s+h) - u(s)| ds.$$

Put $N_2 = \sup_{t \in (0, T_0)} |f(t, u(t))|$. Then, it follows from Theorem 4.11 that

$$I_2 \le C_p \int_0^h (t+h-s)^{-\alpha\gamma-1} |f(s, u(s))| ds$$
$$\le \frac{C_p N_2}{-\alpha\gamma} \big( (t+h)^{-\alpha\gamma} - t^{-\alpha\gamma} \big).$$

Collecting these estimates and using the inequality $(t+h)^{-\alpha\gamma} - t^{-\alpha\gamma} \le h^{-\alpha\gamma}$ ($0 < -\alpha\gamma < 1$) we have

$$|u(t+h) - u(t)|$$
$$\le \frac{C_p N_2 + C_p}{-\alpha\gamma} \big( (t+h)^{-\alpha\gamma} - t^{-\alpha\gamma} \big) + \frac{M'_p}{-\alpha\gamma} T_0^{-\alpha\gamma} h^\kappa$$
$$+ M' C_p \int_0^t (t-s)^{-\alpha\gamma-1} |u(s+h) - u(s)| ds$$
$$\le \frac{C_p N_2 + C_p + M' C_p}{-\alpha\gamma} h^\varsigma + M' C_p \int_0^t (t-s)^{-\alpha\gamma-1} |u(s+h) - u(s)| ds,$$

where $\varsigma = \min\{\kappa, -\alpha\gamma\} > \alpha(\gamma + 1)$. Now, it follows from the usual Gronwall inequality that $u$ has Hölder continuity on $(0, T_0]$. This completes the proof of theorem. $\square$

### 4.5.5 Applications

In this subsection, we present three examples (Examples 4.4-4.6) motivated from physics, which do not aim at generality but indicate how our theorems can be applied to concrete problems. Examples 4.4 and 4.5 are inspired directly from the work of Carvalho, Dlotko and Nescimento, 2008, and they describe anomalous diffusion on fractals (physical objects of fractional dimension, like some amorphous semiconductors or strongly porous materials; see Anh and Leonenko, 2001; Metzler and Klafter, 2000 and references therein). Example 4.4 is the limit problem of certain fractional diffusion equations in complex systems on domains of "dumbbell with a thin handle"(see, e.g., Anh and Leonenko, 2001; Metzler and Klafter, 2000). Example 4.5 displays anomalous dynamical behavior of anomalous transport processes (see, e.g., Anh and Leonenko, 2001; Metzler and Klafter, 2000). Example 4.6 is a modified fractional Schrödinger equation with fractional Laplacians whose physical background is statistical physics and fractional quantum mechanics (see, e.g., Hu and Kallianpur, 2000; Podlubny, 1999). We refer the reader to Kirane, Laskri and Tatar, 2005 and references therein for more research results related to fractional Laplacians.

**Example 4.4.** Consider the system of fractional partial differential equations in the form

$$
\begin{cases}
{}_0^C D_t^\alpha w - \Delta w + w = f(w), & x \in \Omega, \ t > 0, \\[2mm]
\dfrac{\partial w}{\partial n} = 0, & x \in \partial\Omega, \\[2mm]
{}_0^C D_t^\alpha v - \dfrac{1}{g}(g v_x)_x + v = f(v), & x \in (0,1), \\[2mm]
v(0) = w(P_0), v(1) = w(P_1), \\[2mm]
w(x,0) = w_0(x) \quad x \in \Omega, & v(x,0) = v_0(x), \ x \in (0,1),
\end{cases}
\tag{4.63}
$$

where $\Omega = D_1 \cup D_2$ and $D_1$ and $D_2$ are mutually disjoint bounded domains in $\mathbb{R}^N (N \geq 2)$ with smooth boundaries, joined by the line segment $Q_0$, and ${}_0^C D_t^\alpha$, $0 < \alpha < 1$, is the regularized Caputo fractional derivative of order $\alpha$, that is,

$$
({}_0^C D_t^\alpha u)(t,x) = \frac{1}{\Gamma(1-\alpha)} \left( \frac{\partial}{\partial t} \int_0^t (t-s)^{-\alpha} u(s,x)\, ds - t^{-\alpha} u(0,x) \right).
\tag{4.64}
$$

When $\alpha = 1$, we regard (4.63) as the limit problem of (4.40) as $\varepsilon \to 0$, which is described in more detail in Example 4.2. Here, our objective is to show that system (4.63) is well posed in $V_0^p = L^p(\Omega) \oplus L_g^p(0,1)$ $(1 \leq p < \infty)$.

Let the operators $A_0 : D(A_0) \subset V_0^p \mapsto V_0^p$ be defined by

$$
D(A_0) = \{(w,v) \in V_0^P : \ w \in D(\Delta_\Omega), v \in L_g^p(0,1), w(P_0) = v(0), w(P_1) = v(1)\},
$$

$$
A_0(w,v) = \left( -\Delta w + w, -\frac{1}{g}(g v')' + v \right), \quad (w,v) \in V_0^p,
$$

where $\Delta_\Omega$ is the Laplace operator with homogeneous Neumann boundary conditions in $L^p(\Omega)$ and

$$D(\Delta_\Omega) = \left\{ u \in W^{2,p}(\Omega) : \left. \frac{\partial u}{\partial n} \right|_{\partial\Omega} = 0 \right\}.$$

From Example 4.2, if $p > \frac{N}{2}$, then $A_0 \in \Theta_\mu^{-\gamma'}(V_0^p)$ for some $\gamma' \in (0, 1 - \frac{N}{2p})$ and $\mu \in (0, \frac{\pi}{2})$. Therefore, system (4.63) can be seen as an abstract evolution equation in the form

$$\begin{cases} {}_0^C D_t^\alpha u + A_0 u = f(u), & t > 0, \\ u(0) = u_0 = (w_0, v_0) \in V_0^p, \end{cases} \tag{4.65}$$

We assume that the nonlinearity $f : \mathbb{R} \to \mathbb{R}$ is globally Lipschitz continuous. It can define a Nemitskiĭ operator from $V_0^p$ into itself by $f(w, v) = (f_\Omega(w), f_I(v))$ with $f_\Omega(w)(x) = f(w(x))$, $x \in \Omega$ and $f_I(v)(x) = f(v(x))$, $x \in (0, 1)$ such that

$$|f(u) - f(u')|_{V_0^p} \le L''(r)|u - u'|_{V_0^p}.$$

for all $u, u' \in V_0^p$ satisfying $|u|_{V_0^p}, |u'|_{V_0^p} \le r$. Hence, from Remark 4.14, (4.65) (that is, (4.63)) has a unique mild solution provided that $u_0 \in D(A_0^\beta)$ with $\beta > 1 - \gamma'$ (in particular, $u_0 \in D(A_0)$).

**Example 4.5.** Let $\Omega$ be a bounded domain in $\mathbb{R}^N$ ($N \ge 1$) with boundary $\partial\Omega$ of class $C^4$. Consider the fractional initial-boundary value problem of form

$$\begin{cases} ({}_0^C D_t^\alpha u)(t, x) - \Delta u(t, x) = f(u(t, x)), & t > 0, \ x \in \Omega, \\ u|_{\partial\Omega} = 0, \\ u(0, x) = u_0(x), & x \in \Omega, \end{cases} \tag{4.66}$$

in the space $C^l(\overline{\Omega})$ ($0 < l < 1$), where $\Delta$ stands for the Laplacian operator with respect to the spatial variable and ${}_0^C D_t^\alpha$, representing the regularized Caputo fractional derivative of order $\alpha$ ($0 < \alpha < 1$), is given by (4.64). Set

$$\tilde{A} = -\Delta, \quad D(\tilde{A}) = \{u \in C^{2+l}(\overline{\Omega}) : u = 0 \ \text{on} \ \partial\Omega\}.$$

It follows from Example 4.3 that there exist $\nu$, $\varepsilon > 0$ such that $\tilde{A} + \nu \in \Theta_{\frac{\pi}{2}-\varepsilon}^{\frac{l}{2}-1}(C^l(\overline{\Omega}))$. Then, problem (4.66) can be written abstractly as

$$\smash{{}_0^C D_t^\alpha u(t) + \tilde{A} u(t) = f(u), \quad t > 0,}$$

With respect to the nonlinearity $f$, we assume that $f : \mathbb{R} \to \mathbb{R}$ is continuously differentiable and satisfies the condition

$$|f(x) - f(y)| \le \frac{k(r)}{r}|x - y|, \quad |x|, |y| \le r, \tag{4.67}$$

for any $r > 0$. It defines a Nemitskiĭ operator from $C^l(\overline{\Omega})$ into itself by $f(u)(x) = f(u(x))$ with

$$|f(u) - f(v)|_{C^l(\overline{\Omega})} \le k(r)|u - v|_{C^l(\overline{\Omega})}, \quad |v|_{C^l(\overline{\Omega})}, |u|_{C^l(\overline{\Omega})} \le r.$$

Noting $\frac{l}{2} - 1 \in (-1, -\frac{1}{2})$, we then obtain the following conclusion: (i) according to Remark 4.14, (4.66) has a unique mild solution for each $u_0 \in D(\widetilde{A}^\beta)$ with $\beta > \frac{l}{2}$. Moreover, (ii) if $f', f''$ are continuously differentiable functions satisfying the condition (4.67), then one finds that the Nemitskiĭ operator satisfies the assumptions of Theorem 4.18 and Theorem 4.20, which implies that for each $u_0 \in D(\widetilde{A})$ with $\widetilde{A}u_0 \in D(\widetilde{A}^\beta)$ $(\beta > \frac{l}{2})$, the corresponding mild solution to (4.66) is also a unique classical solution.

**Example 4.6.** Consider the following fractional Cauchy problem

$$\begin{cases} ({}^C_0 D^\alpha_t y)(t, x) + (-i\Delta + \sigma)^{1/2} u(t, x) = f(u(t, x)), & t > 0, \ x \in \mathbb{R}^2, \\ u(0, x) = u_0(x), & x \in \mathbb{R}^2, \end{cases} \tag{4.68}$$

in $L^3(\mathbb{R}^2)$, where $\sigma > 0$ is a suitable constant, $i\Delta$ is the Schrödinger operator and ${}^C_0 D^\alpha_t (0 < \alpha < 1)$ is given by (4.64). Let

$$\widehat{A} = (-i\Delta + \sigma)^{1/2}, \quad D(\widehat{A}) = W^{1,3}(\mathbb{R}^2) \quad \text{(a Sobolev space)}.$$

Then $i\Delta$ generates a $\beta$-times integrated semigroup $S^\beta(t)$ with $\beta = \frac{5}{12}$ on $L^3(\mathbb{R}^2)$ such that $\|S^\beta(t)\|_{B(L^3(\mathbb{R}^2))} \leq \widehat{M}t^\beta$ for all $t \geq 0$ and some constants $\widehat{M} > 0$ (see Neerven and Straub, 1998). Therefore, by virtue of Theorem 1.3.5 and Definition 1.3.1 for $C = I$ of Xiao and Liang, 1998, we deduce that the operator $-i\Delta + \sigma$ belongs to $\Theta^{\beta-1}_{\frac{\pi}{2}}(L^3(\mathbb{R}^2))$, which denotes the family of all linear closed operators $A : D(A) \subset L^3(\mathbb{R}^2) \to L^3(\mathbb{R}^2)$ satisfying

$$\sigma(A) \subset S_{\frac{\pi}{2}} = \left\{ z \in \mathbb{C} \setminus \{0\} : \ |\arg z| \leq \frac{\pi}{2} \right\} \cup \{0\},$$

and for every $\frac{\pi}{2} < \mu < \tau$ there exists a constant $C_\mu$ such that $\|R(z; A)\| \leq C_\mu |z|^{\beta-1}$. Thus, it follows from Proposition 3.6 of Periago and Straub, 2002 that $\widehat{A} \in \Theta^{-1+2\beta}_\omega(L^3(\mathbb{R}^2))$ for some $0 < \omega < \frac{2}{\pi}$. Moreover, the system (4.68) can be rewritten as follows:

$$\begin{cases} ({}^C_0 D^\alpha_t y)(t, x) + \widehat{A}u = f(u), & t > 0, \\ u(0, x) = u_0 \in L^3(\mathbb{R}^2). \end{cases}$$

Assume that $f : \mathbb{C} \to \mathbb{C}$ is globally Lipschitz continuous. Then we have a Nemitskiĭ operator from $L^3(\mathbb{R}^2)$ to itself given by $f(u)(x) = f(u(x))$, and for a constant $\widehat{L}(r)$ and all $u, v \in L^3(\mathbb{R}^2)$ such that $|u|_{L^3(\mathbb{R}^2)} \leq r$ and $|v|_{L^3(\mathbb{R}^2)} \leq r$. Consequently, it follows from Remark 4.14 that (4.68) has a unique mild solution provided $u_0 \in D(\widehat{A})^\tau$ with $\tau > \frac{5}{6}$.

## 4.6 Notes and Remarks

The results in Section 4.2 are taken from Zhou, Zhang and Shen, 2013. The material in Section 4.3 due to Zhou, Shen and Zhang, 2013. The main results in Section 4.4 is taken from Wang, Zhou and Fečkan, 2014. The contents of Section 4.5 are adopted from Wang, Chen and Xiao, 2012.

# Chapter 5

# Fractional Impulsive Differential Equations

## 5.1 Introduction

The theory of impulsive differential equations has recently years been an object of increasing interest because of its wide applicability in biology, in medicine and in more and more fields. The reason for this applicability arises from the fact that impulsive differential problems is an appropriate model for describing process which at certain moments change their state rapidly and which cannot be described using the classical differential problems. For a wide bibliography and exposition on this object see for instance the monographs of Benchohra, Henderson and Ntouyas, 2006; Bainov and Simeonov, 1993; Lakshmikantham, Bainov and Simeonov, 1989; Yang, 2001 and the papers of Abada, Benchohra and Hammouche, 2009; Ahmed, 2003 and 2007; Akhmet, 2005; Fan and Li, 2010; Fan, 2010; Liang, Liu and Xiao, 2009; Liu, 1999; Battelli and Fečkan, 1997; Mophou, 2010; Nieto and O'Regan, 2009; Wang, Xiang and Peng, 2009; Wang and Wei, 2010; Wei, Xiang and Peng, 2006; Wei, Hou and Teo, 2006.

Recently, a number of papers have been recently written on Cauchy problems, boundary value problems and nonlocal problems for fractional impulsive differential equations, one can see Ahmad *et al.*, 2009 and 2010; Bench, 2009; Agarwal *et al.*, 2010; Ahmad and Wang, 2010; Balachandran, 2010; Tian and Bai, 2010; Cao and Chen, 2010; Wang *et al.*, 2010; Wang, Ahmad and Zhang, 2010; Wang, 2011; Wang, Zhang and Song, 2011; Yang and Chen, 2011; Cao and Chen, 2012 and the references therein.

However, Fečkan, Zhou and Wang, 2012; Kosmatov, 2012, point out on the error in former solutions for some fractional impulsive differential equations by constructing a counterexample and establish a general framework to seek a natural solution for fractional impulsive differential equations.

Section 5.2 is concerned with the existence and uniqueness of solutions for fractional impulsive initial value equations. In Section 5.3, we give some sufficient conditions for the existence of the solutions for fractional impulsive boundary value equations, and use a new generalized singular Gronwall inequality to obtain the data dependence. In Section 5.4, we establish the existence results of solutions for fractional impulsive Langevin equations by utilizing boundedness, continuity, monotonicity and nonnegative of Mittag-Leffler function and fixed point methods. Section 5.5 is devoted to the existence of $PC$-mild solutions for Cauchy problems and nonlocal problems for fractional impulsive evolution equations.

## 5.2 Impulsive Initial Value Problems

### 5.2.1 *Introduction*

Consider the Cauchy problems for the following fractional impulsive differential equations

$$\begin{cases} {}^{C}_{0}D^{q}_{t}u(t) = f(t, u(t)), & t \in J', \\ u(t^{+}_{k}) = u(t^{-}_{k}) + y_{k}, & y_{k} \in \mathbb{R}, \ k = 1, 2, \dots, m, \\ u(0) = u_{0}, \end{cases} \tag{5.1}$$

where ${}^{C}_{0}D^{q}_{t}$ is Caputo fractional derivative of order $q \in (0,1)$ with the lower limit zero, $u_{0} \in \mathbb{R}$, $f : J \times \mathbb{R} \to \mathbb{R}$ is jointly continuous, and $t_{k}$ satisfy $0 = t_{0} < t_{1} < \cdots < t_{m} < t_{m+1} = T$, $u(t^{+}_{k}) = \lim_{\epsilon \to 0^{+}} u(t_{k} + \epsilon)$ and $u(t^{-}_{k}) = \lim_{\epsilon \to 0^{-}} u(t_{k} + \epsilon)$ represent the right and left limits of $u(t)$ at $t = t_{k}$.

In Subsection 5.2.2, we introduce the definition of a solution of the problem (5.1). Subsection 5.2.3 is concerned with the existence and uniqueness of solutions for (5.1).

### 5.2.2 *Formula of Solutions*

Note that

$$u(t) = u_{0} - \frac{1}{\Gamma(q)} \int_{0}^{a} (a - s)^{q-1} h(s)ds + \frac{1}{\Gamma(q)} \int_{0}^{t} (t - s)^{q-1} h(s)ds,$$

solves the Cauchy problems

$$\begin{cases} {}^{C}_{0}D^{q}_{t}u(t) = h(t), & t \in J, \\ u(0) = u_{0} - \dfrac{1}{\Gamma(q)} \displaystyle\int_{0}^{a} (a - s)^{q-1} h(s)ds. \end{cases}$$

One can obtain the following result immediately.

**Lemma 5.1.** *Let* $q \in (0,1)$ *and* $h : J \to \mathbb{R}$ *be continuous. A function* $u \in C(J, \mathbb{R})$ *is a solution of the fractional integral equation*

$$u(t) = u_{0} - \frac{1}{\Gamma(q)} \int_{0}^{a} (a - s)^{q-1} h(s)ds + \frac{1}{\Gamma(q)} \int_{0}^{t} (t - s)^{q-1} h(s)ds,$$

*if and only if* $u$ *is a solution of the following fractional Cauchy problems*

$$\begin{cases} {}^{C}_{0}D^{q}_{t}u(t) = h(t), & t \in J, \\ u(a) = u_{0}, & a > 0. \end{cases}$$

As a consequence of Lemma 5.1 we have the following result which is useful in what follows.

**Lemma 5.2.** *Let $q \in (0,1)$ and $h : J \to \mathbb{R}$ be continuous. A function $u$ is a solution of the fractional integral equation*

$$u(t) = \begin{cases} u_0 + \dfrac{1}{\Gamma(q)} \displaystyle\int_0^t (t-s)^{q-1} h(s)ds, & \text{for } t \in [0, t_1), \\[2mm] u_0 + y_1 + \dfrac{1}{\Gamma(q)} \displaystyle\int_0^t (t-s)^{q-1} h(s)ds, & \text{for } t \in (t_1, t_2), \\[2mm] u_0 + y_1 + y_2 + \dfrac{1}{\Gamma(q)} \displaystyle\int_0^t (t-s)^{q-1} h(s)ds, & \text{for } t \in (t_2, t_3), \\[2mm] \quad\vdots & \\[2mm] u_0 + \displaystyle\sum_{i=1}^m y_i + \dfrac{1}{\Gamma(q)} \displaystyle\int_0^t (t-s)^{q-1} h(s)ds, & \text{for } t \in (t_m, T], \end{cases} \tag{5.2}$$

*if and only if $u$ is a solution of the following impulsive problem*

$$\begin{cases} {}^C_0 D_t^q u(t) = h(t), & t \in (0, T], \\ u(t_k^+) = u(t_k^-) + y_k, & k = 1, 2, \ldots, m, \\ u(0) = u_0. \end{cases} \tag{5.3}$$

**Proof.** Assume $u$ satisfies (5.3). If $t \in [0, t_1]$, then

$$ {}^C_0 D_t^q u(t) = h(t), \quad t \in (0, t_1] \text{ with } u(0) = u_0. \tag{5.4}$$

Integrating the expression (5.4) from $0$ to $t$ by virtue of Definition 1.1, one can obtain

$$u(t) = u_0 + \frac{1}{\Gamma(q)} \int_0^t (t-s)^{q-1} h(s)ds.$$

If $t \in (t_1, t_2]$, then

$$ {}^C_0 D_t^q u(t) = h(t), \quad t \in (t_1, t_2] \text{ with } u(t_1^+) = u(t_1^-) + y_1.$$

By Lemma 5.1, one obtain

$$u(t) = u(t_1^+) - \frac{1}{\Gamma(q)} \int_0^{t_1} (t_1 - s)^{q-1} h(s)ds + \frac{1}{\Gamma(q)} \int_0^t (t-s)^{q-1} h(s)ds$$

$$= u(t_1^-) + y_1 - \frac{1}{\Gamma(q)} \int_0^{t_1} (t_1 - s)^{q-1} h(s)ds + \frac{1}{\Gamma(q)} \int_0^t (t-s)^{q-1} h(s)ds$$

$$= u_0 + y_1 + \frac{1}{\Gamma(q)} \int_0^t (t-s)^{q-1} h(s)ds.$$

If $t \in (t_2, t_3]$, then using again Lemma 5.1, we get

$$u(t) = u(t_2^+) - \frac{1}{\Gamma(q)} \int_0^{t_2} (t_2 - s)^{q-1} h(s)ds + \frac{1}{\Gamma(q)} \int_0^t (t-s)^{q-1} h(s)ds$$

$$= u(t_2^-) + y_2 - \frac{1}{\Gamma(q)} \int_0^{t_2} (t_2 - s)^{q-1} h(s) ds + \frac{1}{\Gamma(q)} \int_0^t (t - s)^{q-1} h(s) ds$$

$$= u_0 + y_1 + y_2 + \frac{1}{\Gamma(q)} \int_0^t (t - s)^{q-1} h(s) ds.$$

If $t \in (t_k, t_{k+1}]$, then again from Lemma 5.1 we get

$$u(t) = u_0 + \sum_{i=1}^k y_i + \frac{1}{\Gamma(q)} \int_0^t (t - s)^{q-1} h(s) ds.$$

Conversely, assume that $u$ satisfies (5.2). If $t \in (0, t_1]$ then $u(0) = u_0$ and using the fact that $_0^C D_t^q$ is the left inverse of $_0 D_t^{-q}$ we get (5.4). If $t \in (t_k, t_{k+1}]$, $k = 1, 2, \ldots, m$ and using the fact of Caputo fractional derivative of a constant is equal to zero, we obtain $_0^C D_t^q u(t) = h(t)$, $t \in (t_k, t_{k+1}]$ and $u(t_k^+) = u(t_k^-) + y_k$. This completes the proof. $\qquad\square$

**Definition 5.1.** A function $u \in PC^1(J, \mathbb{R})$ is said to be a solution of the problem (5.1) if $u$ satisfies the integral equation

$$u(t) = \begin{cases} u_0 + \dfrac{1}{\Gamma(q)} \displaystyle\int_0^t (t - s)^{q-1} f(s, u(s)) ds, & \text{for } t \in [0, t_1), \\[2ex] u_0 + y_1 + \dfrac{1}{\Gamma(q)} \displaystyle\int_0^t (t - s)^{q-1} f(s, u(s)) ds, & \text{for } t \in (t_1, t_2), \\[2ex] u_0 + y_1 + y_2 + \dfrac{1}{\Gamma(q)} \displaystyle\int_0^t (t - s)^{q-1} f(s, u(s)) ds, & \text{for } t \in (t_2, t_3), \\[2ex] \vdots \\[1ex] u_0 + \displaystyle\sum_{i=1}^m y_i + \dfrac{1}{\Gamma(q)} \displaystyle\int_0^t (t - s)^{q-1} f(s, u(s)) ds, & \text{for } t \in (t_m, T]. \end{cases}$$

**Theorem 5.1.** *(Ye, Gao and Ding, 2007) Suppose $\beta > 0$, $\tilde{a}(t)$ is a nonnegative function locally integrable on $[0, T)$ and $\tilde{g}(t)$ is a nonnegative, nondecreasing continuous function defined on $\tilde{g}(t) \leq M$, $t \in [0, T)$, and suppose $y(t)$ is nonnegative and locally integrable on $[0, T)$ with*

$$y(t) \leq \tilde{a}(t) + \tilde{g}(t) \int_0^t (t - s)^{\beta-1} y(s) ds, \quad t \in [0, T).$$

*Then*

$$y(t) \leq \tilde{a}(t) + \int_0^t \left( \sum_{n=1}^\infty \frac{(\tilde{g}(t)\Gamma(\beta))^n}{\Gamma(n\beta)} (t - s)^{n\beta-1} \tilde{a}(s) \right) ds, \quad t \in [0, T).$$

**Theorem 5.2.** *Under the hypothesis of Theorem 5.1, let $\tilde{a}(t)$ be a nondecreasing function on $[0, T)$. Then we have*

$$y(t) \leq \tilde{a}(t) E_\beta(\tilde{g}(t)\Gamma(\beta)t^\beta),$$

*where $E_\beta$ is the Mittag-Leffler function.*

**Remark 5.1.** There exists a constant $M_{\tilde{g}}^* > 0$ independent of $\tilde{a}$ such that

$$y(t) \leq M_{\tilde{g}}^* \tilde{a} \quad \text{for all } 0 \leq t < T.$$

### 5.2.3  *Existence*

This subsection deals with the existence and uniqueness of solutions for the problem (5.1). Before stating and proving the main results, we introduce the following hypotheses.

**(H1)** $f : J \times \mathbb{R} \to \mathbb{R}$ is jointly continuous;

**(H2)** there exists $q_1 \in (0, q)$ and a real function $m(\cdot) \in L^{\frac{1}{q_1}}(J, \mathbb{R})$ such that $|f(t, u)| \le m(t)$, for all $u \in \mathbb{R}$;

**(H3)** there exists $q_2 \in (0, q)$ and a real function $h(\cdot) \in L^{\frac{1}{q_2}}(J, \mathbb{R})$ such that $|f(t, u_1) - f(t, u_2)| \le h(t)|u_1 - u_2|$, for all $u_1, u_2 \in \mathbb{R}$.

For brevity, let

$$\gamma = \frac{T^q}{\Gamma(q+1)}, \quad \beta = \frac{q-1}{1-q_1}, \quad \alpha = \frac{q-1}{1-q_2}.$$

**Theorem 5.3.** *Assume that (H1)-(H3) hold. If*

$$\frac{T^{(1+\alpha)(1-q_2)} \|h\|_{L^{\frac{1}{q_2}} J}}{\Gamma(q)(1+\alpha)^{1-q_2}} < 1, \qquad (5.5)$$

*then the problem (5.1) has a unique solution on J.*

**Proof.** Transform the problem (5.1) into a fixed point problem. Consider the operator $F : PC(J, \mathbb{R}) \to PC(J, \mathbb{R})$ defined by

$$(Fu)(t) = u_0 + \sum_{i=1}^{k} y_i + \frac{1}{\Gamma(q)} \int_0^t (t-s)^{q-1} f(s, u(s)) ds. \qquad (5.6)$$

It is obvious that $F$ is well defined due to (H1).

**Claim I.** $Fu \in PC(J, \mathbb{R})$ for every $u \in PC(J, \mathbb{R})$.

If $t \in [0, t_1]$, then for every $u \in C([0, t_1], \mathbb{R})$ and any $\delta > 0$, by using Hölder inequality, we get

$$|(Fu)(t + \delta) - (Fu)(t)|$$

$$\le \frac{1}{\Gamma(q)} \left| \int_0^t \left( (t + \delta - s)^{q-1} - (t - s)^{q-1} \right) f(s, u(s)) ds \right|$$

$$+ \frac{1}{\Gamma(q)} \left| \int_t^{t+\delta} (t + \delta - s)^{q-1} f(s, u(s)) ds \right|$$

$$\le \frac{1}{\Gamma(q)} \left( \int_0^t \left( (t + \delta - s)^{q-1} - (t - s)^{q-1} \right)^{\frac{1}{1-q_1}} ds \right)^{1-q_1} \left( \int_0^t (m(s))^{\frac{1}{q_1}} ds \right)^{q_1}$$

$$+ \frac{1}{\Gamma(q)} \left( \int_t^{t+\delta} \left( (t + \delta - s)^{q-1} \right)^{\frac{1}{1-q_1}} ds \right)^{1-q_1} \left( \int_t^{t+\delta} (m(s))^{\frac{1}{q_1}} ds \right)^{q_1},$$

which implies that
$$|Fu(t+\delta) - Fu(t)| \leq \frac{2\delta^{(1+\beta)(1-q_1)}\|m\|_{L^{\frac{1}{q_1}}[0,t_1]}}{\Gamma(q)(1+\beta)^{1-q_1}}.$$
It is easy to see that the right-hand side of the above inequality tends to zero as $\delta \to 0$. Thus, $Fu \in C([0,t_1],\mathbb{R})$.

If $t \in (t_1, t_2]$, then for every $u \in C((t_1, t_2],\mathbb{R})$ and any $\delta > 0$, repeating the same process, one can obtain
$$|Fu(t+\delta) - Fu(t)| \leq \frac{2\delta^{(1+\beta)(1-q_1)}\|m\|_{L^{\frac{1}{q_1}}[t_1,t_2]}}{\Gamma(q)(1+\beta)^{1-q_1}},$$
which implies that $Fu \in C((t_1, t_2],\mathbb{R})$.

If $t \in (t_k, t_{k+1}]$, $k = 1, 2, \ldots, m$, then for every $u \in C((t_k, t_{k+1}],\mathbb{R})$ and any $\delta > 0$, repeating the same process again, one can obtain
$$|Fu(t+\delta) - Fu(t)| \leq \frac{2\delta^{(1+\beta)(1-q_1)}\|m\|_{L^{\frac{1}{q_1}}[t_k,t_{k+1}]}}{\Gamma(q)(1+\beta)^{1-q_1}},$$
which implies that $Fu \in C((t_k, t_{k+1}],\mathbb{R})$.

From the above discussion, we must have $Fu \in PC(J,\mathbb{R})$ for every $u \in PC(J,\mathbb{R})$.

**Claim II.** $F$ is a contraction operator on $PC(J,\mathbb{R})$.

In fact, for arbitrary $u, v \in PC(J,\mathbb{R})$, we get
$$|(Fu)(t) - (Fv)(t)| \leq \frac{1}{\Gamma(q)} \int_0^t (t-s)^{q-1}|f(s,u_1(s)) - f(s,u_2(s))|ds$$
$$\leq \frac{1}{\Gamma(q)} \int_0^t (t-s)^{q-1}h(s)|u_1(s) - u_2(s)|ds$$
$$\leq \frac{\|h\|_{L^{\frac{1}{q_2}}J}}{\Gamma(q)} \left( \int_0^t \left((t-s)^{q-1}\right)^{\frac{1}{1-q_2}} ds \right)^{1-q_2} \|u_1 - u_2\|_{PC}$$
$$= \frac{T^{(1+\alpha)(1-q_2)}\|h\|_{L^{\frac{1}{q_2}}J}}{\Gamma(q)(1+\alpha)^{1-q_2}} \|u_1 - u_2\|_{PC}.$$

Thus, $F$ is a contraction mapping on $PC(J,\mathbb{R})$ due to the condition (5.5). By applying Banach contraction mapping principle we know that the operator $F$ has a unique fixed point on $PC(J,\mathbb{R})$. Therefore, the problem (5.1) has a unique solution

$$u(t) = \begin{cases} u_0 + \dfrac{1}{\Gamma(q)} \displaystyle\int_0^t (t-s)^{q-1}f(s,u(s))ds, & \text{for } t \in [0,t_1), \\[2ex] u_0 + y_1 + \dfrac{1}{\Gamma(q)} \displaystyle\int_0^t (t-s)^{q-1}f(s,u(s))ds, & \text{for } t \in (t_1,t_2), \\[2ex] u_0 + y_1 + y_2 + \dfrac{1}{\Gamma(q)} \displaystyle\int_0^t (t-s)^{q-1}f(s,u(s))ds, & \text{for } t \in (t_2,t_3), \\[1ex] \quad\vdots \\[1ex] u_0 + \displaystyle\sum_{i=1}^m y_i + \dfrac{1}{\Gamma(q)} \displaystyle\int_0^t (t-s)^{q-1}f(s,u(s))ds, & \text{for } t \in (t_m,T]. \end{cases}$$

This completes the proof.       $\square$

The second result is based on Schaefer fixed point theorem.

Now, we replace (H2) into the following linear growth condition:

**(H2)′** there exists a constant $L > 0$ such that

$$|f(t, u)| \leq L(1 + |u|) \quad \text{for each } t \in J \text{ and all } u \in \mathbb{R}.$$

**Theorem 5.4.** *Assume that (H1) and (H2)′ hold. Then the problem (5.1) has at least one solution.*

**Proof.** Transform the problem (5.1) into a fixed point problem. Consider the operator $F : PC(J, \mathbb{R}) \to PC(J, \mathbb{R})$ defined as (5.6). For the sake of convenience, we subdivide the proof into several steps.

**Claim I.** $F$ is continuous.

Let $\{u_n\}$ be a sequence such that $u_n \to u$ in $PC(J, \mathbb{R})$. Then for each $t \in J$, we have

$$
\begin{aligned}
|(Fu_n)(t) - (Fu)(t)| &\leq \frac{1}{\Gamma(q)} \int_0^t (t - s)^{q-1} |f(s, u_n(s)) - f(s, u(s))| ds \\
&\leq \frac{T^q}{\Gamma(q+1)} \|f(\cdot, u_n(\cdot)) - f(\cdot, u(\cdot))\|_{PC}.
\end{aligned}
$$

Due to (H1), $f$ is jointly continuous, then we have

$$\|Fu_n - Fu\|_{PC} \to 0, \quad \text{as } n \to \infty.$$

**Claim II.** $F$ maps bounded sets into bounded sets in $PC(J, \mathbb{R})$.

Indeed, it is enough to show that for any $\eta^* > 0$, there exists a $\ell > 0$ such that for each $u \in B_{\eta^*} = \{y \in PC(J, \mathbb{R}) : \|u\|_{PC} \leq \eta^*\}$, we have $\|Fu\|_{PC} \leq \ell$.

For each $t \in J$, we get

$$
\begin{aligned}
|(Fu)(t)| &\leq |u_0| + \sum_{i=1}^m |y_i| + \frac{1}{\Gamma(\alpha)} \int_0^t (t - s)^{q-1} |f(s, u(s))| ds \\
&\leq |u_0| + \sum_{i=1}^m |y_i| + \frac{LT^q \eta^*}{\Gamma(q+1)},
\end{aligned}
$$

which implies that

$$\|Fy\|_{PC} \leq |u_0| + \sum_{i=1}^m |y_i| + \frac{LT^q \eta^*}{\Gamma(q+1)} =: \ell.$$

**Claim III.** $F$ maps bounded sets into equicontinuous sets of $PC(J, \mathbb{R})$.

For interval $[0, t_1]$, $0 \leq s_1 < s_2 \leq t_1$, $u \in B_{\eta^*}$. Using (H2)′, we have

$$
\begin{aligned}
|(Fu)(s_2) - (Fu)(s_1)| &\leq \frac{1}{\Gamma(q)} \int_0^{s_1} \left( (s_1 - s)^{q-1} - (s_2 - s)^{q-1} \right) |f(s, u(s))| ds \\
&\quad + \frac{1}{\Gamma(q)} \int_{s_1}^{s_2} (s_2 - s)^{q-1} |f(s, y(s))| ds \\
&\leq \frac{L}{\Gamma(q)} \int_0^{s_1} \left( (s_1 - s)^{q-1} - (s_2 - s)^{q-1} \right) (1 + |u(s)|) ds
\end{aligned}
$$

$$+ \frac{L}{\Gamma(q)} \int_{s_1}^{s_2} (s_2 - s)^{q-1} (1 + |u(s)|) ds$$

$$\leq \frac{L(1 + \eta^*)}{\Gamma(q)} \int_0^{s_1} (s_1 - s)^{q-1} - (s_2 - s)^{q-1} ds$$

$$+ \frac{L(1 + \eta^*)}{\Gamma(q)} \int_{s_1}^{s_2} (s_2 - s)^{q-1} ds$$

$$\leq \frac{L(1 + \eta^*)}{\Gamma(q+1)} \left( |s_1^q - s_2^q| + 2(s_2 - s_1)^q \right)$$

$$\leq \frac{3L(1 + \eta^*)(s_2 - s_1)^q}{\Gamma(q+1)}.$$

As $s_2 \to s_1$, the right-hand side of the above inequality tends to zero, therefore $F$ is equicontinuous on interval $[0, t_1]$.

In general, for the time interval $(t_k, t_{k+1}]$, we similarly obtain the following inequality

$$|(Fu)(s_2) - (Fu)(s_1)| \leq \frac{3L(1 + \eta^*)(s_2 - s_1)^q}{\Gamma(q+1)} \to 0, \text{ as } s_2 \to s_1.$$

This yields that $F$ is equicontinuous on interval $(t_k, t_{k+1}]$.

As a consequence of Claim I-III together with $PC$-type Arzela-Ascoli theorem (see Lemma 1.3 in the case of $X = \mathbb{R}$), we can conclude that $F : B_{\eta^*} \to B_{\eta^*}$ is continuous and completely continuous.

**Claim IV.** A priori bound.

Now it remains to show that the set

$$E(F) = \{u \in PC(J, \mathbb{R}) : u = \lambda Fu, \text{ for some } \lambda \in (0, 1)\}$$

is bounded.

Let $u \in E(F)$, then $u = \lambda Fu$ for some $\lambda \in (0, 1)$.

Without loss of generality, for the time interval $t \in (t_k, t_{k+1}]$,

$$|u(t)| \leq |u_0| + \sum_{i=1}^{k} |y_i| + \frac{1}{\Gamma(q)} \int_0^t (t - s)^{q-1} |f(s, u(s))| ds$$

$$\leq |u_0| + \sum_{i=1}^{k} |y_i| + \frac{LT^q}{\Gamma(q+1)} + \frac{L}{\Gamma(q)} \int_0^t (t - s)^{q-1} |u(s)| ds.$$

By Lemma 5.1, there exists a $M_k^* > 0$ such that

$$|u(t)| \leq M_k^*, \quad t \in (t_k, t_{k+1}].$$

Set $M^* = \max_{1 \leq k \leq m} M_k^*$. Thus for every $t \in J$, we have

$$\|u\|_{PC} \leq M^*.$$

This shows that the set $E(F)$ is bounded.

As a consequence of Schaefer fixed point theorem, we deduce that $F$ has a fixed point which is a solution of the problem (5.1). The proof is completed.    □

**Remark 5.2.** Let the assumptions of Theorem 5.4 hold. If $f$ is uniformly Lipschtiz continuous with respect to the second variable, then one can obtain the uniqueness of solutions by virtue of Lemma 5.1 again.

In the following theorem we apply the nonlinear alternative of Leray-Schauder type in which the condition (H2)$'$ is weakened.

**(H2)$''$** There exists a constant $q_3 \in (0, q)$ such that real valued function $\phi_f(t) \in L^{\frac{1}{q_3}}(J, \mathbb{R})$ and there exists a $L^1$-integrable and nondecreasing $\psi : [0, +\infty) \to (0, +\infty)$ such that
$$|f(t, u)| \leq \phi_f(t)\psi(|u|) \text{ for each } t \in J \text{ and all } u \in \mathbb{R};$$
**(H4)** the following inequality
$$r\left(\chi + \frac{\psi(r)T^{q-q_3}(1 - q_3)^{1-q_3}\vartheta}{\Gamma(q)(q - q_3)^{1-q_3}}\right)^{-1} > 1$$
has at least one positive solution, where $\chi = |u_0| + \sum_{i=1}^m |y_i|$ and $\vartheta = \|\phi_f\|_{L^{\frac{1}{q_3}} J}$.

**Theorem 5.5.** *Assume that (H1), (H2)$''$ and (H4) hold. Then the problem (5.1) has at least one solution.*

**Proof.** Consider the operator $F$ defined in Theorem 5.4. It can be easily shown that $F$ is continuous and completely continuous. Repeating the same process in Claim IV in Theorem 5.4, using (H2)$''$ and Hölder inequality again, for each $t \in J$, we have
$$|u(t)| \leq |(Fu)(t)|$$
$$\leq |u_0| + \sum_{i=1}^m |y_i| + \frac{1}{\Gamma(q)} \int_0^t (t - s)^{q-1} \phi_f(s)\psi(|u(s)|)ds$$
$$\leq |u_0| + \sum_{i=1}^m |y_i| + \frac{\psi(\|u\|_{PC})}{\Gamma(q)} \int_0^t (t - s)^{q-1} \phi_f(s)ds$$
$$\leq |u_0| + \sum_{i=1}^m |y_i| + \frac{\psi(\|u\|_{PC})}{\Gamma(q)} \left(\int_0^t (t - s)^{\frac{q-1}{1-q_3}} ds\right)^{1-q_3} \left(\int_0^t (\phi_f(s))^{\frac{1}{q_3}} ds\right)^{q_3}$$
$$\leq \chi + \frac{\psi(\|u\|_{PC})T^{q-q_3}(1 - q_3)^{1-q_3}\vartheta}{\Gamma(q)(q - q_3)^{1-q_3}}.$$
Thus
$$\frac{r}{|u_0| + \sum_{i=1}^m |y_i| + \frac{\psi(r)T^{q-q_3}(1-q_3)^{1-q_3}\vartheta}{\Gamma(q)(q-q_3)^{1-q_3}}} \leq 1.$$
By (H4), there exists a $N^* > 0$ such that $\|u\|_{PC} \neq N^*$.

Let $U = \{u \in PC(J, \mathbb{R}) : \|u\|_{PC} < N^*\}$. The operator $F : \overline{U} \to PC(J, \mathbb{R})$ is continuous and completely continuous. From the choice of $U$, there is no $u \in \partial U$ such that $u = \lambda^* F(u)$, $\lambda^* \in [0, 1]$. As a consequence of the nonlinear alternative of Leray-Schauder type, we deduce that $F$ has a fixed point $u \in \overline{U}$, which implies that the problem (5.1) has at least one solution $u \in PC(J, \mathbb{R})$. This completes the proof. $\square$

## 5.3   Impulsive Boundary Value Problems

### 5.3.1   *Introduction*

In the present section, we consider the boundary value problems for the following fractional impulsive differential equations

$$
\begin{cases}
{}_0^C D_t^q u(t) = f(t, u(t)), & t \in J' := J \setminus \{t_1, \dots, t_m\}, \ J := [0,1], \\
\Delta u(t_k) = y_k, \ \Delta u'(t_k) = \bar{y}_k, & k = 1, 2, \dots, m, \\
u(0) = 0, \ u'(1) = 0,
\end{cases}
\tag{5.7}
$$

where ${}_0^C D_t^q$ is Caputo fractional derivative of order $q \in (1, 2)$ with the lower limit zero, $t_k$ satisfy $0 = t_0 < t_1 < \cdots < t_m < t_{m+1} = 1$, and $y_k, \bar{y}_k \in \mathbb{R}$.

In Subsection 5.3.2, we give a formula of solutions to the problem (5.7). Subsection 5.3.3 is concerned with the existence and uniqueness of solutions for (5.7).

### 5.3.2   *Formula of Solutions*

In this subsection, we give a formula of solutions to boundary problem for impulsive fractional differential equations

$$
\begin{cases}
{}_0^C D_t^q u(t) = h(t), & t \in J', \ q \in (1, 2), \\
\Delta u(t_k) = y_k, \ \Delta u'(t_k) = \bar{y}_k, & k = 1, 2, \dots, m, \\
u(0) = 0, \ u'(1) = 0,
\end{cases}
\tag{5.8}
$$

where $y_k, \bar{y}_k \in \mathbb{R}$.

**Lemma 5.3.** *Let $q \in (1,2)$ and $h : J \to \mathbb{R}$ be continuous. A function $u$ given by*

$$
u(t) = \begin{cases}
\dfrac{1}{\Gamma(q)} \displaystyle\int_0^t (t-s)^{q-1} h(s)\,ds \\
\qquad - \left( \dfrac{1}{\Gamma(q-1)} \displaystyle\int_0^1 (1-s)^{q-2} h(s)\,ds + \displaystyle\sum_{k=1}^m \bar{y}_k \right) t, \quad \text{for } t \in [0, t_1), \\[2mm]
\dfrac{1}{\Gamma(q)} \displaystyle\int_0^t (t-s)^{q-1} h(s)\,ds + \bar{y}_1 (t - t_1) + y_1 \\
\qquad - \left( \dfrac{1}{\Gamma(q-1)} \displaystyle\int_0^1 (1-s)^{q-2} h(s)\,ds + \displaystyle\sum_{k=1}^m \bar{y}_k \right) t, \quad \text{for } t \in (t_1, t_2], \\[2mm]
\quad\vdots \\[2mm]
\dfrac{1}{\Gamma(q)} \displaystyle\int_0^t (t-s)^{q-1} h(s)\,ds + \displaystyle\sum_{i=1}^k \bar{y}_i (t - t_i) + \displaystyle\sum_{i=1}^k y_i \\
\qquad - \left( \dfrac{1}{\Gamma(q-1)} \displaystyle\int_0^1 (1-s)^{q-2} h(s)\,ds + \displaystyle\sum_{k=1}^m \bar{y}_k \right) t, \quad \text{for } t \in (t_k, t_{k+1}], \\[2mm]
\hspace{9cm} k = 1, 2, \dots, m,
\end{cases}
\tag{5.9}
$$

*is a unique solution of the impulsive problem* (5.8)

**Proof.** A general solution $u$ of the first equation of (5.8) on each interval $(t_k, t_{k+1})$ $(k = 0, 1, \ldots, m)$ is given by

$$u(t) = \frac{1}{\Gamma(q)} \int_0^t (t-s)^{q-1} h(s) ds + a_k + b_k t, \quad \text{for } t \in (t_k, t_{k+1}), \qquad (5.10)$$

where $t_0 = 0$ and $t_{m+1} = 1$.

Then, we have

$$u'(t) = \frac{1}{\Gamma(q-1)} \int_0^t (t-s)^{q-2} h(s) ds + b_k, \quad \text{for } t \in (t_k, t_{k+1}). \qquad (5.11)$$

Applying the boundary conditions of (5.8), we find that

$$a_0 = 0, \quad b_m = -\frac{1}{\Gamma(q-1)} \int_0^1 (1-s)^{q-2} h(s) ds.$$

Next, using the right impulsive condition of (5.8), we derive

$$b_k = b_{k-1} + \bar{y}_k, \qquad (5.12)$$

which by (5.3.2) implies

$$b_j = -\frac{1}{\Gamma(q-1)} \int_0^1 (1-s)^{q-2} h(s) ds - \sum_{k=j+1}^m \bar{y}_k, \quad j = 0, 1, 2, \ldots, m-1. \qquad (5.13)$$

Furthermore, using the left impulsive condition of (5.8), we derive

$$a_k + b_k t_k = a_{k-1} + b_{k-1} t_k + y_k,$$

which by (5.12) is equivalent to

$$a_k = a_{k-1} + (b_{k-1} - b_k) t_k + y_k = a_{k-1} + y_k - \bar{y}_k t_k,$$

so by (5.3.2) we obtain

$$a_j = \sum_{k=1}^j (y_k - \bar{y}_k t_k), \quad j = 1, 2, \ldots, m. \qquad (5.14)$$

Hence for $j = 1, 2, \ldots, m$, (5.13) and (5.14) imply

$$a_j + b_j t = \sum_{k=1}^j (y_k - \bar{y}_k t_k) + \left( -\frac{1}{\Gamma(q-1)} \int_0^1 (1-s)^{q-2} h(s) ds - \sum_{k=j+1}^m \bar{y}_k \right) t$$

$$= \sum_{k=1}^j \bar{y}_k (t - t_k) + \sum_{k=1}^j y_k - \left( \frac{1}{\Gamma(q-1)} \int_0^1 (1-s)^{q-2} h(s) ds + \sum_{k=1}^m \bar{y}_k \right) t. \qquad (5.15)$$

Now it is clear that (5.10), (5.3.2) and (5.15) imply (5.9).

Conversely, assume that $u$ satisfies (5.9). By a direct computation, it follows that the solution given by (5.9) satisfies (5.8). This completes the proof. $\square$

Motivated by the above results, we give the following concept of the solution for the problem (5.7).

**Definition 5.2.** A function $u \in PC^1(J, \mathbb{R})$ is said to be a solution of the problem (5.7) if $u$ satisfies the integral equation

$$
u(t) = \begin{cases}
\dfrac{1}{\Gamma(q)} \displaystyle\int_0^t (t-s)^{q-1} h(s)ds \\
\quad - \left( \dfrac{1}{\Gamma(q-1)} \displaystyle\int_0^1 (1-s)^{q-2} f(s,u(s))ds + \displaystyle\sum_{k=1}^m \bar{y}_k \right) t, \quad \text{for } t \in [0, t_1), \\[2em]
\dfrac{1}{\Gamma(q)} \displaystyle\int_0^t (t-s)^{q-1} h(s)ds + \bar{y}_1(t-t_1) + y_1 \\
\quad - \left( \dfrac{1}{\Gamma(q-1)} \displaystyle\int_0^1 (1-s)^{q-2} h(s)ds + \displaystyle\sum_{k=1}^m \bar{y}_k \right) t, \qquad \text{for } t \in (t_1, t_2], \\[2em]
\vdots \\[1em]
\dfrac{1}{\Gamma(q)} \displaystyle\int_0^t (t-s)^{q-1} f(s,u(s))ds + \displaystyle\sum_{i=1}^k \bar{y}_i(t-t_i) + \displaystyle\sum_{i=1}^k y_i \\
\quad - \left( \dfrac{1}{\Gamma(q-1)} \displaystyle\int_0^1 (1-s)^{q-2} f(s,u(s))ds + \displaystyle\sum_{k=1}^m \bar{y}_k \right) t, \quad \text{for } t \in (t_k, t_{k+1}], \\[2em]
\hfill k = 1, 2, \ldots, m.
\end{cases}
$$

Moreover, we need the following known results.

**Lemma 5.4.** *(Wang, Xiang and Peng, 2009) Let $u \in C(J, R)$ satisfy the following inequality:*

$$
|u(t)| \leq a + b \int_0^t |u(\theta)|^{\lambda_1} d\theta + c \int_0^1 |u(\theta)|^{\lambda_2} d\theta, \quad t \in J,
$$

*where $\lambda_1 \in [0,1], \lambda_2 \in [0,1), a, b, c \geq 0$ are constants. Then there exists a constant $M^* > 0$ such that*

$$
|u(t)| \leq M^*.
$$

**Remark 5.3.** For $\lambda_1 < 1$ we can take $M^*$ to be the unique positive solution of $M^* = a + bM^{*\lambda_1} + cM^{*\lambda_2}$. Using the classical Gronwall inequality, for $\lambda_1 = 1$ we can take $M^*$ to be the unique positive solution of $M^* = (a + cM^{*\lambda_2})e^b$.

Using Lemma 5.4, we can obtain the following generalized Gronwall inequality with mixed integral term.

**Lemma 5.5.** *Let $u \in C(J, \mathbb{R})$ satisfy the following inequality:*

$$
|u(t)| \leq a + b \int_0^t (t-s)^{q-1} |u(s)|^{\lambda_1} ds + c \int_0^1 (1-s)^{q-2} |u(s)|^{\lambda_2} ds, \quad (5.16)
$$

*where* $q \in (1,2)$, $a, b, c \geq 0$ *are constants,* $\lambda_1 \in [0, 1 - \frac{1}{p}]$, $\lambda_2 \in [0, 1 - \frac{1}{p})$, *and for some* $p > 1$ *such that* $p(q-2) + 1 > 0$. *Then there exists a constant* $M_* > 0$ *such that*

$$|u(t)| \leq M_*.$$

**Proof.** It follows from (5.16) and Hölder inequality that

$$|u(t)| \leq a + b \left( \int_0^t (t-s)^{p(q-1)} ds \right)^{\frac{1}{p}} \left( \int_0^t |u(s)|^{\frac{\lambda_1 p}{p-1}} ds \right)^{\frac{p-1}{p}}$$
$$+ c \left( \int_0^1 (1-s)^{p(q-2)} ds \right)^{\frac{1}{p}} \left( \int_0^1 |u(s)|^{\frac{\lambda_2 p}{p-1}} ds \right)^{\frac{p-1}{p}}$$
$$\leq a + b \left( \frac{1}{p(q-1)+1} \right)^{\frac{1}{p}} \int_0^t |u(s)|^{\frac{\lambda_1 p}{p-1}} ds + c \left( \frac{1}{p(q-2)+1} \right)^{\frac{1}{p}} \int_0^1 |u(s)|^{\frac{\lambda_2 p}{p-1}} ds$$
$$\leq a + b \int_0^t |u(s)|^{\frac{\lambda_1 p}{p-1}} ds + c \left( \frac{1}{p(q-2)+1} \right)^{\frac{1}{p}} \int_0^1 |u(s)|^{\frac{\lambda_2 p}{p-1}} ds.$$

Applying Lemma 5.4, there exists a constant $M_* > 0$ such that

$$|u(t)| \leq M_*.$$

The proof is completed. □

**Remark 5.4.** Constant $M_*$ can be determined by using Remark 5.3.

### 5.3.3  *Existence*

This subsection deals with the existence and uniqueness of solutions for the problem (5.7).

**Theorem 5.6.** *Let* $f : J \times \mathbb{R} \to \mathbb{R}$ *be a continuous function mapping. Assume that there exists a positive constant* $L$ *such that*

**(A1)** $|f(t,u) - f(t,v)| \leq L|u-v|$, *for all* $t \in J$, $u, v \in \mathbb{R}$,

*with* $L \leq \frac{\Gamma(1+q)}{2(1+q)}$. *Then the problem (5.7) has a unique solution on* $J$.

**Proof.** Setting $\sup_{t \in J} |f(t,0)| = M$ and

$$B_r = \left\{ u \in PC^1(J, \mathbb{R}) : \|u\|_{PC^1} \leq r \right\},$$

where

$$r \geq 2 \left( \frac{1+q}{\Gamma(1+q)} M + \sum_{i=1}^m |\bar{y}_i| + 2 \sum_{i=1}^m |y_i| \right).$$

Define an operator $F : B_r \to PC^1(J, \mathbb{R})$ by

$$(Fu)(t) = \frac{1}{\Gamma(q)} \int_0^t (t-s)^{q-1} f(s, u(s)) ds + \sum_{i=1}^k \bar{y}_i(t - t_i) + \sum_{i=1}^k y_i$$

$$- \left( \frac{1}{\Gamma(q-1)} \int_0^1 (1-s)^{q-2} f(s, u(s)) ds + \sum_{i=1}^{m} \bar{y}_i \right) t,$$

$$t \in (t_k, t_{k+1}], \ k = 0, 1, 2, \ldots, m.$$

It is obvious that $F$ is well defined due to $f : J \times \mathbb{R} \to \mathbb{R}$ is jointly continuous and maps bounded subsets of $J \times \mathbb{R}$ to bounded subsets of $\mathbb{R}$.

**Claim I.** $FB_r \subset B_r$.

For $u \in B_r, t \in J'$, we have

$$|(Fu)(t)| = \left| \frac{1}{\Gamma(q)} \int_0^t (t-s)^{q-1} f(s, u(s)) ds + \sum_{i=1}^{k} \bar{y}_i (t - t_i) + \sum_{i=1}^{k} y_i \right.$$

$$\left. - \left( \frac{1}{\Gamma(q-1)} \int_0^1 (1-s)^{q-2} f(s, u(s)) ds + \sum_{i=1}^{m} \bar{y}_i \right) t \right|$$

$$\leq \left| \frac{1}{\Gamma(q)} \int_0^t (t-s)^{q-1} f(s, u(s)) ds \right| + \left| \frac{1}{\Gamma(q-1)} \int_0^1 (1-s)^{q-2} f(s, u(s)) ds \right|$$

$$+ \sum_{i=1}^{m} |\bar{y}_i| + 2 \sum_{i=1}^{m} |y_i|$$

$$\leq \frac{1}{\Gamma(q)} \int_0^t (t-s)^{q-1} |f(s, u(s)) - f(s, 0)| ds + \frac{1}{\Gamma(q)} \int_0^t (t-s)^{q-1} |f(s, 0)| ds$$

$$+ \frac{1}{\Gamma(q-1)} \int_0^1 (1-s)^{q-2} |f(s, u(s)) - f(s, 0)| ds$$

$$+ \frac{1}{\Gamma(q-1)} \int_0^1 (1-s)^{q-2} |f(s, 0)| ds + \sum_{i=1}^{m} |\bar{y}_i| + 2 \sum_{i=1}^{m} |y_i|$$

$$\leq L \frac{1+q}{\Gamma(1+q)} r + M \frac{1+q}{\Gamma(1+q)} + \sum_{i=1}^{m} |\bar{y}_i| + 2 \sum_{i=1}^{m} |y_i|$$

$$\leq r.$$

**Claim II.** $F$ is a contraction mapping.

For $u, v \in B_r$ and for each $t \in J'$, we obtain

$$|(Fu)(t) - (Fv)(t)|$$

$$= \left| \frac{1}{\Gamma(q)} \int_0^t (t-s)^{q-1} f(s, u(s)) ds - \frac{t}{\Gamma(q-1)} \int_0^1 (1-s)^{q-2} f(s, u(s)) ds \right.$$

$$\left. - \left( \frac{1}{\Gamma(q)} \int_0^t (t-s)^{q-1} f(s, v(s)) ds - \frac{t}{\Gamma(q-1)} \int_0^1 (1-s)^{q-2} f(s, v(s)) ds \right) \right|$$

$$\leq \frac{1}{\Gamma(q)} \int_0^t (t-s)^{q-1} |f(s, u(s)) - f(s, v(s))| ds$$

$$+ \frac{1}{\Gamma(q-1)} \int_0^1 (1-s)^{q-2} |f(s, u(s)) - f(s, v(s))| ds$$

$$\leq \left( \frac{L}{\Gamma(q)} \int_0^t (t-s)^{q-1} ds \right) \|u - v\|_{PC^1} + \left( \frac{L}{\Gamma(q-1)} \int_0^1 (1-s)^{q-2} ds \right) \|u - v\|_{PC^1}$$

$$\leq L \frac{1+q}{\Gamma(1+q)} \|u - v\|_{PC^1}$$

$$\leq \frac{1}{2} \|u - v\|_{PC^1},$$

which implies that

$$\|Fu - Fv\|_{PC^1} \leq \frac{1}{2} \|u - v\|_{PC^1}.$$

Therefore $F$ is a contraction.

Thus, the conclusion of theorem follows by Banach contraction mapping principle. The proof is completed. $\qquad \square$

Now we are ready to state and prove the following existence result.

**Theorem 5.7.** *Let $f : J \times \mathbb{R} \to \mathbb{R}$ be a continuous function mapping with $|f(t, u)| \leq \mu(t)$, for all $(t, u) \in J \times \mathbb{R}$ where $\mu \in L^{\frac{1}{\sigma}}(J, \mathbb{R})$ and $\sigma \in (0, q-1)$. Then the problem (5.7) has at least one solution on $J$.*

**Proof.** Let us choose

$$r \geq \frac{\|\mu\|_{L^{\frac{1}{\sigma}} J}}{\Gamma(q)(\frac{q-\sigma}{1-\sigma})^{1-\sigma}} + \frac{\|\mu\|_{L^{\frac{1}{\sigma}} J}}{\Gamma(q-1)(\frac{q-\sigma-1}{1-\sigma})^{1-\sigma}} + 2 \sum_{i=1}^m |\bar{y}_i| + \sum_{i=1}^m |y_i|,$$

and denote

$$B_r = \{u \in PC^1(J, \mathbb{R}) : \|u\|_{PC^1} \leq r\}.$$

We define the operators $P$ and $Q$ on $B_r$ as

$$(Pu)(t) = \frac{1}{\Gamma(q)} \int_0^t (t-s)^{q-1} f(s, u(s)) ds - \left( \frac{1}{\Gamma(q-1)} \int_0^1 (1-s)^{q-2} f(s, u(s)) ds \right) t,$$

$$(Qu)(t) = \sum_{i=1}^k \bar{y}_i (t - t_i) + \sum_{i=1}^k y_i - \sum_{i=1}^m \bar{y}_i t.$$

For any $u, v \in B_r$ and $t \in J$, using the estimation condition on $f$ and Hölder inequality,

$$\int_0^t |(t-s)^{q-1} f(s, u(s))| ds \leq \left( \int_0^t (t-s)^{\frac{q-1}{1-\sigma}} ds \right)^{1-\sigma} \left( \int_0^t (\mu(s))^{\frac{1}{\sigma}} ds \right)^\sigma$$

$$\leq \frac{\|\mu\|_{L^{\frac{1}{\sigma}} J}}{(\frac{q-\sigma}{1-\sigma})^{1-\sigma}},$$

$$\int_0^1 |(1-s)^{q-2} f(s, u(s))| ds \leq \left( \int_0^1 (1-s)^{\frac{q-2}{1-\sigma}} ds \right)^{1-\sigma} \left( \int_0^1 (\mu(s))^{\frac{1}{\sigma}} ds \right)^\sigma$$

$$\leq \frac{\|\mu\|_{L^{\frac{1}{\sigma}}J}}{(\frac{q-\sigma-1}{1-\sigma})^{1-\sigma}}.$$

Therefore,

$$\|Pu + Qv\|_{PC^1} \leq \frac{\|\mu\|_{L^{\frac{1}{\sigma}}J}}{\Gamma(q)(\frac{q-\sigma}{1-\sigma})^{1-\sigma}} + \frac{\|\mu\|_{L^{\frac{1}{\sigma}}J}}{\Gamma(q-1)(\frac{q-\sigma-1}{1-\sigma})^{1-\sigma}} + 2\sum_{i=1}^{m}|\bar{y}_i| + \sum_{i=1}^{m}|y_i| \leq r.$$

Thus $Pu + Qv \in B_r$. It is obvious that $Q$ is a contraction with the constant zero. On the other hand, the continuity of $f$ implies that the operator $P$ is continuous. Also, $P$ is uniformly bounded on $B_r$ since

$$\|Pu\|_{PC^1} \leq \frac{\|\mu\|_{L^{\frac{1}{\sigma}}J}}{\Gamma(q)(\frac{q-\sigma}{1-\sigma})^{1-\sigma}} + \frac{\|\mu\|_{L^{\frac{1}{\sigma}}J}}{\Gamma(q-1)(\frac{q-\sigma-1}{1-\sigma})^{1-\sigma}} \leq r.$$

Now we need to prove the compactness of the operator $P$.

Letting $\Omega = J \times B_r$, we can define $\sup_{(t,x)\in\Omega}|f(t,u)| = f_{\max}$, and consequently for any $t_k < \tau_2 < \tau_1 \leq t_{k+1}$ we have

$$|(Pu)(\tau_2) - (Pu)(\tau_1)|$$

$$= \left| \frac{1}{\Gamma(q)} \int_0^{\tau_2} (\tau_2 - s)^{q-1} f(s, u(s))ds - \frac{\tau_2}{\Gamma(q-1)} \int_0^1 (1-s)^{q-2} f(s, u(s))ds \right.$$

$$\left. - \left( \frac{1}{\Gamma(q)} \int_0^{\tau_1} (\tau_1 - s)^{q-1} f(s, u(s))ds - \frac{\tau_1}{\Gamma(q-1)} \int_0^1 (1-s)^{q-2} f(s, u(s))ds \right) \right|$$

$$\leq \left| \frac{1}{\Gamma(q)} \int_0^{\tau_2} \left( (\tau_2 - s)^{q-1} - (\tau_1 - s)^{q-1} \right) f(s, u(s))ds \right.$$

$$\left. + \frac{1}{\Gamma(q)} \int_{\tau_2}^{\tau_1} (\tau_1 - s)^{q-1} f(s, u(s))ds \right| + \left| \frac{\tau_2 - \tau_1}{\Gamma(q-1)} \int_0^1 (1-s)^{q-2} f(s, u(s))ds \right|$$

$$\leq f_{\max} \left( \frac{2(\tau_1 - \tau_2)^q + \tau_1^q - \tau_2^q}{\Gamma(1+q)} + \frac{\tau_1 - \tau_2}{\Gamma(q)} \right),$$

which tends to zero as $\tau_2 \to \tau_1$. This yields that $P$ is equicontinuous on interval $(t_k, t_{k+1}]$. So $P$ is relatively compact on $B_r$.

Hence, by $PC$-type Arzela-Ascoli theorem (see Lemma 1.3 in the case of $X = \mathbb{R}$), $P$ is compact on $B_r$. Thus all the assumptions of Theorem 1.7 are satisfied and the conclusion of Theorem 1.7 implies that the problem (5.7) has at least one solution on $J$. The proof is completed. $\square$

In order to obtain the data dependence of solutions, we revise (A1) to the assumption:

(**A2**) there exist $L > 0$ and $\lambda \in (0, 1 - \frac{1}{p})$ where $p(q - 2) + 1 > 0$ with $p > 1$ such that

$$|f(t, u) - f(t, v)| \leq L|u - v|^{\lambda}, \text{ for each } t \in J, \text{ and all } u, v \in \mathbb{R}.$$

Further, we give the following data dependence result.

**Theorem 5.8.** *Assume that the conditions of Theorem 5.7 and the additional condition (A2) hold. Let $v(\cdot)$ be another solution of the problem (5.7) with impulsive conditions $\Delta v(t_k) = y_k$, $\Delta v'(t_k) = \bar{y}_k$, $k = 1, 2, \ldots, m$, and boundary value conditions $v(0) = 0$, $v'(1) = 0$. Then there exists a constant $M_* > 0$ such that $\|u - v\|_{PC^1} \le M_*$.*

**Proof.** By Theorem 5.7, the problem (5.7) has a solution $u(\cdot)$ in $PC^1(J, X)$. Keeping in mind our conditions, $v(\cdot)$ be another solution of the problem (5.7) with impulsive conditions $\Delta v(t_k) = y_k$, $\Delta v'(t_k) = \bar{y}_k$, $k = 1, 2, \ldots, m$, and boundary value conditions $v(0) = 0$, $v'(1) = 0$. Note the condition (A2), we obtain

$$|u(t) - v(t)| \le \frac{L}{\Gamma(q)} \int_0^t (t - s)^{q-1} |u(s) - v(s)|^\lambda ds$$
$$+ \frac{L}{\Gamma(q-1)} \int_0^1 (1 - s)^{q-2} |u(s) - v(s)|^\lambda ds.$$

By Lemma 5.5, we obtain $\|u - v\|_{PC^1} \le M_*$. This completes the proof. $\square$

**Remark 5.5.** Under the assumptions of Theorem 5.8, we do not obtain the uniqueness of the solutions .

**Remark 5.6.** By Remark 5.4 we see that $M_*$ is the unique positive solution of

$$M_* = \frac{L}{\Gamma(q)} M_*^{\frac{\lambda p}{p-1}} + \frac{L}{\Gamma(q-1)} \left( \frac{1}{p(q-2)+1} \right)^{\frac{1}{p}} M_*^{\frac{\lambda p}{p-1}},$$

so

$$M_* = \left( \frac{L}{\Gamma(q)} + \frac{L}{\Gamma(q-1)} \left( \frac{1}{p(q-2)+1} \right)^{\frac{1}{p}} \right)^{\frac{1}{1 - \frac{\lambda p}{p-1}}}.$$

## 5.4 Impulsive Langevin Equations

### 5.4.1 *Introduction*

In 1908, Langevin introduced a concept of a equation of motion of a Brownian particle which is named after Langevin equation, thereafter, Langevin is regarded as one of the founder of the subject of stochastic differential equations. Langevin equations have been widely used to describe stochastic problems in physics, chemistry and electrical engineering. For example, Brownian motion is well described by the Langevin equation when the random fluctuation force is assumed to be white noise. For systems in complex media, standard Langevin equation does not provide the correct description of the dynamics. As a results, various generalizations of Langevin equations have been offered to describe dynamical processes in a fractal medium. One such generalization is the generalized Langevin equation which incorporates the fractal and memory properties with a dissipative memory kernel into the Langevin equation. This gives rise to study Langevin equation with fractional derivatives, see for instance Mainardi and Pironi, 1996; Lutz, 2001; Fa, 2006, 2007; Kobolev and Romanov, 2000; Picozzi and West, 2002; Bazzani, Bassi and Turchetti, 2003; Lim, Li and Teo, 2008; Ahmad and Eloe, 2010; Ahmad, Nieto, Alsaedi et al., 2012; Sandev, Tomovski and Dubbeldam, 2011; Sandev, Metzler and Tomovski, 2012, and the references therein.

It is remarkable that many evolution processes are characterized by the fact that at certain moments of time they experience a change of state abruptly. These processes are subject to short term perturbations whose duration is negligible in comparison with the duration of the processes. Consequently, it is natural to assume that these perturbations act instantaneously, that is in the form of impulses. In particular, when we want to describe fractional Langevin equations subject to abrupt changes as well as other phenomena such as earthquake, it is nature to use fractional impulsive Langevin equations to describe such problems. Thus, we offer to study the following fractional impulsive Langevin equations

$$\begin{cases} {}^{C}_{0}D_{t}^{\beta}({}^{C}_{0}D_{t}^{\alpha} + \lambda)x(t) = f(t, x(t)), & t \in J' := J \setminus \{t_1, \ldots, t_m\}, J := [0, 1], \\ \Delta x(t_k) := x(t_k^+) - x(t_k^-) = I_k, & I_k \in \mathbb{R}, \\ x(0) = 0, \ x(\eta_k) = 0, \ x(1) = 0, & \eta_k \in (t_k, t_{k+1}), \ k = 0, 1, \ldots, m-1, \end{cases} \tag{5.17}$$

where $f : J \times \mathbb{R} \to \mathbb{R}$ is a given function, $0 < \alpha, \beta < 1$ with $0 < \alpha + \beta < 1$, $\lambda > 0$, $0 = t_0 < t_1 < t_2 < \cdots < t_m < t_{m+1} = 1$, $x(t_k^+) = \lim_{\epsilon \to 0^+} x(t_k + \epsilon)$ and $x(t_k^-) = \lim_{\epsilon \to 0^-} x(t_k + \epsilon)$ represent the right and left limits of $x(t)$ at $t = t_k$, the constants $I_k$ denotes the size of the jump.

Moreover, we also study the following nonlinear impulsive problems:

$$\begin{cases} {}^{C}_{0}D_{t}^{\beta}({}^{C}_{0}D_{t}^{\alpha} + \lambda)x(t) = f(t, x(t)), & t \in J', \\ \Delta x(t_k) := x(t_k^+) - x(t_k^-) = I_k(x(t_k^-)), \\ x(0) = 0, \ x(\eta_k) = 0, \ x(1) = 0, & \eta_k \in (t_k, t_{k+1}), \ k = 0, 1, 2, \ldots, m-1. \end{cases} \tag{5.18}$$

where the nonlinear impulsive terms $I_k : \mathbb{R} \to \mathbb{R}$ are specified latter.

Subsection 5.4.2 is devoted to giving the formula of solutions for the linear Langevin equations and some basic properties of classical and generalized Mittag-Leffler functions, then proceed to obtain the general solutions of the linear fractional impulsive Langevin equations. In Subsection 5.4.3, we deal with the existence and uniqueness of solution for the problem (5.17), and extend the existence results to problem (5.18).

### 5.4.2   *Formula of Solutions*

Firstly, we study the following linear Langevin equations with two different fractional derivatives

$$
{}^{C}_{0}D_{t}^{\beta}({}^{C}_{0}D_{t}^{\alpha} + \lambda)x(t) = f(t), \ t \in J_i, \ i = 0, 1, 2, \ldots, m, \tag{5.19}
$$

where $J_0 := [0, t_1]$, $J_i := (t_i, t_{i+1}]$, $i = 1, 2, \ldots, m - 1$, $J_m := (t_m, 1]$.

**Lemma 5.6.** *For $q > 0$, the general solution of fractional differential equation $^C_0 D^q_t u(t) = 0$ is given by*

$$u(t) = c_0 + c_1 t + c_2 t^2 + \cdots + c_{n-1} t^{n-1}$$

*where $c_i \in \mathbb{R}, i = 0, 1, 2, \ldots, n - 1$ $(n = [q] + 1)$ and $[q]$ denotes the integer part of the real number $q$.*

**Remark 5.7.** In view of Lemma 5.6, it follows that

$$_0 D^{-q}_t (^C_0 D^q_t u)(t) = u(t) + c_0 + c_1 t + c_2 t^2 + \cdots + c_{n-1} t^{n-1}$$

where $c_i \in \mathbb{R}, i = 0, 1, 2, \ldots, n - 1$, $n = [q] + 1$.

It is remarkable that Ahmad, Nieto and Alsaedi *et al.*, 2012 studied existence of solutions of linear Langevin equations with two different fractional derivatives by finding a fixed point of a suitable fractional integral equation (see Lemma 2.1 in Ahmad, Nieto and Alsaedi *et al.*), however, the general presentation of solutions for such equations have not been deduced. Here, we try to find a general solution equation (5.19).

**Lemma 5.7.** *A function $u \in C(J_i, \mathbb{R})$, $i = 0, 1, 2, \ldots, m$, is a solution of equation (5.19) if and only if $u$ is a solution of the integral equation*

$$x(t) = E_\alpha(-t^\alpha \lambda) b_i - \frac{1}{\lambda} (1 - E_\alpha(-t^\alpha \lambda)) a_i$$
$$+ \int_0^t (t - z)^{\alpha+\beta-1} E_{\alpha,\alpha+\beta}(-(t - z)^\alpha \lambda) f(z) dz. \tag{5.20}$$

**Proof.** In view of Remark 5.7 and by integrating equation (5.19) from zero to $t$ we have

$$(^C_0 D^\alpha_t + \lambda) x(t) = \int_0^t \frac{(t - s)^{\beta-1}}{\Gamma(\beta)} f(s) ds - a_i, \quad i = 0, 1, \ldots, m,$$

where $a_i$ are constants.

By adopting the same idea and techniques in Zhou and Jiao, 2010, the general solution of

$$(^C_0 D^\alpha_t + \lambda) x(t) = h(t) \tag{5.21}$$

is

$$x(t) = \mathscr{T}(t) b_i + \int_0^t (t - s)^{\alpha-1} \mathscr{S}(t - s) h(s) ds, \tag{5.22}$$

where

$$\mathscr{T}(t) := \int_0^\infty M_\alpha(\theta) e^{-t^\alpha \theta \lambda} d\theta, \quad \mathscr{S}(t) := \alpha \int_0^\infty \theta M_\alpha(\theta) e^{-t^\alpha \theta \lambda} d\theta.$$

Here $M_\alpha$ is the Wtight function (see Definition 1.9). Meanwhile, the solution of equation (5.21) have been considered in the monograph Kilbas, Srivastava and Trujillo, 2006, (see pp. 140-141, (3.1.32)-(3.1.34)), and it is given by the following expression:

$$x(t) = E_\alpha(-t^\alpha \lambda)b_i + \int_0^t (t-s)^{\alpha-1} e_\alpha(-(t-s)^\alpha \lambda)h(s)ds. \tag{5.23}$$

Combined (5.22) and (5.23), we can rewrite $\mathscr{T}(t) = E_\alpha(-t^\alpha \lambda)$, $\mathscr{S}(t) = e_\alpha(-t^\alpha \lambda)$. Note $e_\alpha(z) = \alpha E'_\alpha(z)$ and so

$$(t-s)^{\alpha-1} e_\alpha(-(t-s)^\alpha \lambda) = \frac{d}{ds}\left(\frac{1}{\lambda} E_\alpha(-(t-s)^\alpha \lambda)\right).$$

This yields that

$$\int_0^t (t-s)^{\alpha-1} e_\alpha(-(t-s)^\alpha \lambda)ds = \frac{1}{\lambda}(1 - E_\alpha(-t^\alpha \lambda)).$$

So the final formula of solution of equation (5.19) should be

$$x(t) = E_\alpha(-t^\alpha \lambda)b_i + \int_0^t (t-s)^{\alpha-1} e_\alpha(-(t-s)^\alpha \lambda)$$

$$\times \left(\int_0^s \frac{(s-z)^{\beta-1}}{\Gamma(\beta)} f(z)dz - a_i\right) ds$$

$$= E_\alpha(-t^\alpha \lambda)b_i - \left(\int_0^t (t-s)^{\alpha-1} e_\alpha(-(t-s)^\alpha \lambda)ds\right) a_i$$

$$+ \frac{1}{\Gamma(\beta)} \int_0^t \int_0^s (s-z)^{\beta-1}(t-s)^{\alpha-1} e_\alpha(-(t-s)^\alpha \lambda)f(z)dzds$$

$$= E_\alpha(-t^\alpha \lambda)b_i - \frac{1}{\lambda}(1 - E_\alpha(-t^\alpha \lambda)) a_i$$

$$+ \frac{1}{\Gamma(\beta)} \int_0^t f(z) \int_z^t (t-s)^{\alpha-1}(s-z)^{\beta-1} e_\alpha(-(t-s)^\alpha \lambda)dsdz$$

$$= E_\alpha(-t^\alpha \lambda)b_i - \frac{1}{\lambda}(1 - E_\alpha(-t^\alpha \lambda)) a_i$$

$$+ \int_0^t (t-z)^{\alpha+\beta-1} E_{\alpha,\alpha+\beta}(-(t-z)^\alpha \lambda)f(z)dz,$$

where we use the fact in Theorem 4 of Prabhakar, 1971. Hence

$$x(t) = E_\alpha(-t^\alpha \lambda)b_i - \frac{1}{\lambda}(1 - E_\alpha(-t^\alpha \lambda)) a_i$$

$$+ \int_0^t (t-z)^{\alpha+\beta-1} E_{\alpha,\alpha+\beta}(-(t-z)^\alpha \lambda)f(z)dz,$$

where $E_{\alpha,\alpha+\beta}$ is the generalized Mittag-Leffler function:

$$E_{\alpha,\alpha+\beta}(z) = \sum_{k=0}^\infty \frac{z^k}{\Gamma(k\alpha+\alpha+\beta)} = \frac{1}{\Gamma(\beta)}\sum_{k=0}^\infty \frac{z^k}{\Gamma(k\alpha+\alpha)}B(k\alpha+\alpha,\beta)$$

$$= \frac{1}{\Gamma(\beta)} \sum_{k=0}^{\infty} \frac{z^k}{\Gamma(k\alpha + \alpha)} \int_0^1 u^{k\alpha + \alpha - 1}(1-u)^{\beta-1} du$$

$$= \frac{1}{\Gamma(\beta)} \int_0^1 \sum_{k=0}^{\infty} \frac{(uz)^k}{\Gamma(k\alpha + \alpha)} u^{\alpha-1}(1-u)^{\beta-1} du$$

$$= \frac{1}{\Gamma(\beta)} \int_0^1 u^{\alpha-1}(1-u)^{\beta-1} e_\alpha(uz) du,$$

with $B$ denoting Beta function, which is coincided with

$$E_{\alpha,\alpha+\beta}(-z) = \frac{1}{\Gamma(\beta)} \int_0^1 u^{\alpha-1}(1-u)^{\beta-1} e_\alpha(-uz) du$$

$$= \frac{\alpha}{\Gamma(\beta)} \int_0^1 u^{\alpha-1}(1-u)^{\beta-1} \int_0^{\infty} \theta M_\alpha(\theta) e^{-uz\theta} d\theta du$$

$$= \frac{\alpha}{\Gamma(\beta)} \int_0^{\infty} \theta M_\alpha(\theta) \left( \int_0^1 u^{\alpha-1}(1-u)^{\beta-1} e^{-uz\theta} du \right) d\theta$$

$$= \frac{\Gamma(\alpha+1)}{\Gamma(\alpha+\beta)} \int_0^{\infty} \theta M_\alpha(\theta) {}_1F_1(\alpha, \alpha+\beta, -z\theta) d\theta, \tag{5.24}$$

where

$$_1F_1(\alpha, \alpha+\beta, z) = \frac{1}{B(\alpha,\beta)} \int_0^1 u^{\alpha-1}(1-u)^{\beta-1} e^{uz} du$$

is the hypergeometric function (see, Seaborn, 1991). $\qquad\square$

It is well known that classical and generalized Mittag-Leffler functions have played important role in the study of fractional ordinary differential equation with constant coefficients Bonilla, 2007, and fractional diffusion equations Atkinson and Osseiran, 2011. However, it seems that the known properties of these special functions are not explicit and complete. For example, the literature usually address that the classical and generalized Mittag-Leffler functions are boundedness, but not give an explicit boundedness. Meanwhile, other important properties such as continuity, monotonicity, nonnegative and etc. seems have not been systematically reported. Here, we try to revisit some basic properties of classical and generalized Mittag-Leffler functions by using one-side probability density function.

**Lemma 5.8.** *Let $0 < \alpha, \beta < 1$. The functions $E_\alpha$, $e_\alpha$ and $E_{\alpha,\alpha+\beta}$ are nonnegative and have the following properties:*

**(i)** *For any $\lambda > 0$ and $t \in J$,*

$$E_\alpha(-t^\alpha\lambda) \le 1, \ e_\alpha(-t^\alpha\lambda) \le \frac{1}{\Gamma(\alpha)}, \ E_{\alpha,\alpha+\beta}(-t^\alpha\lambda) \le \frac{1}{\Gamma(\alpha+\beta)}.$$

*Moreover, $E_\alpha(0) = 1$, $e_\alpha(0) = \frac{1}{\Gamma(\alpha)}$, $E_{\alpha,\alpha+\beta}(0) = \frac{1}{\Gamma(\alpha+\beta)}$.*

**(ii)** *For any* $\lambda > 0$ *and* $t_1, t_2 \in J$,

$$E_\alpha(-t_2^\alpha \lambda) \to E_\alpha(-t_1^\alpha \lambda) \text{ as } t_2 \to t_1,$$
$$e_\alpha(-t_2^\alpha \lambda) \to e_\alpha(-t_1^\alpha \lambda) \text{ as } t_2 \to t_1, \qquad (5.25)$$
$$E_{\alpha,\alpha+\beta}(-t_2^\alpha \lambda) \to E_{\alpha,\alpha+\beta}(-t_1^\alpha \lambda) \text{ as } t_2 \to t_1.$$

*Or rather,*

$$|E_\alpha(-t_2^\alpha \lambda) - E_\alpha(-t_1^\alpha \lambda)| := O(|t_2 - t_1|^\alpha) \text{ as } t_2 \to t_1,$$
$$|e_\alpha(-t_2^\alpha \lambda) - e_\alpha(-t_1^\alpha \lambda)| := O(|t_2 - t_1|^\alpha) \text{ as } t_2 \to t_1,$$
$$|E_{\alpha,\alpha+\beta}(-t_2^\alpha \lambda) - E_{\alpha,\alpha+\beta}(-t_1^\alpha \lambda)| := O(|t_2 - t_1|^\alpha) \text{ as } t_2 \to t_1.$$

**(iii)** *For any* $\lambda > 0$, $t_1, t_2 \in J$ *and* $t_1 \le t_2$,

$$E_\alpha(-t_2^\alpha \lambda) \le E_\alpha(-t_1^\alpha \lambda),$$
$$e_\alpha(-t_2^\alpha \lambda) \le e_\alpha(-t_1^\alpha \lambda), \qquad (5.26)$$
$$E_{\alpha,\alpha+\beta}(-t_2^\alpha \lambda) \le E_{\alpha,\alpha+\beta}(-t_1^\alpha \lambda).$$

**(iv)** *For any* $\lambda > 0$ *and* $t_* > 0$,

$$1 - E_\alpha(-t_*^\alpha \lambda) > 0.$$

**Proof.** (i) For any $\lambda > 0$ and $t \in J$,

$$E_\alpha(-t^\alpha \lambda) \le \int_0^\infty M_\alpha(\theta) e^{-t^\alpha \theta \lambda} d\theta \le \int_0^\infty M_\alpha(\theta) d\theta = E_\alpha(0) = 1,$$

$$e_\alpha(-t^\alpha \lambda) \le \alpha \int_0^\infty \theta M_\alpha(\theta) e^{-t^\alpha \theta \lambda} d\theta \le \alpha \int_0^\infty \theta M_\alpha(\theta) d\theta = e_\alpha(0) = \frac{1}{\Gamma(\alpha)},$$

$$E_{\alpha,\alpha+\beta}(-t^\alpha \lambda) \le \frac{1}{\Gamma(\alpha)\Gamma(\beta)} \int_0^1 u^{\alpha-1}(1-u)^{\beta-1} du = E_{\alpha,\alpha+\beta}(0) = \frac{1}{\Gamma(\alpha+\beta)}.$$

(ii) We only check the result (5.25) for $0 \le t_1 < t_2 \le 1$. In fact, using the inequality $t_2^\alpha - t_1^\alpha \le (t_2 - t_1)^\alpha$ and Lagrange mean value theorem we have

$$|e_\alpha(-t_2^\alpha \lambda) - e_\alpha(-t_1^\alpha \lambda)| \le \alpha \int_0^\infty \theta M_\alpha(\theta) \left| e^{-t_2^\alpha \theta \lambda} - e^{-t_1^\alpha \theta \lambda} \right| d\theta$$

$$= \alpha \int_0^\infty \theta M_\alpha(\theta) \left( \int_0^1 e^{-\eta t_2^\alpha \theta \lambda - (1-\eta) t_1^\alpha \theta \lambda} d\eta \right) |t_2^\alpha - t_1^\alpha| \theta \lambda d\theta$$

$$\le |t_2 - t_1|^\alpha \alpha \lambda \int_0^\infty \theta^2 M_\alpha(\theta) d\theta$$

$$\le |t_2 - t_1|^\alpha \frac{2\alpha \lambda}{\Gamma(1+2\alpha)}$$

$$= O(|t_2 - t_1|^\alpha), \text{ as } t_2 \to t_1.$$

Next, one can use the formula (5.24) via the above facts and Beta function to derive

$$|E_{\alpha,\alpha+\beta}(-t_2^\alpha \lambda) - E_{\alpha,\alpha+\beta}(-t_1^\alpha \lambda)|$$

$$\leq \frac{1}{\Gamma(\beta)} \int_0^1 u^{\alpha-1}(1-u)^{\beta-1}|e_\alpha(-t_2^\alpha \lambda u) - e_\alpha(-t_1^\alpha \lambda u)|du$$

$$\leq |t_2 - t_1|^\alpha \frac{2\alpha\lambda B(\alpha,\beta)}{\Gamma(1+2\alpha)\Gamma(\beta)}$$

$$= O(|t_2 - t_1|^\alpha), \quad \text{as } t_2 \to t_1.$$

(iii) Similar to the proof in (ii), for any $t_1, t_2 \in J$ and $t_1 \leq t_2$, we obtain

$$e_\alpha(-t_1^\alpha \lambda) - e_\alpha(-t_2^\alpha \lambda) = \alpha \int_0^\infty \theta M_\alpha(\theta)(e^{-t_1^\alpha \theta \lambda} - e^{-t_2^\alpha \theta \lambda})d\theta$$

$$= \alpha\lambda(t_2^\alpha - t_1^\alpha) \int_0^\infty \theta^2 M_\alpha(\theta)\left(\int_0^1 e^{-\eta t_2^\alpha \theta \lambda - (1-\eta)t_1^\alpha \theta \lambda}d\eta\right)d\theta$$

$$\geq 0.$$

Next, one can use the formula (5.24) via the above facts and Beta function to derive the result (5.26).

(iv) In fact,

$$E_\alpha(-t_*^\alpha \lambda) = \int_0^\infty M_\alpha(\theta)e^{-t_*^\alpha \theta \lambda}d\theta < \int_0^\infty M_\alpha(\theta)d\theta = 1,$$

where we use the fact $e^{-t_*^\alpha \theta \lambda} < 1$ for $\theta \in (0,\infty)$ and $\lambda > 0$. The proof is complete.

$\square$

Secondly, we deduce the general solutions of the following linear fractional impulsive Langevin equations

$$\begin{cases} {}_0^C D_t^\beta ({}_0^C D_t^\alpha + \lambda)x(t) = f(t), & t \in J', \\ \Delta x(t_k) := x(t_k^+) - x(t_k^-) = I_k, & I_k \in \mathbb{R}, \\ x(0) = 0, \ x(\eta_k) = 0, \ x(1) = 0, & \eta_k \in (t_k, t_{k+1}), \ k = 0,1,2,\ldots,m-1. \end{cases} \tag{5.27}$$

For brevity, we denote

$$(Ff)(t) := \int_0^t (t-z)^{\alpha+\beta-1}E_{\alpha,\alpha+\beta}(-(t-z)^\alpha \lambda)f(z)dz, \quad t \in J,$$

$$(T_0 f)(t) := -\frac{1 - E_\alpha(-t^\alpha \lambda)}{1 - E_\alpha(-\eta_0^\alpha \lambda)}(Ff)(\eta_0), \quad t \in J_0,$$

$$(T_i f)(t) := \frac{E_\alpha(-t^\alpha \lambda) - E_\alpha(-\eta_i^\alpha \lambda)}{E_\alpha(-t_i^\alpha \lambda) - E_\alpha(-\eta_i^\alpha \lambda)}[(T_{i-1}f)(t_i)$$

$$+ (Ff)(\eta_i) + I_i] - (Ff)(\eta_i), \quad t \in J_i, \ i = 1,2,\ldots,m-1,$$

$$(T_m f)(t) := \frac{E_\alpha(-t^\alpha \lambda) - E_\alpha(-\lambda)}{E_\alpha(-t_m^\alpha \lambda) - E_\alpha(-\lambda)}[(T_{m-1}f)(t_m)$$

$$+ (Ff)(1) + I_m] - (Ff)(1), \quad t \in J_m.$$

Clearly, by Lemma 5.8(iii), the above symbols are well defined.

**Lemma 5.9.** *A general solution $x$ of equation (5.27) on the interval $J$ is given by*

$$
x(t) = \begin{cases}
(Ff)(t) + (T_0f)(t), & for\ t \in J_0, \\
(Ff)(t) + (T_1f)(t), & for\ t \in J_1, \\
\vdots \\
(Ff)(t) + (T_if)(t), & for\ t \in J_i, \\
\vdots \\
(Ff)(t) + (T_mf)(t), & for\ t \in J_m.
\end{cases}
\tag{5.28}
$$

**Proof.** For $t \in J_0$, integrating both sides of the first equation of (5.27), one can obtain that (see (5.20))

$$
x(t) = E_\alpha(-t^\alpha\lambda)b_0 - \frac{1}{\lambda}\left(1 - E_\alpha(-t^\alpha\lambda)\right)a_0
$$
$$
+ \int_0^t (t-z)^{\alpha+\beta-1} E_{\alpha,\alpha+\beta}(-(t-z)^\alpha\lambda)f(z)dz.
\tag{5.29}
$$

Using the conditions $x(0) = 0$ and $x(\eta_0) = 0$, we get

$$
a_0 = \frac{\lambda}{1 - E_\alpha(-\eta_0^\alpha\lambda)} \int_0^{\eta_0} (\eta_0 - z)^{\alpha+\beta-1} E_{\alpha,\alpha+\beta}(-(\eta_0 - z)^\alpha\lambda)f(z)dz,
$$
$$
b_0 = 0.
\tag{5.30}
$$

Submitting (5.30) to (5.29), we obtain

$$
x(t) = -\frac{1 - E_\alpha(-t^\alpha\lambda)}{1 - E_\alpha(-\eta_0^\alpha\lambda)} \int_0^{\eta_0} (\eta_0 - z)^{\alpha+\beta-1} E_{\alpha,\alpha+\beta}(-(\eta_0 - z)^\alpha\lambda)f(z)dz
$$
$$
+ \int_0^t (t-z)^{\alpha+\beta-1} E_{\alpha,\alpha+\beta}(-(t-z)^\alpha\lambda)f(z)dz,
$$
$$
x(t_1) = (Ff)(t_1) + (T_0f)(t_1).
$$

For $t \in J_1$, integrating both sides of the first equation of (5.27), one can obtain that

$$
x(t) = E_\alpha(-t^\alpha\lambda)b_1 - \frac{1}{\lambda}\left(1 - E_\alpha(-t^\alpha\lambda)\right)a_1
$$
$$
+ \int_0^t (t-z)^{\alpha+\beta-1} E_{\alpha,\alpha+\beta}(-(t-z)^\alpha\lambda)f(z)dz.
$$

Since

$$
x(t_1^+) = E_\alpha(-t_1^\alpha\lambda)b_1 - \frac{1}{\lambda}\left(1 - E_\alpha(-t_1^\alpha\lambda)\right)a_1 + (Ff)(t_1)
$$
$$
x(t_1^-) = (Ff)(t_1) + (T_0f)(t_1),
$$

from $x(t_1^+) = x(t_1^-) + I_1, x(\eta_1) = 0$, it follows

$$
E_\alpha(-t_1^\alpha\lambda)b_1 - \frac{1}{\lambda}\left(1 - E_\alpha(-t_1^\alpha\lambda)\right)a_1 = (T_0f)(t_1) + I_1
$$

$$E_\alpha(-\eta_1^\alpha\lambda)b_1 - \frac{1}{\lambda}\left(1 - E_\alpha(-\eta_1^\alpha\lambda)\right)a_1 + (Ff)(\eta_1) = 0$$

and solving the above equations, we can get

$$a_1 = \frac{\lambda E_\alpha(-\eta_1^\alpha\lambda)}{E_\alpha(-t_1^\alpha\lambda) - E_\alpha(-\eta_1^\alpha\lambda)}[(T_0 f)(t_1) + (Ff)(\eta_1) + I_1] + \lambda(Ff)(\eta_1),$$

$$b_1 = -(Ff)(\eta_1) + \frac{1 - E_\alpha(-\eta_1^\alpha\lambda)}{E_\alpha(-t_1^\alpha\lambda) - E_\alpha(-\eta_1^\alpha\lambda)}[(T_0 f)(t_1) + (Ff)(\eta_1) + I_1].$$

Hence, we obtain

$$x(t) = (Ff)(t) + (T_1 f)(t), \qquad\qquad \text{for } t \in J_2,$$
$$x(t_2) = (Ff)(t_2) + (T_1 f)(t_2).$$

Repeating the above methods on the subinterval $J_i$, $i = 2, 3, \ldots, m - 1$ respectively, one can obtain that $x(t) = (Ff)(t) + (T_i f)(t)$ for $t \in J_i$.

Finally, for $t \in J_m$, integrating both side of the first equation of (5.27) again, one can obtain that

$$x(t) = E_\alpha(-t^\alpha\lambda)b_m - \frac{1}{\lambda}\left(1 - E_\alpha(-t^\alpha\lambda)\right)a_m$$
$$+ \int_0^t (t - z)^{\alpha+\beta-1}E_{\alpha,\alpha+\beta}(-(t - z)^\alpha\lambda)f(z)dz.$$

Note that $x(t_m^+) = x(t_m^-) + I_m = (Fx)(t_m) + (T_{m-1}x)(t_m) + I_m$, $x(1) = 0$, one can obtain

$$a_m = \frac{\lambda E_\alpha(-\lambda)}{E_\alpha(-t_m^\alpha\lambda) - E_\alpha(-\lambda)}[(T_{m-1}f)(t_m) + (Ff)(1) + I_m] + \lambda(Ff)(1),$$

$$b_m = \frac{1 - E_\alpha(-\lambda)}{E_\alpha(-t_m^\alpha\lambda) - E_\alpha(-\lambda)}[(T_{m-1}f)(t_m) + (Ff)(1) + I_m] - (Ff)(1).$$

Then, we get

$$x(t) = (Ff)(t) + (T_m f)(t), \quad \text{for } t \in J_m.$$

This complete the proof. $\qquad\qquad\qquad\qquad\qquad\qquad\qquad\qquad\qquad\square$

**Remark 5.8.** If we denote

$$(T_0 f)(t) := -\frac{1 - E_\alpha(-t^\alpha\lambda)}{1 - E_\alpha(-\eta_0^\alpha\lambda)}(Ff)(\eta_0), \quad t \in J_0,$$

$$(T_i f)(t) := [(T_{i-1}f)(t_i) + I_i + M_i(t_i)(Ff)(\eta_i)]\frac{E_\alpha(-t^\alpha\lambda) - E_\alpha(-\eta_i^\alpha\lambda)}{E_\alpha(-t_i^\alpha\lambda) - E_\alpha(-\eta_i^\alpha\lambda)}$$
$$- M_i(t)(Ff)(\eta_i), \quad t \in J_i, \ i = 1, 2, \ldots, m - 1,$$

$$(T_m f)(t) := [(T_{m-1}f)(t_m) + I_m + M_m(t_m)(Ff)(1)]\frac{E_\alpha(-t^\alpha\lambda) - E_\alpha(-\lambda)}{E_\alpha(-t_m^\alpha\lambda) - E_\alpha(-\lambda)}$$
$$- M_m(t)(Ff)(1), \quad t \in J_m,$$

$$M_i(t) := \frac{1 - E_\alpha(-t^\alpha\lambda)}{1 - E_\alpha(-\eta_i^\alpha\lambda)}, \quad t \in J_i, \ i = 1, 2, \ldots, m - 1,$$

$$M_m(t) := \frac{1 - E_\alpha(-t^\alpha\lambda)}{1 - E_\alpha(-\lambda)},$$

then an alternative formula of general solutions of equation (5.27) on $J$ is given by

$$x(t) = \begin{cases} (Ff)(t) + (T_0f)(t), & \text{for } t \in J_0, \\ (Ff)(t) + (T_1f)(t), & \text{for } t \in J_1, \\ \quad\vdots \\ (Ff)(t) + (T_if)(t), & \text{for } t \in J_i, \\ \quad\vdots \\ (Ff)(t) + (T_mf)(t), & \text{for } t \in J_m. \end{cases} \tag{5.31}$$

By directly computation, one can verify that (5.31) is coincided with (5.28). But, (5.28) seems more suitable than (5.31).

### 5.4.3  *Existence*

This subsection deals with the existence and uniqueness of solution for the problem (5.17). A number of papers have been recently written on fractional impulsive initial and boundary value problems. However, both Fečkan, Zhou and Wang, 2012 and Kosmatov, 2012, point out on the error in former solutions for some impulsive fractional differential equations by construct a counterexample and establish a general framework to seek a nature solution for such problems. Motivated by Fečkan, Zhou and Wang, 2012 and Kosmatov, 2012, we define what it means for the problem (5.17) to have a solution.

**Definition 5.3.** A function $x \in PC(J, \mathbb{R})$ whose Caputo fractional derivative existing on $J$ is said to be a solution of the problem (5.17) if $x(t)$ satisfies

$$x(t) = \begin{cases} (Fx)(t) + (T_0x)(t), & \text{for } t \in J_0, \\ (Fx)(t) + (T_1x)(t), & \text{for } t \in J_1, \\ \quad\vdots \\ (Fx)(t) + (T_ix)(t), & \text{for } t \in J_i, \\ \quad\vdots \\ (Fx)(t) + (T_mx)(t), & \text{for } t \in J_m, \end{cases}$$

where

$$(Fx)(t) = \int_0^t (t - z)^{\alpha+\beta-1} E_{\alpha,\alpha+\beta}(-(t - z)^\alpha\lambda)f(z, x(z))dz, \ t \in J,$$

$$(T_0x)(t) = -\frac{1 - E_\alpha(-t^\alpha\lambda)}{1 - E_\alpha(-\eta_0^\alpha\lambda)}(Fx)(\eta_0), \quad t \in J_0,$$

$$(T_i x)(t) = \frac{E_\alpha(-t^\alpha \lambda) - E_\alpha(-\eta_i^\alpha \lambda)}{E_\alpha(-t_i^\alpha \lambda) - E_\alpha(-\eta_i^\alpha \lambda)}[(T_{i-1}x)(t_i) + (Fx)(\eta_i) + I_i] - (Fx)(\eta_i),$$

$$t \in J_i, i = 1, 2, \ldots, m - 1,$$

$$(T_m x)(t) = \frac{E_\alpha(-t^\alpha \lambda) - E_\alpha(-\lambda)}{E_\alpha(-t_m^\alpha \lambda) - E_\alpha(-\lambda)}[(T_{m-1}x)(t_m) + (Fx)(1) + I_m] - (Fx)(1),$$

$$t \in J_m.$$

Before stating the main results, we introduce the following hypotheses.

**(H1)** $f : J \times \mathbb{R} \to \mathbb{R}$ is jointly continuous;

**(H2)** there exists a function $n(\cdot) \in L^{\frac{1}{q_1}}(J, \mathbb{R}^+)$ such that $|f(t, x)| \le n(t)$ for all $t \in J$ and all $x \in \mathbb{R}$, where $q_1 \in (0, \alpha + \beta)$;

**(H3)** there exists a function $h(\cdot) \in L^{\frac{1}{q_2}}(J, \mathbb{R}^+)$ such that $|f(t, x) - f(t, y)| \le h(t)|x - y|$ for all $t \in J$ and all $x, y \in \mathbb{R}$, where $q_2 \in (0, \alpha + \beta)$.

Consider an operator $N : PC(J, \mathbb{R}) \to PC(J, \mathbb{R})$ defined by

$$(Nx)(t) = \begin{cases} (Fx)(t) + (T_0 x)(t), & \text{for } t \in J_0, \\ (Fx)(t) + (T_1 x)(t), & \text{for } t \in J_1, \\ \vdots \\ (Fx)(t) + (T_i x)(t), & \text{for } t \in J_i, \\ \vdots \\ (Fx)(t) + (T_m x)(t), & \text{for } t \in J_m. \end{cases} \tag{5.32}$$

It is obvious that $N$ is well defined due to (H1). Then, we can transform existence of solutions of the problem (5.17) into a fixed point problem of the operator $N$.

For brevity, denote $p_1 = \frac{\alpha + \beta - 1}{1 - q_1}$, $p_2 = \frac{\alpha + \beta - 1}{1 - q_2}$.

We are ready to state the first existence and uniqueness result in this section.

**Theorem 5.9.** *Assume that (H1)-(H3) hold. If*

$$M_F + M_m < 1, \tag{5.33}$$

*then the problem (5.17) has a unique solution, where*

$$M_F = \frac{\|h\|_{L^{\frac{1}{q_2}} J}}{\Gamma(\alpha + \beta)(1 + p_2)^{1 - q_2}}, \tag{5.34}$$

*and*

$$M_m = \frac{1 - E_\alpha(-\lambda)}{E_\alpha(-t_m^\alpha \lambda) - E_\alpha(-\lambda)}(M_F + M_{m-1}) + M_F, \ldots, M_0 = \frac{M_F}{1 - E_\alpha(-\eta_0^\alpha \lambda)}.$$

**Proof.** We verify that $N$ defined by (5.32) is a contraction mapping. We divide our proof into two steps.

**Claim I.** $Nx \in PC(J, \mathbb{R})$ for every $x \in PC(J, \mathbb{R})$.

If $t \in J_0$, then for every $x \in C(J_0, \mathbb{R})$ and any $\delta > 0$, $0 < t < t + \delta \le t_1$, by (H2), Lemma 5.8 and Hölder inequality, we get

$$
\begin{aligned}
&|(Nx)(t+\delta) - (Nx)(t)| \\
&\le |(Fx)(t+\delta) - (Fx)(t)| + |(T_0 x)(t+\delta) - (T_0 x)(t)| \\
&\le \left| \int_0^{t+\delta} (t+\delta - z)^{\alpha+\beta-1} E_{\alpha,\alpha+\beta}(-(t+\delta-z)^\alpha \lambda) f(z, x(z)) dz \right. \\
&\qquad \left. - \int_0^t (t-z)^{\alpha+\beta-1} E_{\alpha,\alpha+\beta}(-(t-z)^\alpha \lambda) f(z, x(z)) dz \right| \\
&\qquad + \left| \frac{E_\alpha(-(t+\delta)^\alpha \lambda) - E_\alpha(-t^\alpha \lambda)}{1 - E_\alpha(-\eta_0^\alpha \lambda)} (Fx)(\eta_0) \right| \\
&\le \int_0^t (t+\delta-z)^{\alpha+\beta-1} |E_{\alpha,\alpha+\beta}(-(t+\delta-z)^\alpha \lambda) - E_{\alpha,\alpha+\beta}(-(t-z)^\alpha \lambda)| n(z) dz \\
&\qquad + \int_0^t \left| (t+\delta-z)^{\alpha+\beta-1} - (t-z)^{\alpha+\beta-1} \right| |E_{\alpha,\alpha+\beta}(-(t-z)^\alpha \lambda)| n(z) dz \\
&\qquad + \int_t^{t+\delta} (t+\delta-z)^{\alpha+\beta-1} |E_{\alpha,\alpha+\beta}(-(t+\delta-z)^\alpha \lambda)| n(z) dz \\
&\qquad + \frac{|E_\alpha(-(t+\delta)^\alpha \lambda) - E_\alpha(-t^\alpha \lambda)|}{1 - E_\alpha(-\eta_0^\alpha \lambda)} |(Fx)(\eta_0)| \\
&\le O(\delta^\alpha) \left( \int_0^t (t+\delta-z)^{\frac{\alpha+\beta-1}{1-q_1}} dz \right)^{1-q_1} \left( \int_0^t n(z)^{\frac{1}{q_1}} dz \right)^{q_1} \\
&\qquad + \frac{1}{\Gamma(\alpha+\beta)} \left( \int_0^t \left( (t-z)^{\alpha+\beta-1} - (t+\delta-z)^{\alpha+\beta-1} \right)^{\frac{1}{1-q_1}} dz \right)^{1-q_1} \\
&\qquad \times \left( \int_0^t n(z)^{\frac{1}{q_1}} dz \right)^{q_1} + \frac{1}{\Gamma(\alpha+\beta)} \left( \int_t^{t+\delta} (t+\delta-z)^{\frac{\alpha+\beta-1}{1-q_1}} dz \right)^{1-q_1} \\
&\qquad \times \left( \int_t^{t+\delta} n(z)^{\frac{1}{q_1}} dz \right)^{q_1} + O(\delta^\alpha) \frac{|(Fx)(\eta_0)|}{1 - E_\alpha(-\eta_0^\alpha \lambda)} \\
&\le O(\delta^\alpha) \left( \frac{|(Fx)(\eta_0)|}{1 - E_\alpha(-\eta_0^\alpha \lambda)} + \frac{\|n\|_{L^{\frac{1}{q_1}} J_0}}{(1+p_1)^{1-q_1}} \right) + \frac{2\delta^{(1+p_1)(1-q_1)} \|n\|_{L^{\frac{1}{q_1}} J_0}}{\Gamma(\alpha+\beta)(1+p_1)^{1-q_1}} \\
&\to 0, \quad \text{as } \delta \to 0,
\end{aligned}
$$

where we use the facts

$$
\int_0^t (t+\delta-z)^{\frac{\alpha+\beta-1}{1-q_1}} dz = \frac{(t+\delta)^{1+p_1} - \delta^{1+p_1}}{1+p_1} \le \frac{1}{1+p_1},
$$

$$\int_0^t ((t-z)^{\alpha+\beta-1} - (t+\delta-z)^{\alpha+\beta-1})^{\frac{1}{1-q_1}} dz$$

$$\leq \int_0^t ((t-z)^{\frac{\alpha+\beta-1}{1-q_1}} - (t+\delta-z)^{\frac{\alpha+\beta-1}{1-q_1}}) dz$$

$$\leq \frac{\delta^{1+p_1}}{1+p_1},$$

$$\int_t^{t+\delta} (t+\delta-z)^{\frac{\alpha+\beta-1}{1-q_1}} dz = \frac{\delta^{1+p_1}}{1+p_1},$$

and

$$|(Fx)(\eta_0)| \leq \frac{1}{\Gamma(\alpha+\beta)} \left( \int_0^{\eta_0} (\eta_0-z)^{\frac{\alpha+\beta-1}{1-q_1}} dz \right)^{1-q_1} \left( \int_0^{\eta_0} n(z)^{\frac{1}{q_1}} dz \right)^{q_1}$$

$$\leq \frac{\|n\|_{L^{\frac{1}{q_1}} J_0}}{\Gamma(\alpha+\beta)(1+p_1)^{1-q_1}}.$$

Thus we obtain $Nx \in C(J_0, \mathbb{R})$.

If $t \in J_1$, then for every $x \in C(J_1, \mathbb{R})$ and any $\delta > 0$, $t_1 < t < t+\delta \leq t_2$, repeating the above process, one can obtain

$$|(Nx)(t+\delta) - (Nx)(t)|$$

$$\leq O(\delta^\alpha) \left( \frac{(1 - E_\alpha(-\eta_1^\alpha \lambda))(|(Fx)(\eta_1)| + |(T_0 x)(t_1)| + |I_1|)}{E_\alpha(-t_1^\alpha \lambda) - E_\alpha(-\eta_1^\alpha \lambda)} + \frac{\|n\|_{L^{\frac{1}{q_1}} J_1}}{(1+p_1)^{1-q_1}} \right)$$

$$+ \frac{2\delta^{(1+p_1)(1-q_1)} \|n\|_{L^{\frac{1}{q_1}} J_1}}{\Gamma(\alpha+\beta)(1+p_1)^{1-q_1}} \to 0$$

as $\delta > 0$, where we use the fact

$$|(T_0 x)(t_1)| \leq \frac{1}{1 - E_\alpha(-\eta_0^\alpha \lambda)} |(Fx)(\eta_0)|,$$

and

$$|(Fx)(\eta_1)| \leq \frac{\|n\|_{L^{\frac{1}{q_1}} J_1}}{\Gamma(\alpha+\beta)(1+p_1)^{1-q_1}}.$$

Thus $Nx \in C(J_1, \mathbb{R})$.

With the same argument, one can verify that $Nx \in C(J_i, \mathbb{R})$, for every $x \in C(J_i, \mathbb{R})$, $i = 2, \ldots, m$.

From the above fact, we can conclude that $Nx \in PC(J, \mathbb{R})$, for every $x \in PC(J, \mathbb{R})$.

**Claim II.** $N$ is a contraction mapping on $PC(J, \mathbb{R})$.

For arbitrary $x, y \in PC(J, \mathbb{R})$ and $t \in J$, by (H3), Lemma 5.8 and Hölder inequality, we get

$$|(Fx)(t) - (Fy)(t)|$$

$$\leq \frac{1}{\Gamma(\alpha+\beta)} \int_0^t (t-z)^{\alpha+\beta-1}|f(z, x(z)) - f(z, y(z))|dz$$

$$\leq \frac{1}{\Gamma(\alpha+\beta)} \left( \int_0^t (t-z)^{\frac{\alpha+\beta-1}{1-q_2}} dz \right)^{1-q_2} \left( \int_0^t h(z)^{\frac{1}{q_2}} dz \right)^{q_2} \|x-y\|_{PC}$$

$$\leq \frac{\|h\|_{L^{\frac{1}{q_2}} J}}{\Gamma(\alpha+\beta)(1+p_2)^{1-q_2}} \|x-y\|_{PC}$$

$$=: M_F \|x-y\|_{PC}. \tag{5.35}$$

If $t \in J_0$, for arbitrary $x, y \in C(J_0, \mathbb{R})$, we get

$$|(T_0 x)(t) - (T_0 y)(t)| \leq \frac{1}{1 - E_\alpha(-\eta_0^\alpha \lambda)} |(Fx)(\eta_0) - (Fy)(\eta_0)|$$

$$\leq \frac{1}{1 - E_\alpha(-\eta_0^\alpha \lambda)} M_F \|x-y\|_{PC}$$

$$=: M_0 \|x-y\|_{PC}.$$

Thus, $\|Nx - Ny\|_{PC} \leq (M_F + M_0) \|x-y\|_{PC}$.

If $t \in J_1$, for arbitrary $x, y \in C(J_1, \mathbb{R})$, we get

$$|(T_1 x)(t) - (T_1 y)(t)|$$

$$\leq \frac{1 - E_\alpha(-\eta_1^\alpha \lambda)}{E_\alpha(-t_1^\alpha \lambda) - E_\alpha(-\eta_1^\alpha \lambda)} |(Fx)(\eta_1) - (Fy)(\eta_1) + (T_0 x)(t_1) - (T_0 y)(t_1)|$$

$$+ |(Fx)(\eta_1) - (Fy)(\eta_1)|$$

$$\leq \left( \frac{1 - E_\alpha(-\eta_1^\alpha \lambda)}{E_\alpha(-t_1^\alpha \lambda) - E_\alpha(-\eta_1^\alpha \lambda)} (M_F + M_0) + M_F \right) \|x-y\|_{PC}$$

$$=: M_1 \|x-y\|_{PC}.$$

Thus, $\|Nx - Ny\|_{PC} \leq (M_F + M_1)\|x-y\|_{PC}$.

If $t \in J_i$, for arbitrary $x, y \in C(J_i, \mathbb{R})$, $i = 2, \ldots, m-1$, with the same argument, we get $\|Nx - Ny\|_{PC} \leq (M_F + M_i)\|x-y\|_{PC}$.

If $t \in J_m$, for arbitrary $x, y \in C(J_m, \mathbb{R})$, we get

$$|(T_m x)(t) - (T_m y)(t)|$$

$$\leq \frac{1 - E_\alpha(-\lambda)}{E_\alpha(-t_m^\alpha \lambda) - E_\alpha(-\lambda)} |(Fx)(1) - (Fy)(1) + (T_{m-1} x)(t_m) - (T_{m-1} y)(t_m)|$$

$$+ |(Fx)(1) - (Fy)(1)|$$

$$\leq \left( \frac{1 - E_\alpha(-\lambda)}{E_\alpha(-t_m^\alpha \lambda) - E_\alpha(-\lambda)} (M_F + M_{m-1}) + M_F \right) \|x-y\|_{PC}$$

$$=: M_m \|x-y\|_{PC}.$$

Thus, $\|Nx - Ny\|_{PC} \leq (M_F + M_m)\|x-y\|_{PC}$.

Moreover, it is easy to see $M_0 < M_1 < M_2 < \cdots < M_m$. Due to the condition (5.33), $N$ has a unique fixed point on $PC(J, \mathbb{R})$ by Banach contraction mapping principle. This complete the proof. $\qquad\square$

Our second result is based on the well-known fixed point theorem due to Krasnoselskii (see Theorem 1.7).

**Theorem 5.10.** *Assume the conditions (H1)-(H3) hold. If $M_F < 1$, then the problem (5.17) has at least a solution on $PC(J, \mathbb{R})$.*

**Proof.** Setting $B_r = \{x \in PC(J, \mathbb{R}) : \|x\|_{PC} \leq r\}$, where $r \geq \overline{M_F} + \overline{M_m}$, and $\overline{M_m}$, $\overline{M_F}$ are finite positive constants defined by

$$\overline{M_m} = \overline{M_F} + \frac{1 - E_\alpha(-\lambda)}{E_\alpha(-t_m^\alpha \lambda) - E_\alpha(-\lambda)}(|I_m| + \overline{M_{m-1}} + \overline{M_F}), \ldots, \overline{M_0} = \frac{\overline{M_F}}{1 - E_\alpha(-\eta_0^\alpha \lambda)},$$

and

$$\overline{M_F} := \frac{\|n\|_{L^{\frac{1}{q_1}} J}}{\Gamma(\alpha + \beta)(1 + p_1)^{1 - q_1}}.$$

**Claim I.** $(Fx)(t) + (T_i y)(t) \in B_r$ for any $t \in J_i$ and $x, y \in B_r$.

By the Claim I of Theorem 5.9, $(Fx)(t)$ and $(T_i x)(t)$ are obviously continuous in $J_i$ for every $x \in B_r$.

For every $x, y \in B_r$ and $t \in J_0$, by (H2), Lemma 5.8 and Hölder inequality again, we get

$$|(Fx)(t) + (T_0 y)(t)| \leq \left| \int_0^t (t - z)^{\alpha + \beta - 1} E_{\alpha, \alpha + \beta}(-(t - z)^\alpha \lambda) f(z, x(z)) dz \right|$$

$$+ \frac{1 - E_\alpha(-t^\alpha \lambda)}{1 - E_\alpha(-\eta_0^\alpha \lambda)} |(Fy)(\eta_0)|$$

$$\leq \frac{\|n\|_{L^{\frac{1}{q_1}} J_0}}{\Gamma(\alpha + \beta)(1 + p_1)^{1 - q_1}} + \frac{\|n\|_{L^{\frac{1}{q_1}} J_0}}{\Gamma(\alpha + \beta)(1 + p_1)^{1 - q_1}} \frac{1}{1 - E_\alpha(-\eta_0^\alpha \lambda)}$$

$$\leq \overline{M_F} + \overline{M_0} \leq r.$$

For every $x, y \in B_r$ and $t \in J_i$, $i = 1, 2, \ldots, m - 1$, by (H2), Lemma 5.8 and Hölder inequality, we have

$$|(Fx)(t) + (T_i y)(t)|$$

$$\leq \left| \int_0^t (t - z)^{\alpha + \beta - 1} E_{\alpha, \alpha + \beta}(-(t - z)^\alpha \lambda) f(z, x(z)) dz \right|$$

$$+ \left| \frac{E_\alpha(-t^\alpha \lambda) - E_\alpha(-\eta_i^\alpha \lambda)}{E_\alpha(-t_i^\alpha \lambda) - E_\alpha(-\eta_i^\alpha \lambda)} [(T_{i-1} y)(t_i) + (Fy)(\eta_i) + I_i] - (Fy)(\eta_i) \right|$$

$$\leq \frac{\|n\|_{L^{\frac{1}{q_1}} J_i}}{\Gamma(\alpha + \beta)(1 + p_1)^{1 - q_1}}$$

$$+ \frac{1 - E_\alpha(-\eta_i^\alpha \lambda)}{E_\alpha(-t_i^\alpha \lambda) - E_\alpha(-\eta_i^\alpha \lambda)} \left( |I_i| + |(T_{i-1}x)(t_i)| + \frac{\|n\|_{L^{\frac{1}{q_1}} J_i}}{\Gamma(\alpha + \beta)(1 + p_1)^{1-q_1}} \right)$$

$$+ \frac{\|n\|_{L^{\frac{1}{q_1}} J_i}}{\Gamma(\alpha + \beta)(1 + p_1)^{1-q_1}}$$

$$\leq \overline{M_F} + \left( \overline{M_F} + \frac{1 - E_\alpha(-\eta_i^\alpha \lambda)}{E_\alpha(-t_i^\alpha \lambda) - E_\alpha(-\eta_i^\alpha \lambda)} (|I_i| + \overline{M_{i-1}} + \overline{M_F}) \right)$$

$$=: \overline{M_F} + \overline{M_i} \leq r.$$

For every $x, y \in B_r$ and $t \in J_m$, after a similar computation we obtain

$$|(Fx)(t) + (T_m y)(t)| \leq \overline{M_F} + \left( \overline{M_F} + \frac{1 - E_\alpha(-\lambda)}{E_\alpha(-t_m^\alpha \lambda) - E_\alpha(-\lambda)} (|I_m| + \overline{M_{m-1}} + \overline{M_F}) \right)$$

$$:= \overline{M_F} + \overline{M_m} \leq r.$$

Clearly, $\overline{M_m} > \overline{M_{m-1}} > \cdots > \overline{M_0}$. Due to the definition of the ball $B_r$, we must have $Fx + T_i y \in B_r$ for any $t \in J_i$ and $x, y \in B_r$.

**Claim II.** $F$ is a contraction mapping on $B_r$.

By (5.35) we have $\|Fx - Fy\|_{PC} \leq M_F \|x - y\|_{PC}$. The assumption $M_F < 1$ implies that $F$ is a contraction mapping.

**Claim III.** $T_i$ is a completely continuous operator on $B_r|_{J_i}$, $i = 0, 1, 2, \ldots, m$.

Similar to the Claim I of Theorem 5.9, one can easy to verify that $T_i$ is continuous and $\{T_i x : x \in B_r\}$ is an equicontinuous set. Moreover, $\{T_i x : x \in B_r\}$ is uniformly bounded. Thus, $T_i$ is a completely continuous operator on $B_r|_{J_i}$, $i = 0, 1, 2, \ldots, m$ due to Arzela-Ascoli theorem.

Applying Theorem 1.7, the problem (5.17) has at least a solution on $PC(J, \mathbb{R})$. The proof is completed. $\qquad\square$

To end this section, we extend the above existence results to equation (5.18). Now, we denote

$$(Ff)(t) := \int_0^t (t - z)^{\alpha+\beta-1} E_{\alpha,\alpha+\beta}(-(t - z)^\alpha \lambda) f(z, x(z)) dz, \quad t \in J,$$

$$(\overline{T_0 f})(t) := -\frac{1 - E_\alpha(-t^\alpha \lambda)}{1 - E_\alpha(-\eta_0^\alpha \lambda)} (Ff)(\eta_0), \quad t \in J_0,$$

$$(\overline{T_i f})(t) := \frac{E_\alpha(-t^\alpha \lambda) - E_\alpha(-\eta_i^\alpha \lambda)}{E_\alpha(-t_i^\alpha \lambda) - E_\alpha(-\eta_i^\alpha \lambda)} \left[ (\overline{T_{i-1} f})(t_i) \right.$$

$$\left. + (Ff)(\eta_i) + I_i(x(t_k^-)) \right] - (Ff)(\eta_i), \quad t \in J_i, \ i = 1, 2, \ldots, m - 1,$$

$$(\overline{T_m f})(t) := \frac{E_\alpha(-t^\alpha \lambda) - E_\alpha(-\lambda)}{E_\alpha(-t_m^\alpha \lambda) - E_\alpha(-\lambda)} \left[ (\overline{T_{m-1} f})(t_m) \right.$$

$$\left. + (Ff)(1) + I_m(x(t_k^-)) \right] - (Ff)(1). \quad t \in J_m.$$

Using the same method as in Lemma 5.9, one can obtain the following result immediately.

**Lemma 5.10.** *A general solution $x$ of equation* (5.18) *on the interval $J$ is given by*

$$x(t) = \begin{cases} (Ff)(t) + \overline{(T_0 f)}(t), & for\ t \in J_0, \\ (Ff)(t) + \overline{(T_1 f)}(t), & for\ t \in J_1, \\ \vdots \\ (Ff)(t) + \overline{(T_i f)}(t), & for\ t \in J_i, \\ \vdots \\ (Ff)(t) + \overline{(T_m f)}(t), & for\ t \in J_m. \end{cases}$$

We make a necessary assumption on the nonlinear impulsive terms.

**(H4)** There exist constants $L > 0$ and $M > 0$ such that $|I_k(x) - I_k(y)| \le L|x - y|$, with $|I_k(x)| \le M$, for all $x, y \in \mathbb{R}$, $k = 1, 2, \dots, m$.

Now the reader can apply the same methods as in the above theorems to obtain the following existence results. So we omit details of the proof here.

**Theorem 5.11.** *Assume the assumptions (H1)-(H4) hold. If*

$$M_F + \widetilde{M_m} < 1,$$

*then the problem* (5.18) *has a unique solution, where $M_F$ is defined in* (5.34) *and*

$$\widetilde{M_m} = \frac{1 - E_\alpha(-\lambda)}{E_\alpha(-t_m^\alpha \lambda) - E_\alpha(-\lambda)}(M_F + \widetilde{M_{m-1}} + L) + M_F, \dots, M_0$$

$$= \frac{1}{1 - E_\alpha(-\eta_0^\alpha \lambda)} M_F.$$

**Theorem 5.12.** *Assume the assumptions (H1)-(H4) hold. If $M_F < 1$, then the problem* (5.18) *has at least a solution on $PC(J, \mathbb{R})$.*

## 5.5 Impulsive Evolution Equations

### 5.5.1 *Introduction*

Consider the nonlocal Cauchy problems for fractional impulsive evolution equations:

$$\begin{cases} {}^C_0 D^\alpha_t x(t) = Ax(t) + f(t, x(t)), & t \in J,\ t \ne t_k, \\ x(0) = x_0 + g(x), & \\ x(t_k^+) = x(t_k^-) + y_k, & k = 1, 2, \dots, \delta, \end{cases} \tag{5.36}$$

where ${}^C_0 D^\alpha_t$ is Caputo fractional derivative of order $\alpha$, $A: D(A) \subseteq X \to X$ is the generator of a $C_0$-semigroup $\{Q(t)\}_{t \ge 0}$ on a Banach space $X$, $f: J \times X \to X$ is continuous, $x_0$, $y_k$ are the element of $X$, $g$ is a given function, $0 = t_0 < t_1 < t_2 < \cdots < t_\delta < t_{\delta+1} = b$, $x(t_k^+) = \lim_{h \to 0^+} = x(t_k + h)$ and $x(t_k^-) = x(t_k)$ represent respectively the right and left limits of $x(t)$ at $t = t_k$.

In Subsection 5.5.2, we give the definition of mild solution of problem (5.36). Subsection 5.5.3 is devoted to the existence and uniqueness results under the different assumptions on nonlinear term.

## 5.5.2   Cauchy Problems

In this subsection, we introduce a concept of solutions for our problems. We first consider an nonhomogeneous impulsive linear fractional equation of the form

$$\begin{cases} {}^C_0D_t^\alpha x(t) = Ax(t) + h(t), & \alpha \in (0,1), \ t \in J = [0,b], \ t \neq t_k, \\ x(0) = x_0, \\ x(t_k^+) = x(t_k^-) + y_k, & k = 1,2,\ldots,\delta, \end{cases} \tag{5.37}$$

where $h \in PC(J,X)$. We observe that $x(\cdot)$ can be decomposed to $v(\cdot) + w(\cdot)$ where $v$ is the continuous mild solution for

$$\begin{cases} {}^C_0D_t^\alpha v(t) = Av(t) + h(t), & t \in J, \\ v(0) = x_0, \end{cases} \tag{5.38}$$

on $J$, and $w$ is the $PC$-mild solution for

$$\begin{cases} {}^C_0D_t^\alpha w(t) = Aw(t), & t \in J, \ t \neq t_k, \\ w(0) = 0, \\ w(t_k^+) = w(t_k^-) + y_k, & k = 1,2,\ldots,\delta. \end{cases} \tag{5.39}$$

Indeed, by adding together (5.38) with (5.39), it follows (5.37). Note $v$ is continuous, so $v(t_k^+) = v(t_k^-)$, $k = 1,2,\ldots,\delta$. On the other hand, any solution of (5.37) can be decomposed to (5.38) and (5.39).

A mild solution of (5.38) is given by

$$v(t) = S_\alpha(t)x_0 + \int_0^t (t-s)^{\alpha-1} P_\alpha(t-s)h(s)ds, \ t \in J,$$

where

$$S_\alpha(t) = \int_0^\infty M_\alpha(\theta)Q(t^\alpha\theta)d\theta, \quad P_\alpha(t) = \int_0^\infty \alpha\theta M_\alpha(\theta)Q(t^\alpha\theta)d\theta.$$

Now we rewrite system (5.39) in the equivalent integral equation

$$w(t) = \begin{cases} \dfrac{1}{\Gamma(\alpha)} \displaystyle\int_0^t (t-s)^{\alpha-1} Aw(s)ds, & \text{for } t \in [0,t_1], \\[2ex] y_1 + \dfrac{1}{\Gamma(\alpha)} \displaystyle\int_0^t (t-s)^{\alpha-1} Aw(s)ds, & \text{for } t \in (t_1,t_2], \\[2ex] y_1 + y_2 + \dfrac{1}{\Gamma(\alpha)} \displaystyle\int_0^t (t-s)^{\alpha-1} Aw(s)ds, & \text{for } t \in (t_2,t_3], \\[1ex] \quad\vdots \\[1ex] \displaystyle\sum_{i=1}^{\delta} y_i + \dfrac{1}{\Gamma(\alpha)} \displaystyle\int_0^t (t-s)^{\alpha-1} Aw(s)ds, & \text{for } t \in (t_\delta,b]. \end{cases} \tag{5.40}$$

The above equation (5.40) can be expressed as

$$w(t) = \sum_{i=1}^{\delta} \chi_i(t) y_i + \frac{1}{\Gamma(\alpha)} \int_0^t (t-s)^{\alpha-1} Aw(s) ds, \quad \text{for } t \in J, \tag{5.41}$$

where

$$\chi_i(t) = \begin{cases} 0, & \text{for } t \in [0, t_i), \\ 1, & \text{for } t \in [t_i, b] \cup (b, \infty). \end{cases}$$

We adopt the idea used in Section 4.3 and apply the Laplace transform for (5.41) to get

$$u(\lambda) = \sum_{i=1}^{\delta} \frac{e^{-t_i \lambda}}{\lambda} y_i + \frac{1}{\lambda^\alpha} Au(\lambda),$$

which implies

$$u(\lambda) = \sum_{i=1}^{\delta} e^{-t_i \lambda} \lambda^{\alpha-1} \left( \lambda^\alpha I - A \right)^{-1} y_i.$$

Note that the Laplace transform of $S_\alpha(t) y_i$ is $\lambda^{\alpha-1} (\lambda^\alpha I - A)^{-1} y_i$. Thus we can derive the mild solution of (5.39) as

$$w(t) = \sum_{i=1}^{\delta} \chi_i(t) S_\alpha(t - t_i) y_i.$$

Summarizing, the mild solution of (5.37) is given by

$$x(t) = S_\alpha(t) x_0 + \sum_{i=1}^{\delta} \chi_i(t) S_\alpha(t - t_i) y_i + \int_0^t (t-s)^{\alpha-1} P_\alpha(t-s) h(s) ds,$$

i.e.,

$$x(t) = \begin{cases} S_\alpha(t) x_0 + \int_0^t (t-s)^{\alpha-1} P_\alpha(t-s) h(s) ds, & \text{for } t \in [0, t_1], \\[2mm] S_\alpha(t) x_0 + S_\alpha(t - t_1) y_1 + \int_0^t (t-s)^{\alpha-1} P_\alpha(t-s) h(s) ds, & \text{for } t \in (t_1, t_2], \\[2mm] \vdots \\[2mm] S_\alpha(t) x_0 + \sum_{i=1}^{\delta} S_\alpha(t - t_i) y_i + \int_0^t (t-s)^{\alpha-1} P_\alpha(t-s) h(s) ds, & \text{for } t \in (t_\delta, b]. \end{cases}$$

By using the above results, we can introduce the following definition of the mild solution for system (5.36).

**Definition 5.4.** By a *PC*-mild solution of the system (5.36) we mean that a function $x \in PC(J, X)$ which satisfies the following integral equation

$$x(t) = \begin{cases} S_\alpha(t)x_0 + \displaystyle\int_0^t (t-s)^{\alpha-1} P_\alpha(t-s) f(s, x(s)) \, ds, & \text{for } t \in [0, t_1], \\[2mm] S_\alpha(t)x_0 + S_\alpha(t-t_1)y_1 \\[1mm] \quad + \displaystyle\int_0^t (t-s)^{\alpha-1} P_\alpha(t-s) f(s, x(s)) \, ds, & \text{for } t \in (t_1, t_2], \\[2mm] \vdots \\[2mm] S_\alpha(t)x_0 + \displaystyle\sum_{i=1}^\delta S_\alpha(t-t_i)y_i \\[1mm] \quad + \displaystyle\int_0^t (t-s)^{\alpha-1} P_\alpha(t-s) f(s, x(s)) \, ds, & \text{for } t \in (t_\delta, b]. \end{cases}$$

### 5.5.3 *Nonlocal Problems*

In this subsection, we derive some existence and uniqueness results concerning the *PC*-mild solution for system (5.36) under the different assumptions on $f$.

**Case I. $f$ is Lipschitz.**

Let us list the following hypotheses:

**(HA)** $A$ is the infinitesimal generator of a compact semigroup $\{T(t)\}_{t \geq 0}$ in $X$;
**(HF1)** $f: J \times X \to X$ is continuous and there exists a constant $q_1 \in (0, \alpha)$ and a real-valued function $L_f(t) \in L^{\frac{1}{q_1}}(J, \mathbb{R}^+)$ such that

$$|f(t, x) - f(t, y)| \leq L_f(t)|x - y|, \quad t \in J, \ x, y \in X.$$

For brevity, let us take

$$T^* = \left[ \left( \frac{1-q_1}{\alpha - q_1} \right) b^{\frac{\alpha - q_1}{1 - q_1}} \right]^{1-q_1} \|L_f\|_{L^{\frac{1}{q_1}} J}.$$

**Theorem 5.13.** *Let (HA) and (HF1) be satisfied. Then for every $x_0 \in X$, the system (5.36) has a unique PC-mild solution on $J$ provided that*

$$0 < \frac{\alpha M T^*}{\Gamma(1 + \alpha)} < 1. \tag{5.42}$$

**Proof.** Let $x_0 \in X$ be fixed. Define an operator $T$ on $PC(J, X)$ by

$$(Tx)(t) = \begin{cases} S_\alpha(t)x_0 + \displaystyle\int_0^t (t-s)^{\alpha-1} P_\alpha(t-s) f(s, x(s))\, ds, & \text{for } t \in [0, t_1], \\[2mm] S_\alpha(t)x_0 + S_\alpha(t-t_1)y_1 \\[1mm] \quad + \displaystyle\int_0^t (t-s)^{\alpha-1} P_\alpha(t-s) f(s, x(s))\, ds, & \text{for } t \in (t_1, t_2], \\[2mm] \vdots \\[2mm] S_\alpha(t)x_0 + \displaystyle\sum_{i=1}^{\delta} S_\alpha(t-t_i)y_i \\[1mm] \quad + \displaystyle\int_0^t (t-s)^{\alpha-1} P_\alpha(t-s) f(s, x(s))\, ds, & \text{for } t \in (t_\delta, b]. \end{cases}$$

By our assumptions and Lemma 1.5, $T$ is well defined on $PC(J, X)$.

**Claim I.** $Tx \in PC(J, X)$ for $x \in PC(J, X)$.

For $0 \le \tau < t \le t_1$, taking into account the imposed assumptions and applying Proposition 4.5, we obtain

$$|(Tx)(t) - (Tx)(\tau)|$$

$$\le |S_\alpha(t)x_0 - S_\alpha(\tau)x_0| + \int_\tau^t (t-s)^{\alpha-1} |P_\alpha(t-s) f(s, x(s))|\, ds$$

$$+ \int_0^\tau (t-s)^{\alpha-1} |P_\alpha(t-s) f(s, x(s)) - P_\alpha(\tau-s) f(s, x(s))|\, ds$$

$$+ \int_0^\tau |(t-s)^{\alpha-1} - (\tau-s)^{\alpha-1}|\, |P_\alpha(\tau-s) f(s, x(s))|\, ds$$

$$\le \|S_\alpha(t) - S_\alpha(\tau)\|_{B(X)} |x_0| + \frac{\alpha M}{\Gamma(1+\alpha)} \int_\tau^t (t-s)^{\alpha-1} |f(s, x(s))|\, ds$$

$$+ \sup_{s \in [0,\tau]} \|P_\alpha(t-s) - P_\alpha(\tau-s)\|_{B(X)} \int_0^\tau (t-s)^{\alpha-1} |f(s, x(s))|\, ds$$

$$+ \frac{\alpha M \|f\|_{C([0,t_1],X)}}{\Gamma(1+\alpha)} \left| \int_0^\tau (\tau-s)^{\alpha-1} ds - \int_0^\tau (t-s)^{\alpha-1} ds \right|$$

$$\le \|S_\alpha(t) - S_\alpha(\tau)\|_{B(X)} |x_0|$$

$$+ \frac{t_1^\alpha \|f\|_{PC}}{\alpha} \sup_{s \in [0,\tau]} \|P_\alpha(t-s) - P_\alpha(\tau-s)\|_{B(X)}$$

$$+ \frac{3M \|f\|_{PC} (t-\tau)^\alpha}{\Gamma(1+\alpha)},$$

where we use the inequality $t^\alpha - \tau^\alpha \le (t-\tau)^\alpha$. Keeping in mind of Proposition 4.7, the first and second terms tend to zero as $t \to \tau$. Moreover, it is obvious that the last terms tends to zero too as $t \to \tau$. Thus, we can deduce that $Tx \in C([0, t_1], X)$.

For $t_1 \le \tau < t < t_2$, keeping in mind our assumptions and applying Proposition 4.5 again, we have

$$|(Tx)(t) - (Tx)(\tau)|$$

$$\leq \|S_\alpha(t) - S_\alpha(\tau)\| |x_0| + \|S_\alpha(t - t_1) - S_\alpha(\tau - t_1)\|_{B(X)} |y_1|$$

$$+ \frac{t_2^\alpha \|f\|_{PC}}{\alpha} \sup_{s \in [0,\tau]} \|P_\alpha(t - s) - P_\alpha(\tau - s)\|_{B(X)}$$

$$+ \frac{3M \|f\|_{PC}(t - \tau)^\alpha}{\Gamma(1 + \alpha)}.$$

As $t \to \tau$, the right-hand side of the above inequality tends to zero. Thus, we can deduce that $Tx \in C((t_1, t_2], X)$.

Similarly, we can also obtain that $Tx \in C((t_2, t_3], X), \ldots, Tx \in C((t_\delta, b], X)$. That is, $Tx \in PC(J, X)$.

**Claim II.** $T$ is contraction on $PC(J, X)$.

For each $t \in [0, t_1]$, it comes from our assumptions and Proposition 4.5 that

$$|(Tx)(t) - (Ty)(t)|$$

$$\leq \frac{\alpha M}{\Gamma(1 + \alpha)} \int_0^t (t - s)^{\alpha - 1} L_f(s) |x(s) - y(s)| ds$$

$$\leq \frac{\alpha M \|x - y\|_{PC}}{\Gamma(1 + \alpha)} \int_0^t (t - s)^{\alpha - 1} L_f(s) ds$$

$$\leq \frac{\alpha M \|x - y\|_{PC}}{\Gamma(1 + \alpha)} \left( \int_0^t (t - s)^{\frac{\alpha - 1}{1 - q_1}} ds \right)^{1 - q_1} \|L_f\|_{L^{\frac{1}{q_1}}[0, t_1]}$$

$$\leq \frac{\alpha M \|x - y\|_{PC}}{\Gamma(1 + \alpha)} \left[ \left( \frac{1 - q_1}{\alpha - q_1} \right) t_1^{\frac{\alpha - q_1}{1 - q_1}} \right]^{1 - q_1} \|L_f\|_{L^{\frac{1}{q_1}}[0, t_1]}.$$

In general, for each $t \in (t_k, t_{k+1}]$, using our assumptions and Proposition 4.5 again,

$$|(Tx)(t) - (Ty)(t)|$$

$$\leq \frac{\alpha M \|x - y\|_{PC}}{\Gamma(1 + \alpha)} \left[ \left( \frac{1 - q_1}{\alpha - q_1} \right) t_{k+1}^{\frac{\alpha - q_1}{1 - q_1}} \right]^{1 - q_1} \|L_f\|_{L^{\frac{1}{q_1}}[t_k, t_{k+1}]}.$$

Thus,

$$\|Tx - Ty\|_{PC} \leq \frac{\alpha M T^*}{\Gamma(1 + \alpha)} \|x - y\|_{PC}.$$

Hence, the condition (5.42) allows us to conclude in view of Banach contraction mapping principle, that $T$ has a unique fixed point $x \in PC(J, X)$ which is just the unique $PC$-mild solution of system (5.36). $\square$

**Case II.** $f$ is not Lipschitz.

We make the following assumptions.

**(C1)** $f : J \times X \to X$ is continuous and maps a bounded set into a bounded set;

**(C2)** for each $x_0 \in X$, there exists a constant $r > 0$ such that

$$M \left( |x_0| + \sum_{k=1}^\delta |y_k| + \frac{b^\alpha}{\Gamma(1 + \alpha)} \sup_{s \in J, \phi \in Y_\Gamma} |f(s, \phi(s))| \right) \leq r,$$

where
$$Y_\Gamma = \{\phi \in PC(J, X) \mid \|\phi\| \leq r \text{ for } t \in J\}.$$

**Theorem 5.14.** *Suppose that (HA), (C1) and (C2) are satisfied. Then for every* $x_0 \in X$, *the system* (5.36) *has at least a PC-mild solution on* $J$.

**Proof.** Let $x_0 \in X$ be fixed. We introduce that map
$$T : PC(J, X) \to PC(J, X)$$
by
$$(Tv)(t) = (T_1 v)(t) + (T_2 v)(t)$$
where
$$(T_1 v)(t) = S_\alpha(t) x_0 + \int_0^t (t - s)^{\alpha-1} P_\alpha(t - s) f(s, v(s)) \, ds, \quad t \in J \setminus \{t_1, t_2, \ldots, t_\delta\},$$
and

$$(T_2 v)(t) = \begin{cases} 0, & t \in [0, t_1], \\ \displaystyle\sum_{i=1}^{k} S_\alpha(t - t_i) y_i, & t \in (t_k, t_{k+1}], \ k = 1, \ldots, \delta. \end{cases} \tag{5.43}$$

For each $t \in [0, t_1]$, $v \in Y_\Gamma$,
$$|(Tv)(t)| \leq |(T_1 v)(t)| + |(T_2 v)(t)|$$
$$\leq M|x_0| + \frac{b^\alpha M}{\Gamma(1 + \alpha)} \sup_{s \in J, \phi \in Y_\Gamma} |f(s, \phi(s))|.$$

For each $t \in (t_k, t_{k+1}]$, $v \in Y_\Gamma$,
$$|(Tv)(t)| \leq |(T_1 v)(t)| + |(T_2 v)(t)|$$
$$\leq M|x_0| + M \sum_{k=1}^{\delta} |y_k| + \frac{b^\alpha M}{\Gamma(1 + \alpha)} \sup_{s \in J, \phi \in Y_\Gamma} |f(s, \phi(s))|.$$

Noting that the condition (C2), we see that $T : Y_\Gamma \to Y_\Gamma$.

**Claim I.** $T$ is a continuous mapping from $Y_\Gamma$ to $Y_\Gamma$.

In order to derive the continuity of $T$, we only check that $T_1$ and $T_2$ are all continuous.

For this purpose, we assume that $v_n \to v$ in $Y_\Gamma$. It comes from the continuity of $f$ that $(\cdot - s)^{\alpha-1} f(s, v_n(s)) \to (\cdot - s)^{\alpha-1} f(s, v(s))$, as $n \to \infty$. Noting that
$$(t - s)^{\alpha-1} |f(s, v_n(s)) - f(s, v(s))| \leq (t - s)^{\alpha-1} \sup_{s \in J, \phi \in Y_\Gamma} |f(s, \phi(s))|,$$
for $s \in [0, t]$, $t \in J$, by means of Lebesgue dominated convergence theorem, we obtain that
$$\int_0^t (t - s)^{\alpha-1} |f(s, v_n(s)) - f(s, v(s))| \, ds \to 0, \text{ as } n \to \infty.$$

It is easy to see that for each $t \in J$,
$$|(T_1 v_n)(t) - (T_1 v)(t)| \leq \frac{\alpha M}{\Gamma(1 + \alpha)} \int_0^t (t - s)^{\alpha-1} |f(s, v_n(s)) - f(s, v(s))| \, ds$$
$$\to 0, \text{ as } n \to \infty.$$

Thus, $T_1$ is continuous. On the other hand, it is obvious that $T_2$ is continuous. Since $T_1$ and $T_2$ are continuous, $T$ is continuous.

**Claim II.** $T$ is a compact operator, or $T_1$ and $T_2$ are compact operators.

The compactness of $T_2$ is clear since it is a constant map (see (5.43)).

Now we prove the compactness of $T_1$. For each $t \in J$, the set $\{S_\alpha(t)x_0\}$ is precompact in $X$ since $S_\alpha(t)$, $t > 0$ is compact.

Also, for each $t \in J$, arbitrary $b > h > 0$, $\varepsilon > 0$, the set

$$\left\{ T(h^\alpha \varepsilon) \int_0^{t-h} (t-s)^{q-1} \left( \alpha \int_\varepsilon^\infty \theta M_\alpha(\theta) T((t-s)^\alpha \theta - h^\alpha \varepsilon) d\theta \right) \right.$$
$$\left. \times f(s, v(s)) \, ds \mid v \in Y_\Gamma \right\}$$
$$= \left\{ \alpha \int_0^{t-h} \int_\varepsilon^\infty \theta(t-s)^{\alpha-1} M_\alpha(\theta) T((t-s)^\alpha \theta) f(s, v(s)) \, d\theta ds \,\bigg|\, v \in Y_\Gamma \right\}$$

is precompact in $X$ since $T(h^\alpha \varepsilon)$ is compact.

Proceeding as in the proof of Theorem 3.1 in our previous work Zhou and Jiao, 2010b, one can obtain

$$\alpha \int_0^{t-h} \int_\varepsilon^\infty \theta(t-s)^{\alpha-1} M_\alpha(\theta) T((t-s)^\alpha \theta) f(s, v(s)) \, d\theta ds$$
$$\to \alpha \int_0^t \int_0^\infty \theta(t-s)^{\alpha-1} M_\alpha(\theta) T((t-s)^\alpha \theta) f(s, v(s)) \, d\theta ds,$$

as $h \to 0$, $\varepsilon \to 0$.

Thus, we can conclude that

$$\left\{ \int_0^t (t-s)^{\alpha-1} P_\alpha(t-s) f(s, v(s)) \, ds \mid v \in Y_\Gamma \right\}$$
$$= \left\{ \alpha \int_0^t \int_0^\infty \theta(t-s)^{\alpha-1} M_\alpha(\theta) T((t-s)^\alpha \theta) f(s, v(s)) \, d\theta ds \mid v \in Y_\Gamma \right\}$$

is precompact in $X$.

Therefore, the set

$$\left\{ S_\alpha(t)x_0 + \sum_{i=1}^k S_\alpha(t-t_i)y_i + \int_0^t (t-s)^{\alpha-1} P_\alpha(t-s) f(s, v(s)) \, ds \,\bigg|\, v \in Y_\Gamma \right\}$$

is precompact in $X$.

Thus, for each $t \in J$, $\{(T_1 v)(t) \mid v \in Y_\Gamma\}$ is precompact in $X$.

Next, we show the equicontinuity of $\mathcal{M} = \{(T_1 v)(\cdot) \mid v \in Y_\Gamma\}$.

The equicontinuity of $\{S_\alpha(t)x_0 \mid t \in J \setminus \{t_1, t_2, \ldots, t_\delta\}\}$, can be shown using the fact of $S_\alpha(\cdot)$ is continuous.

Now, we only need to check the equicontinuity of the second term in $\mathcal{M}$.

For $t \in J$, let $0 \le t' < t'' \le t_1$, set

$$I_1 = \left| \int_{t'}^{t''} (t''-s)^{\alpha-1} P_\alpha(t''-s) f(s, v(s)) ds \right|,$$
$$I_2 = \left| \int_0^{t'} ((t''-s)^{\alpha-1} - (t'-s)^{\alpha-1}) P_\alpha(t''-s) f(s, v(s)) ds \right|,$$

$$I_3 = \left| \int_0^{t'} (t' - s)^{\alpha-1} \left( P_\alpha(t'' - s) - P_\alpha(t' - s) \right) f(s, v(s)) ds \right|.$$

After some computation, we have

$$\left| \int_0^{t''} (t'' - s)^{\alpha-1} P_\alpha(t'' - s) f(s, v(s)) ds - \int_0^{t'} (t'' - s)^{\alpha-1} P_\alpha(t' - s) f(s, v(s)) ds \right|$$

$$\leq I_1 + I_2 + I_3.$$

Now repeating the previous discussion in Theorem 3.1 of Zhou and Jiao, 2010 we derive that $I_1, I_2, I_2$ tend to zero as $t'' \to t'$.

Accordingly, we see that the functions in $\mathcal{M}$ are equicontinuous. Therefore, $T_1$ is a compact operator by Arzela-Ascoli theorem, and hence $T$ is also a compact operator. Now, Schauder fixed point theorem implies that $T$ has a fixed point, which gives rise to a $PC$-mild solution. □

To end this section, we make the following assumptions.

**(D1)** $f \colon J \times X \to X$ is continuous and there exists a function $m(\cdot) \in L^\infty(J, \mathbb{R}^+)$ such that

$$|f(t, x)| \leq m(t), \text{ for all } x \in X \text{ and } t \in J.$$

**Theorem 5.15.** *Suppose that (HA) and (D1) are satisfied. Then system (5.36) has at least a PC-mild solution on $J$.*

**Proof.** We defined that $T : PC(J, X) \to PC(J, X)$ as in Theorem 5.14 by $(Tv)(t) = (T_1 v)(t) + (T_2 v)(t)$. Then we proceed in several steps.

**Claim I.** $T$ is a continuous mapping from $PC(J, X)$ to $PC(J, X)$.

Let $\{v_n\}$ be a sequence in $PC(J, X)$ such that $v_n \to v$ in $PC(J, X)$. It comes from (D1) that $(\cdot - s)^{\alpha-1} f(s, v_n(s)) \to (\cdot - s)^{\alpha-1} f(s, v(s))$, as $n \to \infty$, and note that

$$(t - s)^{\alpha-1} |f(s, v_n(s)) - f(s, v(s))| \leq 2m(s)(t - s)^{\alpha-1} \in L^1(J, \mathbb{R}^+),$$

for $s \in [0, t]$, $t \in J$. Similar to the discussion in Theorem 5.14, one can prove that $T$ is a continuous mapping from $PC(J, X)$ to $PC(J, X)$.

**Claim II.** $T$ maps bounded sets into bounded sets in $PC(J, X)$.

So, let us prove that for any $r > 0$ there exits a $M^* > 0$ such that for each $v \in B_r = \{v \in PC(J, X) \mid \|v\|_{PC} \leq r\}$, we have $\|Tv\|_{PC} \leq M^*$.

Indeed, for any $v \in B_r$,

$$|(Tv)(t)| \leq |(T_1 v)(t)| + |(T_2 v)(t)|$$

$$\leq M|x_0| + M \sum_{i=1}^{\delta} |y_i| + \frac{b^\alpha M}{\Gamma(1 + \alpha)} \|m\|_{L^\infty} J,$$

which implies

$$\|Tv\|_{PC} \leq M|x_0| + M \sum_{i=1}^{\delta} |y_i| + \frac{b^\alpha M}{\Gamma(1 + \alpha)} \|m\|_{L^\infty} J \equiv M^*.$$

**Claim III.** $T$ is a compact operator.

In order to verify that $T$ is a compact operator, one can repeat the same process in Claim II of Theorem 5.14 only need replace $\sup_{s \in J, \phi \in Y_\Gamma} \|f(s, \phi(s))\|$ by $\|m\|_{L^\infty J}$.

**Claim IV.** The set $\Theta = \{x \in PC(J, X) \mid x = \lambda Tx, \lambda \in [0, 1]\}$ is bounded.

Let $v \in \Theta$. Then $v(t) = \lambda(Tv)(t)$ for some $\lambda \in [0, 1]$. Thus, for $t \in J$, directly calculation implies that $\|Tv\|_{PC} \leq M^*$. Hence, we deduce that $\Theta$ is a bounded set.

Since we have already proven that $T$ is continuous and compact, thanks to the Schaefer fixed point theorem, $T$ has a fixed point which is a $PC$-mild solution of system (5.36) on $J$. $\qquad \square$

**Remark 5.9.** In the assumption (D1), the condition $m(\cdot) \in L^\infty(J, \mathbb{R}^+)$ can be replaced by $m(\cdot) \in L^{\frac{1}{q_2}}(J, \mathbb{R}^+)$ where $\frac{1}{q_2} \in [0, \alpha)$.

## 5.6  Notes and Remarks

The material in Sections 5.2 due to Fečkan, Zhou and Wang, 2012. The results in Section 5.3 are adopted from Wang, Zhou and Fečkan, 2012. The main results of Section 5.4 are from Wang, Fečkan and Zhou, 2013. The material in Sections 5.5 due to Wang, Fečkan and Zhou, 2011.

# Chapter 6

# Fractional Boundary Value Problems

## 6.1 Introduction

Critical point theory and variational methods are crucial in the study of many mathematical models of real-world problems. We realized that critical point theory, which has been mostly developed by specialist in ordinary differential equations, partial differential equations, differential topology, optimization, should be made more popular among people working in fractional differential equations.

The main purpose of this chapter is to present a new approach via critical point theory to study the existence of solutions for the boundary value problem of fractional differential equations. In Section 6.2, we consider the existence of solutions for fractional boundary value problems by using the critical point theory. Section 6.3 is devoted to the existence of multiple solutions to the boundary value problem which arises from studying the steady fractional advection dispersion equation. In Section 6.4, according to variational methods, we investigate the multiplicity results for the solutions for boundary value problem.

## 6.2 Solution for BVP with Left and Right Fractional Integrals

### 6.2.1 *Introduction*

In this section, we consider the fractional boundary value problem (BVP for short) of the following form

$$
\begin{cases}
\dfrac{d}{dt}\left( \dfrac{1}{2}\, {}_0D_t^{-\beta}(u'(t)) + \dfrac{1}{2}\, {}_tD_T^{-\beta}(u'(t)) \right) + \nabla F(t, u(t)) = 0, & \text{a.e. } t \in [0, T], \\
u(0) = u(T) = 0,
\end{cases}
\tag{6.1}
$$

where ${}_0D_t^{-\beta}$ and ${}_tD_T^{-\beta}$ are the left and right Riemann-Liouville fractional integrals of order $0 \le \beta < 1$, respectively, $F: [0, T] \times \mathbb{R}^N \to \mathbb{R}$ is a given function satisfying some assumptions and $\nabla F(t, x)$ is the gradient of $F$ at $x$. In particular, if $\beta = 0$, BVP (6.1) reduces to the standard second order BVP.

Physical models containing fractional differential operators have recently renewed attention from scientists which is mainly due to applications as models for

physical phenomena exhibiting anomalous diffusion. A strong motivation for investigating the BVP (6.1) comes from fractional advection dispersion equation (ADE
for short). A fractional ADE is a generalization of the classical ADE in which the
second-order derivative is replaced with a fractional-order derivative. In contrast to
the classical ADE, the fractional ADE has solutions that resemble the highly skewed
and heavy-tailed breakthrough curves observed in field and laboratory studies (see
Benson, Schumer, Meerschaert *et al.*, 2001; Benson, Wheatcraft and Meerschaert,
2000a), in particular in contaminant transport of ground-water flow (see Benson,
Wheatcraft and Meerschaert, 2000b). Benson *et al.* stated that solutes moving
through a highly heterogeneous aquifer violations violates the basic assumptions of
local second-order theories because of large deviations from the stochastic process
of Brownian motion.

Let $\phi(t,x)$ represents the concentration of a solute at a point $x$ at time $t$ in an
arbitrary bounded connected set $\Omega \subset \mathbb{R}^N$. According to Benson, Wheatcraft and
Meerschaert, 2000a; Fix and Roop, 2004, the $N$-dimensional form of the fractional
ADE can be written as

$$\frac{\partial \phi}{\partial t} = -\nabla(v\phi) - \nabla(\nabla^{-\beta}(-k\nabla\phi)) + f, \quad \text{in } \Omega, \tag{6.2}$$

where $v$ is a constant mean velocity, $k$ is a constant dispersion coefficient, $v\phi$ and
$-k\nabla\phi$ denote the mass flux from advection and dispersion respectively. The components of $\nabla^{-\beta}$ in (6.2) are linear combination of the left and right Riemann-Liouville
fractional integral operators

$$(\nabla^{-\beta}(-k\nabla\phi))_i = (q \,_{-\infty}D_{x_i}^{-\beta} + (1-q) \,_{x_i}D_{+\infty}^{-\beta})\left(-k\frac{\partial \phi}{\partial x_i}\right), \quad i = 1, \ldots, N, \tag{6.3}$$

where $q \in [0,1]$ describes the skewness of the transport process, and $\beta \in [0,1)$ is the
order of the left and right Liouville-Weyl fractional integral operators on the real
line (see Definition 1.4). This equation may be interpreted as stating that the mass
flux of a particle is related to the negative gradient via a combination of the left
and right fractional integrals. Equation (6.3) is physically interpreted as a Fick's
law for concentrations of particles with a strong nonlocal interaction.

For discussions of equation (6.2), see Benson, Wheatcraft and Meerschaert,
2000b; Fix and Roop, 2004. When $\beta = 0$, the dispersion operators in (6.2) are
identical and the classical ADE is recovered. In a more general version of (6.2), $k$ is
replaced by a symmetric positive definite matrix.

A special case of the fractional ADE (equation (6.2)) describes symmetric transitions. In this case, $\nabla^{-\beta}$ is equivalent to the symmetric operator

$$(\nabla^{-\beta})_i = \frac{1}{2} \,_{-\infty}D_{x_i}^{-\beta} + \frac{1}{2} \,_{x_i}D_{+\infty}^{-\beta}, \quad i = 1, \ldots, N. \tag{6.4}$$

Combining (6.2) and (6.4) gives the mass balance equation for advection and symmetric fractional dispersion.

The fractional ADE has been studied in one dimension (see, e.g., Benson,
Wheatcraft and Meerschaert, 2000b), and in three dimension (see Lu, Molz and

Fix, 2002), over infinite domains by using the Fourier transform of fractional differential operators to determine a classical solution. Variational methods, especially the Galerkin approximation has been investigated to find the solutions of BVP (see, e.g., Fix and Roop, 2004) and fractional ADE (see, e.g., Ervin and Roop, 2006) on a finite domain by establishing some suitable fractional derivative spaces. A Lagrangian structure for some partial differential equations is obtained by using the fractional embedding theory of continuous Lagrangian systems (see, Cresson, 2010).

We note that for nonlinear BVP, some fixed point theorems were already applied successfully to investigate the existence of solutions (see, e.g., Agarwal, Benchohra and Hamani, 2010; Ahmad and Nieto, 2009; Benchohra, Hamani and Ntouyas, 2009; Zhang, 2010). However, it seems that fixed point theorem is not appropriate for discussing BVP (6.1) since the equivalent integral equation is not easy to be obtained. On the other hand, there is another effective approach, calculus of variation, which proved to be very useful in determining the existence of solutions for integer order differential equation provided that equation with certain boundary conditions possesses a variational structure on some suitable Sobolev spaces, for example, one can refer to Corvellec, Motreanu and Saccon, 2010; Li, Liang and Zhang, 2005; Mawhin and Willem, 1989; Rabinowitz, 1986; Tang and Wu, 2010 and the references therein for detailed discussions.

However, to the best of author's knowledge, there are few results on the solutions to BVP which were established by the critical point theory, since it is often very difficult to establish a suitable space and variational functional for fractional differential equations with some boundary conditions. These difficulties are mainly caused by the following properties of fractional integral and fractional derivative operators. These are:

(i) the composition rule in general fails to be satisfied by fractional integral and fractional derivative operators (e.g., Lemma 2.21 in Kilbas, Srivastava and Trujillo, 2006);

(ii) the fractional integral is a singular integral operator and fractional derivative operator is non-local (see Definitions 1.1, 1.2 and 1.3), and

(iii) the adjoint of a fractional differential operator is not the negative of itself (e.g., Lemma 2.7 in Kilbas, Srivastava and Trujillo, 2006).

It should be mentioned here that the fractional variational principles were started to be investigated deeply. The fractional calculus of variations was introduced by Riewe, 1996, where he presented a new approach to mechanics that allows one to obtain the equations for a nonconservative system using certain functionals. Klimek, 2002, gave another approach by considering fractional derivatives, and corresponding Euler-Lagrange equations were obtained, using both Lagrangian and Hamiltonian formalisms. Agrawal, 2002, presented Euler-Lagrange equations for unconstrained and constrained fractional variational problems, and as a continuation of Agrawal's work, the generalized mechanics are considered to obtain the

Hamiltonian formulation for the Lagrangian depending on fractional derivative of coordinates (see Rabei, Nawafleh, Hijjawi *et al.*, 2007). The recent book by Malinowska and Torres, 2012, provides a broad introduction to the important subject of fractional calculus of variations.

In Section 6.2, we investigate the existence of solutions for BVP (6.1). The technical tool is the critical point theory. In Subsection 6.2.2, we develop a fractional derivative space and some propositions are proven which aid in our analysis, and in Subsection 6.2.3, we shall exhibit a variational structure for BVP (6.1). The results presented in Subsections 6.2.2 and 6.2.3 are basic, but crucial to limpidly reveal that under some suitable assumptions, the critical points of the variational functional defined on a suitable Hilbert space are the solutions of BVP (6.1). In Subsection 6.2.4, we introduce some critical point theorems. Also, various criteria on the existence of solutions for BVP (6.1) is established.

As it was already mentioned, if $\beta = 0$, then BVP (6.1) reduces to the standard second order BVP of the following form

$$\begin{cases} u''(t) + \nabla F(t, u(t)) = 0, & \text{a.e. } t \in [0, T], \\ u(0) = u(T) = 0, \end{cases}$$

where $F: [0, T] \times \mathbb{R}^N \to \mathbb{R}$ is a given function and $\nabla F(t, x)$ is the gradient of $F$ at $x$. Although many excellent results have been worked out on the existence of solutions for second order BVP (e.g., Li, Liang and Zhang, 2005; Rabinowitz, 1986), it seems that no similar results were obtained in the literature for fractional BVP. The present results in Section 6.2 are to show that the critical point theory is an effective approach to tackle the existence of solutions for fractional BVP.

### 6.2.2  *Fractional Derivative Space*

Let us recall that for any fixed $t \in [0, T]$ and $1 \leq p < \infty$,

$$\|u\|_{L^p[0,t]} = \left( \int_0^t |u(\xi)|^p d\xi \right)^{\frac{1}{p}}, \quad \|u\|_{L^p} = \left( \int_0^T |u(t)|^p dt \right)^{\frac{1}{p}} \quad \text{and} \quad \|u\| = \max_{t \in [0,T]} |u(t)|.$$

The following result yields the boundedness of the Riemann-Liouville fractional integral operators from the space $L^p([0, T], \mathbb{R}^N)$ to the space $L^p([0, T], \mathbb{R}^N)$, where $1 \leq p < \infty$. It should be mentioned here that the similar results have been presented in Fix and Roop, 2004; Kilbas, Srivastava and Trujillo, 2006; Samko, Kilbas and Marichev, 1993.

**Lemma 6.1.** *Let $0 < \alpha \leq 1$ and $1 \leq p < \infty$. For any $f \in L^p([0, T], \mathbb{R}^N)$, we have*

$$\|_0D_\xi^{-\alpha} f\|_{L^p[0,t]} \leq \frac{t^\alpha}{\Gamma(\alpha + 1)} \|f\|_{L^p[0,t]}, \quad \text{for } \xi \in [0, t], \quad t \in [0, T]. \tag{6.5}$$

**Proof.** Inspired by the proof of Young theorem in Adams, 1975, we can prove (6.5). In fact, if $p = 1$, we have

$$\|_0D_\xi^{-\alpha} f\|_{L^1[0,t]} = \frac{1}{\Gamma(\alpha)} \left| \int_0^t \int_0^\xi (\xi - \tau)^{\alpha-1} f(\tau) d\tau d\xi \right|$$

$$\leq \frac{1}{\Gamma(\alpha)} \int_0^t \int_0^\xi (\xi - \tau)^{\alpha-1} |f(\tau)| d\tau d\xi$$

$$= \frac{1}{\Gamma(\alpha)} \int_0^t |f(\tau)| d\tau \int_\tau^t (\xi - \tau)^{\alpha-1} d\xi$$

$$= \frac{1}{\Gamma(\alpha+1)} \int_0^t |f(\tau)|(t - \tau)^\alpha d\tau$$

$$\leq \frac{t^\alpha}{\Gamma(\alpha+1)} \|f\|_{L^1[0,t]}, \quad \text{for } t \in [0,T]. \tag{6.6}$$

Now, suppose that $1 < p < \infty$ and $g \in L^q([0,T], \mathbb{R}^N)$, where $\frac{1}{p} + \frac{1}{q} = 1$. We have

$$\left| \int_0^t g(\xi) \int_0^\xi (\xi - \tau)^{\alpha-1} f(\tau) d\tau d\xi \right|$$

$$= \left| \int_0^t g(\xi) \int_0^\xi \tau^{\alpha-1} f(\xi - \tau) d\tau d\xi \right|$$

$$\leq \int_0^t |g(\xi)| \int_0^\xi \tau^{\alpha-1} |f(\xi - \tau)| d\tau d\xi$$

$$= \int_0^t \tau^{\alpha-1} d\tau \int_\tau^t |g(\xi)| |f(\xi - \tau)| d\xi$$

$$\leq \int_0^t \tau^{\alpha-1} d\tau \left( \int_\tau^t |g(\xi)|^q d\xi \right)^{\frac{1}{q}} \left( \int_\tau^t |f(\xi - \tau)|^p d\xi \right)^{\frac{1}{p}}$$

$$\leq \frac{t^\alpha}{\alpha} \|f\|_{L^p[0,t]} \|g\|_{L^q[0,t]}, \quad \text{for } t \in [0,T]. \tag{6.7}$$

For any fixed $t \in [0,T]$, consider the functional $H_{\xi * f} : L^q([0,T], \mathbb{R}^N) \to \mathbb{R}$

$$H_{\xi * f}(g) = \int_0^t \left( \int_0^\xi (\xi - \tau)^{\alpha-1} f(\tau) d\tau \right) g(\xi) d\xi. \tag{6.8}$$

According to (6.7), it is obvious that $H_{\xi * f} \in (L^q([0,T], \mathbb{R}^N))^*$, where $(L^q([0,T], \mathbb{R}^N))^*$ denotes the dual space of $L^q([0,T], \mathbb{R}^N)$. Therefore, by (6.7), (6.8) and Riesz representation theorem, there exists $h \in L^p([0,T], \mathbb{R}^N)$ such that

$$\int_0^t h(\xi) g(\xi) d\xi = \int_0^t \left( \int_0^\xi (\xi - \tau)^{\alpha-1} f(\tau) d\tau \right) g(\xi) d\xi \tag{6.9}$$

and

$$\|h\|_{L^p[0,t]} \leq \frac{t^\alpha}{\alpha} \|f\|_{L^p[0,t]} \tag{6.10}$$

for all $g \in L^q([0,T], \mathbb{R}^N)$. Hence, we have by (6.9)

$$\frac{1}{\Gamma(\alpha)} h(\xi) = \frac{1}{\Gamma(\alpha)} \int_0^\xi (\xi - \tau)^{\alpha-1} f(\tau) d\tau = {}_0 D_\xi^{-\alpha} f(\xi), \quad \text{for } \xi \in [0,t],$$

which means that

$$\|_0 D_\xi^{-\alpha} f\|_{L^p[0,t]} = \frac{1}{\Gamma(\alpha)} \|h\|_{L^p[0,t]} \leq \frac{t^\alpha}{\Gamma(\alpha+1)} \|f\|_{L^p[0,t]} \tag{6.11}$$

according to (6.10). Combining (6.6) and (6.11), we obtain the inequality (6.5). $\square$

In order to establish a variational structure for BVP (6.1), it is necessary to construct appropriate function spaces. Denote by $C_0^\infty([0,T],\mathbb{R}^N)$ the set of all functions $h \in C^\infty([0,T],\mathbb{R}^N)$ with $h(0) = h(T) = 0$. According to Lemma 6.1, for any $h \in C_0^\infty([0,T],\mathbb{R}^N)$ and $1 < p < \infty$, we have $h \in L^p([0,T],\mathbb{R}^N)$ and $_0^C D_t^\alpha h \in L^p([0,T],\mathbb{R}^N)$. Therefore, one can construct a set of space $E_0^{\alpha,p}$, which depend on $L^p$-integrability of Caputo fractional derivative of a function.

**Definition 6.1.** Let $0 < \alpha \le 1$ and $1 < p < \infty$. The fractional derivative space $E_0^{\alpha,p}$ is defined by the closure of $C_0^\infty([0,T],\mathbb{R}^N)$ with respect to the norm

$$\|u\|_{\alpha,p} = \left( \int_0^T |u(t)|^p dt + \int_0^T |_0^C D_t^\alpha u(t)|^p dt \right)^{\frac{1}{p}}, \quad \forall u \in E_0^{\alpha,p}.$$

**Remark 6.1.**

(i) It is obvious that the fractional derivative space $E_0^{\alpha,p}$ is the space of functions $u \in L^p([0,T],\mathbb{R}^N)$ having an $\alpha$-order Caputo fractional derivative $_0^C D_t^\alpha u \in L^p([0,T],\mathbb{R}^N)$ and $u(0) = u(T) = 0$.

(ii) For any $u \in E_0^{\alpha,p}$, noting the fact that $u(0) = 0$, we have $_0^C D_t^\alpha u(t) = {}_0 D_t^\alpha u(t)$, $t \in [0,T]$ according to Proposition 1.1.

(iii) It is easy to verify that $E_0^{\alpha,p}$ is a reflexive and separable Banach space.

**Proposition 6.1.** *Let $0 < \alpha \le 1$ and $1 < p < \infty$. The fractional derivative space $E_0^{\alpha,p}$ is a reflexive and separable Banach space.*

**Proof.** In fact, owing to $L^p([0,T],\mathbb{R}^N)$ be reflexive and separable, the Cartesian product space

$$L_2^p([0,T],\mathbb{R}^N) = L^p([0,T],\mathbb{R}^N) \times L^p([0,T],\mathbb{R}^N)$$

is also a reflexive and separable Banach space with respect to the norm

$$\|v\|_{L_2^p} = \left( \sum_{i=1}^2 \|v_i\|_{L^p[0,T]}^p \right)^{\frac{1}{p}}, \tag{6.12}$$

where $v = (v_1, v_2) \in L_2^p([0,T],\mathbb{R}^N)$.

Consider the space $\Omega = \{(u, {}_0^C D_t^\alpha u) : u \in E_0^{\alpha,p}\}$, which is a closed subset of $L_2^p([0,T],\mathbb{R}^N)$ as $E_0^{\alpha,p}$ is closed. Therefore, $\Omega$ is also a reflexive and separable Banach space with respect to the norm (6.12) for $v = (v_1, v_2) \in \Omega$.

We form the operator $A : E_0^{\alpha,p} \to \Omega$ as follows

$$A : u \to (u, {}_0^C D_t^\alpha u), \quad \forall u \in E_0^{\alpha,p}.$$

It is obvious that

$$\|u\|_{\alpha,p} = \|Au\|_{L_2^p},$$

which means that the operator $A : u \to (u, {}_0^C D_t^\alpha u)$ is an isometric isomorphic mapping and the space $E_0^{\alpha,p}$ is isometric isomorphic to the space $\Omega$. Thus $E_0^{\alpha,p}$ is a reflexive and separable Banach space, and this completes the proof. $\square$

Applying Proposition 1.9 and Lemma 6.1, we now can give the following useful estimates.

**Proposition 6.2.** *Let $0 < \alpha \leq 1$ and $1 < p < \infty$. For all $u \in E_0^{\alpha,p}$, we have*

$$\|u\|_{L^p[0,T]} \leq \frac{T^\alpha}{\Gamma(\alpha+1)} \|_0^C D_t^\alpha u\|_{L^p[0,T]}. \tag{6.13}$$

*Moreover, if $\alpha > \frac{1}{p}$ and $\frac{1}{p} + \frac{1}{q} = 1$, then*

$$\|u\| \leq \frac{T^{\alpha-\frac{1}{p}}}{\Gamma(\alpha)((\alpha-1)q+1)^{\frac{1}{q}}} \|_0^C D_t^\alpha u\|_{L^p[0,T]}. \tag{6.14}$$

**Proof.** For any $u \in E_0^{\alpha,p}$, according to (1.11) and noting the fact that $u(0) = 0$, we have that

$$_0D_t^{-\alpha}(_0^C D_t^\alpha u(t)) = u(t), \quad t \in [0,T].$$

Therefore, in order to prove inequalities (6.13) and (6.14), we only need to prove that

$$\|_0D_t^{-\alpha}(_0^C D_t^\alpha u)\|_{L^p[0,T]} \leq \frac{T^\alpha}{\Gamma(\alpha+1)} \|_0^C D_t^\alpha u\|_{L^p[0,T]}, \tag{6.15}$$

where $0 < \alpha \leq 1$ and $1 < p < \infty$, and

$$\|_0D_t^{-\alpha}(_0^C D_t^\alpha u)\| \leq \frac{T^{\alpha-\frac{1}{p}}}{\Gamma(\alpha)((\alpha-1)q+1)^{\frac{1}{q}}} \|_0^C D_t^\alpha u\|_{L^p[0,T]}, \tag{6.16}$$

where $\alpha > \frac{1}{p}$ and $\frac{1}{p} + \frac{1}{q} = 1$.

Firstly, we note that $_0^C D_t^\alpha u \in L^p([0,T], \mathbb{R}^N)$, the inequality (6.15) follows from (6.5) directly.

We are now in a position to prove (6.16). For $\alpha > \frac{1}{p}$, choose $q$ such that $\frac{1}{p} + \frac{1}{q} = 1$. $\forall u \in E_0^{\alpha,p}$, we have

$$\begin{aligned}
|_0D_t^{-\alpha}(_0^C D_t^\alpha u(t))| &= \frac{1}{\Gamma(\alpha)} \left| \int_0^t (t-s)^{\alpha-1}\, _0^C D_s^\alpha u(s) ds \right| \\
&\leq \frac{1}{\Gamma(\alpha)} \left( \int_0^t (t-s)^{(\alpha-1)q} ds \right)^{\frac{1}{q}} \|_0^C D_t^\alpha u\|_{L^p[0,T]} \\
&\leq \frac{T^{\frac{1}{q}+\alpha-1}}{\Gamma(\alpha)((\alpha-1)q+1)^{\frac{1}{q}}} \|_0^C D_t^\alpha u\|_{L^p[0,T]} \\
&= \frac{T^{\alpha-\frac{1}{p}}}{\Gamma(\alpha)((\alpha-1)q+1)^{\frac{1}{q}}} \|_0^C D_t^\alpha u\|_{L^p[0,T]},
\end{aligned}$$

and this completes the proof. $\qquad \square$

According to (6.13), we can consider $E_0^{\alpha,p}$ with respect to the norm

$$\|u\|_{\alpha,p} = \|_0^C D_t^\alpha u\|_{L^p[0,T]} = \left( \int_0^T |_0^C D_t^\alpha u(t)|^p dt \right)^{\frac{1}{p}} \tag{6.17}$$

in the following analysis.

**Proposition 6.3.** *Let $0 < \alpha \leq 1$ and $1 < p < \infty$. Assume that $\alpha > \frac{1}{p}$ and the sequence $\{u_k\}$ converges weakly to $u$ in $E_0^{\alpha,p}$, i.e., $u_k \rightharpoonup u$. Then $u_k \to u$ in $C([0,T], \mathbb{R}^N)$, i.e., $\|u - u_k\| \to 0$, as $k \to \infty$.*

**Proof.** If $\alpha > \frac{1}{p}$, then by (6.14) and (6.17), the injection of $E_0^{\alpha,p}$ into $C([0,T], \mathbb{R}^N)$, with its natural norm $\| \cdot \|$, is continuous, i.e., if $u_k \to u$ in $E_0^{\alpha,p}$, then $u_k \to u$ in $C([0,T], \mathbb{R}^N)$.

Since $u_k \rightharpoonup u$ in $E_0^{\alpha,p}$, it follows that $u_k \rightharpoonup u$ in $C([0,T], \mathbb{R}^N)$. In fact, For any $h \in (C([0,T], \mathbb{R}^N))^*$, if $u_k \to u$ in $E_0^{\alpha,p}$, then $u_k \to u$ in $C([0,T], \mathbb{R}^N)$, and thus $h(u_k) \to h(u)$. Therefore, $h \in (E_0^{\alpha,p})^*$, which means that $(C([0,T], \mathbb{R}^N))^* \subseteq (E_0^{\alpha,p})^*$.

Hence, if $u_k \rightharpoonup u$ in $E_0^{\alpha,p}$, then for any $h \in (C([0,T], \mathbb{R}^N))^*$, we have $h \in (E_0^{\alpha,p})^*$, and thus $h(u_k) \to h(u)$, i.e., $u_k \rightharpoonup u$ in $C([0,T], \mathbb{R}^N)$.

By the Banach–Steinhaus theorem, $\{u_k\}$ is bounded in $E_0^{\alpha,p}$ and, hence, in $C([0,T], \mathbb{R}^N)$. We are now in a position to prove that the sequence $\{u_k\}$ is equi-uniformly continuous.

Let $\frac{1}{p} + \frac{1}{q} = 1$ and $0 \leq t_1 < t_2 \leq T$. For every $f \in L^p([0,T], \mathbb{R}^N)$, by using Hölder inequality and noting that $\alpha > \frac{1}{p}$, we have

$$|_0D_{t_1}^{-\alpha}f(t_1) - {}_0D_{t_2}^{-\alpha}f(t_2)|$$

$$= \frac{1}{\Gamma(\alpha)}\left| \int_0^{t_1}(t_1 - s)^{\alpha-1}f(s)ds - \int_0^{t_2}(t_2 - s)^{\alpha-1}f(s)ds \right|$$

$$\leq \frac{1}{\Gamma(\alpha)}\left| \int_0^{t_1}(t_1 - s)^{\alpha-1}f(s)ds - \int_0^{t_1}(t_2 - s)^{\alpha-1}f(s)ds \right|$$

$$+ \frac{1}{\Gamma(\alpha)}\left| \int_{t_1}^{t_2}(t_2 - s)^{\alpha-1}f(s)ds \right|$$

$$\leq \frac{1}{\Gamma(\alpha)}\int_0^{t_1}\left((t_1 - s)^{\alpha-1} - (t_2 - s)^{\alpha-1}\right)|f(s)|ds$$

$$+ \frac{1}{\Gamma(\alpha)}\int_{t_1}^{t_2}(t_2 - s)^{\alpha-1}|f(s)|ds \tag{6.18}$$

$$\leq \frac{1}{\Gamma(\alpha)}\left( \int_0^{t_1}\left((t_1 - s)^{\alpha-1} - (t_2 - s)^{\alpha-1}\right)^q ds \right)^{\frac{1}{q}}\|f\|_{L^p[0,T]}$$

$$+ \frac{1}{\Gamma(\alpha)}\left( \int_{t_1}^{t_2}(t_2 - s)^{(\alpha-1)q}ds \right)^{\frac{1}{q}}\|f\|_{L^p[0,T]}$$

$$\leq \frac{1}{\Gamma(\alpha)} \left( \int_0^{t_1} (t_1 - s)^{(\alpha-1)q} - (t_2 - s)^{(\alpha-1)q} ds \right)^{\frac{1}{q}} \|f\|_{L^p[0,T]}$$

$$+ \frac{1}{\Gamma(\alpha)} \left( \int_{t_1}^{t_2} (t_2 - s)^{(\alpha-1)q} ds \right)^{\frac{1}{q}} \|f\|_{L^p[0,T]}$$

$$= \frac{\|f\|_{L^p[0,T]}}{\Gamma(\alpha)(1 + (\alpha-1)q)^{\frac{1}{q}}} \left( t_1^{(\alpha-1)q+1} - t_2^{(\alpha-1)q+1} + (t_2 - t_1)^{(\alpha-1)q+1} \right)^{\frac{1}{q}}$$

$$+ \frac{\|f\|_{L^p[0,T]}}{\Gamma(\alpha)(1 + (\alpha-1)q)^{\frac{1}{q}}} \left( (t_2 - t_1)^{(\alpha-1)q+1} \right)^{\frac{1}{q}}$$

$$\leq \frac{2\|f\|_{L^p[0,T]}}{\Gamma(\alpha)(1 + (\alpha-1)q)^{\frac{1}{q}}} (t_2 - t_1)^{\alpha - 1 + \frac{1}{q}}$$

$$= \frac{2\|f\|_{L^p[0,T]}}{\Gamma(\alpha)(1 + (\alpha-1)q)^{\frac{1}{q}}} (t_2 - t_1)^{\alpha - \frac{1}{p}}.$$

Therefore, the sequence $\{u_k\}$ is equi-uniformly continuous since, for $0 \leq t_1 < t_2 \leq T$, by applying (6.18) and in view of (6.17), we have

$$|u_k(t_1) - u_k(t_2)| = \left| {}_0D_{t_1}^{-\alpha} \left( {}_0^C D_{t_1}^\alpha u_k(t_1) \right) - {}_0D_{t_2}^{-\alpha} \left( {}_0^C D_{t_2}^\alpha u_k(t_2) \right) \right|$$

$$\leq \frac{2(t_2 - t_1)^{\alpha - \frac{1}{p}}}{\Gamma(\alpha)(1 + (\alpha-1)q)^{\frac{1}{q}}} \left\| {}_0^C D_t^\alpha u_k \right\|_{L^p[0,T]}$$

$$= \frac{2(t_2 - t_1)^{\alpha - \frac{1}{p}}}{\Gamma(\alpha)(1 + (\alpha-1)q)^{\frac{1}{q}}} \|u_k\|_{\alpha,p}$$

$$\leq c(t_2 - t_1)^{\alpha - \frac{1}{p}},$$

where $\frac{1}{p} + \frac{1}{q} = 1$ and $c \in \mathbb{R}^+$ is a constant. By Arzela-Ascoli theorem, $\{u_k\}$ is relatively compact in $C([0,T], \mathbb{R}^N)$. By the uniqueness of the weak limit in $C([0,T], \mathbb{R}^N)$, every uniformly convergent subsequence of $\{u_k\}$ converges uniformly on $[0,T]$ to $u$. The proof is completed. $\qquad \square$

### 6.2.3 *Variational Structure*

In this section, we establish a variational structure which enables us to reduce the existence of solutions of BVP (6.1) to the one of critical points of corresponding functional defined on the space $E_0^{\alpha,p}$ with $p = 2$ and $\frac{1}{2} < \alpha \leq 1$.

First of all, making use of Proposition 1.4, for any $u \in AC([0,T], \mathbb{R}^N)$, BVP

(6.1) transforms to

$$
\begin{cases}
\dfrac{d}{dt}\left( \dfrac{1}{2}\, {_0D_t^{-\frac{\beta}{2}}}({_0D_t^{-\frac{\beta}{2}}}u'(t)) + \dfrac{1}{2}\, {_tD_T^{-\frac{\beta}{2}}}({_tD_T^{-\frac{\beta}{2}}}u'(t)) \right) + \nabla F(t, u(t)) = 0, \\
u(0) = u(T) = 0,
\end{cases}
\tag{6.19}
$$

for almost every $t \in [0, T]$, where $\beta \in [0, 1)$.

Furthermore, in view of Definition 1.3 and Proposition 1.2, it is obvious that $u \in AC([0, T], \mathbb{R}^N)$ is a solution of BVP (6.19) if and only if $u$ is a solution of the following problem

$$
\begin{cases}
\dfrac{d}{dt}\left( \dfrac{1}{2}\, {_0D_t^{\alpha-1}}({_0^CD_t^{\alpha}}u(t)) - \dfrac{1}{2}\, {_tD_T^{\alpha-1}}({_t^CD_T^{\alpha}}u(t)) \right) + \nabla F(t, u(t)) = 0, \\
u(0) = u(T) = 0,
\end{cases}
\tag{6.20}
$$

for almost every $t \in [0, T]$, where $\alpha = 1 - \frac{\beta}{2} \in (\frac{1}{2}, 1]$. Therefore, we seek a solution $u$ of BVP (6.20) which, of course, corresponds to the solution $u$ of BVP (6.1) provided that $u \in AC([0, T], \mathbb{R}^N)$.

Let us denote by

$$
D^{\alpha}(u(t)) = \frac{1}{2}\, {_0D_t^{\alpha-1}}({_0^CD_t^{\alpha}}u(t)) - \frac{1}{2}\, {_tD_T^{\alpha-1}}({_t^CD_T^{\alpha}}u(t)).
\tag{6.21}
$$

We are now in a position to give a definition of the solution of BVP (6.20).

**Definition 6.2.** A function $u \in AC([0, T], \mathbb{R}^N)$ is called a solution of BVP (6.20) if

**(i)** $D^{\alpha}(u(t))$ is derivable for almost every $t \in [0, T]$, and
**(ii)** $u$ satisfies (6.20).

In the sequel, we treat BVP (6.20) in the Hilbert space $E^{\alpha} = E_0^{\alpha,2}$ with the corresponding norm $\|u\|_{\alpha} = \|u\|_{\alpha,2}$ which we defined in (6.17).

Consider the functional $u \to -\int_0^T ({_0^CD_t^{\alpha}}u(t), {_t^CD_T^{\alpha}}u(t))dt$ on $E^{\alpha}$. The following estimate is useful for our further discussion.

**Proposition 6.4.** *If* $\frac{1}{2} < \alpha \leq 1$, *then for any* $u \in E^{\alpha}$, *we have*

$$
|\cos(\pi\alpha)|\|u\|_{\alpha}^2 \leq -\int_0^T ({_0^CD_t^{\alpha}}u(t), {_t^CD_T^{\alpha}}u(t))dt \leq \frac{1}{|\cos(\pi\alpha)|}\|u\|_{\alpha}^2.
\tag{6.22}
$$

**Proof.** Let $u \in E^{\alpha}$ and $\tilde{u}$ be the extension of $u$ by zero on $\mathbb{R} \setminus [0, T]$. Then $\mathrm{supp}(\tilde{u}) \subseteq [0, T]$. However, as the left and right fractional derivatives are nonlocal,

$$
\mathrm{supp}({_{-\infty}D_t^{\alpha}}\tilde{u}) \subseteq [0, \infty) \quad \text{and} \quad \mathrm{supp}({_tD_{+\infty}^{\alpha}}\tilde{u}) \subseteq (-\infty, T].
$$

Nonetheless, the product $({_{-\infty}D_t^{\alpha}}\tilde{u}, {_tD_{+\infty}^{\alpha}}\tilde{u})$ has support in $[0, T]$.

On the other hand, according to Theorem 2.3 and Lemma 2.4 in Ervin and Roop, 2006, we have

$$
\int_{-\infty}^{\infty} ({_{-\infty}D_t^{\alpha}}\tilde{u}(t), {_tD_{+\infty}^{\alpha}}\tilde{u}(t))dt = \cos(\pi\alpha) \int_{-\infty}^{\infty} |{_{-\infty}D_t^{\alpha}}\tilde{u}(t)|^2 dt
$$

$$
= \cos(\pi\alpha) \int_{-\infty}^{\infty} |{_tD_{+\infty}^{\alpha}}\tilde{u}(t)|^2 dt,
\tag{6.23}
$$

where $_{-\infty}D_t^\alpha$ and $_tD_{+\infty}^\alpha$ are Liouville-Weyl fractional derivatives on the real line (see Definition 1.4). Helpful in establishing (6.23) is the Fourier transform of Liouville-Weyl fractional derivative on the real line (see Podlubny, 1999). Hence, according to Remark 6.1, (6.23) and noting that $\cos(\pi\alpha) \in [-1, 0)$ as $\alpha \in (\frac{1}{2}, 1]$, we have

$$-\int_0^T ({}_0^C D_t^\alpha u(t), {}_t^C D_T^\alpha u(t))dt = -\int_0^T ({}_0 D_t^\alpha u(t), {}_t D_T^\alpha u(t))dt$$

$$= -\int_0^T ({}_{-\infty}D_t^\alpha \tilde{u}(t), {}_t D_{+\infty}^\alpha \tilde{u}(t))dt$$

$$= -\int_{-\infty}^\infty ({}_{-\infty}D_t^\alpha \tilde{u}(t), {}_t D_{+\infty}^\alpha \tilde{u}(t))dt$$

$$= -\cos(\pi\alpha) \int_{-\infty}^\infty |{}_{-\infty}D_t^\alpha \tilde{u}(t)|^2 dt$$

$$= -\cos(\pi\alpha) \int_0^\infty |{}_0 D_t^\alpha \tilde{u}(t)|^2 dt$$

$$\geq -\cos(\pi\alpha) \int_0^T |{}_0 D_t^\alpha u(t)|^2 dt$$

$$= |\cos(\pi\alpha)| \int_0^T |{}_0^C D_t^\alpha u(t)|^2 dt$$

$$= |\cos(\pi\alpha)| \|u\|_\alpha^2. \tag{6.24}$$

On the other hand, by using Young inequality, we obtain

$$\left| \int_0^T ({}_0^C D_t^\alpha u(t), {}_t^C D_T^\alpha u(t))dt \right| = \left| \int_0^T ({}_0 D_t^\alpha u(t), {}_t D_T^\alpha u(t))dt \right|$$

$$\leq \int_0^T \frac{1}{\sqrt{2\varepsilon}} |{}_0 D_t^\alpha u(t)| \sqrt{2\varepsilon} \, |{}_t D_T^\alpha u(t)| dt$$

$$\leq \frac{1}{4\varepsilon} \int_0^T |{}_0 D_t^\alpha u(t)|^2 dt + \varepsilon \int_0^T |{}_t D_T^\alpha u(t)|^2 dt$$

$$= \frac{1}{4\varepsilon} \int_0^T |{}_0^C D_t^\alpha u(t)|^2 dt + \varepsilon \int_0^\infty |{}_t D_{+\infty}^\alpha \tilde{u}(t)|^2 dt$$

$$\leq \frac{1}{4\varepsilon} \|u\|_\alpha^2 + \varepsilon \int_{-\infty}^\infty |{}_t D_{+\infty}^\alpha \tilde{u}(t)|^2 dt$$

$$= \frac{1}{4\varepsilon} \|u\|_\alpha^2 + \frac{\varepsilon}{|\cos(\pi\alpha)|} \left| \int_{-\infty}^\infty ({}_{-\infty}D_t^\alpha \tilde{u}(t), {}_t D_{+\infty}^\alpha \tilde{u}(t))dt \right|$$

$$= \frac{1}{4\varepsilon} \|u\|_\alpha^2 + \frac{\varepsilon}{|\cos(\pi\alpha)|} \left| \int_0^T ({}_0 D_t^\alpha u(t), {}_t D_T^\alpha u(t))dt \right|$$

$$= \frac{1}{4\varepsilon} \|u\|_\alpha^2 + \frac{\varepsilon}{|\cos(\pi\alpha)|} \left| \int_0^T ({}_0^C D_t^\alpha u(t), {}_t^C D_T^\alpha u(t))dt \right|.$$

Therefore, by taking $\varepsilon = |\cos(\pi\alpha)|/2$, we have

$$\left| \int_0^T ({}_0^C D_t^\alpha u(t), {}_t^C D_T^\alpha u(t))dt \right| \leq \frac{1}{|\cos(\pi\alpha)|} \|u\|_\alpha^2. \tag{6.25}$$

The inequality (6.22) follows then from (6.24) and (6.25), and the proof is complete. □

**Remark 6.2.** According to (6.22) and (6.23), for any $u \in E^\alpha$, it is obvious that

$$\int_0^T |{}_t^C D_T^\alpha u(t)|^2 dt \leq \int_{-\infty}^\infty |{}_t D_{+\infty}^\alpha \tilde{u}(t)|^2 dt$$

$$= -\int_0^T \frac{({}_0^C D_t^\alpha u(t), {}_t^C D_T^\alpha u(t))}{|\cos(\pi\alpha)|} dt$$

$$\leq \frac{1}{|\cos(\pi\alpha)|^2} \|u\|_\alpha^2,$$

which means that ${}_t^C D_T^\alpha u \in L^2([0,T], \mathbb{R}^N)$.

In the following, we establish a variational structure on $E^\alpha$ with $\alpha \in (\frac{1}{2}, 1]$. Also, we show that the critical points of that functional are indeed solutions of BVP (6.20), and therefore, are solutions of BVP (6.1).

**Theorem 6.1.** *Let $L : [0,T] \times \mathbb{R}^N \times \mathbb{R}^N \times \mathbb{R}^N \to \mathbb{R}$ be defined by*

$$L(t, x, y, z) = -\frac{1}{2}(y, z) - F(t, x),$$

*where $F : [0,T] \times \mathbb{R}^N \to \mathbb{R}$ satisfies the following assumption:*

**(A)** *$F(t, x)$ is measurable in $t$ for each $x \in \mathbb{R}^N$, continuously differentiable in $x$ for almost every $t \in [0,T]$ and there exist $m_1 \in C(\mathbb{R}^+, \mathbb{R}^+)$ and $m_2 \in L^1([0,T], \mathbb{R}^+)$ such that*

$$|F(t, x)| \leq m_1(|x|)m_2(t), \quad |\nabla F(t, x)| \leq m_1(|x|)m_2(t)$$

*for all $x \in \mathbb{R}^N$ and a.e. in $t \in [0,T]$.*

*If $\frac{1}{2} < \alpha \leq 1$, then the functional defined by*

$$\varphi(u) = \int_0^T L(t, u(t), {}_0^C D_t^\alpha u(t), {}_t^C D_T^\alpha u(t)) dt$$

$$= \int_0^T \left( -\frac{1}{2}({}_0^C D_t^\alpha u(t), {}_t^C D_T^\alpha u(t)) - F(t, u(t)) \right) dt \qquad (6.26)$$

*is continuously differentiable on $E^\alpha$, and $\forall u, v \in E^\alpha$, we have*

$$\langle \varphi'(u), v \rangle = \int_0^T \left( D_x L(t, u(t), {}_0^C D_t^\alpha u(t), {}_t^C D_T^\alpha u(t)), v(t) \right) dt$$

$$+ \int_0^T \left( D_y L(t, u(t), {}_0^C D_t^\alpha u(t), {}_t^C D_T^\alpha u(t)), {}_0^C D_t^\alpha v(t) \right) dt$$

$$+ \int_0^T \left( D_z L(t, u(t), {}_0^C D_t^\alpha u(t), {}_t^C D_T^\alpha u(t)), {}_t^C D_T^\alpha v(t) \right) dt$$

$$= -\int_0^T \frac{1}{2} \left( ({}_0^C D_t^\alpha u(t), {}_t^C D_T^\alpha v(t)) + ({}_t^C D_T^\alpha u(t), {}_0^C D_t^\alpha v(t)) \right) dt$$

$$- \int_0^T (\nabla F(t, u(t)), v(t)) dt. \qquad (6.27)$$

**Proof.** First, we note that for a.e. $t \in [0, T]$ and every $[x, y, z] \in \mathbb{R}^N \times \mathbb{R}^N \times \mathbb{R}^N$, one has

$$|L(t, x, y, z)| \leq m_1(|x|)m_2(t) + \frac{1}{4}(|y|^2 + |z|^2), \tag{6.28}$$

$$|D_x L(t, x, y, z)| \leq m_1(|x|)m_2(t), \tag{6.29}$$

$$|D_y L(t, x, y, z)| \leq \frac{1}{2}|z| \quad \text{and} \quad |D_z L(t, x, y, z)| \leq \frac{1}{2}|y|. \tag{6.30}$$

Then, inspired by the proof of Theorem 1.4 in Mawhin and Willem, 1989, it suffices to prove that at every point $u, \varphi$ has a directional derivative $\varphi'(u) \in (E^\alpha)^*$ given by (6.27) and that the mapping

$$\varphi' : E^\alpha \to (E^\alpha)^*, \quad u \to \varphi'(u)$$

is continuous.

1) It follows easily from Remark 6.2 and (6.28) that $\varphi$ is everywhere finite on $E^\alpha$. Let us define, for $u$ and $v$ fixed in $E^\alpha$, $t \in [0, T]$, $\lambda \in [-1, 1]$,

$$G(\lambda, t) = L(t, u(t) + \lambda v(t), {}_0^C D_t^\alpha u(t) + \lambda\, {}_0^C D_t^\alpha v(t), {}_t^C D_T^\alpha u(t) + \lambda\, {}_t^C D_T^\alpha v(t))$$

and

$$\psi(\lambda) = \int_0^T G(\lambda, t)dt = \varphi(u + \lambda v).$$

We shall apply Leibniz formula of differentiation under integral sign to $\psi$. By (6.29) and (6.30), we have

$$|D_\lambda G(\lambda, t)|$$
$$= \left|(D_x L(t, u(t) + \lambda v(t), {}_0^C D_t^\alpha u(t) + \lambda\, {}_0^C D_t^\alpha v(t), {}_t^C D_T^\alpha u(t) + \lambda\, {}_t^C D_T^\alpha v(t)), v(t))\right|$$
$$+ \left|(D_y L(t, u(t) + \lambda v(t), {}_0^C D_t^\alpha u(t) + \lambda\, {}_0^C D_t^\alpha v(t),\right.$$
$$\left. {}_t^C D_T^\alpha u(t) + \lambda\, {}_t^C D_T^\alpha v(t)), {}_0^C D_t^\alpha v(t))\right|$$
$$+ \left|(D_z L(t, u(t) + \lambda v(t), {}_0^C D_t^\alpha u(t) + \lambda\, {}_0^C D_t^\alpha v(t),\right.$$
$$\left. {}_t^C D_T^\alpha u(t) + \lambda\, {}_t^C D_T^\alpha v(t)), {}_t^C D_T^\alpha v(t))\right|$$
$$\leq m_1(|u(t) + \lambda v(t)|)m_2(t)|v(t)| + \frac{1}{2}|{}_t^C D_T^\alpha u(t) + \lambda\, {}_t^C D_T^\alpha v(t)||{}_0^C D_t^\alpha v(t)|$$
$$+ \frac{1}{2}|{}_0^C D_t^\alpha u(t) + \lambda\, {}_0^C D_t^\alpha v(t)||{}_t^C D_T^\alpha v(t)|$$
$$\leq m_0 m_2(t)|v(t)| + \frac{1}{2}|{}_t^C D_T^\alpha u(t)||{}_0^C D_t^\alpha v(t)| + \frac{1}{2}|{}_0^C D_t^\alpha u(t)||{}_t^C D_T^\alpha v(t)|$$
$$+ |{}_0^C D_t^\alpha v(t)||{}_t^C D_T^\alpha v(t)|,$$

where

$$m_0 = \max_{(\lambda, t) \in [-1, 1] \times [0, T]} m_1(|u(t) + \lambda v(t)|).$$

Since $m_2 \in L^1([0, T], \mathbb{R}^+)$, $v$ is continuous on $[0, T]$, and in view of Remark 6.2, we have

$$|D_\lambda G(\lambda, t)| \leq d(t),$$

where $d \in L^1([0,T], \mathbb{R}^+)$. Thus Leibniz formula is applicable and

$$\frac{d}{d\lambda}\psi(0) = \int_0^T D_\lambda G(0,t)dt$$

$$= \int_0^T (D_x L(t, u(t), {}_0^C D_t^\alpha u(t), {}_t^C D_T^\alpha u(t)), v(t))dt$$

$$+ \int_0^T (D_y L(t, u(t), {}_0^C D_t^\alpha u(t), {}_t^C D_T^\alpha u(t)), {}_0^C D_t^\alpha v(t))dt$$

$$+ \int_0^T (D_z L(t, u(t), {}_0^C D_t^\alpha u(t), {}_t^C D_T^\alpha u(t)), {}_t^C D_T^\alpha v(t))dt.$$

Moreover,

$$\left| D_x L(t, u(t), {}_0^C D_t^\alpha u(t), {}_t^C D_T^\alpha u(t)) \right| \le m_1(|u(t)|)m_2(t),$$

$$\left| D_y L(t, u(t), {}_0^C D_t^\alpha u(t), {}_t^C D_T^\alpha u(t)) \right| \le \frac{1}{2} v|_t^C D_T^\alpha u(t)|$$

and

$$\left| D_z L(t, u(t), {}_0^C D_t^\alpha u(t), {}_t^C D_T^\alpha u(t)) \right| \le \frac{1}{2} |_0^C D_t^\alpha u(t)|.$$

Thus, by Remark 6.2 and (6.14),

$$\langle \varphi'(u), v \rangle = \int_0^T (D_x L(t, u(t), {}_0^C D_t^\alpha u(t), {}_t^C D_T^\alpha u(t)), v(t))dt$$

$$+ \int_0^T (D_y L(t, u(t), {}_0^C D_t^\alpha u(t), {}_t^C D_T^\alpha u(t)), {}_0^C D_t^\alpha v(t))dt$$

$$+ \int_0^T (D_z L(t, u(t), {}_0^C D_t^\alpha u(t), {}_t^C D_T^\alpha u(t)), {}_t^C D_T^\alpha v(t))dt$$

$$\le c_1\|v\| + c_2\|_0^C D_t^\alpha v(t)\|_{L^2} + c_3\|_t^C D_T^\alpha v(t)\|_{L^2}$$

$$\le c_1\|v\| + c_2\|v\|_\alpha + \frac{c_3}{|\cos(\pi\alpha)|}\|v\|_\alpha$$

$$\le c_4\|v\|_\alpha,$$

where $c_1, c_2, c_3$ and $c_4$ are some positive constants. Therefore, $\varphi$ has, at $u$, a directional derivative $\varphi'(u) \in (E^\alpha)^*$ given by (6.27).

2) By a theorem of Krasnoselskii, (6.29) and (6.30) imply that the mapping from $E^\alpha$ into $L^1([0,T], \mathbb{R}^N) \times L^2([0,T], \mathbb{R}^N) \times L^2([0,T], \mathbb{R}^N)$ defined by

$$u \to (D_x L(\cdot, u, {}_0^C D_t^\alpha u, {}_t^C D_T^\alpha u), D_y L(\cdot, u, {}_0^C D_t^\alpha u, {}_t^C D_T^\alpha u), D_z L(\cdot, u, {}_0^C D_t^\alpha u, {}_t^C D_T^\alpha u))$$

is continuous, so that $\varphi'$ is continuous from $E^\alpha$ into $(E^\alpha)^*$, and the proof is completed. $\square$

**Theorem 6.2.** *Let $\frac{1}{2} < \alpha \le 1$ and $\varphi$ be defined by (6.26). If condition (A) is satisfied and $u \in E^\alpha$ is a solution of corresponding Euler equation $\varphi'(u) = 0$, then $u$ is a solution of BVP (6.20) which, of course, corresponding to the solution of BVP (6.1).*

**Proof.** By Theorem 6.1 and Proposition 1.10, we have

$$
\begin{aligned}
0 &= \langle \varphi'(u), v \rangle \\
&= -\int_0^T \frac{1}{2}[({}_0^C D_t^\alpha u(t), {}_t^C D_T^\alpha v(t)) + ({}_t^C D_T^\alpha u(t), {}_0^C D_t^\alpha v(t))]dt \\
&\quad - \int_0^T (\nabla F(t, u(t)), v(t))dt \\
&= \int_0^T \left( \frac{1}{2}({}_0 D_t^{\alpha-1}({}_0^C D_t^\alpha u(t)), v'(t)) - \frac{1}{2}({}_t D_T^{\alpha-1}({}_t^C D_T^\alpha u(t)), v'(t)) \right) dt \\
&\quad - \int_0^T (\nabla F(t, u(t)), v(t))dt
\end{aligned}
\tag{6.31}
$$

for all $v \in E^\alpha$.

Let us define $w \in C([0, T], \mathbb{R}^N)$ by

$$
w(t) = \int_0^t \nabla F(s, u(s))ds, \quad t \in [0, T],
$$

so that

$$
\int_0^T (w(t), v'(t))dt = \int_0^T \left( \int_0^t (\nabla F(s, u(s)), v'(t))ds \right) dt.
$$

By the Fubini theorem and noting that $v(T) = 0$, we obtain

$$
\begin{aligned}
\int_0^T (w(t), v'(t))dt &= \int_0^T \left( \int_s^T (\nabla F(s, u(s)), v'(t))dt \right) ds \\
&= \int_0^T (\nabla F(s, u(s)), v(T) - v(s))ds \\
&= -\int_0^T (\nabla F(s, u(s)), v(s))ds.
\end{aligned}
$$

Hence, by (6.31) we have, for every $v \in E^\alpha$,

$$
\int_0^T \left( \frac{1}{2} {}_0 D_t^{\alpha-1}({}_0^C D_t^\alpha u(t)) - \frac{1}{2} {}_t D_T^{\alpha-1}({}_t^C D_T^\alpha u(t)) + w(t), v'(t) \right) dt = 0. \tag{6.32}
$$

If $(e_j)$ denotes the canonical basis of $\mathbb{R}^N$, we can choose $v \in E^\alpha$ such that

$$
v(t) = \sin \frac{2k\pi t}{T} e_j \quad \text{or} \quad v(t) = e_j - \cos \frac{2k\pi t}{T} e_j, \quad k = 1, 2, \ldots \quad \text{and } j = 1, \ldots, N.
$$

The theory of Fourier series and (6.32) imply that

$$
\frac{1}{2} {}_0 D_t^{\alpha-1}({}_0^C D_t^\alpha u(t)) - \frac{1}{2} {}_t D_T^{\alpha-1}({}_t^C D_T^\alpha u(t)) + w(t) = C,
$$

a.e. $t \in [0, T]$, for some $C \in \mathbb{R}^N$. According to the definition of $w \in C([0, T], \mathbb{R}^N)$, we have

$$
\frac{1}{2} {}_0 D_t^{\alpha-1}({}_0^C D_t^\alpha u(t)) - \frac{1}{2} {}_t D_T^{\alpha-1}({}_t^C D_T^\alpha u(t)) = -\int_0^t \nabla F(s, u(s))ds + C,
$$

a.e. $t \in [0,T]$, for some $C \in \mathbb{R}^N$.

In view of $\nabla F(\cdot, u(\cdot)) \in L^1([0,T], \mathbb{R}^N)$, we shall identify the equivalence class $D^\alpha(u(t))$ given by (6.21) and its continuous representation

$$D^\alpha(u(t)) = \frac{1}{2} \, _0D_t^{\alpha-1}(_0^C D_t^\alpha u(t)) - \frac{1}{2} \, _tD_T^{\alpha-1}(_t^C D_T^\alpha u(t))$$

$$= -\int_0^t \nabla F(s, u(s))ds + C \tag{6.33}$$

for $t \in [0,T]$.

Therefore, it follows from (6.33) and a classical result of Lebesgue theory that $-\nabla F(\cdot, u(\cdot))$ is the classical derivative of $D^\alpha(u(t))$ a.e. on $[0,T]$ which means that (i) in Definition 6.2 is verified.

Since $u \in E^\alpha$ implies that $u \in AC([0,T], \mathbb{R}^N)$, it remains to show that $u$ satisfies (6.20). In fact, according to (6.33), we can get that

$$\frac{d}{dt} D^\alpha(u(t)) = \frac{d}{dt}\left(\frac{1}{2} \, _0D_t^{\alpha-1}(_0^C D_t^\alpha u(t)) - \frac{1}{2} \, _tD_T^{\alpha-1}(_t^C D_T^\alpha u(t))\right) = -\nabla F(t, u(t)).$$

Moreover, $u \in E^\alpha$ implies that $u(0) = u(T) = 0$, and therefore (6.1) is verified. The proof is completed. $\qquad\square$

From now on, $\varphi$ given by (6.26) is considered as a functional on $E^\alpha$ with $\frac{1}{2} < \alpha \le 1$.

### 6.2.4 *Existence Under Ambrosetti-Rabinowitz Condition*

According to Theorem 6.2, we know that in order to find solutions of BVP (6.1), it suffices to obtain the critical points of functional $\varphi$ given by (6.26). We need to use some critical point theorems.

First, we use Theorem 1.12 to consider the existence of solutions for BVP (6.1). Assume that condition (A) is satisfied. Recall that, in our setting in (6.26), the corresponding functional $\varphi$ on $E^\alpha$ given by

$$\varphi(u) = \int_0^T \left(-\frac{1}{2}(_0^C D_t^\alpha u(t), _t^C D_T^\alpha u(t)) - F(t, u(t))\right)dt$$

is continuously differentiable according to Theorem 6.1 and is also weakly lower semi-continuous functional on $E^\alpha$ as the sum of a convex continuous function (see Theorem 1.2 in Mawhin and Willem, 1989) and of a weakly continuous one (see Proposition 1.2 in Mawhin and Willem, 1989).

In fact, according to Proposition 6.3, if $u_k \rightharpoonup u$ in $E^\alpha$, then $u_k \to u$ in $C([0,T], \mathbb{R}^N)$. Therefore, $F(t, u_k(t)) \to F(t, u(t))$ a.e. $t \in [0,T]$. By Lebesgue dominated convergence theorem, we have $\int_0^T F(t, u_k(t))dt \to \int_0^T F(t, u(t))dt$, which means that the functional $u \to \int_0^T F(t, u(t))dt$ is weakly continuous on $E^\alpha$. Moreover, the following lemma implies that the functional $u \to -\int_0^T [(_0^C D_t^\alpha u(t), _t^C D_T^\alpha u(t))/2]dt$ is convex and continuous on $E^\alpha$.

**Lemma 6.2.** *Let $\frac{1}{2} < \alpha \le 1$ and condition (A) be satisfied. If $u \in E^\alpha$, then the functional $H : E^\alpha \to \mathbb{R}^N$ denoted by*

$$H(u) = -\frac{1}{2}\int_0^T ({}_0^C D_t^\alpha u(t), {}_t^C D_T^\alpha u(t))dt$$

*is convex and continuous on $E^\alpha$.*

**Proof.** The continuity follows from (6.22) and (6.17) directly. We are now in a position to prove the convexity of $H$.

Let $\lambda \in (0,1)$, $u, v \in E^\alpha$ and $\tilde{u}, \tilde{v}$ be the extension of $u$ and $v$ by zero on $\mathbb{R}/[0,T]$ respectively. Since Caputo fractional derivative operator is linear operator, we have by Remark 6.2 and (6.23) that

$$H((1-\lambda)u + \lambda v)$$

$$= -\frac{1}{2}\int_0^T ({}_0^C D_t^\alpha ((1-\lambda)u(t) + \lambda v(t)), {}_t^C D_T^\alpha ((1-\lambda)u(t) + \lambda v(t)))dt$$

$$= -\frac{1}{2}\int_0^T ({}_0 D_t^\alpha ((1-\lambda)u(t) + \lambda v(t)), {}_t D_T^\alpha ((1-\lambda)u(t) + \lambda v(t)))dt$$

$$= -\frac{1}{2}\int_{-\infty}^\infty ({}_{-\infty} D_t^\alpha ((1-\lambda)\tilde{u}(t) + \lambda \tilde{v}(t)), {}_t D_{+\infty}^\alpha ((1-\lambda)\tilde{u}(t) + \lambda \tilde{v}(t)))dt$$

$$= \frac{|\cos(\pi\alpha)|}{2}\int_{-\infty}^\infty |{}_{-\infty} D_t^\alpha ((1-\lambda)\tilde{u}(t) + \lambda \tilde{v}(t))|^2 dt$$

$$\le \frac{|\cos(\pi\alpha)|}{2}\int_{-\infty}^\infty ((1-\lambda)|{}_{-\infty} D_t^\alpha \tilde{u}(t)|^2 + \lambda|{}_{-\infty} D_t^\alpha \tilde{v}(t)|^2)\, dt$$

$$= \int_{-\infty}^\infty \left(-\frac{1-\lambda}{2}({}_{-\infty} D_t^\alpha \tilde{u}(t), {}_t D_{+\infty}^\alpha \tilde{u}(t)) - \frac{\lambda}{2}({}_{-\infty} D_t^\alpha \tilde{v}(t), {}_t D_{+\infty}^\alpha \tilde{v}(t))\right) dt$$

$$= \int_0^T \left(-\frac{1-\lambda}{2}({}_0^C D_t^\alpha u(t), {}_t^C D_T^\alpha u(t)) - \frac{\lambda}{2}({}_0^C D_t^\alpha v(t), {}_t^C D_T^\alpha v(t))\right) dt$$

$$= (1-\lambda)H(u) + \lambda H(v),$$

which implies that $H$ is a convex functional defined on $E^\alpha$. This completes the proof. $\square$

According to the arguments above, if $\varphi$ is coercive, by Theorem 1.12, $\varphi$ has a minimum so that BVP (6.1) is solvable. It remains to find conditions under which $\varphi$ is coercive on $E^\alpha$, i.e. $\lim_{\|u\|_\alpha \to \infty} \varphi(u) = +\infty$, for $u \in E^\alpha$. We shall see that it suffices to require that $F(t,x)$ is bounded by a function for a.e., $t \in [0,T]$ and all $x \in \mathbb{R}^N$.

**Theorem 6.3.** *Let $\alpha \in (\frac{1}{2}, 1]$ and assume that $F$ satisfies condition (A). If*

$$|F(t,x)| \le \bar{a}|x|^2 + \bar{b}(t)|x|^{2-\gamma} + \bar{c}(t), \quad t \in [0,T],\ x \in \mathbb{R}^N, \qquad (6.34)$$

*where $\bar{a} \in [0, |\cos(\pi\alpha)|\Gamma^2(\alpha+1)/2T^{2\alpha})$, $\gamma \in (0,2)$, $\bar{b} \in L^{2/\gamma}([0,T],\mathbb{R})$, and $\bar{c} \in L^1([0,T],\mathbb{R})$, then BVP (6.1) has at least one solution which minimizes $\varphi$ on $E^\alpha$.*

**Proof.** According to arguments above, our problem reduces to prove that $\varphi$ is coercive on $E^\alpha$. For $u \in E^\alpha$, it follows from (6.22), (6.34) and (6.13) that

$$\varphi(u) = -\frac{1}{2} \int_0^T ({}_0^C D_t^\alpha u(t), {}_t^C D_T^\alpha u(t)) dt - \int_0^T F(t, u(t)) dt$$

$$\geq \frac{|\cos(\pi\alpha)|}{2} \int_0^T |{}_0^C D_t^\alpha u(t)|^2 dt - \bar{a} \int_0^T |u(t)|^2 dt$$

$$- \int_0^T \bar{b}(t)|u(t)|^{2-\gamma} dt - \int_0^T \bar{c}(t) dt$$

$$= \frac{|\cos(\pi\alpha)|}{2} \|u\|_\alpha^2 - \bar{a}\|u\|_{L^2}^2 - \int_0^T \bar{b}(t)|u(t)|^{2-\gamma} dt - \bar{c}_1$$

$$\geq \frac{|\cos(\pi\alpha)|}{2} \|u\|_\alpha^2 - \bar{a}\|u\|_{L^2}^2 - \left(\int_0^T |\bar{b}(t)|^{2/\gamma} dt\right)^{\gamma/2} \left(\int_0^T |u(t)|^2 dt\right)^{1-\gamma/2} - \bar{c}_1$$

$$= \frac{|\cos(\pi\alpha)|}{2} \|u\|_\alpha^2 - \bar{a}\|u\|_{L^2}^2 - \bar{b}_1\|u\|_{L^2}^{2-\gamma} - \bar{c}_1$$

$$\geq \frac{|\cos(\pi\alpha)|}{2} \|u\|_\alpha^2 - \frac{\bar{a}T^{2\alpha}}{\Gamma^2(\alpha+1)} \|u\|_\alpha^2 - \bar{b}_1 \left(\frac{T^\alpha}{\Gamma(\alpha+1)}\right)^{2-\gamma} \|u\|_\alpha^{2-\gamma} - \bar{c}_1$$

$$= \left(\frac{|\cos(\pi\alpha)|}{2} - \frac{\bar{a}T^{2\alpha}}{\Gamma^2(\alpha+1)}\right) \|u\|_\alpha^2 - \bar{b}_1 \left(\frac{T^\alpha}{\Gamma(\alpha+1)}\right)^{2-\gamma} \|u\|_\alpha^{2-\gamma} - \bar{c}_1,$$

where $\bar{b}_1 = \left(\int_0^T |\bar{b}(t)|^{2/\gamma} dt\right)^{\gamma/2}$ and $\bar{c}_1 = \int_0^T \bar{c}(t) dt$.

Noting that $\bar{a} \in [0, |\cos(\pi\alpha)|\Gamma^2(\alpha+1)/2T^{2\alpha})$ and $\gamma \in (0, 2)$, we have

$$\varphi(u) = +\infty \quad \text{as} \quad \|u\|_\alpha \to \infty,$$

and hence $\varphi$ is coercive, which completes the proof. $\square$

Our task is now to use Theorem 1.13 (Mountain pass theorem) to find a nonzero critical point of functional $\varphi$ on $E^\alpha$.

**Theorem 6.4.** *Let $\alpha \in (\frac{1}{2}, 1]$ and suppose that $F$ satisfies condition (A). If*

**(A1)** *$F \in C([0, T] \times \mathbb{R}^N, \mathbb{R})$ and there exists $\mu \in [0, \frac{1}{2})$ and $M > 0$ such that $0 < F(t, x) \leq \mu(\nabla F(t, x), x)$ for all $x \in \mathbb{R}^N$ with $|x| \geq M$ and $t \in [0, T]$;*

**(A2)** *$\limsup_{|x| \to 0} F(t, x)/|x|^2 < |\cos(\pi\alpha)|\Gamma^2(\alpha+1)/2T^{2\alpha}$ uniformly for $t \in [0, T]$ and $x \in \mathbb{R}^N$;*

*are satisfied, then BVP (6.1) has at least one nonzero solution on $E^\alpha$.*

**Proof.** We will verify that $\varphi$ satisfies all conditions of Theorem 1.13.

First, we will prove that $\varphi$ satisfies (PS) condition. Since $F(t, x) - \mu(\nabla F(t, x), x)$ is continuous for $t \in [0, T]$ and $|x| \leq M$, there exists $c \in \mathbb{R}^+$, such that

$$F(t, x) \leq \mu(\nabla F(t, x), x) + c, \quad t \in [0, T], \quad |x| \leq M.$$

By condition (A1), we obtain

$$F(t, x) \leq \mu(\nabla F(t, x), x) + c, \quad t \in [0, T], \quad x \in \mathbb{R}^N. \tag{6.35}$$

Let $\{u_k\} \subset E^\alpha$, $|\varphi(u_k)| \leq K$, $k = 1, 2, \ldots$, $\varphi'(u_k) \to 0$. Notice that

$$\langle \varphi'(u_k), u_k \rangle = -\int_0^T [({}_0^C D_t^\alpha u_k(t), {}_t^C D_T^\alpha u_k(t)) + (\nabla F(t, u_k(t)), u_k(t))]dt. \quad (6.36)$$

It follows from (6.35), (6.36) and (6.22) that

$$K \geq \varphi(u_k) = -\frac{1}{2}\int_0^T ({}_0^C D_t^\alpha u_k(t), {}_t^C D_T^\alpha u_k(t))dt - \int_0^T F(t, u_k(t))dt$$

$$\geq -\frac{1}{2}\int_0^T ({}_0^C D_t^\alpha u_k(t), {}_t^C D_T^\alpha u_k(t))dt - \mu\int_0^T (\nabla F(t, u_k(t)), u_k(t))dt - cT$$

$$= \left(\mu - \frac{1}{2}\right)\int_0^T ({}_0^C D_t^\alpha u_k(t), {}_t^C D_T^\alpha u_k(t))dt + \mu\langle \varphi'(u_k), u_k \rangle - cT$$

$$\geq \left(\frac{1}{2} - \mu\right)|\cos(\pi\alpha)|\|u_k\|_\alpha^2 - \mu\|\varphi'(u_k)\|_\alpha\|u_k\|_\alpha - cT, \quad k = 1, 2, \ldots.$$

Since $\varphi'(u_k) \to 0$, there exists $N_0 \in \mathbb{N}$ such that

$$K \geq \left(\frac{1}{2} - \mu\right)|\cos(\pi\alpha)|\|u_k\|_\alpha^2 - \|u_k\|_\alpha - cT, \quad k > N_0,$$

and this implies that $\{u_k\} \subset E^\alpha$ is bounded. Since $E^\alpha$ is a reflexive space, going to a subsequence if necessary, we may assume that $u_k \rightharpoonup u$ weakly in $E^\alpha$, thus we have

$$\langle \varphi'(u_k) - \varphi'(u), u_k - u \rangle = \langle \varphi'(u_k), u_k - u \rangle - \langle \varphi'(u), u_k - u \rangle$$

$$\leq \|\varphi'(u_k)\|_\alpha\|u_k - u\|_\alpha - \langle \varphi'(u), u_k - u \rangle$$

$$\to 0, \quad \text{as } k \to \infty. \quad (6.37)$$

Moreover, according to (6.14) and Proposition 6.3, we have $u_k$ is bounded in $C([0, T], \mathbb{R}^N)$ and $\|u_k - u\| \to 0$ as $k \to \infty$. Hence, we have

$$\int_0^T \nabla F(t, u_k(t))dt \to \int_0^T \nabla F(t, u(t))dt \quad \text{as } k \to \infty. \quad (6.38)$$

Noting that

$$\langle \varphi'(u_k) - \varphi'(u), u_k - u \rangle$$

$$= -\int_0^T ({}_0^C D_t^\alpha(u_k(t) - u(t)), {}_t^C D_T^\alpha(u_k(t) - u(t)))dt$$

$$- \int_0^T \Big((\nabla F(t, u_k(t)) - \nabla F(t, u(t))), (u_k(t) - u(t))\Big)dt$$

$$\geq |\cos(\pi\alpha)|\|u_k - u\|_\alpha^2 - \left|\int_0^T (\nabla F(t, u_k(t)) - \nabla F(t, u(t)))dt\right|\|u_k - u\|.$$

Combining (6.37) and (6.38), it is easy to verify that $\|u_k - u\|_\alpha^2 \to 0$ as $k \to \infty$, and hence that $u_k \to u$ in $E^\alpha$. Thus, we obtain the desired convergence property.

From $\limsup_{|x| \to 0} F(t, x)/|x|^2 < |\cos(\pi\alpha)|\Gamma^2(\alpha + 1)/2T^{2\alpha}$ uniformly for $t \in [0, T]$, there exists $\epsilon \in (0, |\cos(\pi\alpha)|)$ and $\delta > 0$ such that $F(t, x) \leq (|\cos(\pi\alpha)| - \epsilon)(\Gamma^2(\alpha + 1)/2T^{2\alpha})|x|^2$ for all $t \in [0, T]$ and $x \in \mathbb{R}^N$ with $|x| \leq \delta$.

Let $\rho = \frac{\Gamma(\alpha)((\alpha-1)/2+1)^{\frac{1}{2}}}{T^{\alpha-\frac{1}{2}}}\delta$ and $\sigma = \epsilon\rho^2/2 > 0$. Then it follows from (6.14) that

$$\|u\| \leq \frac{T^{\alpha-\frac{1}{2}}}{\Gamma(\alpha)((\alpha-1)/2+1)^{\frac{1}{2}}}\|u\|_\alpha = \delta$$

for all $u \in E^\alpha$ with $\|u\|_\alpha = \rho$. Therefore, we have

$$\begin{aligned}
\varphi(u) &= -\frac{1}{2}\int_0^T ({}^C_0D_t^\alpha u(t), {}^C_tD_T^\alpha u(t))dt - \int_0^T F(t, u(t))dt \\
&\geq \frac{|\cos(\pi\alpha)|}{2}\|u\|_\alpha^2 - (|\cos(\pi\alpha)| - \epsilon)\frac{\Gamma^2(\alpha+1)}{2T^{2\alpha}}\int_0^T |u(t)|^2 dt \\
&\geq \frac{|\cos(\pi\alpha)|}{2}\|u\|_\alpha^2 - \frac{1}{2}(|\cos(\pi\alpha)| - \epsilon)\|u\|_\alpha^2 \\
&= \frac{1}{2}\epsilon\|u\|_\alpha^2 \\
&= \sigma
\end{aligned}$$

for all $u \in E^\alpha$ with $\|u\|_\alpha = \rho$. This implies (ii) in Theorem 1.13 is satisfied.

It is obvious from the definition of $\varphi$ and (A2) that $\varphi(0) = 0$, and therefore, it suffices to show that $\varphi$ satisfies (iii) in Theorem 1.13.

Since $0 < F(t, x) \leq \mu(\nabla F(t, x), x)$ for all $x \in \mathbb{R}^N$ and $|x| \geq M$, a simple regularity argument then shows that there exists $r_1, r_2 > 0$ such that

$$F(t, x) \geq r_1|x|^{1/\mu} - r_2, \quad x \in \mathbb{R}^N, \quad t \in [0, T].$$

For any $u \in E^\alpha$ with $u \neq 0$, $\kappa > 0$ and noting that $\mu \in [0, \frac{1}{2})$ and (6.22), we have

$$\begin{aligned}
\varphi(\kappa u) &= -\frac{1}{2}\int_0^T ({}^C_0D_t^\alpha \kappa u(t), {}^C_tD_T^\alpha \kappa u(t))dt - \int_0^T F(t, \kappa u(t))dt \\
&\leq \frac{\kappa^2}{2|\cos(\pi\alpha)|}\|u\|_\alpha^2 - r_1\int_0^T |\kappa u(t)|^{1/\mu}dt + r_2 T \\
&= \frac{\kappa^2}{2|\cos(\pi\alpha)|}\|u\|_\alpha^2 - r_1\kappa^{1/\mu}\|u\|_{L^{1/\mu}}^{1/\mu} + r_2 T \\
&\to -\infty
\end{aligned}$$

as $\kappa \to \infty$. Then there exists a sufficiently large $\kappa_0$ such that $\varphi(\kappa_0 u) \leq 0$. Hence (iii) in Theorem 1.13 holds.

Lastly noting that $\varphi(0) = 0$ while for our critical point $u$, $\varphi(u) \geq \sigma > 0$. Hence $u$ is a nontrivial weak solution of BVP (6.1), and this completes the proof. $\square$

**Corollary 6.1.** $\forall \alpha \in (\frac{1}{2}, 1]$, *suppose that $F$ satisfies conditions (A) and (A1). If*

**(A2)′** $F(t, x) = o(|x|^2)$, *as $|x| \to 0$ uniformly for $t \in [0, T]$ and $x \in \mathbb{R}^N$*

*is satisfied, then BVP (6.1) has at least one nonzero solution on $E^\alpha$.*

### 6.2.5   *Superquadratic Case*

Under the usual Ambrosetti-Rabinowitz condition, it is easy to show that the energy functional associated with the system has the Mountain Pass geometry and satisfies the (PS) condition. However, the A.R. condition is so strong that many potential functions cannot satisfy it, then the problem becomes more delicate and complicated.

Assume that $F : [0, T] \times \mathbb{R}^N \to \mathbb{R}$ satisfies the condition (A) which is assumed as in Subsection 6.2.3.

In the following, we introduce the function space $E^\alpha$, where $\alpha \in (\frac{1}{2}, 1]$. For $u \in E^\alpha$, where

$$E^\alpha := \left\{ u \in L^2(0, T; \mathbb{R}^N) : \, {}^C_0D_t^\alpha u \in L^2(0, T; \mathbb{R}^N) \right\}$$

is a reflexive Banach space with the norm defined by

$$\|u\|_\alpha = \|{}^C_0D_t^\alpha u\|_{L^2}$$

and

$$\|u\| = \max_{t \in [0, T]} |u(t)|.$$

It follows from Theorem 6.1 that the functional $\varphi$ on $E^\alpha$ given by

$$\varphi(u) = \int_0^T \left( -\frac{1}{2} \left( {}^C_0D_t^\alpha u(t), \, {}^C_tD_T^\alpha u(t) \right) - F(t, u(t)) \right) dt$$

is continuously differentiable on $E^\alpha$. Moreover, we have

$$\langle \varphi'(u), v \rangle = -\int_0^T \frac{1}{2} \left( \left( {}^C_0D_t^\alpha u(t), \, {}^C_tD_T^\alpha v(t) \right) + \left( {}^C_tD_T^\alpha u(t), \, {}^C_0D_t^\alpha v(t) \right) \right) dt$$

$$- \int_0^T \left( \nabla F(t, u(t)), v(t) \right) dt.$$

Recall that a sequence $\{u_n\} \subset E^\alpha$ is said to be a (C) sequence of $\varphi$ if $\varphi(u_n)$ is bounded and $(1 + \|u_n\|_\alpha)\|\varphi(u_n)\|_\alpha \to 0$ as $n \to \infty$. The functional $\varphi$ satisfies condition (C) if every (C) sequence of $\varphi$ has a convergent subsequence. This condition is due to Cerami, 1978.

For the superquadratic case, we make the following assumptions:

**(A3)** $\lim_{|x| \to 0} \frac{F(t,x)}{|x|^2} = 0$, $\quad \liminf_{|x| \to \infty} \frac{F(t,x)}{|x|^2} \geq L > \frac{\pi^2}{|\cos(\pi\alpha)|\Gamma^2(2-\alpha)T^{2\alpha}(3-2\alpha)}$
uniformly for some $L > 0$ and a.e. $t \in [0, T]$;

**(A4)** $\limsup_{|x| \to +\infty} \frac{F(t,x)}{|x|^r} \leq M < +\infty$ uniformly for some $M > 0$ and a.e. $t \in [0, T]$;

**(A5)** $\liminf_{|x| \to +\infty} \frac{(\nabla F(t,x),x) - 2F(t,x)}{|x|^\mu} \geq Q > 0$ uniformly for some $Q > 0$ and a.e. $t \in [0, T]$, where $r > 2$ and $\mu > r - 2$.

We will first establish the following lemma.

**Lemma 6.3.** *Assume (A), (A4), (A5) hold, then the functional $\varphi$ satisfies condition (C).*

**Proof.** Let $\{u_n\} \subset E^\alpha$ is a (C) sequence of $\varphi$, that is $\varphi(u_n)$ is bounded and $(1 + \|u_n\|_\alpha)\|\varphi'(u_n)\|_\alpha \to 0$ as $n \to \infty$. Then there exists $M_0$ such that

$$|\varphi(u_n)| \le M_0 \quad \text{and} \quad (1 + \|u_n\|_\alpha)\|\varphi'(u_n)\|_\alpha \le M_0, \tag{6.39}$$

for all $n \in \mathbb{N}$.

By (A4), there exist positive constants $B_1$ and $M_1$ such that

$$F(t, x) \le B_1 |x|^r$$

for all $|x| \ge M_1$ and a.e. $t \in [0, T]$.

It follows from (A) that

$$|F(t, x)| \le \max_{s \in [0, M_1]} a(s)b(t)$$

for all $|x| \le M_1$ and a.e. $t \subset [0, T]$. Therefore, we obtain

$$F(t, x) \le B_1 |x|^r + \max_{s \in [0, M_1]} a(s)b(t), \tag{6.40}$$

for all $x \in \mathbb{R}^N$ and a.e. $t \in [0, T]$.

Combining (6.22) and (6.40), we get

$$\frac{|\cos(\pi\alpha)|}{2} \|u_n\|_\alpha^2 \le \varphi(u_n) + \int_0^T F(t, u_n(t))dt$$

$$\le M_0 + \max_{s \in [0, M_1]} a(s) \int_0^T b(t)dt + B_1 \int_0^T |u_n(t)|^r dt. \tag{6.41}$$

On the other hand, by (A5), there exist $\eta > 0$ and $M_2 > 0$ such that

$$\big(\nabla F(t, x), x\big) - 2F(t, x) \ge \eta |x|^\mu$$

for a.e. $t \in [0, T]$ and $|x| \ge M_2$.

By (A), we have

$$|\big(\nabla F(t, x), x\big) - 2F(t, x)| \le (2 + M_2) \max_{s \in [0, M_2]} a(s)b(t)$$

for all $|x| \le M_2$ and a.e. $t \in [0, T]$.

Therefore, we obtain

$$\big(\nabla F(t, x), x\big) - 2F(t, x) \ge \eta |x|^\mu - (2 + M_2) \max_{s \in [0, M_2]} a(s)b(t), \tag{6.42}$$

for all $x \in \mathbb{R}^N$ and a.e. $t \in [0, T]$.

It follows from (6.39) and (6.42) that

$$3M_0 \ge 2\varphi(u_n) - \langle\varphi'(u_n), u_n\rangle$$

$$= 2 \int_0^T \Big[-\frac{1}{2}({}_0^C D_t^\alpha u_n(t), {}_t^C D_T^\alpha u_n(t)) - F(t, u_n(t))\Big]dt$$

$$- \int_0^T \Big[-({}_0^C D_t^\alpha u_n(t), {}_t^C D_T^\alpha u_n(t)) - (\nabla F(t, u_n(t)), u_n(t))\Big]dt$$

$$= \int_0^T \left[ (\nabla F(t, u_n(t)), u_n(t)) - 2F(t, u_n(t)) \right] dt$$

$$\geq \eta \int_0^T |u_n(t)|^\mu dt - (2 + M_2) \max_{s \in [0, M_2]} a(s) \int_0^T b(t) dt,$$

thus, $\int_0^T |u_n(t)|^\mu dt$ is bounded.

If $\mu > r$, then

$$\int_0^T |u_n(t)|^r dt \leq T^{\frac{\mu-r}{\mu}} \left( \int_0^T |u_n(t)|^\mu dt \right)^{r/\mu},$$

which combining (6.41) implies that $\|u_n\|_\alpha$ is bounded.

If $\mu \leq r$, then

$$\int_0^T |u_n(t)|^r dt \leq \|u_n\|_\infty^{r-\mu} \int_0^T |u_n(t)|^\mu dt \leq C_1^{r-\mu} \|u_n\|_\alpha^{r-\mu} \int_0^T |u_n(t)|^\mu dt,$$

where

$$C_1 := \frac{T^{\alpha - \frac{1}{2}}}{\Gamma(\alpha)(2\alpha - 1)^{\frac{1}{2}}}$$

by (6.14).

Since $\mu > r - 2$, it follows from (6.41) that $\|u_n\|_\alpha$ is bounded too. Thus $\|u_n\|_\alpha$ is bounded in $E^\alpha$.

By Proposition 6.3, the sequence $\{u_n\}$ has a subsequence, also denoted by $\{u_n\}$, such that

$$u_n \rightharpoonup u \text{ in } E^\alpha \quad \text{and} \quad u_n \to u \text{ in } C([0, T], \mathbb{R}^N).$$

Then we obtain $u_n \to u$ in $E^\alpha$ by use of the same argument of Theorem 6.4. The proof of Lemma 6.3 is completed. □

We state our first existence result as follows.

**Theorem 6.5.** *Assume that (A3)-(A5) hold and that $F(t, x)$ satisfies the condition (A). Then BVP (6.1) has at least one solution on $E^\alpha$.*

**Proof.** By (A3), there exist $\epsilon_1 \in (0, |\cos(\pi\alpha)|)$ and $\delta > 0$ such that

$$F(t, x) \leq (|\cos(\pi\alpha)| - \epsilon_1) \frac{\Gamma^2(\alpha + 1)}{2T^{2\alpha}} |x|^2$$

for a.e. $t \in [0, T]$ and $x \in \mathbb{R}^N$ with $|x| \leq \delta$.

Let

$$\rho = \frac{\Gamma(\alpha)(2(\alpha - 1) + 1)^{\frac{1}{2}}}{T^{\alpha - \frac{1}{2}}} \delta \quad \text{and} \quad \sigma = \frac{\epsilon_1 \rho^2}{2} > 0.$$

Then it follows from (6.14) that

$$\|u\| \leq \frac{T^{\alpha - \frac{1}{2}}}{\Gamma(\alpha)(2(\alpha - 1) + 1)^{\frac{1}{2}}} \|u\|_\alpha = \delta$$

for all $u \in E^\alpha$ with $\|u\|_\alpha = \rho$.

Therefore, we have

$$
\begin{aligned}
\varphi(u) &= \int_0^T \left[ -\frac{1}{2} \left( {}_0^C D_t^\alpha u(t), {}_t^C D_T^\alpha u(t) \right) - F(t, u(t)) \right] dt \\
&\geq \frac{|\cos(\pi\alpha)|}{2} \|u\|_\alpha^2 - (|\cos(\pi\alpha)| - \epsilon_1) \frac{\Gamma^2(\alpha+1)}{2T^{2\alpha}} \int_0^T |u(t)|^2 dt \\
&\geq \frac{|\cos(\pi\alpha)|}{2} \|u\|_\alpha^2 - \frac{(|\cos(\pi\alpha)| - \epsilon_1)}{2} \|u\|_\alpha^2 \\
&= \frac{\epsilon_1}{2} \|u\|_\alpha^2 \\
&= \sigma
\end{aligned}
$$

for all $u \in E^\alpha$ with $\|u\|_\alpha = \rho$. This implies that (ii) in Theorem 1.13 is satisfied.

It is obvious from the definition of $\varphi$ and (A3) that $\varphi(0) = 0$, and therefore, it suffices to show that $\varphi$ satisfies (iii) in Theorem 1.13.

By (A3), there exist $\epsilon_2 > 0$ and $M_3 > 0$ such that

$$
F(t, x) > \left( \frac{\pi^2}{|\cos(\pi\alpha)|\Gamma^2(2-\alpha)T^{2\alpha}(3-2\alpha)} + \epsilon_2 \right) |x|^2
$$

for all $|x| \geq M_3$ and a.e. $t \in [0, T]$.

It follows from (A) that

$$
|F(t, x)| \leq \max_{s \in [0, M_3]} a(s)b(t),
$$

for all $|x| \leq M_3$ and a.e. $t \in [0, T]$.

Therefore, we obtain

$$
\begin{aligned}
F(t, x) \geq & \left( \frac{\pi^2}{|\cos(\pi\alpha)|\Gamma^2(2-\alpha)T^{2\alpha}(3-2\alpha)} + \epsilon_2 \right) (|x|^2 - M_3^2) \\
& - \max_{s \in [0, M_3]} a(s)b(t),
\end{aligned}
\tag{6.43}
$$

for all $x \in \mathbb{R}^N$ and a.e. $t \in [0, T]$.

Choosing $u_0 = (\frac{T}{\pi} \sin \frac{\pi t}{T}, 0, \dots, 0) \in E^\alpha$, then

$$
\|u_0\|_{L^2}^2 = \frac{T^3}{2\pi^2} \quad \text{and} \quad \|u_0\|_\alpha^2 \leq \frac{T^{3-2\alpha}}{\Gamma^2(2-\alpha)(3-2\alpha)}.
\tag{6.44}
$$

For $\varsigma > 0$ and noting that (6.43) and (6.44), we have

$$\varphi(\varsigma u_0) = \int_0^T \left[ -\frac{1}{2} \left( {}_0^C D_t^\alpha \varsigma u_0(t), {}_t^C D_T^\alpha \varsigma u_0(t) \right) - F(t, \varsigma u_0(t)) \right] dt$$

$$\leq \frac{\varsigma^2}{2|\cos(\pi\alpha)|} \|u_0\|_\alpha^2$$

$$- \left( \frac{\varsigma^2 \pi^2}{|\cos(\pi\alpha)|T^{2\alpha}\Gamma^2(2-\alpha)(3-2\alpha)} + \varsigma^2 \epsilon_2 \right) \int_0^T |u_0(t)|^2 dt + C_2$$

$$\leq \frac{\varsigma^2}{2|\cos(\pi\alpha)|} \cdot \frac{T^{3-2\alpha}}{\Gamma^2(2-\alpha)(3-2\alpha)}$$

$$- \frac{\varsigma^2 \pi^2}{|\cos(\pi\alpha)|T^{2\alpha}\Gamma^2(2-\alpha)(3-2\alpha)} \cdot \frac{T^3}{2\pi^2} - \frac{\varsigma^2 \epsilon_2 T^3}{2\pi^2} + C_2$$

$$\to -\infty$$

as $\varsigma \to \infty$, where $C_2$ is a positive constant. Then there exists a sufficiently large $\varsigma_0$ such that $\varphi(\varsigma_0 u_0) \leq 0$. Hence (iii) in Theorem 1.13 holds.

Finally, noting that $\varphi(0) = 0$ while for critical point $u$, $\varphi(u) \geq \sigma > 0$. Hence $u$ is a nontrivial solution of BVP (6.1), and this completes the proof. □

We give an example to illustrate our results.

**Example 6.1.** In BVP (6.1), let
$$F(t, x) = \ln(1 + 2|x|^2)|x|^2.$$
These show that all conditions of Theorem 6.5 are satisfied, where
$$r = 2.5, \quad \mu = 2.$$
By Theorem 6.5, BVP (6.1) has at least one solution $u \in E^\alpha$.

### 6.2.6 *Asymptotically Quadratic Case*

For the asymptotically quadratic case, we assume:

**(A4)'** $\limsup_{|x|\to+\infty} \frac{F(t,x)}{|x|^2} \leq M < +\infty$ uniformly for some $M > 0$ and a.e. $t \in [0, T]$;

**(A6)** there exists $\tau(t) \in L^1([0, T], \mathbb{R}^+)$ such that $(\nabla F(t, x), x) - 2F(t, x) \geq \tau(t)$ for all $x \in \mathbb{R}^N$ and a.e. $t \in [0, T]$;

**(A7)** $\lim_{|x|\to+\infty}[(\nabla F(t, x), x) - 2F(t, x)] = +\infty$ for a.e. $t \in [0, T]$.

**Theorem 6.6.** *Assume that $F(t, x)$ satisfies (A), (A3), (A4)', (A6) and (A7). Then BVP (6.1) has at least one solution on $E^\alpha$.*

The following lemmas are needed in the proof of Theorem 6.6.

**Lemma 6.4.** *Assume that (A7) holds. Then for any $\varepsilon > 0$, there exists a subset $E_\varepsilon \subset [0, T]$ with $\alpha([0, T]\backslash E_\varepsilon) < \varepsilon$ such that*
$$\lim_{|x|\to\infty} [(\nabla F(t, x), x) - 2F(t, x)] = +\infty$$
*uniformly for $t \in E_\varepsilon$.*

The proof is similar to that of Lemma 2 in Tang and Wu, 2001, and is omitted.

**Lemma 6.5.** *Assume that (A), (A4)′, (A6) and (A7) hold. Then the functional $\varphi$ satisfies condition (C).*

**Proof.** Suppose that $\{u_n\} \subset E^\alpha$ is a (C) sequence of $\varphi$, that is $\varphi(u_n)$ is bounded and $(1 + \|u_n\|_\alpha)\|\varphi'(u_n)\|_\alpha \to 0$ as $n \to \infty$. Then we have

$$\liminf_{n\to\infty}[\langle \varphi'(u_n), u_n \rangle - 2\varphi(u_n)] > -\infty,$$

which implies that

$$\limsup_{n\to\infty} \int_0^T [(\nabla F(t, u_n), \; u_n) - 2F(t, u_n)]dt < +\infty. \tag{6.45}$$

We only need to show that $\{u_n\}$ is bounded in $E^\alpha$. If $\{u_n\}$ is unbounded, we may assume, without loss of generality, that $\|u_n\|_\alpha \to \infty$ as $n \to \infty$. Put $z_n = \frac{u_n}{\|u_n\|_\alpha}$, we then have $\|z_n\|_\alpha = 1$. Going to a sequence if necessary, we assume that $z_n \rightharpoonup z$ in $E^\alpha$, $z_n \to z$ in $C([0,T], \mathbb{R}^N)$ and $L^2([0,T], \mathbb{R}^N)$.

By (A2)′, it follows that there exist constants $B_2 > 0$ and $M_4 > 0$ such that

$$F(t, x) \leq B_2|x|^2$$

for all $|x| \geq M_4$ and a.e. $t \in [0, T]$.

By condition (A), it follows that

$$|F(t, x)| \leq \max_{s \in [0, M_4]} a(s)b(t)$$

for all $|x| \leq M_4$ and a.e. $t \in [0, T]$. Therefore, we obtain

$$F(t, x) \leq B_2|x|^2 + \max_{s \in [0, M_4]} a(s)b(t)$$

for all $x \in \mathbb{R}^N$ and a.e. $t \in [0, T]$. Therefore, we have

$$\varphi(u) = \int_0^T \left[ -\frac{1}{2} \left({}_0^c D_t^\alpha u(t), {}_t^c D_T^\alpha u(t)\right) - F(t, u(t)) \right]dt$$

$$\geq \frac{|\cos(\pi\alpha)|}{2}\|u\|_\alpha^2 - B_2 \int_0^T |u|^2 dt - \max_{s\in[0,M_4]} a(s) \int_0^T b(t)dt,$$

from which, it follows that

$$\frac{\varphi(u_n)}{\|u_n\|_\alpha^2} \geq \frac{|\cos(\pi\alpha)|}{2} - B_2\|z_n\|_{L_2}^2 - \frac{1}{\|u_n\|_\alpha^2} \max_{s\in[0,M_4]} a(s) \int_0^T b(t)dt.$$

Passing to the limit in the last inequality, we get

$$\frac{|\cos(\pi\alpha)|}{2} - B_2\|z\|_{L_2}^2 \leq 0,$$

which yields $z \neq 0$. Therefore, there exists a subset $E \subset [0, T]$ with $\alpha(E) > 0$ such that $z(t) \neq 0$ on $E$.

By virtue of Lemma 6.4, for $\varepsilon = \frac{1}{2}\alpha(E) > 0$, we can choose a subset $E_\varepsilon \subset [0,T]$ with $\alpha([0,T]\backslash E_\varepsilon) < \varepsilon$ such that

$$\lim_{|x|\to\infty} [(\nabla F(t,x),x) - 2F(t,x)] = +\infty, \tag{6.46}$$

uniformly for $t \in E_\varepsilon$.

We assert that $\alpha(E \cap E_\varepsilon) > 0$. If not, $\alpha(E \cap E_\varepsilon) = 0$.

Since $E = (E \cap E_\varepsilon) \cup (E\backslash E_\varepsilon)$, it follows that

$$0 < \alpha(E) = \alpha(E \cap E_\varepsilon) + \alpha(E\backslash E_\varepsilon)$$
$$\leq \alpha([0,T]\backslash E_\varepsilon)$$
$$< \varepsilon = \frac{1}{2}\alpha(E),$$

which leads to a contradiction and establishes the assertion.

By (A6), we obtain

$$\int_0^T [(\nabla F(t,u_n),u_n) - 2F(t,u_n)]dt$$

$$= \int_{E \cap E_\varepsilon} [(\nabla F(t,u_n),u_n) - 2F(t,u_n)]dt$$

$$+ \int_{[0,T]\backslash(E \cap E_\varepsilon)} [(\nabla F(t,u_n),u_n) - 2F(t,u_n)]dt$$

$$\geq \int_{E \cap E_\varepsilon} [(\nabla F(t,u_n),u_n) - 2F(t,u_n)]dt - \int_0^T |\tau(t)|dt. \tag{6.47}$$

By (6.46), (6.47) and Fatou lemma, it follows that

$$\lim_{n\to\infty} \int_0^T [(\nabla F(t,u_n),u_n) - 2F(t,u_n)]dt = +\infty,$$

which contradicts (6.45). This contradiction shows that $\|u_n\|_\alpha$ is bounded in $E^\alpha$ and this completes the proof. $\qquad\square$

**Theorem 6.7.** *Assume that $F(t,x)$ satisfies (A), (A3), (A4)′ and the following conditions:*

**(A6)′** *there exists $\tau(t) \in L^1(0,T;\mathbb{R}^+)$ such that $(\nabla F(t,x),x) - 2F(t,x) \leq \tau(t)$ for all $x \in \mathbb{R}^N$ and a.e. $t \in [0,T]$;*

**(A7)′** $\lim_{|x|\to+\infty} [(\nabla F(t,x),x) - 2F(t,x)] = -\infty$ *for a.e. $t \in [0,T]$.*

*Then BVP (6.1) has at least one solution on $E^\alpha$.*

By virtue of Lemma 6.4 and Lemma 6.5, similar to Theorem 6.5, we can complete the proof of Theorem 6.6 by using the similar proof of Theorem 6.5. Theorem 6.7 can be proved similarly.

We give an example to illustrate our results.

**Example 6.2.** In BVP (6.1), let $T = 2\pi$ and $F(t,x) = \kappa f(x)(2 + \sin t)\arctan|x|^2$, where $\kappa > 0$ and $f(x)$ will be specified below.

Let $f(x) = |x|^2 + \ln(1 + |x|^2)$. Noting that $0 \le \ln(1 + |x|^2) \le |x|^2$, we see that (A) and (A4)' hold. It is also easy to see that (A3) hold for

$$\kappa > \frac{(2\pi)^{1-2\alpha}}{|\cos(\pi\alpha)|\Gamma^2(2-\alpha)(3-2\alpha)}.$$

Furthermore, we have

$$(\nabla f(x), x) - 2f(x) = \frac{2|x|^2}{1+|x|^2} - 2\ln(1+|x|^2) \to -\infty$$

as $|x| \to +\infty$. Therefore, we have

$$(\nabla F(t,x), x) - 2F(t,x)$$

$$= \kappa \frac{2|x|^2}{1+|x|^4} f(x)(2 + \sin t) + \kappa[(\nabla f(x), x) - 2f(x)](2 + \sin t)\arctan|x|^2$$

$$\to -\infty$$

uniformly for all $t \in [0, 2\pi]$ as $|x| \to +\infty$. Thus (A6)' and (A7)' hold. By virtue of Theorem 6.7, we conclude that BVP (6.1) has at least one solution on $E^\alpha$.

If $f(x) = |x|^2 - \ln(1 + |x|^2)$, then exact the same conclusions as above hold true by Theorem 6.6.

## 6.3 Multiple Solutions for BVP with Parameters

### 6.3.1 *Introduction*

In this section, we study the existence of three solutions to BVP of the form

$$\begin{cases} \dfrac{d}{dt}\left(\dfrac{1}{2}\,_0D_t^{-\beta}(u'(t)) + \dfrac{1}{2}\,_tD_T^{-\beta}(u'(t))\right) + \lambda\nabla F(t, u(t)) = 0, & t \in [0, T], \\ u(0) = u(T) = 0, \end{cases} \tag{6.48}$$

where $T > 0$, $\lambda > 0$ is a parameter, $0 \le \beta < 1$, $_0D_t^{-\beta}$ and $_tD_T^{-\beta}$ are the left and right Riemann-Liouville fractional integrals of order $\beta$, respectively, $N \ge 1$ is an integer, $F : [0, T] \times \mathbb{R}^N \to \mathbb{R}$ is a given function such that $F(t, \mathbf{x})$ is measurable in $t$ for each $\mathbf{x} = (x_1, \ldots, x_N) \in \mathbb{R}^N$ and continuously differentiable in $\mathbf{x}$ for a.e. $t \in [0, T]$, $F(t, 0, \ldots, 0) \equiv 0$ on $[0, T]$, and $\nabla F(t, \mathbf{x}) = (\partial F/\partial x_1, \ldots, \partial F/\partial x_N)$ is the gradient of $F$ at $\mathbf{x}$. By a solution of (6.48), we mean an absolutely continuous function $u : [0, T] \to \mathbb{R}^N$ such that $u(t)$ satisfies both equation for a.e. $t \in [0, T]$ and the boundary conditions in (6.48). We notice that when $\beta = 0$, problem (6.48) has the form

$$\begin{cases} u''(t) + \lambda\nabla F(t, u(t)) = 0, & t \in [0, T], \\ u(0) = u(T) = 0, \end{cases} \tag{6.49}$$

which has been extensively studied.

The equation in (6.48) is motivated by the steady fractional advection dispersion equation studied in Ervin and Roop, 2006,

$$- D\, a\, (p\,_0 D_t^{-\beta} + q\,_t D_T^{-\beta}) Du + b(t) Du + c(t) u = f, \tag{6.50}$$

where $D$ represents a single spatial derivative, $0 \leq p, q \leq 1$ satisfying $p + q = 1$, $a > 0$ is a constant, and $b$, $c$, $f$ are functions satisfying some suitable conditions. The interest in (6.50) arises from its application as a model for physical phenomena exhibiting anomalous diffusion; i.e., diffusion not accurately modeled by the usual advection dispersion equation. Anomalous diffusion has been used in modeling turbulent flow (see Carreras, Lynch and Zaslavsky, 2001; Shlesinger, West and Klafter, 1987) , and chaotic dynamics of classical conservative systems (see, Zaslavsky, Stevens and Weitzner, 1993). The reader may find more background information and applications on (6.50) in Benson, Wheatcraft and Meerschaert, 2000a; Ervin and Roop, 2006.

**Example 6.3.** When $N = 1$, problem (6.48) reduces to the scalar BVP

$$\begin{cases} \dfrac{d}{dt} \left( \dfrac{1}{2}\,_0 D_t^{-\beta}(u'(t)) + \dfrac{1}{2}\,_t D_T^{-\beta}(u'(t)) \right) + \lambda f(t, u(t)) = 0, \quad t \in [0, T], \\ u(0) = u(T) = 0, \end{cases} \tag{6.51}$$

where $f : [0, T] \times \mathbb{R} \to \mathbb{R}$ is such that $f(t, x)$ is measurable in $t$ for each $x \in \mathbb{R}$ and continuous in $x$ for a.e. $t \in [0, T]$.

It is clear that the equation in (6.51) is of the special form of (6.50) with $D = d/dt$, $a = 1$, $p = q = \frac{1}{2}$, $b(t) = c(t) = 0$, and $f = \lambda f(t, u)$.

We also notice that since (6.50) is the steady fractional advection dispersion equation, it has no dependence on the time variable and it just depends on the space variable $t$ (here, the notation $t$ stands for the space variable in (6.50)). Since the space we studied is one dimensional and has the form of an interval, say $[0, T]$, the boundary conditions in the space reduce to the conditions at the two endpoints $t = 0$ and $t = T$ of the interval. In Subsection 6.3.2, we discuss the existence of Dirichlet type boundary conditions.

## 6.3.2 *Existence*

For $0 \leq \beta < 1$ given in (6.48), let $\alpha = 1 - \frac{\beta}{2} \in (\frac{1}{2}, 1]$ and define

$$\rho_\alpha = \frac{16N}{T^2 \Gamma^2(2 - \alpha)} \left( \frac{1}{3 - 2\alpha} \left( \frac{T}{4} \right)^{3 - 2\alpha} + \int_{T/4}^{3T/4} g^2(t)\, dt + \int_{3T/4}^{T} h^2(t)\, dt \right), \tag{6.52}$$

where

$$g(t) = t^{1-\alpha} - (t - T/4)^{1-\alpha}, \tag{6.53}$$
$$h(t) = t^{1-\alpha} - (t - T/4)^{1-\alpha} - (t - 3T/4)^{1-\alpha}. \tag{6.54}$$

In the remainder of this section, for some $c, d, l, m, p \in \mathbb{R}$, let the bold letters $\mathbf{c}$, $\mathbf{d}$, $\mathbf{l}$, $\mathbf{m}$, and $\mathbf{p}$ be the constant vectors in $\mathbb{R}^N$ defined by

$$\mathbf{c} = (c, \ldots, c), \quad \mathbf{d} = (d, \ldots, d), \quad \mathbf{l} = (l, \ldots, l), \quad \mathbf{m} = (m, \ldots, m), \quad \mathbf{p} = (p, \ldots, p),$$

and any other bold letter, such as $\mathbf{x}$, is used to denote an arbitrary vector in $\mathbb{R}^N$.

Let $E^\alpha$ be the space of functions $u \in L^2([0, T], \mathbb{R}^N)$ having an $\alpha$-order Caputo fractional derivatives ${}_0^C D_t^\alpha u \in L^2([0, T], \mathbb{R}^N)$ and $u(0) = u(T) = 0$. Then, by Remark 6.1 (i) and Proposition 6.1, $E^\alpha$ is a reflexive and separable Banach space with the norm

$$\|u\|_\alpha = \left( \int_0^T |u(t)|^2 dt + \int_0^T |{}_0^C D_t^\alpha u(t)|^2 dt \right)^{\frac{1}{2}} \qquad \text{for any } u \in E^\alpha.$$

We see that the norm $\|u\|_\alpha$ is equivalent to the norm defined as the follow

$$\|u\|_\alpha = \left( \int_0^T |{}_0^C D_t^\alpha u(t)|^2 dt \right)^{\frac{1}{2}} \qquad \text{for any } u \in E^\alpha.$$

We recall the norms

$$\|u\|_{L^2} = \left( \int_0^T |u(t)|^2 dt \right)^{\frac{1}{2}} \qquad \text{and} \qquad \|u\| = \max_{t \in [0,T]} |u(t)|.$$

For $u \in E^\alpha$, let the functionals $\Phi$ and $\Psi$ be defined as follows

$$\Phi(u) = -\frac{1}{2} \int_0^T \left( {}_0^C D_t^\alpha u(t), {}_t^C D_T^\alpha u(t) \right) dt, \tag{6.55}$$

$$\Psi(u) = \int_0^T F(t, u(t)) dt. \tag{6.56}$$

Then, by Theorem 6.1, we see that $\Phi$ and $\Psi$ are continuously differentiable, and for any $u, v \in E^\alpha$, we have

$$\langle \Phi'(u), v \rangle = -\frac{1}{2} \int_0^T \left[ \left( {}_0^C D_t^\alpha u(t), {}_t^C D_T^\alpha v(t) \right) + \left( {}_t^C D_T^\alpha u(t), {}_0^C D_t^\alpha v(t) \right) \right] dt, \tag{6.57}$$

$$\langle \Psi'(u), v \rangle = \int_0^T \left( \nabla F(t, u(t)), v(t) \right) dt.$$

Parts (i) and (ii) of Lemma 6.6 below are taken from Lemma 6.2 and Theorem 6.2, respectively.

**Lemma 6.6.** *We have that*

**(i)** *The functional $\Phi$ is convex and continuous on $E^\alpha$.*

**(ii)** *If $u \in E^\alpha$ is a critical point of the functional $\Phi - \lambda\Psi$, then $u$ is a solution of BVP (6.48).*

We now state the results of this section.

**Theorem 6.8.** *Assume that there exist four positive constants $c, d, l$ and $m$, with*

$$d < m \quad \text{and} \quad c < \frac{T^{\alpha - \frac{1}{2}} \rho_\alpha^{\frac{1}{2}} d}{\Gamma(\alpha)(2\alpha - 1)^{\frac{1}{2}}} < |\cos(\pi\alpha)| l < |\cos(\pi\alpha)| m, \tag{6.58}$$

*such that*

$$F(t, \mathbf{x}) \geq 0, \quad \text{for } (t, \mathbf{x}) \in [0, T] \times [-m, m]^N, \tag{6.59}$$

$$\max_{|\mathbf{x}| \leq c} F(t, \mathbf{x}) \leq F(t, \mathbf{c}), \quad \max_{|\mathbf{x}| \leq l} F(t, \mathbf{x}) \leq F(t, \mathbf{l}), \quad \max_{|\mathbf{x}| \leq m} F(t, \mathbf{x}) \leq F(t, \mathbf{m}), \tag{6.60}$$

$$\frac{\int_0^T F(t, \mathbf{c}) dt}{c^2} < \frac{\Gamma^2(\alpha) \cos^2(\pi\alpha)(2\alpha - 1)}{T^{2\alpha - 1} \rho_\alpha d^2} \left( \int_{T/4}^{3T/4} F(t, \mathbf{d}) dt - \int_0^T F(t, \mathbf{c}) dt \right), \tag{6.61}$$

$$\frac{\int_0^T F(t, \mathbf{l}) dt}{l^2} < \frac{\Gamma^2(\alpha) \cos^2(\pi\alpha)(2\alpha - 1)}{T^{2\alpha - 1} \rho_\alpha d^2} \left( \int_{T/4}^{3T/4} F(t, \mathbf{d}) dt - \int_0^T F(t, \mathbf{c}) dt \right), \tag{6.62}$$

$$\frac{\int_0^T F(t, \mathbf{m}) dt}{m^2 - l^2} < \frac{\Gamma^2(\alpha) \cos^2(\pi\alpha)(2\alpha - 1)}{T^{2\alpha - 1} \rho_\alpha d^2} \left( \int_{T/4}^{3T/4} F(t, \mathbf{d}) dt - \int_0^T F(t, \mathbf{c}) dt \right). \tag{6.63}$$

*Then, for each* $\lambda \in (\underline{\lambda}, \bar{\lambda})$*, the system* (6.48) *has at least three solutions* $u_1$, $u_2$ *and* $u_3$ *such that* $\max_{t \in [0,T]} |u_1(t)| < c$*,* $\max_{t \in [0,T]} |u_2(t)| < l$*, and* $\max_{t \in [0,T]} |u_3(t)| < m$*, where*

$$\underline{\lambda} = \frac{\rho_\alpha d^2}{2|\cos(\pi\alpha)|(\int_{T/4}^{3T/4} F(t, \mathbf{d}) dt - \int_0^T F(t, \mathbf{c}) dt)} \tag{6.64}$$

*and*

$$\bar{\lambda} = \min \left\{ \frac{\Gamma^2(\alpha)(2\alpha - 1)|\cos(\pi\alpha)|c^2}{2T^{2\alpha - 1} \int_0^T F(t, \mathbf{c}) dt}, \frac{\Gamma^2(\alpha)(2\alpha - 1)|\cos(\pi\alpha)|l^2}{2T^{2\alpha - 1} \int_0^T F(t, \mathbf{l}) dt}, \right.$$
$$\left. \frac{\Gamma^2(\alpha)(2\alpha - 1)|\cos(\pi\alpha)|(m^2 - l^2)}{2T^{2\alpha - 1} \int_0^T F(t, \mathbf{m}) dt} \right\}. \tag{6.65}$$

**Proof.** For any $x \in \mathbb{R}$, let $p(x) = \max\{-m, \min\{x, m\}\}$. For any $\mathbf{x} = (x_1, \ldots, x_N) \in E^\alpha$, let $\tilde{F}(t, \mathbf{x}) = F(t, \tilde{\mathbf{x}})$, where $\tilde{\mathbf{x}} = (p(x_1), \ldots, p(x_N))$. Then, $\tilde{F}(t, \mathbf{x})$ is measurable in $t$ for each $\mathbf{x} \in \mathbb{R}^N$ and continuously differentiable in $\mathbf{x}$ for a.e. $t \in [0, T]$, and $\tilde{F}(t, 0, \ldots, 0) = 0$ on $[0, T]$. Note that $-m \leq p(u_i) \leq m$ for any $u = (u_1, \ldots, u_N) \in E^\alpha$ and $i = 1, \ldots, N$. Then, (6.59) implies that

$$\tilde{F}(t, u) \geq 0, \quad \text{for } (t, u) \in [0, T] \times E^\alpha. \tag{6.66}$$

Note that $d < m$ and $c < l < m$ by (6.58). Then, we have

$$\tilde{F}(t, \mathbf{x}) = F(t, \mathbf{x}), \quad \text{for } (t, \mathbf{x}) \in [0, T] \times \mathbb{R}^N \text{ with } |\mathbf{x}| < m,$$
$$\tilde{F}(t, \mathbf{c}) = F(t, \mathbf{c}), \quad \tilde{F}(t, \mathbf{d}) = F(t, \mathbf{d}), \tag{6.67}$$
$$\tilde{F}(t, \mathbf{l}) = F(t, \mathbf{l}), \quad \tilde{F}(t, \mathbf{m}) = F(t, \mathbf{m}).$$

Let the continuously differentiable functional $\Phi$ be given by (6.55) and the functional $\tilde{\Psi}$ be defined by

$$\tilde{\Psi}(u) = \int_0^T \tilde{F}(t, u(t)) dt, \quad \text{for } u \in E^\alpha. \tag{6.68}$$

Then, by Proposition 6.4 and (6.55), we have

$$\frac{1}{2}|\cos(\pi\alpha)|\,\|u\|_\alpha^2 \le \Phi(u) \le \frac{1}{2|\cos(\pi\alpha)|}\|u\|_\alpha^2, \quad \text{for } u \in E^\alpha. \tag{6.69}$$

Moreover, $\tilde{\Psi}$ is continuously differentiable, and for any $u, v \in E^\alpha$, in view of (6.66), we have

$$\tilde{\Psi}(u) \ge 0 \quad \text{and} \quad \langle \tilde{\Psi}'(u), v \rangle = \int_0^T \left( \nabla \tilde{F}(t, u(t)), v(t) \right) dt. \tag{6.70}$$

In the following, we will apply Lemma 1.6 with $X = E^\alpha$ to the functionals $\Phi$ and $\tilde{\Psi}$.

We first show that some basic assumptions of Lemma 1.6 are satisfied. The convexity and coercivity of $\Phi$ follow from Lemma 6.6(i) and (6.69), respectively. For any $u, v \in E^\alpha$, from Proposition 6.4 and (6.57),

$$\langle \Phi'(u) - \Phi'(v), u - v \rangle$$

$$= -\frac{1}{2}\int_0^T [({}_0^C D_t^\alpha u(t), {}_t^C D_T^\alpha(u(t) - v(t))) + ({}_t^C D_T^\alpha u(t), {}_0^C D_t^\alpha(u(t) - v(t)))]\,dt$$

$$+ \frac{1}{2}\int_0^T [({}_0^C D_t^\alpha v(t), {}_t^C D_T^\alpha(u(t) - v(t))) + ({}_t^C D_T^\alpha v(t), {}_0^C D_t^\alpha(u(t) - v(t)))]\,dt$$

$$= -\int_0^T ({}_0^C D_t^\alpha(u(t) - v(t)), {}_t^C D_T^\alpha(u(t) - v(t)))\,dt$$

$$\ge |\cos(\pi\alpha)|\,\|u - v\|_\alpha^2.$$

Thus, $\Phi'$ is uniformly monotone. Hence, by Theorem 26.A(d) in Zeidler, 1990, $(\Phi')^{-1} : (E^\alpha)^* \to E^\alpha$ exists and is continuous. Suppose that $u_n \rightharpoonup u \in E^\alpha$. Then, by Proposition 6.3 $u_n \to u$ in $C([0, T], \mathbb{R}^N)$. Since $\tilde{F}(t, \mathbf{x})$ is continuously differentiable in $\mathbf{x}$ for a.e. $t \in [0, 1]$, from the derivative formula in (6.70), we have $\tilde{\Psi}'(u_n) \to \tilde{\Psi}'(u)$, i.e., $\tilde{\Psi}'$ is strongly continuous. Therefore, $\tilde{\Psi}'$ is a compact operator by Proposition 26.2 in Zeidler, 1990.

Next, note that the facts that $\tilde{F}(t, 0, \dots, 0) = 0$ on $[0, T]$ and the inequality in (6.70), from Proposition 6.4, (6.55) and (6.68), we see that conditions (i) and (ii) of Lemma 1.6 are satisfied.

Now, we show that condition (iii) of Lemma 1.6 holds. For $i = 1, \dots, N$, let

$$w_i(t) = \begin{cases} \dfrac{4d}{T}t, & t \in [0, T/4), \\[2mm] d, & t \in [T/4, 3T/4], \\[2mm] \dfrac{4d}{T}(T - t), & t \in (3T/4, T], \end{cases}$$

and $w(t) = (w_1(t), \dots, w_N(t))$. Then, $w \in E^\alpha$ and

$$_0^C D_t^\alpha w_i(t) = \frac{4d}{T\Gamma(2 - \alpha)} \begin{cases} t^{1-\alpha}, & t \in [0, T/4), \\[2mm] g(t), & t \in [T/4, 3T/4], \\[2mm] h(t), & t \in (3T/4, T], \end{cases} \tag{6.71}$$

where $g(t)$ and $h(t)$ are defined by (6.53) and (6.54). From (6.52) and (6.71),

$$\int_0^T |{}_0^C D_t^\alpha w(t)|^2 dt$$

$$= N\left(\int_0^T |{}_0^C D_t^\alpha w_1(t)|^2 dt + \int_{T/4}^{3T/4} |{}_0^C D_t^\alpha w_1(t)|^2 dt + \int_{3T/4}^T |{}_0^C D_t^\alpha w_1(t)|^2 dt\right)$$

$$= \frac{16Nd^2}{T^2 \Gamma^2(2-\alpha)}\left(\int_0^{T/4} t^{2-2\alpha} dt + \int_{T/4}^{3T/4} g^2(t) dt + \int_{3T/4}^T |h(t)|^2 dt\right)$$

$$= \frac{16Nd^2}{T^2 \Gamma^2(2-\alpha)}\left(\frac{1}{3-2\alpha}\left(\frac{T}{4}\right)^{3-2\alpha} + \int_{T/4}^{3T/4} g^2(t) dt + \int_{3T/4}^T h^2(t) dt\right)$$

$$= \rho_\alpha d^2.$$

Then, $\|w\|_\alpha^2 = \rho_\alpha d^2$. Thus, from (6.69) with $u = w$,

$$\frac{1}{2}|\cos(\pi\alpha)|\rho_\alpha d^2 \leq \Phi(w) \leq \frac{1}{2|\cos(\pi\alpha)|}\rho_\alpha d^2. \tag{6.72}$$

Let

$$r_1 = \frac{\Gamma^2(\alpha)(2\alpha-1)|\cos(\pi\alpha)|}{2T^{2\alpha-1}}c^2, \quad r_2 = \frac{\Gamma^2(\alpha)(2\alpha-1)|\cos(\pi\alpha)|}{2T^{2\alpha-1}}l^2,$$

$$r_3 = \frac{\Gamma^2(\alpha)(2\alpha-1)|\cos(\pi\alpha)|}{2T^{2\alpha-1}}(m^2 - l^2). \tag{6.73}$$

Then, from (6.58) and (6.72), we have $r_1 < \Phi(w) < r_2$ and $r_3 > 0$. For any $u \in E^\alpha$, from the first inequality in (6.69), we see that $\|u\|_\alpha^2 \leq 2\Phi(u)/|\cos(\pi\alpha)|$. Then, by (6.14) and (6.17), we have

$$\|u\|^2 \leq \frac{T^{2\alpha-1}}{\Gamma^2(\alpha)(2\alpha-1)}\|u\|_\alpha^2 \leq \frac{2T^{2\alpha-1}\Phi(u)}{\Gamma^2(\alpha)(2\alpha-1)|\cos(\pi\alpha)|}.$$

Thus, by (6.73), we have the following implications:

$$\Phi(u) < r_1 \Rightarrow \|u\| < c,$$
$$\Phi(u) < r_2 \Rightarrow \|u\| < l, \tag{6.74}$$
$$\Phi(u) < r_2 + r_3 \Rightarrow \|u\| < m.$$

This, together with (6.60) and (6.67), implies

$$\sup_{u\in\Phi^{-1}(-\infty,r_1)} \int_0^T \tilde{F}(t, u(t)) dt \leq \int_0^T \max_{|\mathbf{x}|\leq c} F(t, \mathbf{x}) dt \leq \int_0^T F(t, \mathbf{c}) dt,$$

$$\sup_{u\in\Phi^{-1}(-\infty,r_2)} \int_0^T \tilde{F}(t, u(t)) dt \leq \int_0^T \max_{|\mathbf{x}|\leq l} F(t, \mathbf{x}) dt \leq \int_0^T F(t, \mathbf{l}) dt, \tag{6.75}$$

$$\sup_{u\in\Phi^{-1}(-\infty,r_2+r_3)} \int_0^T \tilde{F}(t, u(t)) dt \leq \int_0^T \max_{|\mathbf{x}|\leq m} F(t, \mathbf{x}) dt \leq \int_0^T F(t, \mathbf{m}) dt.$$

Let $\varphi$, $\beta$, $\gamma$ and $\alpha$ be defined by (1.16)-(1.19). Then, taking into account the fact that $0 \in \Phi^{-1}(-\infty, r_i)$, $i = 1, 2$, from (6.68) and (6.73), it follows that

$$\varphi(r_1) \leq \frac{\sup_{u\in\Phi^{-1}(-\infty,r_1)} \tilde{\Psi}(u)}{r_1} \leq \frac{2T^{2\alpha-1}\int_0^T F(t, \mathbf{c}) dt}{\Gamma^2(\alpha)(2\alpha-1)|\cos(\pi\alpha)|c^2}, \tag{6.76}$$

$$\varphi(r_2) \leq \frac{\sup_{u \in \Phi^{-1}(-\infty, r_2)} \tilde{\Psi}(u)}{r_2} \leq \frac{2T^{2\alpha - 1} \int_0^T F(t, 1) dt}{\Gamma^2(\alpha)(2\alpha - 1)|\cos(\pi\alpha)|l^2}, \qquad (6.77)$$

$$\gamma(r_2, r_3) = \frac{\sup_{u \in \Phi^{-1}(-\infty, r_2 + r_3)} \tilde{\Psi}(u)}{r_3} \leq \frac{2T^{2\alpha - 1} \int_0^T F(t, \mathbf{m}) dt}{\Gamma^2(\alpha)(2\alpha - 1)|\cos(\pi\alpha)|(m^2 - l^2)}. \qquad (6.78)$$

On the other hand, in view of the fact that $w(t) = \mathbf{d} < \mathbf{m}$ on $[T/4, 3T/4]$ and from (6.66) and (6.67),

$$\int_0^T \tilde{F}(t, w(t)) dt \geq \int_{T/4}^{3T/4} \tilde{F}(t, w(t)) dt = \int_{T/4}^{3T/4} \tilde{F}(t, \mathbf{d}) dt.$$

Note that $w \in \Phi^{-1}[r_1, r_2)$, from (1.17) and (6.75), we obtain

$$\beta(r_1, r_2) \geq \inf_{u \in \Phi^{-1}(-\infty, r_1)} \frac{\tilde{\Psi}(w) - \tilde{\Psi}(u)}{\Phi(w) - \Phi(u)}$$

$$\geq \inf_{u \in \Phi^{-1}(-\infty, r_1)} \frac{\tilde{\Psi}(w) - \tilde{\Psi}(u)}{\Phi(w)}$$

$$\geq \frac{\int_{T/4}^{3T/4} \tilde{F}(t, \mathbf{d}) dt - \int_0^T \tilde{F}(t, \mathbf{c}) dt}{\Phi(w)}.$$

By (6.72), $1/\Phi(w) \geq 2|\cos(\pi\alpha)|/(\rho_\alpha d^2)$. Then

$$\beta(r_1, r_2) \geq \frac{2|\cos(\pi\alpha)|}{\rho_\alpha d^2} \left( \int_{T/4}^{3T/4} \tilde{F}(t, \mathbf{d}) dt - \int_0^T \tilde{F}(t, \mathbf{c}) dt \right). \qquad (6.79)$$

For $\underline{\lambda}$ and $\bar{\lambda}$ defined by (6.64) and (6.65), from (6.61)-(6.63) and (6.76)-(6.79), we have

$$\varphi(r_1) < \frac{1}{\bar{\lambda}} < \frac{1}{\underline{\lambda}} < \beta(r_1, r_2),$$

$$\varphi(r_2) < \frac{1}{\bar{\lambda}} < \frac{1}{\underline{\lambda}} < \beta(r_1, r_2),$$

$$\gamma(r_2, r_3) < \frac{1}{\bar{\lambda}} < \frac{1}{\underline{\lambda}} < \beta(r_1, r_2).$$

In view of (1.19), $\alpha(r_1, r_2, r_3) < 1/\bar{\lambda} < 1/\underline{\lambda} < \beta(r_1, r_2)$; i.e., condition (iii) of Lemma 1.6 holds. Hence, all the assumptions of Lemma 1.6 are satisfied. Then, by Lemma 1.6, for each $\lambda \in (\underline{\lambda}, \bar{\lambda})$, the functional $\Phi - \lambda\tilde{\Psi}$ has three distinct critical points $u_1$, $u_2$ and $u_3$ such that $u_1 \in \Phi^{-1}(-\infty, r_1)$, $u_2 \in \Phi^{-1}[r_1, r_2)$, and $u_3 \in \Phi^{-1}(-\infty, r_2 + r_3)$. From (6.74), we have

$$\|u_1\| < c, \quad \|u_2\| < l, \quad \|u_3\| < m.$$

Then, in view of (6.65), (6.67) and (6.68), we have $\tilde{\Psi}(u) = \Psi(u)$. Therefore, $u_1$, $u_2$ and $u_3$ are three distinct critical points of the functional $\Phi - \lambda\Psi$. Thus, by Proposition 6.6 (ii), $u_1$, $u_2$ and $u_3$ are three distinct solutions of (6.48). This completes the proof of the theorem. $\qquad \square$

The following results are consequences of Theorem 6.8. In particular, Corollaries 6.2 and 6.4 give some conditions for the system (6.49) to have at least three solutions, and Corollary 6.3 provide some relatively simpler existence criteria for the system (6.48).

**Corollary 6.2.** *Assume that there exist four positive constants $c$, $d$, $l$ and $m$, with*

$$c < (8N)^{\frac{1}{2}}d < l < m,$$

*such that (6.59) and (6.60) hold, and*

$$\frac{\int_0^T F(t, \mathbf{c})dt}{c^2} < \frac{1}{8Nd^2}\left(\int_{T/4}^{3T/4} F(t, \mathbf{d})dt - \int_0^T F(t, \mathbf{c})dt\right),$$

$$\frac{\int_0^T F(t, \mathbf{l})dt}{l^2} < \frac{1}{8Nd^2}\left(\int_{T/4}^{3T/4} F(t, \mathbf{d})dt - \int_0^T F(t, \mathbf{c})dt\right),$$

$$\frac{\int_0^T F(t, \mathbf{m})dt}{m^2 - l^2} < \frac{1}{8Nd^2}\left(\int_{T/4}^{3T/4} F(t, \mathbf{d})dt - \int_0^T F(t, \mathbf{c})dt\right).$$

*Then, for each $\lambda \in (\underline{\lambda}_1, \bar{\lambda}_1)$, system (6.49) has at least three solutions $u_1$, $u_2$, and $u_3$ such that $\max_{t\in[0,T]}|u_1(t)| < c$, $\max_{t\in[0,T]}|u_2(t)| < l$, and $\max_{t\in[0,T]}|u_3(t)| < m$, where*

$$\underline{\lambda}_1 = \frac{4Nd^2}{T(\int_{T/4}^{3T/4} F(t, \mathbf{d})dt - \int_0^T F(t, \mathbf{c})dt)},$$

$$\bar{\lambda}_1 = \min\left\{\frac{c^2}{2T\int_0^T F(t, \mathbf{c})dt}, \frac{l^2}{2T\int_0^T F(t, \mathbf{l})dt}, \frac{m^2 - l^2}{2T\int_0^T F(t, \mathbf{m})dt}\right\}.$$

**Proof.** When $\alpha = 1$, from (6.52), we have $\rho_\alpha = 8N/T$. Then, under the assumptions of Corollary 6.2, it is easy to see that all the conditions of Theorem 6.8 hold for $\alpha = 1$. Note that the system (6.49) is a special case of the system (6.48) with $\alpha = 1$. The conclusion then follows directly from Theorem 6.8. The proof is completed. $\square$

**Corollary 6.3.** *Assume that there exist three positive constants $c$, $d$ and $p$, with*

$$d < p \quad and \quad c < \frac{T^{\alpha - \frac{1}{2}}\rho_\alpha^{\frac{1}{2}}d}{\Gamma(\alpha)(2\alpha - 1)^{\frac{1}{2}}} < \frac{|\cos(\pi\alpha)|p}{\sqrt{2}}, \tag{6.80}$$

*such that*

$$F(t, \mathbf{x}) \geq 0 \quad for (t, \mathbf{x}) \in [0, T] \times [-p, p]^N, \tag{6.81}$$

$$\max_{|\mathbf{x}|\leq c} F(t, \mathbf{x}) \leq F(t, \mathbf{c}), \quad \max_{|\mathbf{x}|\leq p/\sqrt{2}} F(t, \mathbf{x}) \leq F(t, \frac{\mathbf{p}}{\sqrt{2}}), \quad \max_{|\mathbf{x}|\leq p} F(t, \mathbf{x}) \leq F(t, \mathbf{p}), \tag{6.82}$$

$$\frac{\int_0^T F(t, \mathbf{c})dt}{c^2} < \frac{\Gamma^2(\alpha)\cos^2(\pi\alpha)(2\alpha - 1)}{T^{2\alpha-1}\rho_\alpha d^2(1 + \cos^2(\pi\alpha))}\int_{T/4}^{3T/4} F(t, \mathbf{d})dt, \tag{6.83}$$

$$\frac{\int_0^T F(t,\mathbf{p})dt}{p^2} < \frac{\Gamma^2(\alpha)\cos^2(\pi\alpha)(2\alpha-1)}{2T^{2\alpha-1}\rho_\alpha d^2(1+\cos^2(\pi\alpha))} \int_{T/4}^{3T/4} F(t,\mathbf{d})dt. \tag{6.84}$$

*Then, for each* $\lambda \in (\underline{\lambda}_2, \bar{\lambda}_2)$*, system* (6.48) *has at least three solutions* $u_1$, $u_2$, *and* $u_3$ *such that* $\max_{t\in[0,T]}|u_1(t)| < c$, $\max_{t\in[0,T]}|u_2(t)| < p/\sqrt{2}$, *and* $\max_{t\in[0,T]}|u_3(t)| < p$, *where*

$$\underline{\lambda}_2 = \frac{\rho_\alpha d^2(1+\cos^2(\pi\alpha))}{2|\cos(\pi\alpha)|\int_{T/4}^{3T/4} F(t,\mathbf{d})dt}, \tag{6.85}$$

$$\bar{\lambda}_2 = \min\left\{\frac{\Gamma^2(\alpha)(2\alpha-1)|\cos(\pi\alpha)|c^2}{2T^{2\alpha-1}\int_0^T F(t,\mathbf{c})dt}, \frac{\Gamma^2(\alpha)(2\alpha-1)|\cos(\pi\alpha)|p^2}{4T^{2\alpha-1}\int_0^T F(t,\mathbf{p})dt}\right\}. \tag{6.86}$$

**Proof.** Let $l = p/\sqrt{2}$ and $m = p$. Then, from (6.80)-(6.82), we see that (6.58)-(6.60) hold. By (6.82) and (6.84), we have

$$\frac{\int_0^T F(t,\mathbf{l})dt}{l^2} = \frac{2\int_0^T F(t,\mathbf{p}/\sqrt{2})dt}{p^2} \le \frac{2\int_0^T F(t,\mathbf{p})dt}{p^2}$$
$$< \frac{\Gamma^2(\alpha)\cos^2(\pi\alpha)(2\alpha-1)}{T^{2\alpha-1}\rho_\alpha d^2(1+\cos^2(\pi\alpha))}\int_{T/4}^{3T/4} F(t,\mathbf{d})dt, \tag{6.87}$$

and

$$\frac{\int_0^T F(t,\mathbf{m})dt}{m^2-l^2} = \frac{2\int_0^T F(t,\mathbf{p})dt}{p^2}$$
$$< \frac{\Gamma^2(\alpha)\cos^2(\pi\alpha)(2\alpha-1)}{T^{2\alpha-1}\rho_\alpha d^2(1+\cos^2(\pi\alpha))}\int_{T/4}^{3T/4} F(t,\mathbf{d})dt. \tag{6.88}$$

Note from (6.80) it follows that

$$\frac{\Gamma^2(\alpha)(2\alpha-1)}{T^{2\alpha-1}\rho_\alpha d^2} < \frac{1}{c^2}.$$

Combining this inequality with (6.83), we obtain

$$\frac{\Gamma^2(\alpha)\cos^2(\pi\alpha)(2\alpha-1)}{T^{2\alpha-1}\rho_\alpha d^2}\left(\int_{T/4}^{3T/4} F(t,\mathbf{d})dt - \int_0^T F(t,\mathbf{c})dt\right)$$
$$> \frac{\Gamma^2(\alpha)\cos^2(\pi\alpha)(2\alpha-1)}{T^{2\alpha-1}\rho_\alpha d^2}\int_{T/4}^{3T/4} F(t,\mathbf{d})dt - \frac{\cos^2(\pi\alpha)}{c^2}\int_0^T F(t,\mathbf{c})dt$$
$$> \frac{\Gamma^2(\alpha)\cos^2(\pi\alpha)(2\alpha-1)}{T^{2\alpha-1}\rho_\alpha d^2}\int_{T/4}^{3T/4} F(t,\mathbf{d})dt$$
$$- \frac{\Gamma^2(\alpha)\cos^4(\pi\alpha)(2\alpha-1)}{T^{2\alpha-1}\rho_\alpha d^2(1+\cos^2(\pi\alpha))}\int_{T/4}^{3T/4} F(t,\mathbf{c})dt$$
$$= \frac{\Gamma^2(\alpha)\cos^2(\pi\alpha)(2\alpha-1)}{T^{2\alpha-1}\rho_\alpha d^2(1+\cos^2(\pi\alpha))}\int_{T/4}^{3T/4} F(t,\mathbf{d})dt. \tag{6.89}$$

By (6.83) and (6.87)-(6.89), we see that (6.61)-(6.62) hold. From (6.64), (6.65), (6.85), (6.86) and (6.89), we have $\underline{\lambda} < \underline{\lambda}_2$ and $\bar{\lambda} = \bar{\lambda}_2$. Therefore, the conclusion now follows from Theorem 6.8. The proof is completed. $\qquad\square$

**Corollary 6.4.** *Assume that there exist three positive constants c, d and p, with*

$$c < (8N)^{\frac{1}{2}}d < \frac{p}{\sqrt{2}}, \tag{6.90}$$

*such that (6.81) and (6.82) hold, and*

$$\frac{\int_0^T F(t,\mathbf{c})dt}{c^2} < \frac{1}{16Nd^2} \int_{T/4}^{3T/4} F(t,\mathbf{d})dt, \tag{6.91}$$

*and*

$$\frac{\int_0^T F(t,\mathbf{p})dt}{p^2} < \frac{1}{32Nd^2} \int_{T/4}^{3T/4} F(t,\mathbf{d})dt. \tag{6.92}$$

*Then, for each $\lambda \in (\underline{\lambda}_3, \bar{\lambda}_3)$, system (6.49) has at least three solutions $u_1$, $u_2$, and $u_3$ such that $\max_{t\in[0,T]} |u_1(t)| < c$, $\max_{t\in[0,T]} |u_2(t)| < p/\sqrt{2}$, and $\max_{t\in[0,T]} |u_3(t)| < p$, where*

$$\underline{\lambda}_3 = \frac{8Nd^2}{T \int_{T/4}^{3T/4} F(t,\mathbf{d})dt},$$

$$\bar{\lambda}_3 = \min\left\{ \frac{c^2}{2T \int_0^T F(t,\mathbf{c})dt}, \frac{p^2}{4T \int_0^T F(t,\mathbf{p})dt} \right\}.$$

**Proof.** When $\alpha = 1$, from (6.52), we have $\rho_\alpha = 8N/T$. Under the assumptions of Corollary 6.4, it is easy to see that all the conditions of Corollary 6.3 hold for $\alpha = 1$. Note that system (6.49) is a special case of system (6.48) with $\alpha = 1$. The conclusion then follows directly from Corollary 6.3. The proof is completed. $\qquad\square$

**Remark 6.3.** We want to point out that when $F$ does not depend on $t$, (6.91) and (6.92) reduce to

$$\frac{F(\mathbf{c})}{c^2} < \frac{F(\mathbf{d})}{32Nd^2} \quad \text{and} \quad \frac{F(\mathbf{p})}{p^2} < \frac{F(\mathbf{d})}{64Nd^2}, \tag{6.93}$$

and $\underline{\lambda}_3$ and $\bar{\lambda}_3$ become

$$\underline{\lambda}_3 = \frac{16Nd^2}{T^2F(\mathbf{d})} \quad \text{and} \quad \bar{\lambda}_3 = \min\left\{ \frac{c^2}{2T^2F(\mathbf{c})}, \frac{p^2}{4T^2F(\mathbf{p})} \right\}. \tag{6.94}$$

**Remark 6.4.** We observe that, in our results, no asymptotic condition on $F$ is needed and only local conditions on $F$ are imposed to guarantee the existence of solutions. Moreover, in the conclusions of the above results, one of the three solutions may be trivial since $\nabla F(t, 0, \ldots, 0)$ may be zero.

In the remainder of this subsection, we give two examples to illustrate the applicability of our results.

**Example 6.4.** Let $T > 0$. For $(t, x, y) \in [0, T] \times \mathbb{R}^2$, let $F(t, x, y) = tG(x, y)$, where $G : \mathbb{R}^2 \to \mathbb{R}$ satisfies that $G(-x, -y) = G(x, y)$, and that for $x \in [0, \infty)$ and $y \in \mathbb{R}$,

$$
G(x, y) = \begin{cases}
x^3 + |y|^3, & 0 \le x \le 1,\ 0 \le |y| \le 1, \\
x^3 + 2|y|^{3/2} - 1, & 0 \le x \le 1,\ |y| > 1, \\
2x^{3/2} + |y|^3 - 1, & x > 1,\ 0 \le |y| \le 1, \\
2x^{3/2} + 2|y|^{3/2} - 2, & x > 1,\ |y| > 1.
\end{cases} \tag{6.95}
$$

It is easy to verify that $F : [0, T] \times \mathbb{R}^2 \to \mathbb{R}$ is measurable in $t$ for $(x, y) \in \mathbb{R}^2$ and continuously differentiable in $x$ and $y$ for $t \in [0, T]$, and $F(t, 0, 0) \equiv 0$ on $[0, T]$.

Let $0 \le \beta < 1$, $\alpha = 1 - \frac{\beta}{2} \in (\frac{1}{2}, 1]$, $\rho_\alpha$ be defined by (6.52), and $u(t) = (u_1(t), u_2(t))$. We claim that for each

$$
\lambda \in \left( \frac{\rho_\alpha (1 + \cos^2(\pi\alpha))}{T^2 |\cos(\pi\alpha)|}, \infty \right),
$$

the system

$$
\begin{cases}
\dfrac{d}{dt} \left( \dfrac{1}{2}\, {}_0 D_t^{-\beta}(u'(t)) + \dfrac{1}{2}\, {}_t D_T^{-\beta}(u'(t)) \right) + \lambda \nabla F(t, u(t)) = 0, & t \in [0, T], \\
u(0) = u(T) = 0
\end{cases} \tag{6.96}
$$

has at least three solutions.

In fact, system (6.96) is a special case of system (6.48) with $N = 2$. For $0 < c < 1$ and $p > 1$, in view of (6.95), we have

$$
\frac{\int_0^T F(t, c, c)dt}{c^2} = \frac{2c^3 \int_0^T t\,dt}{c^2} = T^2 c, \tag{6.97}
$$

$$
\frac{\int_0^T F(t, p, p)dt}{p^2} = \frac{(4p^{3/2} - 2) \int_0^T t\,dt}{p^2} = \frac{T^2(2p^{3/2} - 1)}{p^2}. \tag{6.98}
$$

Choose $d = 1$. Then,

$$
\int_{T/4}^{3T/4} F(t, d, d)dt = 2 \int_{T/4}^{3T/4} t\,dt = \frac{1}{2}T^2. \tag{6.99}
$$

By (6.97)-(6.99), we see that there exist $0 < c^* < 1$ and $p^* > 1$ such that (6.80), (6.83) and (6.84) hold for any $0 < c < c^*$ and $p > p^*$. Moreover, (6.81) and (6.82) hold for any $c, p > 0$. Finally, note from (6.85) and (6.86) that

$$
\lambda_2 = \frac{\rho_\alpha (1 + \cos^2(\pi\alpha))}{T^2 |\cos(\pi\alpha)|},
$$

$\bar{\lambda}_2 \to \infty$ as $c \to 0^+$ and $p \to \infty$.

Then, the claim follows from Corollary 6.3.

**Example 6.5.** Let $F : \mathbb{R}^2 \to \mathbb{R}$ satisfy that $F(-x, -y) = F(x, y)$, and that for $x \in [0, \infty)$ and $y \in \mathbb{R}$,

$$F(x, y) = \begin{cases} x^3, & 0 \leq x \leq 1, \ 0 \leq |y| \leq 1, \\ x^3 + 2|y|^{3/2} - 3|y| + 1, & 0 \leq x \leq 1, \ |y| > 1, \\ 2x^{3/2} - 1, & x > 1, \ 0 \leq |y| \leq 1, \\ 2x^{3/2} + 2|y|^{3/2} - 3|y|, & x > 1, \ |y| > 1. \end{cases} \tag{6.100}$$

It is easy to verify that $F : \mathbb{R}^2 \to \mathbb{R}$ is continuously differentiable in $x$ and $y$ and $F(0, 0) = 0$.

Let $T > 0$ and $u(t) = (u_1(t), u_2(t))$. We claim that for each $\lambda \in (32/T^2, \infty)$, the system

$$\begin{cases} u''(t) + \lambda \nabla F(u(t)) = 0, & t \in [0, T], \\ u(0) = u(T) = 0 \end{cases} \tag{6.101}$$

has at least three solutions. In fact, the system (6.101) is a special case of the system (6.49) with $N = 2$. For $0 < c < 1$ and $p > 1$, from (6.100), we have

$$\frac{F(c, c)}{c^2} = \frac{c^3}{c^2} = c, \tag{6.102}$$

$$\frac{F(p, p)}{p^2} = \frac{4p^{3/2} - 3p}{p^2} = \frac{4p^{\frac{1}{2}} - 3}{p}. \tag{6.103}$$

Choose $d = 1$. Then

$$\frac{F(d, d)}{32 N d^2} = \frac{1}{64} \quad \text{and} \quad \frac{F(d, d)}{64 N d^2} = \frac{1}{128}. \tag{6.104}$$

By (6.102)-(6.104), we see that there exist $0 < c^* < 1$ and $p^* > 1$ such that (6.90) and (6.93) hold for any $0 < c < c^*$ and $p > p^*$. Moreover, (6.81) and (6.82) hold for any $c, p > 0$. Finally, note from (6.94) that

$$\lambda_3 = \frac{32}{T^2} \quad \text{and} \quad \bar{\lambda}_3 \to \infty, \quad \text{as } c \to 0^+ \text{ and } p \to \infty.$$

Then, the claim follows from Corollary 6.4 and Remark 6.3.

**Remark 6.5.** As noted in Remark 6.4, one of the three solutions in the conclusions of the above examples may be trivial.

## 6.4 Infinite Solutions for BVP with Left and Right Fractional Integrals

### 6.4.1 *Introduction*

In this section, we consider BVP (6.1), i.e.,

$$\begin{cases} \frac{d}{dt} \left( \frac{1}{2} \, _0D_t^{-\beta}(u'(t)) + \frac{1}{2} \, _tD_T^{-\beta}(u'(t)) \right) + \nabla F(t, u(t)) = 0, & \text{a.e. } t \in [0, T], \\ u(0) = u(T) = 0, \end{cases}$$

where $_0D_t^{-\beta}$ and $_tD_T^{-\beta}$ are the left and right Riemann-Liouville fractional integrals of order $0 \le \beta < 1$ respectively. Assume that $F : [0, T] \times \mathbb{R}^N \to \mathbb{R}$ satisfies the condition (A) which is assumed as in Subsection 6.2.3.

In particular, if $\beta = 0$, BVP (6.1) reduces to the standard second-order BVP.

In the Subsection 6.4.2, using variational methods we prove the multiplicity results for the solutions of problem (6.1).

### 6.4.2  *Existence*

Making use of the Proposition 1.4 and Definition 1.3, for any $u \in AC([0, T], \mathbb{R}^N)$, BVP (6.1) is equivalent to (6.20).

In the following, we will treat BVP (6.20) in the Hilbert space $E^\alpha = E_0^{\alpha,2}$ with the corresponding norm $\|u\|_\alpha = \|u\|_{\alpha,2}$.

As $E^\alpha$ is a reflexive and separable Banach space, then there are $e_j \in E^\alpha$ and $e_j^* \in (E^\alpha)^*$ such that

$$E^\alpha = \overline{\text{span}\{e_j : j = 1, 2, \ldots\}} \quad \text{and} \quad (E^\alpha)^* = \overline{\text{span}\{e_j^* : j = 1, 2, \ldots\}}.$$

For $k = 1, 2, \ldots$, denote

$$X_j := \text{span}\{e_j\}, \quad Y_k := \bigoplus_{j=1}^k X_j, \quad Z_k := \overline{\bigoplus_{j=k}^\infty X_j}.$$

**Theorem 6.9.** *Assume that $F(t, x)$ satisfies the condition (A), and suppose the following conditions hold:*

**(A1)** *there exist $\kappa > 2$ and $r > 0$ such that*

$$\kappa F(t, x) \le (\nabla F(t, x), x)$$

*for a.e. $t \in [0, T]$ and all $|x| \ge r$ in $\mathbb{R}^N$;*

**(A2)** *there exist positive constants $\mu > 2$ and $Q > 0$ such that*

$$\limsup_{|x| \to +\infty} \frac{F(t, x)}{|x|^\mu} \le Q$$

*uniformly for a.e. $t \in [0, T]$;*

**(A3)** *there exist $\mu' > 2$ and $Q' > 0$ such that*

$$\limsup_{|x| \to +\infty} \frac{F(t, x)}{|x|^{\mu'}} \ge Q'$$

*uniformly for a.e. $t \in [0, T]$;*

**(A4)** *$F(t, x) = F(t, -x)$ for $t \in [0, T]$ and all $x$ in $\mathbb{R}^N$.*

*Then BVP (6.1) has infinite solutions $\{u_n\}$ on $E^\alpha$ for every positive integer $n$ such that $\|u_n\|_\infty \to \infty$, as $n \to \infty$.*

**Proof.** Let $\{u_n\} \subset E^\alpha$ such that $\varphi(u_n)$ is bounded and $\varphi'(u_n) \to 0$ as $n \to \infty$, where $\varphi(u)$ and $\varphi'(u)$ are defined by (6.26) and (6.27) respectively. First we prove $\{u_n\}$ is a bounded sequence, otherwise, $\{u_n\}$ would be unbounded sequence, passing to a subsequence, still denoted by $\{u_n\}$, such that $\|u_n\|_\alpha \geq 1$ and $\|u_n\|_\alpha \to \infty$, as $n \to \infty$.

Noting that

$$\langle \varphi'(u_n), u_n \rangle = -\int_0^T \left( ({}^C_0 D_t^\alpha u_n(t), {}^C_t D_T^\alpha u_n(t)) + (\nabla F(t, u_n(t)), u_n(t)) \right) dt.$$

In view of the condition (A1) and (6.22) that

$$\varphi(u_n) - \frac{1}{\kappa} \langle \varphi'(u_n), u_n \rangle$$

$$= \left( \frac{1}{\kappa} - \frac{1}{2} \right) \int_0^T ({}^C_0 D_t^\alpha u_n(t), {}^C_t D_T^\alpha u_n(t)) dt$$

$$+ \int_0^T \left( \frac{1}{\kappa} (\nabla F(t, u_n(t)), u_n(t)) - F(t, u_n(t)) \right) dt$$

$$\geq \left( \frac{1}{2} - \frac{1}{\kappa} \right) |\cos(\pi\alpha)| \|u_n\|_\alpha^2$$

$$+ \left( \int_{\Omega_1} + \int_{\Omega_2} \right) \left( \frac{1}{\kappa} (\nabla F(t, u_n(t)), u_n(t)) - F(t, u_n(t)) \right) dt$$

$$\geq \left( \frac{1}{2} - \frac{1}{\kappa} \right) |\cos(\pi\alpha)| \|u_n\|_\alpha^2 - C_1,$$

where $\Omega_1 := \{t \in [0, T] : |u_n(t)| \leq r\}$, $\Omega_2 := [0, T] \backslash \Omega_1$ and $C_1$ is a positive constant.

Since $\varphi(u_n)$ is bounded, there exists a positive constant $C_2$, such that $|\varphi(u_n)| \leq C_2$. Hence, we have

$$C_2 \geq \varphi(u_n) \geq \left( \frac{1}{2} - \frac{1}{\kappa} \right) |\cos(\pi\alpha)| \|u_n\|_\alpha^2 + \frac{1}{\kappa} \langle \varphi'(u_n), u_n \rangle - C_1$$

$$\geq \left( \frac{1}{2} - \frac{1}{\kappa} \right) |\cos(\pi\alpha)| \|u_n\|_\alpha^2 - \frac{1}{\kappa} \|\varphi'(u_n)\|_\alpha \|u_n\|_\alpha - C_1, \quad (6.105)$$

so $\{u_n\}$ is a bounded sequence in $E^\alpha$ by (6.105).

Since $E^\alpha$ is a reflexive space, going to a subsequence if necessary, we may assume that $u_n \rightharpoonup u$ weakly in $E^\alpha$, thus we have

$$\langle \varphi'(u_n) - \varphi'(u), u_n - u \rangle = \langle \varphi'(u_n), u_n - u \rangle - \langle \varphi'(u), u_n - u \rangle$$

$$\leq \|\varphi'(u_n)\|_\alpha \|u_n - u\|_\alpha - \langle \varphi'(u), u_n - u \rangle$$

$$\to 0, \quad \text{as } n \to \infty. \quad (6.106)$$

Moreover, according to (6.14) and Proposition 6.3, we have $u_n$ is bounded in $C([0, T], \mathbb{R}^N)$ and $\|u_n - u\| \to 0$ as $n \to \infty$.

Observing that

$$\langle \varphi'(u_n) - \varphi'(u), u_n - u \rangle$$

$$= -\int_0^T (^C_0D_t^\alpha(u_n(t) - u(t)), ^C_tD_T^\alpha(u_n(t) - u(t)))dt$$

$$- \int_0^T (\nabla F(t, u_n(t)) - \nabla F(t, u(t)), u_n(t) - u(t))dt$$

$$\geq |\cos(\pi\alpha)|\|u_n(t) - u(t)\|_\alpha^2$$

$$- \left| \int_0^T (\nabla F(t, u_n(t)) - \nabla F(t, u(t)))dt \right| \|u_n(t) - u(t)\|.$$

Combining this with (6.106), it is easy to verify that $\|u_n(t) - u(t)\|_\alpha \to 0$ as $n \to \infty$, and hence that $u_n \to u$ in $E^\alpha$. Thus, $\{u_n\}$ admits a convergent subsequence.

For any $u \in Y_k$, let

$$\|u\|_* := \left( \int_0^T |u(t)|^{\mu'} dt \right)^{1/\mu'}, \tag{6.107}$$

and it is easy to verify that $\|\cdot\|_*$ define by (6.107) is a norm of $Y_k$. Since all the norms of a finite dimensional normed space are equivalent, so there exists positive constant $C_3$ such that

$$C_3\|u\|_\alpha \leq \|u\|_* \quad \text{for } u \in Y_k. \tag{6.108}$$

In view of (A3), there exist two positive constants $M_1$ and $C_4$ such that

$$F(t, x) \geq M_1|x|^{\mu'}, \tag{6.109}$$

for a.e. $t \in [0, T]$ and $|x| \geq C_4$.

It follows from (6.22), (6.108) and (6.109) that

$$\varphi(u) = -\int_0^T \frac{1}{2}(^C_0D_t^\alpha u(t), ^C_tD_T^\alpha u(t))dt - \int_0^T F(t, u(t))dt$$

$$\leq \frac{1}{|2\cos(\pi\alpha)|}\|u\|_\alpha^2 - \int_{\Omega_3} F(t, u(t))dt - \int_{\Omega_4} F(t, u(t))dt$$

$$\leq \frac{1}{|2\cos(\pi\alpha)|}\|u\|_\alpha^2 - M_1\int_{\Omega_3} |u(t)|^{\mu'} dt - \int_{\Omega_4} F(t, u(t))dt$$

$$= \frac{1}{|2\cos(\pi\alpha)|}\|u\|_\alpha^2 - M_1\int_0^T |u(t)|^{\mu'} dt + M_1\int_{\Omega_4} |u(t)|^{\mu'} dt - \int_{\Omega_4} F(t, u(t))dt$$

$$\leq \frac{1}{|2\cos(\pi\alpha)|}\|u\|_\alpha^2 - C_3^{\mu'} M_1\|u\|_\alpha^{\mu'} + C_5,$$

where $\Omega_3 := \{t \in [0, T] : |u(t)| \geq C_4\}$, $\Omega_4 := [0, T]\backslash\Omega_3$ and $C_5$ is a positive constant.

Since $\mu' > 2$, then there exist positive constants $d_k$ such that

$$\varphi(u) \leq 0, \quad \text{for } u \in Y_k, \quad \text{and } \|u\|_\alpha \geq d_k. \tag{6.110}$$

For any $u \in Z_k$, let

$$\|u\|_\mu := \left( \int_0^T |u(t)|^\mu dt \right)^{1/\mu} \quad \text{and} \quad \beta_k := \sup_{\substack{u \in Z_k \\ \|u\|_\alpha = 1}} \|u\|_\mu,$$

then we conclude $\beta_k \to 0$ as $k \to \infty$.

In fact, it is obvious that $\beta_k \geq \beta_{k+1} > 0$, so $\beta_k \to \beta$. For every $k \in \mathbb{N}$, there exists $u_k \in Z_k$ such that

$$\|u_k\|_\alpha = 1 \quad \text{and} \quad \|u_k\|_\mu > \beta_k/2. \tag{6.111}$$

As $E^\alpha$ is reflexive, $\{u_k\}$ has a weakly convergent subsequence, still denoted by $\{u_k\}$, such that $u_k \rightharpoonup u$. We claim $u = 0$.

In fact, for any $f_m \in \{f_n : n = 1, 2, \ldots\}$, we have $f_m(u_k) = 0$, when $k > m$, so

$$f_m(u_k) \to 0, \quad \text{as } k \to \infty$$

for any $f_m \in \{f_n : n = 1, 2, \ldots\}$, therefore $u = 0$.

By Proposition 6.3, when $u_k \rightharpoonup 0$ in $E^\alpha$, then $u_k \to 0$ strongly in $C([0,T], \mathbb{R}^N)$. So we conclude $\beta = 0$ by (6.111).

In view of (A2), there exist two positive constants $M_2$ and $C_6$ such that

$$F(t, x) \leq M_2 |x|^\mu$$

uniformly for a.e. $t \in [0, T]$ and $|x| \geq C_6$. We have

$$\varphi(u) = -\int_0^T \frac{1}{2} ({}_0^C D_t^\alpha u(t), {}_t^C D_T^\alpha u(t)) dt - \int_0^T F(t, u(t)) dt$$

$$\geq \frac{|\cos(\pi\alpha)|}{2} \|u\|_\alpha^2 - \int_{\Omega_5} F(t, u(t)) dt - \int_{\Omega_6} F(t, u(t)) dt$$

$$\geq \frac{|\cos(\pi\alpha)|}{2} \|u\|_\alpha^2 - M_2 \int_{\Omega_5} |u(t)|^\mu dt - \int_{\Omega_6} F(t, u(t)) dt$$

$$= \frac{|\cos(\pi\alpha)|}{2} \|u\|_\alpha^2 - M_2 \int_0^T |u(t)|^\mu dt + M_2 \int_{\Omega_6} |u(t)|^\mu dt - \int_{\Omega_6} F(t, u(t)) dt$$

$$\geq \frac{|\cos(\pi\alpha)|}{2} \|u\|_\alpha^2 - M_2 \beta_k^\mu \|u\|_\alpha^\mu - C_7,$$

where $\Omega_5 := \{t \in [0, T] : |u(t)| \geq C_6\}$, $\Omega_6 := [0, T] \backslash \Omega_5$ and $C_7$ is a positive constant.

Choosing $r_k = 1/\beta_k$, it is obvious that $r_k \to \infty$ as $k \to \infty$, then

$$b_k := \inf_{\substack{u \in Z_k \\ \|u\|_\alpha = \rho_k}} \varphi(u) \to \infty \quad \text{as } k \to \infty,$$

that is, the condition (H3) in Theorem 1.14 is satisfied.

In view of (6.110), let $\rho_k := \max\{d_k, r_k + 1\}$, then

$$a_k := \max_{\substack{u \in Y_k \\ \|u\|_\alpha = \rho_k}} \varphi(u) \leq 0,$$

and this shows the condition of (H2) in Theorem 1.14 is satisfied.

We have proved the functional $\varphi$ satisfies all the conditions of Theorem 1.14, then $\varphi$ has an unbounded sequence of critical values $c_n = \varphi(u_n)$ by Theorem 1.14. We only need to show $\|u_n\| \to \infty$ as $n \to \infty$.

In fact, since $u_n$ is a critical point of the functional $\varphi$, that is

$$\langle \varphi'(u_n), u_n \rangle = - \int_0^T \left[ ({}_0^C D_t^\alpha u_n(t), {}_t^C D_T^\alpha u_n(t)) + (\nabla F(t, u_n(t)), u_n(t)) \right] dt = 0.$$

Hence, we have

$$\begin{aligned}
c_n = \varphi(u_n) &= - \int_0^T \frac{1}{2} \left( {}_0^C D_t^\alpha u_n(t), {}_t^C D_T^\alpha u_n(t) \right) dt - \int_0^T F(t, u_n(t)) dt, \\
&= \frac{1}{2} \int_0^T (\nabla F(t, u_n(t)), u_n(t)) dt - \int_0^T F(t, u_n(t)) dt, \\
&\leq \frac{1}{2} \int_0^T |\nabla F(t, u_n(t))| |u_n(t)| dt + \int_0^T |F(t, u_n(t))| dt, \quad (6.112)
\end{aligned}$$

since $c_n \to \infty$, we conclude

$$\|u_n\| \to \infty, \quad \text{as } n \to \infty$$

by (6.112). In fact, if not, going to a subsequence if necessary, we may assume that

$$\|u_n\| \leq M_3,$$

for all $n \in \mathbb{N}$ and some positive constant $M_3$.

Combining condition (A) and (6.112), we have

$$\begin{aligned}
c_n &\leq \frac{1}{2} \int_0^T |\nabla F(t, u_n(t))| |u_n(t)| dt + \int_0^T |F(t, u_n(t))| dt \\
&\leq \frac{1}{2} (M_3 + 1) \max_{0 \leq s \leq M_3} m_1(s) \int_0^T m_2(t) dt,
\end{aligned}$$

which contradicts the unboundedness of $c_n$. This completes the proof of Theorem 6.6. $\qquad \square$

**Example 6.6.** In BVP (6.1), let $F(t, x) = |x|^4$, and choose

$$\kappa = 4, \quad r = 2, \quad \mu = \mu' = 4 \text{ and } Q = Q' = 1,$$

so it is easy to verify that all the conditions (A1)-(A4) are satisfied. Then by Theorem 6.9, BVP (6.1) has infinite solutions $\{u_k\}$ on $E^\alpha$ for every positive integer $k$ such that $\|u_k\| \to \infty$, as $k \to \infty$.

**Theorem 6.10.** *Assume that $F(t, x)$ satisfies the following assumption:*

**(A5)** $F(t, x) := a(t)|x|^\gamma$, *where $a(t) \in L^\infty([0, T], \mathbb{R}^+)$ and $1 < \gamma < 2$ is a constant.*

*Then BVP (6.1) has infinite solutions $\{u_n\}$ on $E^\alpha$ for every positive integer $n$ with $\|u_n\|_\alpha$ bounded.*

**Proof.** Let us show that $\varphi$ satisfies conditions in Theorem 1.15 item by item. First, we show that $\varphi$ satisfies the (PS)$^*_c$ condition for every $c \in \mathbb{R}$.

Suppose $n_j \to \infty$, $u_{n_j} \in Y_{n_j}$, $\varphi(u_n) \to c$ and $(\varphi|_{Y_{n_j}})'(u_{n_j}) \to 0$, then $\{u_{n_j}\}$ is a bounded sequence, otherwise, $\{u_{n_j}\}$ would be unbounded sequence, passing to a subsequence, still denoted by $\{u_{n_j}\}$ such that $\|u_{n_j}\|_\alpha \geq 1$ and $\|u_{n_j}\|_\alpha \to \infty$. Note that

$$\langle \varphi'(u_{n_j}), u_{n_j} \rangle - \gamma\varphi(u_{n_j}) = \left(-1 + \frac{\gamma}{2}\right)\int_0^T ({}_0^C D_t^\alpha u_{n_j}(t), {}_t^C D_T^\alpha u_{n_j}(t))dt. \qquad (6.113)$$

However, from (6.113), we have

$$-\gamma\varphi(u_{n_j}) \geq \left(1 - \frac{\gamma}{2}\right)|\cos(\pi\alpha)|\|u_{n_j}\|_\alpha^2 - \|(\varphi|_{Y_{n_j}})'(u_{n_j})\|\|u_{n_j}\|_\alpha,$$

thus $\|u_{n_j}\|_\alpha$ is a bounded sequence in $E^\alpha$. Going, if necessary, to a subsequence, we can assume that $u_{n_j} \rightharpoonup u$ in $E^\alpha$. As $E^\alpha = \overline{\bigcup_{n_j} Y_{n_j}}$, we can choose $v_{n_j} \in Y_{n_j}$ such that $v_{n_j} \to u$.

Hence

$$\lim_{n_j \to \infty} \langle \varphi'(u_{n_j}), u_{n_j} - u \rangle$$

$$= \lim_{n_j \to \infty} \langle \varphi'(u_{n_j}), u_{n_j} - v_{n_j} \rangle + \lim_{n_j \to \infty} \langle \varphi'(u_{n_j}), v_{n_j} - u \rangle$$

$$= \lim_{n_j \to \infty} \langle (\varphi|_{Y_{n_j}})'(u_{n_j}), u_{n_j} - v_{n_j} \rangle$$

$$= 0.$$

So we have

$$\lim_{n_j \to \infty} \langle \varphi'(u_{n_j}) - \varphi'(u), u_{n_j} - u \rangle$$

$$= \lim_{n_j \to \infty} \langle \varphi'(u_{n_j}), u_{n_j} - u \rangle - \lim_{n_j \to \infty} \langle \varphi'(u), u_{n_j} - u \rangle$$

$$= 0,$$

and

$$\langle \varphi'(u_{n_j}) - \varphi'(u), u_{n_j} - u \rangle$$

$$= -\int_0^T ({}_0^C D_t^\alpha(u_{n_j}(t) - u(t)), {}_t^C D_T^\alpha(u_{n_j}(t) - u(t)))dt$$

$$- \int_0^T (\nabla F(t, u_{n_j}(t)) - \nabla F(t, u(t)), u_{n_j}(t) - u(t))dt$$

$$\geq |\cos(\pi\alpha)|\|u_{n_j}(t) - u(t)\|_\alpha^2$$

$$- \left|\int_0^T (\nabla F(t, u_{n_j}(t)) - \nabla F(t, u(t)))dt\right|\|u_{n_j}(t) - u(t)\|,$$

we can conclude $u_{n_j} \to u$ in $E^\alpha$, furthermore, we have $\varphi'(u_{n_j}) \to \varphi'(u)$.

Let us prove $\varphi'(u) = 0$ below. Taking arbitrarily $w_k \in Y_k$, notice when $n_j \geq k$, we have

$$\langle \varphi'(u), w_k \rangle = \langle \varphi'(u) - \varphi'(u_{n_j}), w_k \rangle + \langle \varphi'(u_{n_j}), w_k \rangle$$

$$= \langle \varphi'(u) - \varphi'(u_{n_j}), w_k \rangle + \langle (\varphi|_{Y_{n_j}})'(u_{n_j}), w_k \rangle.$$

Let $n_j \to \infty$ in the right side of above equation. Then

$$\langle \varphi'(u), w_k \rangle = 0, \quad \forall \, w_k \in Y_k,$$

so $\varphi'(u) = 0$, this shows that $\varphi$ satisfies the $(PS)_c^*$ for every $c \in \mathbb{R}$.

For any finite dimensional subspace $E \subset E^\alpha$, there exists $\varepsilon > 0$ such that

$$\alpha\{t \in [0, T] : a(t)|u(t)|^\gamma \geq \varepsilon \|u\|_\alpha^\gamma\} \geq \varepsilon, \quad \forall \, u \in E \backslash \{0\}. \tag{6.114}$$

Otherwise, for any positive integer $n$, there exists $u_n \in E \backslash \{0\}$ such that

$$\alpha\left\{t \in [0, T] : a(t)|u_n(t)|^\gamma \geq \frac{1}{n}\|u_n\|_\alpha^\gamma\right\} < \frac{1}{n}.$$

Set $v_n := \frac{u_n(t)}{\|u_n\|_\alpha} \in E \backslash \{0\}$, then $\|v_n\|_\alpha = 1$ for all $n \in \mathbb{N}$ and

$$\alpha\left\{t \in [0, T] : a(t)|v_n(t)|^\gamma \geq \frac{1}{n}\right\} < \frac{1}{n}. \tag{6.115}$$

Since $\dim E < \infty$, it follows from the compactness of the unit sphere of $E$ that there exists a subsequence, denoted also by $\{v_n\}$, such that $\{v_n\}$ converges to some $v_0$ in $E$. It is obvious that $\|v_0\|_\alpha = 1$.

By the equivalence of the norms on the finite dimensional space, we have $v_n \to v_0$ in $L^2([0, T], \mathbb{R}^N)$, i.e.,

$$\int_0^T |v_n - v_0|^2 dt \to 0, \quad \text{as } n \to \infty. \tag{6.116}$$

By (6.116) and Hölder inequality, we have

$$\int_0^T a(t)|v_n - v_0|^\gamma dt \leq \left(\int_0^T a(t)^{\frac{2}{2-\gamma}} dt\right)^{\frac{2-\gamma}{2}} \left(\int_0^T |v_n - v_0|^2 dt\right)^{\frac{\gamma}{2}}$$

$$= \|a\|_{\frac{2}{2-\gamma}} \left(\int_0^T |v_n - v_0|^2 dt\right)^{\frac{\gamma}{2}}$$

$$\to 0, \quad \text{as } n \to \infty. \tag{6.117}$$

Thus, there exist $\xi_1, \xi_2 > 0$ such that

$$\alpha\{t \in [0, T] : a(t)|v_0(t)|^\gamma \geq \xi_1\} \geq \xi_2. \tag{6.118}$$

In fact, if not, we have

$$\alpha\left\{t \in [0, T] : a(t)|v_0(t)|^\gamma \geq \frac{1}{n}\right\} = 0$$

for all positive integer $n$.

It implies that

$$0 \leq \int_0^T a(t)|v_0|^{\gamma+2} dt < \frac{T}{n}\|v_0\|^2 \leq \frac{C_8^2 T}{n}\|v_0\|_\alpha^2 \to 0,$$

as $n \to \infty$, where

$$C_8 := \frac{T^{\alpha - \frac{1}{2}}}{\Gamma(\alpha)(2\alpha - 1)^{\frac{1}{2}}}$$

by (6.14). Hence $v_0 = 0$ which contradicts that $\|v_0\|_\alpha = 1$. Therefore, (6.118) holds. Now let

$$\Omega_0 = \{t \in [0,T] : a(t)|v_0(t)|^\gamma \geq \xi_1\}, \quad \Omega_n = \left\{t \in [0,T] : a(t)|v_n(t)|^\gamma < \frac{1}{n}\right\},$$

and $\Omega_n^c = [0,T] \setminus \Omega_n = \{t \in [0,T] : a(t)|v_n(t)|^\gamma \geq \frac{1}{n}\}$.

By (6.115) and (6.118), we have

$$\alpha(\Omega_n \cap \Omega_0) = \alpha(\Omega_0 \setminus (\Omega_n^c \cap \Omega_0))$$
$$\geq \alpha(\Omega_0) - \alpha(\Omega_n^c \cap \Omega_0)$$
$$\geq \xi_2 - \frac{1}{n}$$

for all positive integer $n$. Let $n$ be large enough such that

$$\xi_2 - \frac{1}{n} \geq \frac{1}{2}\xi_2 \quad \text{and} \quad \frac{1}{2^{\gamma-1}}\xi_1 - \frac{1}{n} \geq \frac{1}{2^\gamma}\xi_1,$$

then we have

$$\int_0^T a(t)|v_n - v_0|^\gamma dt \geq \int_{\Omega_n \cap \Omega_0} a(t)|v_n - v_0|^\gamma dt$$
$$\geq \frac{1}{2^{\gamma-1}} \int_{\Omega_n \cap \Omega_0} a(t)|v_0|^\gamma dt - \int_{\Omega_n \cap \Omega_0} a(t)|v_n|^\gamma dt$$
$$\geq \left(\frac{1}{2^{\gamma-1}}\xi_1 - \frac{1}{n}\right) \alpha(\Omega_n \cap \Omega_0)$$
$$\geq \frac{\xi_1}{2^\gamma} \frac{\xi_2}{2} = \frac{\xi_1 \xi_2}{2^{\gamma+1}} > 0$$

for all large $n$, which is a contradiction to (6.117). Therefore, (6.114) holds.

For any $u \in Z_k$, let

$$\|u\|_2 := \left(\int_0^T |u(t)|^2 dt\right)^{\frac{1}{2}} \quad \text{and} \quad \gamma_k := \sup_{\substack{u \in Z_k \\ \|u\|_\alpha = 1}} \|u\|_2,$$

then we conclude $\gamma_k \to 0$ as $k \to \infty$ in the same way as in the proof of Theorem 6.6.

$$\varphi(u) = -\int_0^T \frac{1}{2}(_0^C D_t^\alpha u(t), _t^C D_T^\alpha u(t))dt - \int_0^T F(t, u(t))dt$$
$$\geq \frac{1}{2}|\cos(\pi\alpha)|\|u\|_\alpha^2 - \int_0^T a(t)|u(t)|^\gamma dt$$
$$\geq \frac{1}{2}|\cos(\pi\alpha)|\|u\|_\alpha^2 - \left(\int_0^T a(t)^{\frac{2}{2-\gamma}} dt\right)^{\frac{2-\gamma}{2}} \|u\|_2^\gamma$$
$$\geq \frac{1}{2}|\cos(\pi\alpha)|\|u\|_\alpha^2 - \left(\int_0^T a(t)^{\frac{2}{2-\gamma}} dt\right)^{\frac{2-\gamma}{2}} \gamma_k^\gamma \|u\|_\alpha^\gamma. \quad (6.119)$$

Let $\rho_k := \left(\frac{4c\gamma_k^\gamma}{|\cos(\pi\alpha)|}\right)^{\frac{1}{2-\gamma}}$, where $c := \left(\int_0^T a(t)^{\frac{2}{2-\gamma}} dt\right)^{\frac{1}{2-\gamma}}$, it is obvious that $\rho_k \to 0$, as $k \to \infty$.

In view of (6.119), we conclude

$$\inf_{\substack{u \in Z_k \\ \|u\|_\alpha = \rho_k}} \varphi(u) \geq \frac{|\cos(\pi\alpha)|}{4}\rho_k^2 > 0,$$

so the condition (H7) in Theorem 1.15 is satisfied.

Furthermore, by (6.119), for any $u \in Z_k$ with $\|u\|_\alpha \leq \rho_k$, we have

$$\varphi(u) \geq -c\gamma_k^\alpha \|u\|_\alpha^\gamma.$$

Therefore,

$$-c\gamma_k^\gamma \rho_k^\gamma \leq \inf_{\substack{u \in Z_k \\ \|u\|_\alpha \leq \rho_k}} \varphi(u) \leq 0.$$

So we have

$$\inf_{\substack{u \in Z_k \\ \|u\|_\alpha \leq \rho_k}} \varphi(u) \to 0,$$

for $\rho_k, \gamma_k \to 0$, as $k \to \infty$. Hence (H5) in Theorem 1.15 is satisfied.

For any $u \in Y_k \setminus \{0\}$,

$$\varphi(u) = -\int_0^T \frac{1}{2}({}_0^C D_t^\alpha u(t), {}_{,t}^C D_T^\alpha u(t))dt - \int_0^T F(t, u(t))dt$$

$$\leq \frac{1}{2|\cos(\pi\alpha)|}\|u\|_\alpha^2 - \int_0^T a(t)|u(t)|^\gamma dt$$

$$\leq \frac{1}{2|\cos(\pi\alpha)|}\|u\|_\alpha^2 - \varepsilon\|u\|_\alpha^\gamma a(\Omega_u)$$

$$\leq \frac{1}{2|\cos(\pi\alpha)|}\|u\|_\alpha^2 - \varepsilon^2\|u\|_\alpha^\gamma,$$

where $\varepsilon$ is given in (6.114), and $\Omega_u := \{t \in [0,T] : a(t)|u(t)|^\gamma \geq \varepsilon\|u\|_\alpha^\gamma\}$.

Choosing $0 < r_k < \min\{\rho_k, (|\cos(\pi\alpha)|\varepsilon^2)^{\frac{1}{2-\gamma}}\}$, we conclude

$$i_k := \max_{\substack{u \in Y_k \\ \|u\|_\alpha = r_k}} \varphi(u) < -\frac{1}{2|\cos(\pi\alpha)|}r_k^2 < 0, \quad \forall\, k \in \mathbb{N},$$

that is, the condition (H6) in Theorem 1.15 is satisfied.

We have proved the functional $\varphi$ satisfies all the conditions of Theorem 1.15, then $\varphi$ has a bounded sequence of negative critical values $c_n = \varphi(u_n)$ converging to 0 by Theorem 1.15, we only need to show $\|u_n\|_\alpha$ is bounded as for every positive integer $n$. Since

$$c_n = \varphi(u_n) = -\int_0^T \frac{1}{2}({}_0^C D_t^\alpha u_n(t), {}_{,t}^C D_T^\alpha u_n(t))dt - \int_0^T F(t, u_n(t))dt$$

$$= -\int_0^T \frac{1}{2}({}_0^C D_t^\alpha u_n(t), {}_{,t}^C D_T^\alpha u_n(t))dt - \int_0^T a(t)|u_n(t)|^\gamma dt$$

$$\geq \frac{|\cos(\pi\alpha)|}{2}\|u_n\|_\alpha^2 - a_0\|u_n\|^\gamma T$$

$$\geq \frac{|\cos(\pi\alpha)|}{2}\|u_n\|_\alpha^2 - a_0 T C_8^\gamma\|u_n\|_\alpha^\gamma, \tag{6.120}$$

where $a_0 = \operatorname{ess\,sup}\{a(t) : t \in [0, T]\}$, by Theorem 1.15, $c_n \to 0$ as $n \to \infty$. If $\|u_n\|_\alpha$ has an unbounded sequence, then $c_n$ is unbounded by (6.120), which is a contradiction. The proof is completed. $\qquad\square$

**Example 6.7.** In BVP (6.1), let $F(t, x) = a(t)|x|^{\frac{3}{2}}$, where

$$a(t) = \begin{cases} T, & t = 0, \\ t, & 0 < t \leq T. \end{cases}$$

By Theorem 6.10, BVP (6.1) has infinite solutions $\{u_k\}$ on $E^\alpha$ for every positive integer $k$ with $\|u_k\|_\alpha$ bounded.

## 6.5 Solutions for BVP with Left and Right Fractional Derivatives

### 6.5.1 *Introduction*

In Section 6.5, we consider the BVP of the following form

$$\begin{cases} {}_tD_T^\alpha({}_0D_t^\alpha u(t)) = \nabla F(t, u(t)), & \text{a.e. } t \in [0, T], \\ u(0) = u(T) = 0, \end{cases} \tag{6.121}$$

where ${}_tD_T^\alpha$ and ${}_0D_t^\alpha$ are the right and left Riemann-Liouville fractional derivatives of order $0 < \alpha \leq 1$ respectively, $F: [0, T] \times \mathbb{R}^N \to \mathbb{R}$ is a given function satisfying some assumptions and $\nabla F(t, x)$ is the gradient of $F$ at $x$.

In particular, if $\alpha = 1$, BVP (6.121) reduces to the standard second order boundary value problem of the following form

$$\begin{cases} u''(t) + \nabla F(t, u(t)) = 0, & \text{a.e. } t \in [0, T], \\ u(0) = u(T) = 0, \end{cases}$$

where $F: [0, T] \times \mathbb{R}^N \to \mathbb{R}$ is a given function and $\nabla F(t, x)$ is the gradient of $F$ at $x$. Although many excellent results have been worked out on the existence of solutions for second order BVP (e.g., Li, Liang and Zhang, 2005; Nieto and O'Regan, 2009; Rabinowitz, 1986), it seems that no similar results were obtained in the literature for fractional BVP.

According to Benson, Wheatcraft and Meerschaert, 2000a, the one-dimensional form of the fractional ADE can be written as

$$\frac{\partial \mathcal{C}}{\partial t} = -v\frac{\partial \mathcal{C}}{\partial x} + \mathcal{D}j\frac{\partial^\gamma \mathcal{C}}{\partial x^\gamma} + \mathcal{D}(1 - j)\frac{\partial^\gamma \mathcal{C}}{\partial(-x)^\gamma}, \tag{6.122}$$

where $\mathcal{C}$ is the expected concentration, $t$ is time, $v$ is a constant mean velocity, $x$ is distance in the direction of mean velocity, $\mathcal{D}$ is a constant dispersion coefficient, $0 \leq j \leq 1$ describes the skewness of the transport process, and $\gamma$ is the order of left and right fractional differential operators. For discussions of this equation, see Benson, Wheatcraft and Meerschaert, 2000b; Fix and Roop, 2004, when $\gamma = 2$, the dispersion operators are identical and the classical ADE is recovered. Fundamental (Green function) solutions are Lévy's $\gamma$-stable densities.

A special case of the fractional ADE (equation (6.122)) describes symmetric transitions, where $j = \frac{1}{2}$. Defining the symmetric operator equivalent to the Riesz potential in Samko, Kilbas and Marichev, 1993,

$$2\nabla^\gamma \equiv D_+^\gamma + D_-^\gamma$$

gives the mass balance equation for advection and symmetric fractional dispersion

$$\frac{\partial \mathcal{C}}{\partial t} = -v\nabla\mathcal{C} + \mathcal{D}\nabla^\gamma\mathcal{C}.$$

In Subsection 6.5.2, we shall establish a variational structure for BVP (6.121). We show that under some suitable assumptions, the critical points of the variational functional defined on a suitable Hilbert space are the solutions of BVP (6.121). In Subsection 6.5.3, the existence of weak solutions for BVP (6.121) with $\frac{1}{2} < \alpha \le 1$ will be established, where $\alpha$ is the order of fractional derivative in BVP (6.121). In Subsection 6.5.4, we will give some existence results of solutions for BVP (6.121).

### 6.5.2 *Variational Structure*

**Proposition 6.5.** *Let $0 < \alpha \le 1$ and $1 < p < \infty$. For all $u \in E_0^{\alpha,p}$, if $\alpha > \frac{1}{p}$, we have $_0D_t^{-\alpha}(_0D_t^\alpha u(t)) = u(t)$. Moreover, we can get that $E_0^{\alpha,p} \in C_0([0,T], \mathbb{R}^N)$.*

**Proof.** Let $\frac{1}{p} + \frac{1}{q} = 1$ and $0 \le t_1 < t_2 \le T$. For every $f \in L^p([0,T], \mathbb{R}^N)$, by using Hölder inequality and noting that $\alpha > \frac{1}{p}$, we have

$$|_0D_{t_1}^{-\alpha}f(t_1) - _0D_{t_2}^{-\alpha}f(t_2)|$$

$$= \frac{1}{\Gamma(\alpha)}\left|\int_0^{t_1}(t_1-s)^{\alpha-1}f(s)ds - \int_0^{t_2}(t_2-s)^{\alpha-1}f(s)ds\right|$$

$$= \frac{1}{\Gamma(\alpha)}\left|\int_0^{t_1}(t_1-s)^{\alpha-1}f(s)ds - \int_0^{t_1}(t_2-s)^{\alpha-1}f(s)ds\right.$$
$$\left. + \frac{1}{\Gamma(\alpha)}\left|\int_{t_1}^{t_2}(t_2-s)^{\alpha-1}f(s)ds\right|\right.$$

$$\le \frac{1}{\Gamma(\alpha)}\int_0^{t_1}\left((t_1-s)^{\alpha-1} - (t_2-s)^{\alpha-1}\right)|f(s)|ds$$
$$+ \frac{1}{\Gamma(\alpha)}\int_{t_1}^{t_2}(t_2-s)^{\alpha-1}|f(s)|ds$$

$$\le \frac{1}{\Gamma(\alpha)}\left(\int_0^{t_1}\left((t_1-s)^{\alpha-1} - (t_2-s)^{\alpha-1}\right)^q ds\right)^{\frac{1}{q}}\|f\|_{L^p[0,T]}$$
$$+ \frac{1}{\Gamma(\alpha)}\left(\int_{t_1}^{t_2}(t_2-s)^{(\alpha-1)q}ds\right)^{\frac{1}{q}}\|f\|_{L^p[0,T]}$$

$$\le \frac{1}{\Gamma(\alpha)}\left(\int_0^{t_1}(t_1-s)^{(\alpha-1)q} - (t_2-s)^{(\alpha-1)q}ds\right)^{\frac{1}{q}}\|f\|_{L^p[0,T]} \qquad (6.123)$$

$$+ \frac{1}{\Gamma(\alpha)} \left( \int_{t_1}^{t_2} (t_2 - s)^{(\alpha-1)q} ds \right)^{\frac{1}{q}} \|f\|_{L^p[0,T]}$$

$$= \frac{\|f\|_{L^p[0,T]}}{\Gamma(\alpha)(1 + (\alpha-1)q)^{\frac{1}{q}}} (t_1^{(\alpha-1)q+1} - t_2^{(\alpha-1)q+1} + (t_2 - t_1)^{(\alpha-1)q+1})^{\frac{1}{q}}$$

$$+ \frac{\|f\|_{L^p[0,T]}}{\Gamma(\alpha)(1 + (\alpha-1)q)^{\frac{1}{q}}} ((t_2 - t_1)^{(\alpha-1)q+1})^{\frac{1}{q}}$$

$$\leq \frac{2\|f\|_{L^p[0,T]}}{\Gamma(\alpha)(1 + (\alpha-1)q)^{\frac{1}{q}}} (t_2 - t_1)^{\alpha-1+\frac{1}{q}}$$

$$= \frac{2\|f\|_{L^p[0,T]}}{\Gamma(\alpha)(1 + (\alpha-1)q)^{\frac{1}{q}}} (t_2 - t_1)^{\alpha-\frac{1}{p}}.$$

For any $u \in E_0^{\alpha,p}$, as $_0D_t^\alpha u(t) \in L^p([0,T], \mathbb{R}^N)$, we apply (6.123) to obtain the continuity of the function $_0D_t^{-\alpha}(_0D_t^\alpha u(t))$ on $[0,T]$. We complete the argument by using Propositions 1.6-1.7, and we have

$$_0D_t^{-\alpha}(_0D_t^\alpha u(t)) = u(t) + Ct^{\alpha-1}, \quad t \in [0,T],$$

where $C \in \mathbb{R}^N$.

Since $u(0) = 0$ and $_0D_t^{-\alpha}(_0D_t^\alpha u(t))$ is continuous in $[0,T]$, we can get that $C = 0$, which means that $_0D_t^{-\alpha}(_0D_t^\alpha u(t)) = u(t)$ and $u$ is continuous in $[0,T]$. $\square$

**Remark 6.6.** In the case that $1 - \alpha \geq \frac{1}{p}$, for any $u \in E_0^{\alpha,p}$, we also have $_0D_t^{-\alpha}(_0D_t^\alpha u(t)) = u(t)$. In fact, set $f(t) = _0D_t^{\alpha-1}u(t)$. According to Propositions 1.6-1.7, we only need to prove that $f(0) = [_0D_t^{\alpha-1}u(t)]_{t=0} = 0$. Noting that $1 - \alpha \geq \frac{1}{p}$, by using Hölder inequality, Lemma 6.1 and the similar method in the proof of Lemma 7 in Fix and Roop, 2004, we can obtain the desired result, i.e. $f(0) = 0$. We skip the proof since it is similar to Lemma 7 in Fix and Roop, 2004.

If $\alpha > \frac{1}{p}$, the following theorem is useful for us to establish the variational structure on the space $E_0^{\alpha,p}$ for BVP (6.121).

**Theorem 6.11.** *Let $1 < p < \infty$, $\frac{1}{p} < \alpha \leq 1$ and $L : [0,T] \times \mathbb{R}^N \times \mathbb{R}^N \to \mathbb{R}$, $(t, x, y) \to L(t, x, y)$ be measurable in $t$ for each $[x, y] \in \mathbb{R}^N \times \mathbb{R}^N$ and continuously differentiable in $[x, y]$ for almost every $t \in [0,T]$. If there exist $m_1 \in C(\mathbb{R}^+, \mathbb{R}^+)$, $m_2 \in L^1([0,T], \mathbb{R}^+)$ and $m_3 \in L^q([0,T], \mathbb{R}^+)$, $1 < q < \infty$, such that, for a.e. $t \in [0,T]$ and every $[x, y] \in \mathbb{R}^N \times \mathbb{R}^N$, one has*

$$|L(t, x, y)| \leq m_1(|x|)(m_2(t) + |y|^p),$$
$$|D_x L(t, x, y)| \leq m_1(|x|)(m_2(t) + |y|^p),$$
$$|D_y L(t, x, y)| \leq m_1(|x|)(m_3(t) + |y|^{p-1}),$$

*where $\frac{1}{p} + \frac{1}{q} = 1$, then the functional $\varphi$ defined by*

$$\varphi(u) = \int_0^T L(t, u(t), _0D_t^\alpha u(t)) dt$$

*is continuously differentiable on* $E_0^{\alpha,p}$, *and* $\forall u, v \in E_0^{\alpha,p}$, *we have*

$$\langle \varphi'(u), v \rangle = \int_0^T [(D_x L(t, u(t), {}_0D_t^\alpha u(t)), v(t))$$
$$+ (D_y L(t, u(t), {}_0D_t^\alpha u(t)), {}_0D_t^\alpha v(t))]dt. \tag{6.124}$$

**Proof.** It suffices to prove that $\varphi$ has at every point $u$ a directional derivative $\varphi'(u) \in (E_0^{\alpha,p})^*$ given by (6.124) and that the mapping

$$\varphi' : E_0^{\alpha,p} \to (E_0^{\alpha,p})^*, \quad u \to \varphi'(u)$$

is continuous.

We omit the rather technical proof which is similar to the proof of Theorem 1.4 in Mawhin and Willem, 1989. In fact, the only change we need is to replace the weak derivatives for $u$ and $v$ of Theorem 1.5 in Mawhin and Willem, 1989, by ${}_0D_t^\alpha u$ and ${}_0D_t^\alpha v$ respectively. The proof is completed. $\square$

We are now in a position to give the definition for the solution of BVP (6.121).

**Definition 6.3.** A function $u : [0, T] \to \mathbb{R}^N$ is called a solution of BVP (6.121) if

**(i)** ${}_tD_T^{\alpha-1}({}_0D_t^\alpha u(t))$ and ${}_0D_t^{\alpha-1}u(t)$ are differentiable for almost every $t \in [0, T]$, and

**(ii)** $u$ satisfies (6.121).

For a solution $u \in E^\alpha$ of BVP (6.121) such that $\nabla F(\cdot, u(\cdot)) \in L^1([0, T], \mathbb{R}^N)$, multiplying (6.121) by $v \in C_0^\infty([0, T], \mathbb{R}^N)$ yields

$$\int_0^T [({}_tD_T^\alpha({}_0D_t^\alpha u(t)), v(t)) - (\nabla F(t, u(t)), v(t))]dt$$
$$= \int_0^T [({}_0D_t^\alpha u(t), {}_0D_t^\alpha v(t)) - (\nabla F(t, u(t)), v(t))]dt$$
$$= 0 \tag{6.125}$$

after applying (1.13) and Definition 6.3. Therefore, we can give the definition of weak solution for BVP (6.121) as follows.

**Definition 6.4.** By the weak solution of BVP (6.121), we mean that the function $u \in E^\alpha$ such that $\nabla F(\cdot, u(\cdot)) \in L^1([0, T], \mathbb{R}^N)$ and satisfies (6.125) for all $v \in C_0^\infty([0, T], \mathbb{R}^N)$.

Any solution $u \in E^\alpha$ of BVP (6.121) is a weak solution provided that $\nabla F(\cdot, u(\cdot)) \in L^1([0, T], \mathbb{R}^N)$. Our task is now to establish a variational structure on $E^\alpha$ with $\alpha \in (\frac{1}{2}, 1]$, which enables us to reduce the existence of weak solutions of BVP (6.121) to the one of finding critical points of corresponding functional.

**Corollary 6.5.** *Let* $L : [0, T] \times \mathbb{R}^N \times \mathbb{R}^N \to \mathbb{R}$ *be defined by*

$$L(t, x, y) = \frac{1}{2}|y|^2 - F(t, x),$$

where $F : [0, T] \times \mathbb{R}^N \to \mathbb{R}$ *satisfies the following condition (A) which is assumed as in Subsection 6.2.3.*

*If $\frac{1}{2} < \alpha \leq 1$ and $u \in E^\alpha$ is a solution of corresponding Euler equation $\varphi'(u) = 0$, where $\varphi$ is defined as*

$$\varphi(u) = \int_0^T \left( \frac{1}{2} |_0 D_t^\alpha u(t)|^2 - F(t, u(t)) \right) dt, \quad for \ u \in E^\alpha, \tag{6.126}$$

*then $u$ is a weak solution of BVP (6.121) with $\frac{1}{2} < \alpha \leq 1$.*

**Proof.** By Theorem 6.11, we have

$$0 = \langle \varphi'(u), v \rangle = \int_0^T [(_0 D_t^\alpha u(t), {}_0 D_t^\alpha v(t)) - (\nabla F(t, u(t)), v(t))] dt$$

for all $v \in E^\alpha$ and hence for all $v \in C_0^\infty([0, T], \mathbb{R}^N)$. Thus, according to Definition 6.4, $u$ is a weak solution of BVP (6.121). The proof is completed. $\square$

**Remark 6.7.** Generally speaking, a critical point $u$ of $\varphi$ on $E^\alpha$ will be a weak solution of BVP (6.121). However, we shall show that every weak solution is also a solution of BVP (6.121).

### 6.5.3 *Existence of Weak Solutions*

According to Corollary 6.5, we know that in order to find weak solutions of BVP (6.121), it suffices to obtain the critical points of functional $\varphi$ given by (6.126). We need to use some critical point theorems.

First, we use Theorem 1.12 to consider the existence of weak solutions for BVP (6.121). Assume that the condition (A) is satisfied. Recall that, in our setting in (6.126), the corresponding functional $\varphi$ on $E^\alpha$ is continuously differentiable according to Corollary 6.5 and is also weakly lower semi-continuous functional on $E^\alpha$ as the sum of a convex continuous function (see Theorem 1.2 in Mawhin and Willem, 1989) and of a weakly continuous one (see Proposition 1.2 in Mawhin and Willem, 1989).

In fact, according to Proposition 6.3, if $u_k \rightharpoonup u$ in $E^\alpha$, then $u_k \to u$ in $C([0, T], \mathbb{R}^N)$. Therefore, $F(t, u_k(t)) \to F(t, u(t))$ a.e. $t \in [0, T]$. By Lebesgue dominated convergence theorem, we have $\int_0^T F(t, u_k(t)) dt \to \int_0^T F(t, u(t)) dt$, which means that the functional $u \to \int_0^T F(t, u(t)) dt$ is weakly continuous on $E^\alpha$. Moreover, since fractional derivative operator is linear operator, the functional $u \to \int_0^T (|_0 D_t^\alpha u(t)|^2/2) dt$ is convex and continuous on $E^\alpha$.

If $\varphi$ is coercive, by Theorem 1.12, $\varphi$ has a minimum so that BVP (6.121) is solvable. It remains to find conditions under which $\varphi$ is coercive on $E^\alpha$, i.e. $\lim_{\|u\|_\alpha \to \infty} \varphi(u) = +\infty$, for $u \in E^\alpha$. We shall see that it suffices to require that $F(t, x)$ is bounded by a function for a.e. $t \in [0, T]$ and all $x \in \mathbb{R}^N$.

**Theorem 6.12.** *Let $\alpha \in (\frac{1}{2}, 1]$ and assume that $F$ satisfies (A). If*

$$|F(t, x)| \leq \bar{a}|x|^2 + \bar{b}(t)|x|^{2-\gamma} + \bar{c}(t), \quad a.e. \ t \in [0, T], \ x \in \mathbb{R}^N, \tag{6.127}$$

*where* $\bar{a} \in [0, \Gamma^2(\alpha+1)/2T^{2\alpha})$, $\gamma \in (0,2)$, $\bar{b} \in L^{2/\gamma}([0,T], \mathbb{R})$, *and* $\bar{c} \in L^1([0,T], \mathbb{R})$, *then BVP* (6.121) *has at least one weak solution which minimizes* $\varphi$ *on* $E^\alpha$.

**Proof.** According to the arguments above, our problem reduces to prove that $\varphi$ is coercive on $E^\alpha$. For $u \in E^\alpha$, it follows from (6.127) and (6.13) that

$$
\begin{aligned}
\varphi(u) &= \frac{1}{2} \int_0^T |_0 D_t^\alpha u(t)|^2 dt - \int_0^T F(t, u(t)) dt \\
&\geq \frac{1}{2} \int_0^T |_0 D_t^\alpha u(t)|^2 dt - \bar{a} \int_0^T |u(t)|^2 dt - \int_0^T \bar{b}(t)|u(t)|^{2-\gamma} dt - \int_0^T \bar{c}(t) dt \\
&= \frac{1}{2} \|u\|_\alpha^2 - \bar{a}\|u\|_{L^2}^2 - \int_0^T \bar{b}(t)|u(t)|^{2-\gamma} dt - \bar{c}_1 \\
&\geq \frac{1}{2} \|u\|_\alpha^2 - \bar{a}\|u\|_{L^2}^2 - \left( \int_0^T |\bar{b}(t)|^{2/\gamma} dt \right)^{\gamma/2} \left( \int_0^T |u(t)|^2 dt \right)^{1-\gamma/2} - \bar{c}_1 \\
&= \frac{1}{2} \|u\|_\alpha^2 - \bar{a}\|u\|_{L^2}^2 - \bar{b}_1 \|u\|_{L^2}^{2-\gamma} - \bar{c}_1 \\
&\geq \frac{1}{2} \|u\|_\alpha^2 - \frac{\bar{a}T^{2\alpha}}{\Gamma^2(\alpha+1)} \|u\|_\alpha^2 - \bar{b}_1 \left( \frac{T^\alpha}{\Gamma(\alpha+1)} \right)^{2-\gamma} \|u\|_\alpha^{2-\gamma} - \bar{c}_1 \\
&= \left( \frac{1}{2} - \frac{\bar{a}T^{2\alpha}}{\Gamma^2(\alpha+1)} \right) \|u\|_\alpha^2 - \bar{b}_1 \left( \frac{T^\alpha}{\Gamma(\alpha+1)} \right)^{2-\gamma} \|u\|_\alpha^{2-\gamma} - \bar{c}_1,
\end{aligned}
$$

where $\bar{b}_1 = (\int_0^T |\bar{b}(t)|^{2/\gamma} dt)^{\gamma/2}$ and $\bar{c}_1 = \int_0^T \bar{c}(t) dt$. Noting that $\bar{a} \in [0, \Gamma^2(\alpha+1)/2T^{2\alpha})$ and $\gamma \in (0,2)$, we have $\varphi(u) = +\infty$ as $\|u\|_\alpha \to \infty$, and hence $\varphi$ is coercive, which completes the proof. $\square$

Let $a_0 = \min_{\lambda \in [\frac{1}{2}, 1]} \{\Gamma^2(\lambda+1)/2T^{2\lambda}\}$. The following result follows immediately from Theorem 6.12.

**Corollary 6.6.** $\forall \alpha \in (\frac{1}{2}, 1]$ *and if* $F$ *satisfies the condition* (A) *and* (6.127) *with* $a \in [0, a_0)$, *then BVP* (6.121) *has at least one weak solution which minimizes* $\varphi$ *on* $E^\alpha$.

Our task is now to use Theorem 1.13 (Mountain pass theorem) to find a nonzero critical point of functional $\varphi$ on $E^\alpha$.

**Theorem 6.13.** *Let* $\alpha \in (\frac{1}{2}, 1]$ *and suppose that* $F$ *satisfies the condition* (A). *If*

**(A1)** $F \in C([0,T] \times \mathbb{R}^N, \mathbb{R})$ *and there exists* $\mu \in [0, \frac{1}{2})$ *and* $M > 0$ *such that*
$$0 < F(t, x) \leq \mu(\nabla F(t, x), x) \text{ for all } x \in \mathbb{R}^N \text{ with } |x| \geq M \text{ and } t \in [0,T];$$
**(A2)** $\limsup_{|x| \to 0} F(t, x)/|x|^2 < \Gamma^2(\alpha+1)/2T^{2\alpha}$ *uniformly for* $t \in [0,T]$ *and* $x \in \mathbb{R}^N$

*are satisfied, then BVP* (6.121) *has at least one nonzero weak solution on* $E^\alpha$.

**Proof.** We will verify that $\varphi$ satisfies all the conditions of Theorem 1.13.

First, we will prove that $\varphi$ satisfies (PS) condition. Since $F(t,x) - \mu(\nabla F(t,x), x)$ is continuous for $t \in [0,T]$ and $|x| \le M$, there exists $c \in \mathbb{R}^+$, such that

$$F(t,x) \le \mu(\nabla F(t,x), x) + c, \quad t \in [0,T], \ |x| \le M.$$

By assumption (A1), we obtain

$$F(t,x) \le \mu(\nabla F(t,x), x) + c, \quad t \in [0,T], \ x \in \mathbb{R}^N. \tag{6.128}$$

Let $\{u_k\} \subset E^\alpha$, $|\varphi(u_k)| \le K$, $k = 1, 2, \ldots$, $\varphi'(u_k) \to 0$. Notice that

$$\langle \varphi'(u_k), u_k \rangle = \int_0^T [({}_0D_t^\alpha u_k(t), {}_0D_t^\alpha u_k(t)) - (\nabla F(t, u_k(t)), u_k(t))]dt$$

$$= \|u_k\|_\alpha^2 - \int_0^T (\nabla F(t, u_k(t)), u_k(t))dt. \tag{6.129}$$

It follows from (6.128) and (6.129) that

$$K \ge \varphi(u_k) = \frac{1}{2}\|u_k\|_\alpha^2 - \int_0^T F(t, u_k(t))dt$$

$$\ge \frac{1}{2}\|u_k\|_\alpha^2 - \mu \int_0^T (\nabla F(t, u_k(t)), u_k(t))dt - cT$$

$$= \left(\frac{1}{2} - \mu\right)\|u_k\|_\alpha^2 + \mu\langle \varphi'(u_k), u_k \rangle - cT$$

$$\ge \left(\frac{1}{2} - \mu\right)\|u_k\|_\alpha^2 - \mu\|\varphi'(u_k)\|_\alpha\|u_k\|_\alpha - cT, \quad k = 1, 2, \ldots .$$

Since $\varphi'(u_k) \to 0$, there exists $N_0 \in \mathbb{N}$ such that

$$K \ge \left(\frac{1}{2} - \mu\right)\|u_k\|_\alpha^2 - \|u_k\|_\alpha - cT, \quad k > N_0,$$

and this implies that $\{u_k\} \subset E^\alpha$ is bounded. Since $E^\alpha$ is a reflexive space, going to a subsequence if necessary, we may assume that $u_k \rightharpoonup u$ weakly in $E^\alpha$, thus we have

$$\langle \varphi'(u_k) - \varphi'(u), u_k - u \rangle = \langle \varphi'(u_k), u_k - u \rangle - \langle \varphi'(u), u_k - u \rangle$$

$$\le \|\varphi'(u_k)\|_\alpha\|u_k - u\|_\alpha - \langle \varphi'(u), u_k - u \rangle$$

$$\to 0, \quad \text{as } k \to \infty. \tag{6.130}$$

Moreover, according to (6.14) and Proposition 6.3, we get that $u_k$ is bounded in $C([0,T], \mathbb{R}^N)$ and $\|u_k - u\| = 0$ as $k \to \infty$. Hence, we have

$$\int_0^T \nabla F(t, u_k(t))dt \to \int_0^T \nabla F(t, u(t))dt, \quad \text{as } k \to \infty. \tag{6.131}$$

Noting that

$$\langle \varphi'(u_k) - \varphi'(u), u_k - u \rangle$$

$$= \int_0^T ({}_0D_t^\alpha u_k(t) - {}_0D_t^\alpha u(t))^2 dt - \int_0^T (\nabla F(t, u_k(t)) - \nabla F(t, u(t)))(u_k(t) - u(t))dt$$

$$\geq \|u_k - u\|_\alpha^2 - \left| \int_0^T (\nabla F(t, u_k(t)) - \nabla F(t, u(t))) dt \right| \|u_k - u\|.$$

Combining (6.130) and (6.131), it is easy to verify that $\|u_k - u\|_\alpha^2 \to 0$ as $k \to \infty$, and hence $u_k \to u$ in $E_\alpha$. Thus, we obtain the desired convergence property.

From $\limsup_{|x| \to 0} F(t, x)/|x|^2 < \Gamma^2(\alpha+1)/2T^{2\alpha}$ uniformly for $t \in [0, T]$, there exists $\epsilon \in (0, 1)$ and $\delta > 0$ such that $F(t, x) \leq (1 - \epsilon)(\Gamma^2(\alpha+1)/2T^{2\alpha})|x|^2$ for all $t \in [0, T]$ and $x \in \mathbb{R}^N$ with $|x| \leq \delta$.

Let $\rho = \frac{\Gamma(\alpha)((\alpha-1)/2+1)^{\frac{1}{2}}}{T^{\alpha-\frac{1}{2}}} \delta$ and $\sigma = \epsilon \rho^2 / 2 > 0$. Then it follows from (6.14) that

$$\|u\| \leq \frac{T^{\alpha-\frac{1}{2}}}{\Gamma(\alpha)((\alpha-1)/2+1)^{\frac{1}{2}}} \|u\|_\alpha = \delta$$

for all $u \in E^\alpha$ with $\|u\|_\alpha = \rho$. Therefore, we have

$$\varphi(u) = \frac{1}{2}\|u\|_\alpha^2 - \int_0^T F(t, u(t)) dt$$

$$\geq \frac{1}{2}\|u\|_\alpha^2 - (1 - \epsilon)\frac{\Gamma^2(\alpha+1)}{2T^{2\alpha}} \int_0^T |u(t)|^2 dt$$

$$\geq \frac{1}{2}\|u\|_\alpha^2 - \frac{1}{2}(1 - \epsilon)\|u\|_\alpha^2$$

$$= \frac{1}{2}\epsilon\|u\|_\alpha^2 = \sigma$$

for all $u \in E^\alpha$ with $\|u\|_\alpha = \rho$. This implies (ii) in Theorem 1.13 is satisfied.

It is obvious from the definition of $\varphi$ and (A2) that $\varphi(0) = 0$, and therefore, it suffices to show that $\varphi$ satisfies (iii) in Theorem 1.13.

Since $0 < F(t, x) \leq \mu(\nabla F(t, x), x)$ for all $x \in \mathbb{R}^N$ and $|x| \geq M$, a simple regularity argument then shows that there exists $r_1, r_2 > 0$ such that

$$F(t, x) \geq r_1 |x|^{1/\mu} - r_2, \quad x \in \mathbb{R}^N, \quad t \in [0, T].$$

For any $u \in E^\alpha$ with $u \neq 0$, $\kappa > 0$ and noting that $\mu \in [0, \frac{1}{2})$, we have

$$\varphi(\kappa u) = \frac{1}{2}\|\kappa u\|_\alpha^2 - \int_0^T F(t, \kappa u(t)) dt$$

$$\leq \frac{1}{2}\kappa^2\|u\|_\alpha^2 - r_1 \int_0^T |\kappa u(t)|^{1/\mu} dt + r_2 T$$

$$= \frac{1}{2}\kappa^2\|u\|_\alpha^2 - r_1 \kappa^{1/\mu}\|u\|_{L^{1/\mu}}^{1/\mu} + r_2 T$$

$$\to -\infty \quad \text{as } \kappa \to \infty.$$

Then there exists a sufficiently large $\kappa_0$ such that $\varphi(\kappa_0 u) \leq 0$. Hence (iii) holds.

Lastly noting that $\varphi(0) = 0$ while for our critical point $u$, $\varphi(u) \geq \sigma > 0$. Hence $u$ is a nontrivial weak solution of BVP (6.121), and this completes the proof.  $\square$

**Theorem 6.14.** $\forall \alpha \in (\frac{1}{2}, 1]$, *suppose that $F$ satisfies conditions (A) and (A1). If*

**(A2)′** $F(t,x) = o(|x|^2)$, as $|x| \to 0$ *uniformly for* $t \in [0,T]$ *and* $x \in \mathbb{R}^N$

*is satisfied, then BVP* (6.121) *has at least one nonzero weak solution on* $E^\alpha$.

**Remark 6.8.** The assumptions in Theorem 6.12 and Theorem 6.13 are classical and the examples can be found in many papers which use critical point theory to discuss differential equations, see, e.g., Li, Liang and Zhang, 2005; Mawhin and Willem, 1989; Rabinowitz, 1986 and references therein.

### 6.5.4 Existence of Solutions

We firstly give the following lemma which is useful for our further discussion.

**Lemma 6.7.** *Let* $0 < \alpha \le 1$. *If* $u \in E^\alpha$ *is a weak solution of BVP* (6.121), *then there exists a constant* $C \in \mathbb{R}^N$ *such that*

$$_0D_t^\alpha u(t) = {}_tD_T^{-\alpha}\nabla F(t,u(t)) + C(T-t)^{\alpha-1}, \quad \text{a.e. } t \in [0,T].$$

**Proof.** Since $u \in E^\alpha$ is a weak solution of BVP (6.121), i.e. $\forall h \in C_0^\infty([0,T],\mathbb{R}^N)$, we have

$$\int_0^T [({}_0D_t^\alpha u(t), {}_0D_t^\alpha h(t)) - (\nabla F(t,u(t)), h(t))]dt = 0. \tag{6.132}$$

Noting that $\nabla F(\cdot, u(\cdot)) \in L^1([0,T],\mathbb{R}^N)$, and applying a similar argument as that for (6.5) in the proof of Lemma 6.1, we get that $_tD_T^{-\alpha}\nabla F(\cdot, u(\cdot)) \in L^1([0,T],\mathbb{R}^N)$. Let us define $w \in L^1([0,T],\mathbb{R}^N)$ by

$$w(t) = {}_tD_T^{-\alpha}\nabla F(t,u(t)), \quad t \in [0,T],$$

so that

$$\begin{aligned}
\int_0^T (w(t), {}_0D_t^\alpha h(t))dt &= \int_0^T ({}_tD_T^\alpha w(t), h(t))dt \\
&= \int_0^T ({}_tD_T^\alpha({}_tD_T^{-\alpha}\nabla F(t,u(t))), h(t))dt \\
&= \int_0^T (\nabla F(t,u(t)), h(t))dt
\end{aligned}$$

by applying (1.13) and Proposition 1.5.

Hence, by (6.132) we have, for every $h \in C_0^\infty([0,T],\mathbb{R}^N)$,

$$\int_0^T ({}_0D_t^\alpha u(t) - w(t), {}_0D_t^\alpha h(t))dt = 0. \tag{6.133}$$

According to Proposition 1.1 and in view of $h \in C_0^\infty([0,T],\mathbb{R}^N)$, we have $_0D_t^\alpha h(t) = {}_0D_t^{\alpha-1}h'(t)$. Since $_0D_t^\alpha u \in L^2([0,T],\mathbb{R}^N)$ and $w \in L^1([0,T],\mathbb{R}^N)$, using (1.12) and (6.133), we get that

$$\int_0^T \left({}_tD_T^{\alpha-1}({}_0D_t^\alpha u(t) - w(t)), h'(t)\right) dt = 0.$$

If $(e_j)$ denotes the Canonical basis of $\mathbb{R}^N$, we can choose

$$h(t) = \sin \frac{2k\pi t}{T} e_j \quad \text{or} \quad h(t) = e_j - \cos \frac{2k\pi t}{T} e_j, \quad k = 1, \dots \text{ and } j = 1, \dots, N.$$

In view of $_tD_T^{\alpha-1}(_0D_t^\alpha u - w) \in L^1([0,T], \mathbb{R}^N)$, and the theory of Fourier series implies that

$$_tD_T^{\alpha-1}(_0D_t^\alpha u(t) - w(t)) = \tilde{C} \tag{6.134}$$

a.e. on $[0,T]$ for some $\tilde{C} \in \mathbb{R}^N$. Using Proposition 1.5 and Proposition 1.3, we can get that

$$_0D_t^\alpha u(t) = w(t) + C(T-t)^{\alpha-1}, \quad \text{a.e. } t \in [0,T],$$

for some $C \in \mathbb{R}^N$ and this completes the proof. $\qquad\square$

**Remark 6.9.**

**(i)** According to (6.134) and Proposition 1.4, we have

$$_tD_T^{\alpha-1}(_0D_t^\alpha u(t)) = _tD_T^{\alpha-1}(_tD_T^{-\alpha}\nabla F(t, u(t))) + \tilde{C} = _tD_T^{-1}\nabla F(t, u(t)) + \tilde{C}$$

a.e. on $[0,T]$ for some $\tilde{C} \in \mathbb{R}^N$. In view of Definition 1.1 and $\nabla F(\cdot, u(\cdot)) \in L^1([0,T], \mathbb{R}^N)$, we shall identify the equivalence class $_tD_T^{\alpha-1}(_0D_t^\alpha u)$ and its continuous representant

$$_tD_T^{\alpha-1}(_0D_t^\alpha u(t)) = \int_t^T \nabla F(s, u(s))ds + \tilde{C} \tag{6.135}$$

for $t \in [0,T]$.

**(ii)** It follows from (6.135) and a classical result of Lebesgue theory that $-\nabla F(\cdot, u(\cdot))$ is the classical derivative of $_tD_T^{\alpha-1}(_0D_t^\alpha u)$ a.e. on $[0,T]$.

We are now in a position to show that every weak solution of BVP (6.121) is also a solution of BVP (6.121).

**Theorem 6.15.** *Let $0 < \alpha \leq 1$. If $u \in E^\alpha$ is a weak solution of BVP (6.121), then $u$ is also a solution of BVP (6.121).*

**Proof.** Firstly, We notice that $_0D_t^{\alpha-1}u(t)$ is derivative for almost every $t \in [0,T]$ and $(_0D_t^{\alpha-1}u(t))' = {_0D_t^\alpha u(t)} \in L^2([0,T], \mathbb{R}^N)$ as $u \in E^\alpha$. On the other hand, Remark 6.9 implies that $_tD_T^{\alpha-1}(_0D_t^\alpha u(t))$ is derivative a.e. on $[0,T]$ and $(_tD_T^{\alpha-1}(_0D_t^\alpha u(t)))' \in L^1([0,T], \mathbb{R}^N)$. Therefore, (i) in Definition 6.3 is verified.

It remains to show that $u$ satisfies (6.121). In fact, according to Definition 1.2 and (6.135), we can get that

$$_tD_T^\alpha(_0D_t^\alpha u(t)) = -(_tD_T^{\alpha-1}(_0D_t^\alpha u(t)))' = \nabla F(t, u(t)), \quad \text{a.e. } t \in [0,T].$$

Moreover, $u \in E^\alpha$ implies that $u(0) = u(T) = 0$, and therefore (6.121) is verified. The proof is completed. $\qquad\square$

The conclusions in Subsection 6.5.3 and Theorem 6.15 imply that BVP (6.121) with $\alpha \in (\frac{1}{2}, 1]$ possesses at least one solution if $F$ satisfies some hypotheses. However, we would like to consider the existence of solutions for BVP (6.121) with $\alpha = \frac{1}{2}$ under the same hypotheses.

For any given $\epsilon_0 \in (0, \frac{1}{2})$, let $\epsilon \in (0, \epsilon_0)$ and $\delta = \delta(\epsilon) = \frac{1}{2} + \epsilon$. According to Corollary 6.6 and Theorem 6.14, if (A) and (6.127) with $a \in [0, a_0)$, or (A), (A1) and (A2)$'$ are satisfied, then $\forall \epsilon \in (0, \epsilon_0)$, the following BVP

$$\begin{cases} {}_tD_T^\delta({}_0D_t^\delta u(t)) = \nabla F(t, u(t)), & \text{a.e. } t \in [0, T], \\ u(0) = u(T) = 0 \end{cases} \tag{6.136}$$

has at least a weak solution $u_\epsilon \in E^\delta$. Moreover, according to Theorem 6.15, $u_\epsilon$ is also the solution of BVP (6.136). Now, our idea is to obtain the solution of BVP (6.121) with $\delta = \frac{1}{2}$ by considering the approximation of $u_\epsilon$ as $\epsilon \to 0$.

**Theorem 6.16.** *Assume that there exists $\epsilon_0 \in (0, \frac{1}{2})$ such that $\forall \epsilon \in (0, \epsilon_0)$ and $\delta = \delta(\epsilon) = \frac{1}{2} + \epsilon$, BVP (6.136) possesses a weak solution $u_\epsilon \in E^\delta$. Moreover, if the following conditions are satisfied*

**(A3)** *there exist $\beta > 2$ and $m \in L^\beta([0, T], \mathbb{R}^+)$ such that $|\nabla F(t, u_\epsilon(t))| \leq m(t)$;*
**(A4)** *there exists $\beta_1 > 1/(\frac{1}{2} - \epsilon_0)$ such that ${}_0D_t^\delta u_\epsilon \in L^{\beta_1}([0, T], \mathbb{R}^N)$.*

*Then there exists a sequence $\{\epsilon_n\}$ such that $\epsilon_1 > \epsilon_2 > \cdots > \epsilon_n \to 0$ as $n \to \infty$, $u(t) = \lim_{n\to\infty} u_{\epsilon_n}(t)$ exists uniformly on $[0, T]$ and $u$ is a solution of BVP (6.121) with $\alpha = \frac{1}{2}$.*

**Proof.** According to Theorem 6.15, $u_\epsilon$ is also a solution of BVP (6.136). Thus, we have

$$ {}_tD_T^\delta({}_0D_t^\delta u_\epsilon(t)) = \nabla F(t, u_\epsilon(t)), \quad \text{a.e. } t \in [0, T]. \tag{6.137}$$

Propositions 1.6-1.7 implies that equation (6.137) is equivalent to the integral equation

$$ {}_0D_t^\delta u_\epsilon(t) = {}_tD_T^{-\delta}(\nabla F(t, u_\epsilon(t))) + C(T - t)^{\delta-1}, \quad \text{a.e. } t \in [0, T], \tag{6.138}$$

where $C = (1/\Gamma(\delta))[{}_tD_T^{\delta-1}({}_0D_t^\delta u_\epsilon(t))]_{t=T}$. Noting that ${}_0D_t^\delta u_\epsilon \in L^{\beta_1}([0, T], \mathbb{R}^N)$ according to (A4), direct calculation gives that

$$\begin{aligned} \left| {}_tD_T^{\delta-1}({}_0D_t^\delta u_\epsilon(t)) \right| &\leq \frac{1}{\Gamma(1-\delta)} \int_t^T (s-t)^{-\delta} |{}_0D_s^\delta u_\epsilon(s)| ds \\ &\leq \frac{1}{\Gamma(1-\delta)} \left( \int_t^T (s-t)^{\frac{-\delta\beta_1}{\beta_1-1}} ds \right)^{1-1/\beta_1} \|{}_0D_t^\delta u_\epsilon\|_{L^{\beta_1}} \\ &\leq c(T-t)^{1-\delta-1/\beta_1} \|{}_0D_t^\delta u_\epsilon\|_{L^{\beta_1}}, \end{aligned}$$

where $c \in \mathbb{R}^+$ is a constant. It is obvious that $1-\delta-1/\beta_1 > 0$ since $\beta_1 > 1/(\frac{1}{2}-\epsilon_0) > 1/(1-\delta)$, then we have $C = (1/\Gamma(\delta))[{}_tD_T^{\delta-1}({}_0D_t^\delta u_\epsilon(t))]_{t=T} = 0$. Therefore, (6.138) can be written as

$$ {}_0D_t^\delta u_\epsilon(t) = {}_tD_T^{-\delta}(\nabla F(t, u_\epsilon(t))), \quad \text{a.e. } t \in [0, T]. \tag{6.139}$$

According to Proposition 6.5 and in view of the continuity of $u_\epsilon \in E^\delta$, (6.139) is equivalent to the integral equation

$$u_\epsilon(t) = {}_0D_t^{-\delta}({}_tD_T^{-\delta}\nabla F(t, u_\epsilon(t))), \quad t \in [0, T]. \tag{6.140}$$

On the other hand, we observe that $m \in L^\beta([0, T], \mathbb{R}^+)$ and $\beta > 2$ in (A3) imply that

$$\begin{aligned}
\left|{}_tD_T^{-\delta}m(t)\right| &\leq \frac{1}{\Gamma(\delta)} \int_t^T (s-t)^{\delta-1}|m(s)|ds \\
&\leq \frac{1}{\Gamma(\delta)} \left( \int_t^T (s-t)^{\frac{(\delta-1)\beta}{\beta-1}}ds \right)^{1-1/\beta} \|m\|_{L^\beta} \\
&\leq c_1 T^{\delta-1/\beta}\|m\|_{L^\beta} \\
&\leq c_1\|m\|_{L^\beta} \max_{\lambda \in [\frac{1}{2}, 1]} \{T^{\lambda-1/\beta}\}, \quad t \in [0, T],
\end{aligned}$$

where $c_1 \in \mathbb{R}^+$ is a constant. Therefore, there exists a constant $M \in \mathbb{R}^+$ such that $\|{}_tD_T^{-\delta}m\| \leq M$, which means that $|{}_tD_T^{-\delta}\nabla F(t, u_\epsilon(t))| \leq M$ on $[0, T]$ since $|\nabla F(t, u_\epsilon(t))| \leq m(t)$.

Set $G(t, u_\epsilon(t)) = {}_tD_T^{-\delta}\nabla F(t, u_\epsilon(t))$, and we have by (6.140)

$$\begin{aligned}
|u_\epsilon(t)| &\leq \frac{1}{\Gamma(\delta)} \int_0^t (t-s)^{\delta-1}|G(s, u_\epsilon(s))|ds \\
&\leq \frac{M}{\Gamma(\delta)} \int_0^t (t-s)^{\delta-1}ds \\
&\leq \frac{M}{\Gamma(\delta+1)}T^\delta \\
&\leq M \max_{\lambda \in [\frac{1}{2}, 1]} \left\{ \frac{T^\lambda}{\Gamma(\lambda+1)} \right\}, \quad t \in [0, T].
\end{aligned} \tag{6.141}$$

The last inequality follows from the continuity of $T^\lambda/\Gamma(\lambda+1)$ with respect to $\lambda > 0$ and the fact that $\Gamma(\lambda) > 0$ for $\lambda > 0$. Furthermore, letting $0 \leq t_1 < t_2 \leq T$, we see that

$$\begin{aligned}
&|u_\epsilon(t_1) - u_\epsilon(t_2)| \\
&= \frac{1}{\Gamma(\delta)} \left| \int_0^{t_1} (t_1-s)^{\delta-1}G(s, u_\epsilon(s))ds - \int_0^{t_2} (t_2-s)^{\delta-1}G(s, u_\epsilon(s))ds \right| \\
&= \frac{1}{\Gamma(\delta)} \left| \int_0^{t_1} \left((t_1-s)^{\delta-1} - (t_2-s)^{\delta-1}\right) G(s, u_\epsilon(s))ds \right. \\
&\quad \left. + \int_{t_1}^{t_2} (t_2-s)^{\delta-1}G(s, u_\epsilon(s))ds \right| \\
&\leq \frac{M}{\Gamma(\delta)} \left| \int_0^{t_1} (t_1-s)^{\delta-1} - (t_2-s)^{\delta-1}ds + \int_{t_1}^{t_2} (t_2-s)^{\delta-1}ds \right| \\
&= \frac{M}{\Gamma(\delta+1)} \left(2(t_2-t_1)^\delta + t_1^\delta - t_2^\delta\right)
\end{aligned} \tag{6.142}$$

$$\leq \frac{2M}{\Gamma(\delta+1)}(t_2-t_1)^\delta$$

$$\leq 2M \max_{\lambda\in[\frac{1}{2},1]}\left\{\frac{(t_2-t_1)^\lambda}{\Gamma(\lambda+1)}\right\}.$$

It then follows from (6.141) and (6.142) that the family $\{u_\epsilon\}$ forms an equicontinuous and uniformly bounded functions. Application of Arzela-Ascoli theorem shows the existence of a sequence $\{\epsilon_n\}$ such that $\epsilon_1 > \epsilon_2 > \ldots > \epsilon_n \to 0$ as $n \to \infty$, and $u(t) = \lim_{n\to\infty} u_{\epsilon_n}(t)$ exists uniformly on $[0,T]$. Since the continuity and boundedness of $\nabla F(t,\cdot)$ imply the continuity of ${}_tD_T^{-\delta}\nabla F(t,\cdot)$, we obtain that

$${}_tD_T^{-\delta}\nabla F(t,u_{\epsilon_n}(t)) \to {}_tD_T^{-\delta}\nabla F(t,u(t)), \quad \text{as } n \to \infty,$$

and combining (6.139) yields

$$u(t) = {}_0D_t^{-\delta}({}_tD_T^{-\delta}\nabla F(t,u(t))), \quad t \in [0,T].$$

This proves that $u$ is a solution of BVP (6.121) by using the Proposition 1.5 and Lemma 6.1. The proof is completed. $\qquad\square$

**Example 6.8.** Set $F(t,x) = m(t)\sin(|x|)$, where $m \in L^\beta([0,T],\mathbb{R}^+)$ and $x \in \mathbb{R}^N$. Then (A3) is verified since $|F(t,x)| \leq m(t)$ for $x \in \mathbb{R}^N$. If for any $\epsilon \in (0,\epsilon_0)$, we have $u_\epsilon \in AC([0,T],\mathbb{R}^N)$ and $u_\epsilon' \in L^{\beta_1}([0,T],\mathbb{R}^N)$, then ${}_0D_t^\delta u_\epsilon \in L^{\beta_1}([0,T],\mathbb{R}^N)$ by using Proposition 1.1 and (6.5). Thus, (A4) is satisfied.

## 6.6    Notes and Remarks

The results in Subsections 6.2.1-6.2.4 are taken from Jiao and Zhou, 2011. The material in Subsections 6.2.5-6.2.6 due to Chen and Tang, 2012. The results in Section 6.3 are adopted from Kong, 2013. The main results of Section 6.4 are from Chen and Tang, 2013. The material in Sections 6.5 due to Jiao and Zhou, 2012.

# Chapter 7

# Fractional Partial Differential Equations

## 7.1 Introduction

Fractional calculus is recognized as one of the best tools to describe long-memory processes. Such models are interesting for engineers and physicists but also for pure mathematicians. The most important among such models are those described by partial differential equations (PDEs) containing fractional derivatives. Their evolutions behave in a much more complex way than in the classical integer-order case and the study of the corresponding dynamics is a hugely demanding task. Although some results of qualitative analysis for fractional partial differential equations (FPDEs) can be similarly obtained, many classical PDEs' methods are hardly applicable directly to FPDEs. New theories and methods are thus required to be specifically developed for FPDEs, whose investigation becomes more challenging. Comparing with PDEs' classical theory, the researches on FPDEs are only on their initial stage of development.

The main objective of this chapter is to investigate the existence and regularity for a variety of time-fractional partial differential equations with applications. Section 7.2 is devoted to study of global and local existence, regularity of mild solutions for Navier-Stokes equations. In Section 7.3, we investigate the existence of a weak solution for Euler-Lagrange equations. In Section 7.4, we discuss existence and regularity of fractional diffusion equations. And in the last section, we present some results on existence of mild solution for fractional Schrödinger equations.

## 7.2 Fractional Navier-Stokes Equations

### 7.2.1 *Introduction*

The Navier-Stokes equations describe the motion of the incompressible Newtonian fluid flows ranging from large scale atmospheric motions to the lubrication of ball bearings, and express the conservation of mass and momentum. For more details we refer to the monographs of Cannone, 1995 and Varnhorn, 1994. We find this system which is so rich in phenomena that the whole power of mathematical theory is needed to discuss the existence, regularity and boundary conditions; see, e.g.,

Lemarié-Rieusset, 2002 and von Wahl, 2013.

It is worth mentioning that Leray carried out an initial study that a boundary-value problem for the time-dependent Navier-Stokes equations possesses a unique smooth solution on some intervals of time provided the data are sufficiently smooth. Since then many results on the existence for weak, mild and strong solutions for the Navier-Stokes equations have been investigated intensively by many authors; see, e.g., de Almeida and Ferreira, 2013; Heck, Kim and Kozono, 2013; Iwabuchi and Takada, 2013; Koch *et al.*, 2009; Masmoudi and Wong, 2015 and Weissler, 1980. Moreover, one can find results on regularity of weak and strong solution from Amrouche and Rejaiba, 2014; Chemin and Gallagher, 2010; Chemin, Gallagher and Paicu, 2011; Choe, 2015; Danchin, 2000; Giga and Yoshikazu, 1991; Kozono, 1998; Raugel and Sell, 1993 and the references therein.

Theoretical analysis and experimental data have shown that classical diffusion equation fails to describe diffusion phenomenon in heterogeneous porous media that exhibits fractal characteristics. How is the classical diffusion equation modified to make it appropriate to depict anomalous diffusion phenomena? This problem is interesting for researchers. Fractional calculus have been found effective in modelling anomalous diffusion processes since it has been recognized as one of the best tools to characterize the long memory processes. Consequently, it is reasonable and significant to propose the generalized Navier-Stokes equations with Caputo fractional derivative operator, which can be used to simulate anomalous diffusion in fractal media. Its evolutions behave in a much more complex way than in classical inter-order case and the corresponding investigation becomes more challenging.

The main effort on time-fractional Navier-Stokes equations has been put into attempts to derive numerical solutions and analytical solutions; see Ganji *et al.*, 2010; El-Shahed and Salem, 2004 and Momani and Zaid, 2006. However, to the best of our knowledge, there are very few results on the existence and regularity of mild solutions for time-fractional Navier-Stokes equations. Recently, De Carvalho-Neto and Gabriela, 2015 dealt with the existence and uniqueness of global and local mild solutions for the time-fractional Navier-Stokes equations.

Motivated by above discussion, in this section we study the following time-fractional Navier-Stokes equations in an open set $\Omega \subset \mathbb{R}^n$ $(n \geq 3)$:

$$\begin{cases} \partial_t^\alpha u - \nu \Delta u + (u \cdot \nabla)u = -\nabla p + f, & t > 0, \\ \nabla \cdot u = 0, \\ u|_{\partial\Omega} = 0, \\ u(0, x) = a, \end{cases} \tag{7.1}$$

where $\partial_t^\alpha$ is Caputo fractional derivative of order $\alpha \in (0,1)$, $u = (u_1(t,x), u_2(t,x), \ldots, u_n(t,x))$ represents the velocity field at a point $x \in \Omega$ and time $t > 0$, $p = p(t,x)$ is the pressure, $\nu$ the viscosity, $f = f(t,x)$ is the external force and $a = a(x)$ is the initial velocity. From now on, we assume that $\Omega$ has a smooth boundary.

Firstly, we get rid of the pressure term by applying Helmholtz projector $P$ to equation (7.1), which converts equation (7.1) to

$$\begin{cases} \partial_t^\alpha u - \nu P \Delta u + P(u \cdot \nabla)u = Pf, & t > 0, \\ \nabla \cdot u = 0, \\ u|_{\partial\Omega} = 0, \\ u(0, x) = a. \end{cases}$$

The operator $-\nu P \Delta$ with Dirichlet type boundary conditions is, basically, the Stokes operator $A$ in the divergence-free function space under consideration. Then we rewrite (7.1) as the following abstract form

$$\begin{cases} {}_0^C D_t^\alpha u = -Au + F(u, u) + Pf, & t > 0, \\ u(0) = a, \end{cases} \tag{7.2}$$

where $F(u, v) = -P(u \cdot \nabla)v$. If one can give sense to the Helmholtz projector $P$ and the Stokes operator $A$, then the solution of equation (7.2) is also the solution of equation (7.1).

The objective of this section is to establish the existence and uniqueness of global and local mild solutions of problem (7.2) in $H^{\beta,q}$. Further, we prove the regularity results which state essentially that if $Pf$ is Hölder continuous then there is a unique classical solution $u(t)$ such that $Au$ and ${}_0^C D_t^\alpha u(t)$ are Hölder continuous in $J_q$.

In Subsection 7.2.2, we recall some notations, definitions, and preliminary facts. Subsection 7.2.3 is devoted to the existence and uniqueness of global mild solution in $H^{\beta,q}$ of problem (7.2). In Subsection 7.2.4, we proceed to study the local mild solution in $H^{\beta,q}$ and use the iteration method to obtain the existence and uniqueness of local mild solution in $J_q$ of problem (7.2). Finally, Subsection 7.2.5 is concerned with the existence and regularity of classical solution in $J_q$ of problem (7.2).

### 7.2.2 Preliminaries

In this Subsection, we introduce notations, definitions, and preliminary facts which are used throughout this section.

Let $\Omega = \{(x_1, \dots, x_n) : x_n > 0\}$ be open subset of $\mathbb{R}^n$, where $n \geq 3$. Let $1 < q < \infty$. Then there is a bounded projection $P$ called the on $(L^q(\Omega))^n$, whose range is the closure of

$$C_\sigma^\infty(\Omega) := \{u \in (C^\infty(\Omega))^n : \nabla \cdot u = 0, \ u \text{ has compact support in } \Omega\},$$

and whose null space is the closure of

$$\{u \in (C^\infty(\Omega))^n : u = \nabla\phi, \ \phi \in C^\infty(\Omega)\}.$$

For notational convenience, let $J_q := \overline{C_\sigma^\infty(\Omega)}^{|\cdot|_q}$, which is a closed subspace of $(L^q(\Omega))^n$. $(W^{m,q}(\Omega))^n$ is a Sobolev space with the norm $|\cdot|_{m,q}$.

$A = -\nu P \Delta$ denotes the Stokes operator in $J_q$ whose domain is $D_q(A) = D_q(\Delta) \cap J_q$; here,

$$D_q(\Delta) = \{u \in (W^{2,q}(\Omega))^n : u|_{\partial\Omega} = 0\}.$$

It is known that $-A$ is a closed linear operator and generates the bounded analytic semigroup $\{e^{-tA}\}$ on $J_q$.

So as to state our results, we need to introduce the definitions of the fractional power spaces associated with $-A$. For $\beta > 0$ and $u \in J_q$, define

$$A^{-\beta}u = \frac{1}{\Gamma(\beta)} \int_0^\infty t^{\beta-1} e^{-tA} u \, dt.$$

Then $A^{-\beta}$ is a bounded, one-to-one operator on $J_q$. Let $A^\beta$ be the inverse of $A^{-\beta}$. For $\beta > 0$, we denote the space $H^{\beta,q}$ by the range of $A^{-\beta}$ with the norm

$$|u|_{H^{\beta,q}} = |A^\beta u|_q.$$

It is easy to check that $e^{-tA}$ extends (or restricts) to a bounded analytic semigroup on $H^{\beta,q}$. For more details, we refer to von Wahl, 2013.

Let $X$ be a Banach space and $J$ be an interval of $\mathbb{R}$. $C(J, X)$ denotes the set of all continuous $X$-valued functions. For $0 < \vartheta < 1$, $C^\vartheta(J, X)$ stands for the set of all functions which are Hölder continuous with the exponent $\vartheta$.

Let $\alpha \in (0, 1]$ and $u : [0, \infty) \times \mathbb{R}^n \to \mathbb{R}^n$, Caputo fractional derivative with respect to time of the function $u$ can be written as

$$\partial_t^\alpha u(t, x) = \partial_t \left( \int_0^t g_{1-\alpha}(t-s)\big(u(t, x) - u(0, x)\big) ds \right), \quad t > 0.$$

For more details, we refer the reader to Kilbas, Srivastava and Trujillo, 2006.

Let us introduce the generalized Mittag-Leffler functions:

$$E_\alpha(-t^\alpha A) = \int_0^\infty M_\alpha(s) e^{-st^\alpha A} ds, \quad e_\alpha(-t^\alpha A) = \int_0^\infty \alpha s M_\alpha(s) e^{-st^\alpha A} ds,$$

where $M_\alpha$ is the Wright function (see Definition 1.9).

In the following, we give some properties of the generalized Mittag-Leffler functions:

**Proposition 7.1.**

(i) $e_\alpha(-t^\alpha A) = \frac{1}{2\pi i} \int_{\Gamma_\theta} e_\alpha(-\mu t^\alpha)(\mu I + A)^{-1} d\mu$;

(ii) $A^\gamma e_\alpha(-t^\alpha A) = \frac{1}{2\pi i} \int_{\Gamma_\theta} \mu^\gamma e_\alpha(-\mu t^\alpha)(\mu I + A)^{-1} d\mu$.

**Proof.** (i) In view of $\int_0^\infty \alpha s M_\alpha(s) e^{-st} ds = e_\alpha(-t)$ and Fubini theorem, we get

$$e_\alpha(-t^\alpha A) = \int_0^\infty \alpha s M_\alpha(s) e^{-st^\alpha A} ds$$

$$= \frac{1}{2\pi i} \int_0^\infty \alpha s M_\alpha(s) \int_{\Gamma_\theta} e^{-\mu s t^\alpha}(\mu I + A)^{-1} d\mu \, ds$$

$$= \frac{1}{2\pi i} \int_{\Gamma_\theta} e_\alpha(-\mu t^\alpha)(\mu I + A)^{-1} d\mu,$$

where $\Gamma_\theta$ is a suitable integral path.

(ii) A similar argument shows that

$$A^\gamma e_\alpha(-t^\alpha A) = \int_0^\infty \alpha s M_\alpha(s) A^\gamma e^{-st^\alpha A} ds$$

$$= \frac{1}{2\pi i} \int_0^\infty \alpha s M_\alpha(s) \int_{\Gamma_\theta} \mu^\gamma e^{-\mu st^\alpha} (\mu I + A)^{-1} d\mu ds$$

$$= \frac{1}{2\pi i} \int_{\Gamma_\theta} \mu^\gamma e_\alpha(-\mu t^\alpha)(\mu I + A)^{-1} d\mu. \qquad \Box$$

Moreover, we have the following results.

**Lemma 7.1.** *(Wang, Chen and Xiao, 2012) For $t > 0$, $E_\alpha(-t^\alpha A)$ and $e_\alpha(-t^\alpha A)$ are continuous in the uniform operator topology. Moreover, for every $r > 0$, the continuity is uniform on $[r, \infty)$.*

**Lemma 7.2.** *(Wang, Chen and Xiao, 2012) Let $0 < \alpha < 1$. Then*

(i) *for all $u \in X$, $\lim_{t \to 0^+} E_\alpha(-t^\alpha A)u = u$;*
(ii) *for all $u \in D(A)$ and $t > 0$, ${}_0^C D_t^\alpha E_\alpha(-t^\alpha A)u = -AE_\alpha(-t^\alpha A)u$;*
(iii) *for all $u \in X$, $E_\alpha'(-t^\alpha A)u = -t^{\alpha-1} A e_\alpha(-t^\alpha A)u$;*
(iv) *for $t > 0$, $E_\alpha(-t^\alpha A)u = I_t^{1-\alpha}\left(t^{\alpha-1} e_\alpha(-t^\alpha A)u\right)$.*

Before presenting the definition of mild solution of problem (7.2), we give the following lemma for a given function $h : [0, \infty) \to X$. For more details we refer to Zhou, 2014, 2016.

**Lemma 7.3.** *If*

$$u(t) = a + \frac{1}{\Gamma(\alpha)} \int_0^t (t - s)^{\alpha-1}\left(Au(s) + h(s)\right) ds, \text{ for } t \geq 0 \qquad (7.3)$$

*holds, then we have*

$$u(t) = E_\alpha(-t^\alpha A)a + \int_0^t (t - s)^{\alpha-1} e_\alpha(-(t - s)^\alpha A) h(s) ds.$$

We rewrite (7.2) as

$$u(t) = a + \frac{1}{\Gamma(\alpha)} \int_0^t (t - s)^{\alpha-1}\left(Au(s) + F(u(s), u(s)) + Pf(s)\right) ds, \text{ for } t \geq 0.$$

Inspired by above discussion, we adopt the following concepts of mild solution to problem (7.2).

**Definition 7.1.** A function $u : [0, \infty) \to H^{\beta,q}$ is called a global mild solution of problem (7.2) in $H^{\beta,q}$, if $u \in C([0, \infty), H^{\beta,q})$ and for $t \in [0, \infty)$

$$u(t) = E_\alpha(-t^\alpha A)a + \int_0^t (t - s)^{\alpha-1} e_\alpha(-(t - s)^\alpha A) F(u(s), u(s)) ds$$

$$+ \int_0^t (t - s)^{\alpha-1} e_\alpha(-(t - s)^\alpha A) Pf(s) ds. \qquad (7.4)$$

**Definition 7.2.** Let $0 < T < \infty$. A function $u : [0,T] \to H^{\beta,q}$ (or $J_q$) is called a local mild solution of problem (7.2) in $H^{\beta,q}$ (or $J_q$), if $u \in C([0,T], H^{\beta,q})$ (or $C([0,T], J_q)$) and $u$ satisfies (7.4) for $t \in [0,T]$.

For convenience, we define two operators $\Phi$ and $\mathcal{G}$ as follows:

$$\Phi(t) = \int_0^t (t-s)^{\alpha-1} e_\alpha(-(t-s)^\alpha A) f(s) ds,$$

$$\mathcal{G}(u,v)(t) = \int_0^t (t-s)^{\alpha-1} e_\alpha(-(t-s)^\alpha A) F(u(s), v(s)) ds.$$

In subsequent proof we use the following fixed point result.

**Lemma 7.4.** *(Cannone, 1995) Let $(X, |\cdot|_X)$ be a Banach space, $G : X \times X \to X$ a bilinear operator and $L$ a positive real number such that*

$$|G(u,v)|_X \le L|u|_X|v|_X, \ \forall \ u, \ v \in X.$$

*Then for any $u_0 \in X$ with $|u_0|_X < \frac{1}{4L}$, the equation $u = u_0 + G(u,u)$ has a unique solution $u \in X$.*

### 7.2.3    Global Existence

Our main purpose in this subsection is to establish sufficient conditions for existence and uniqueness of mild solution to problem (7.2) in $H^{\beta,q}$. To this end we assume that:

(f) $Pf$ is continuous for $t > 0$ and $|Pf(t)|_q = o(t^{-\alpha(1-\beta)})$ as $t \to 0$ for $0 < \beta < 1$.

**Lemma 7.5.** *(Galdi, 1998; Weissler, 1980) Let $1 < q < \infty$ and $\beta_1 \le \beta_2$. Then there is a constant $C = C(\beta_1, \beta_2)$ such that*

$$|e^{-tA}v|_{H^{\beta_2,q}} \le Ct^{-(\beta_2-\beta_1)} |v|_{H^{\beta_1,q}}, \quad t > 0$$

*for $v \in H^{\beta_1,q}$. Furthermore, $\lim_{t\to 0} t^{(\beta_2-\beta_1)} |e^{-tA}v|_{H^{\beta_2,q}} = 0$.*

Now, we study an important technical lemma, that helps us to prove the main theorems of this subsection.

**Lemma 7.6.** *Let $1 < q < \infty$ and $\beta_1 \le \beta_2$. Then for any $T > 0$, there exists a constant $C_1 = C_1(\beta_1, \beta_2) > 0$ such that*

$$|E_\alpha(-t^\alpha A)v|_{H^{\beta_2,q}} \le C_1 t^{-\alpha(\beta_2-\beta_1)} |v|_{H^{\beta_1,q}}$$

*and*

$$|e_\alpha(-t^\alpha A)v|_{H^{\beta_2,q}} \le C_1 t^{-\alpha(\beta_2-\beta_1)} |v|_{H^{\beta_1,q}}$$

*for all $v \in H^{\beta_1,q}$ and $t \in (0,T]$. Furthermore,*

$$\lim_{t\to 0} t^{\alpha(\beta_2-\beta_1)} |E_\alpha(-t^\alpha A)v|_{H^{\beta_2,q}} = 0.$$

**Proof.** Let $v \in H^{\beta_1,q}$. By Lemma 7.5, we estimate

$$|E_\alpha(-t^\alpha A)v|_{H^{\beta_2,q}} \leq \int_0^\infty M_\alpha(s)|e^{-st^\alpha A}v|_{H^{\beta_2,q}}ds$$

$$\leq \left(C\int_0^\infty M_\alpha(s)s^{-(\beta_2-\beta_1)}ds\right)t^{-\alpha(\beta_2-\beta_1)}|v|_{H^{\beta_1,q}}$$

$$\leq C_1 t^{-\alpha(\beta_2-\beta_1)}|v|_{H^{\beta_1,q}}.$$

More precisely, Lebesgue dominated convergence theorem shows

$$\lim_{t\to0} t^{\alpha(\beta_2-\beta_1)}|E_\alpha(-t^\alpha A)v|_{H^{\beta_2,q}} \leq \int_0^\infty M_\alpha(s)\lim_{t\to0} t^{\alpha(\beta_2-\beta_1)}|e^{-st^\alpha A}v|_{H^{\beta_2,q}}ds = 0.$$

Similarly,

$$|e_\alpha(-t^\alpha A)v|_{H^{\beta_2,q}} \leq \int_0^\infty \alpha s M_\alpha(s)|e^{-st^\alpha A}v|_{H^{\beta_2,q}}ds$$

$$\leq \left(\alpha C\int_0^\infty M_\alpha(s)s^{1-(\beta_2-\beta_1)}ds\right)t^{-\alpha(\beta_2-\beta_1)}|v|_{H^{\beta_1,q}}$$

$$\leq C_1 t^{-\alpha(\beta_2-\beta_1)}|v|_{H^{\beta_1,q}},$$

where the constant $C_1 = C_1(\alpha,\beta_1,\beta_2)$ is such that

$$C_1 \geq C\max\left\{\frac{\Gamma(1-\beta_2+\beta_1)}{\Gamma(1+\alpha(\beta_1-\beta_2))}, \frac{\alpha\Gamma(2-\beta_2+\beta_1)}{\Gamma(1+\alpha(1+\beta_1-\beta_2))}\right\}. \qquad \square$$

For convenience, we denote

$$M(t) = \sup_{s\in(0,t]}\{s^{\alpha(1-\beta)}|Pf(s)|_q\},$$

$$B_1 = C_1\max\{B(\alpha(1-\beta), 1-\alpha(1-\beta)),\ B(\alpha(1-\gamma), 1-\alpha(1-\beta))\},$$

$$L \geq MC_1\max\{B(\alpha(1-\beta), 1-2\alpha(\gamma-\beta)),\ B(\alpha(1-\gamma), 1-2\alpha(\gamma-\beta))\},$$

where $M$ will be given later.

**Theorem 7.1.** *Let $1 < q < \infty$, $0 < \beta < 1$ and (f) hold. For every $a \in H^{\beta,q}$, suppose that*

$$C_1|a|_{H^{\beta,q}} + B_1 M_\infty < \frac{1}{4L}, \qquad (7.5)$$

*where $M_\infty := \sup_{s\in(0,\infty)}\{s^{\alpha(1-\beta)}Pf(s)\}$. If $\frac{n}{2q} - \frac{1}{2} < \beta$, then there exist a $\gamma > \max\{\beta, \frac{1}{2}\}$ and a unique function $u : [0,\infty) \to H^{\beta,q}$ satisfying:*

(a) *$u : [0,\infty) \to H^{\beta,q}$ is continuous and $u(0) = a$;*
(b) *$u : (0,\infty) \to H^{\gamma,q}$ is continuous and $\lim_{t\to0} t^{\alpha(\gamma-\beta)}|u(t)|_{H^{\gamma,q}} = 0$;*
(c) *$u$ satisfies (7.4) for $t \in [0,\infty)$.*

**Proof.** Let $\gamma = \frac{(1+\beta)}{2}$. Define $X_\infty = X[\infty]$ as the space of all curves $u : (0, \infty) \to H^{\beta,q}$ such that:

(i) $u : [0, \infty) \to H^{\beta,q}$ is bounded and continuous;
(ii) $u : (0, \infty) \to H^{\gamma,q}$ is bounded and continuous, moreover,

$$\lim_{t \to 0} t^{\alpha(\gamma-\beta)} |u(t)|_{H^{\gamma,q}} = 0$$

with its natural norm

$$\|u\|_{X_\infty} = \max \left\{ \sup_{t \geq 0} |u(t)|_{H^{\beta,q}}, \ \sup_{t \geq 0} t^{\alpha(\gamma-\beta)} |u(t)|_{H^{\gamma,q}} \right\}.$$

It is obvious that $X_\infty$ is a non-empty complete metric space.

From an argument of Weissler, 1980, we know that $F : H^{\gamma,q} \times H^{\gamma,q} \to J_q$ is a bounded bilinear map, then there exists $M$ such that for $u, v \in H^{\gamma,q}$

$$\begin{aligned} |F(u, v)|_q &\leq M |u|_{H^{\gamma,q}} |v|_{H^{\gamma,q}}, \\ |F(u, u) - F(v, v)|_q &\leq M(|u|_{H^{\gamma,q}} + |v|_{H^{\gamma,q}})|u - v|_{H^{\gamma,q}}. \end{aligned} \tag{7.6}$$

**Claim I.** The operator $\mathcal{G}(u(t), v(t))$ belongs to $C([0, \infty), H^{\beta,q})$ as well as $C((0, \infty), H^{\gamma,q})$ for $u, v \in X_\infty$. For arbitrary $t_0 \geq 0$ fixed and $\varepsilon > 0$ enough small, consider $t > t_0$ (the case $t < t_0$ follows analogously), we have

$$|\mathcal{G}(u(t), v(t)) - \mathcal{G}(u(t_0), v(t_0))|_{H^{\beta,q}}$$

$$\leq \int_{t_0}^{t} (t - s)^{\alpha-1} |e_\alpha(-(t - s)^\alpha A) F(u(s), v(s))|_{H^{\beta,q}} ds$$

$$+ \int_{0}^{t_0} \left| ((t - s)^{\alpha-1} - (t_0 - s)^{\alpha-1}) e_\alpha(-(t - s)^\alpha A) F(u(s), v(s)) \right|_{H^{\beta,q}} ds$$

$$+ \int_{0}^{t_0-\varepsilon} (t_0 - s)^{\alpha-1} \left| \left( e_\alpha(-(t - s)^\alpha A) - e_\alpha(-(t_0 - s)^\alpha A) \right) F(u(s), v(s)) \right|_{H^{\beta,q}} ds$$

$$+ \int_{t_0-\varepsilon}^{t_0} (t_0 - s)^{\alpha-1} \left| \left( e_\alpha(-(t - s)^\alpha A) - e_\alpha(-(t_0 - s)^\alpha A) \right) F(u(s), v(s)) \right|_{H^{\beta,q}} ds$$

$$=: I_{11}(t) + I_{12}(t) + I_{13}(t) + I_{14}(t).$$

We estimate each of the four terms separately. For $I_{11}(t)$, in view of Lemma 7.6, we obtain

$$I_{11}(t) \leq C_1 \int_{t_0}^{t} (t - s)^{\alpha(1-\beta)-1} |F(u(s), v(s))|_q ds$$

$$\leq MC_1 \int_{t_0}^{t} (t - s)^{\alpha(1-\beta)-1} |u(s)|_{H^{\gamma,q}} |v(s)|_{H^{\gamma,q}} ds$$

$$\leq MC_1 \int_{t_0}^{t} (t - s)^{\alpha(1-\beta)-1} s^{-2\alpha(\gamma-\beta)} ds \ \sup_{s \in [0,t]} \{ s^{2\alpha(\gamma-\beta)} |u(s)|_{H^{\gamma,q}} |v(s)|_{H^{\gamma,q}} \}$$

$$= MC_1 \int_{t_0/t}^{1} (1 - s)^{\alpha(1-\beta)-1} s^{-2\alpha(\gamma-\beta)} ds \ \sup_{s \in [0,t]} \{ s^{2\alpha(\gamma-\beta)} |u(s)|_{H^{\gamma,q}} |v(s)|_{H^{\gamma,q}} \}.$$

By the properties of the Beta function, there exists $\delta > 0$ small enough such that for $0 < t - t_0 < \delta$,

$$\int_{t_0/t}^{1} (1-s)^{\alpha(1-\beta)-1} s^{-2\alpha(\gamma-\beta)} ds \to 0,$$

which follows that $I_{11}(t)$ tends to 0 as $t - t_0 \to 0$. For $I_{12}(t)$, since

$$I_{12}(t) \le C_1 \int_0^{t_0} \left( (t_0-s)^{\alpha-1} - (t-s)^{\alpha-1} \right)(t-s)^{-\alpha\beta} |F(u(s), v(s))|_q ds$$

$$\le MC_1 \int_0^{t_0} \left( (t_0-s)^{\alpha-1} - (t-s)^{\alpha-1} \right)(t-s)^{-\alpha\beta} s^{-2\alpha(\gamma-\beta)} ds$$

$$\times \sup_{s \in [0,t_0]} \{ s^{2\alpha(\gamma-\beta)} |u(s)|_{H^{\gamma,q}} |v(s)|_{H^{\gamma,q}} \},$$

noting that

$$\int_0^{t_0} \left| (t_0-s)^{\alpha-1} - (t-s)^{\alpha-1} \right| (t-s)^{-\alpha\beta} s^{-2\alpha(\gamma-\beta)} ds$$

$$\le \int_0^{t_0} (t-s)^{\alpha-1}(t-s)^{-\alpha\beta} s^{-2\alpha(\gamma-\beta)} ds + \int_0^{t_0} (t_0-s)^{\alpha-1}(t-s)^{-\alpha\beta} s^{-2\alpha(\gamma-\beta)} ds$$

$$\le 2 \int_0^{t_0} (t_0-s)^{\alpha(1-\beta)-1} s^{-2\alpha(\gamma-\beta)} ds$$

$$= 2B(\alpha(1-\beta), 1 - 2\alpha(\gamma-\beta)),$$

then by Lebesgue dominated convergence theorem, we have

$$\int_0^{t_0} \left( (t_0-s)^{\alpha-1} - (t-s)^{\alpha-1} \right)(t-s)^{-\alpha\beta} s^{-2\alpha(\gamma-\beta)} ds \to 0, \text{ as } t \to t_0,$$

one deduces that $\lim_{t \to t_0} I_{12}(t) = 0$. For $I_{13}(t)$, since

$$I_{13}(t) \le \int_0^{t_0-\varepsilon} (t_0-s)^{\alpha-1} \left| \left( e_\alpha(-(t-s)^\alpha A) + e_\alpha(-(t_0-s)^\alpha A) \right) F(u(s), v(s)) \right|_{H^{\beta,q}} ds$$

$$\le \int_0^{t_0-\varepsilon} (t_0-s)^{\alpha-1} \left( (t-s)^{-\alpha\beta} + (t_0-s)^{-\alpha\beta} \right) |F(u(s), v(s))|_q ds$$

$$\le 2MC_1 \int_0^{t_0-\varepsilon} (t_0-s)^{\alpha(1-\beta)-1} s^{-2\alpha(\gamma-\beta)} ds$$

$$\times \sup_{s \in [0,t_0]} \{ s^{2\alpha(\gamma-\beta)} |u(s)|_{H^{\gamma,q}} |v(s)|_{H^{\gamma,q}} \},$$

using Lebesgue dominated convergence theorem again, the fact from the uniform continuity of the operator $e_\alpha(-t^\alpha A)$ due to Lemma 7.1 shows

$$\lim_{t \to t_0} I_{13}(t) = \int_0^{t_0-\varepsilon} (t_0-s)^{\alpha-1}$$

$$\times \lim_{t \to t_0} \left| \left( e_\alpha(-(t-s)^\alpha A) - e_\alpha(-(t_0-s)^\alpha A) \right) F(u(s), v(s)) \right|_{H^{\beta,q}} ds$$

$$= 0.$$

For $I_{14}(t)$, by immediate calculation, we estimate

$$I_{14}(t) \leq \int_{t_0-\varepsilon}^{t_0} (t_0 - s)^{\alpha-1}\big((t - s)^{-\alpha\beta} + (t_0 - s)^{-\alpha\beta}\big)|F(u(s), v(s))|_q ds$$

$$\leq 2MC_1 \int_{t_0-\varepsilon}^{t_0} (t_0 - s)^{\alpha(1-\beta)-1} s^{-2\alpha(\gamma-\beta)} ds$$

$$\times \sup_{s\in[t_0-\varepsilon,t_0]} \{s^{2\alpha(\gamma-\beta)}|u(s)|_{H^{\gamma,q}}|v(s)|_{H^{\gamma,q}}\}$$

$$\to 0, \text{ as } \varepsilon \to 0$$

according to the properties of the Beta function. Thenceforth, it follows

$$|\mathcal{G}(u(t), v(t)) - \mathcal{G}(u(t_0), v(t_0))|_{H^{\beta,q}} \to 0, \text{ as } t \to t_0.$$

The continuity of the operator $\mathcal{G}(u, v)$ evaluated in $C((0, \infty), H^{\gamma,q})$ follows by the similar discussion as above. So, we omit the details.

**Claim II.** The operator $\mathcal{G} : X_\infty \times X_\infty \to X_\infty$ is a continuous bilinear operator. By Lemma 7.6, we have

$$|\mathcal{G}(u(t), v(t))|_{H^{\beta,q}} \leq \left| \int_0^t (t - s)^{\alpha-1} e_\alpha(-(t - s)^\alpha A) F(u(s), v(s)) ds \right|_{H^{\beta,q}}$$

$$\leq C_1 \int_0^t (t - s)^{\alpha(1-\beta)-1} |F(u(s), v(s))|_q ds$$

$$\leq MC_1 \int_0^t (t - s)^{\alpha(1-\beta)-1} s^{-2\alpha(\gamma-\beta)} ds$$

$$\times \sup_{s\in[0,t]} \{s^{2\alpha(\gamma-\beta)}|u(s)|_{H^{\gamma,q}}|v(s)|_{H^{\gamma,q}}\}$$

$$\leq MC_1 B(\alpha(1 - \beta), 1 - 2\alpha(\gamma - \beta))\|u\|_{X_\infty}\|v\|_{X_\infty}$$

and

$$|\mathcal{G}(u(t), v(t))|_{H^{\gamma,q}} \leq \left| \int_0^t (t - s)^{\alpha-1} e_\alpha(-(t - s)^\alpha A) F(u(s), v(s)) ds \right|_{H^{\gamma,q}}$$

$$\leq C_1 \int_0^t (t - s)^{\alpha(1-\gamma)-1} |F(u(s), v(s))|_q ds$$

$$\leq MC_1 \int_0^t (t - s)^{\alpha(1-\gamma)-1} s^{-2\alpha(\gamma-\beta)} ds$$

$$\times \sup_{s\in[0,t]} \left\{s^{2\alpha(\gamma-\beta)}|u(s)|_{H^{\gamma,q}}|v(s)|_{H^{\gamma,q}}\right\}$$

$$\leq MC_1 t^{-\alpha(\gamma-\beta)} B(\alpha(1 - \gamma), 1 - 2\alpha(\gamma - \beta))\|u\|_{X_\infty}\|v\|_{X_\infty},$$

it follows that

$$\sup_{t\in[0,\infty)} t^{\alpha(\gamma-\beta)}|\mathcal{G}(u(t), v(t))|_{H^{\gamma,q}} \leq MC_1 B(\alpha(1 - \gamma), 1 - 2\alpha(\gamma - \beta))\|u\|_{X_\infty}\|v\|_{X_\infty}.$$

More precisely,

$$\lim_{t \to 0} t^{\alpha(\gamma-\beta)} |\mathcal{G}(u(t), v(t))|_{H^{\gamma,q}} = 0.$$

Hence, $\mathcal{G}(u, v) \in X_\infty$ and $\|\mathcal{G}(u(t), v(t))\|_{X_\infty} \leq L\|u\|_{X_\infty}\|v\|_{X_\infty}$.

**Claim III.** Condition (c) holds. Let $0 < t_0 < t$. Since

$$|\Phi(t) - \Phi(t_0)|_{H^{\beta,q}}$$

$$\leq \int_{t_0}^t (t - s)^{\alpha-1} |e_\alpha(-(t - s)^\alpha A) P f(s)|_{H^{\beta,q}} ds$$

$$+ \int_0^{t_0} ((t_0 - s)^{\alpha-1} - (t - s)^{\alpha-1}) |e_\alpha(-(t - s)^\alpha A) P f(s)|_{H^{\beta,q}} ds$$

$$+ \int_0^{t_0-\varepsilon} (t_0 - s)^{\alpha-1} \left| \left( e_\alpha(-(t - s)^\alpha A) - e_\alpha(-(t_0 - s)^\alpha A) \right) P f(s) \right|_{H^{\beta,q}} ds$$

$$+ \int_{t_0-\varepsilon}^{t_0} (t_0 - s)^{\alpha-1} \left| \left( e_\alpha(-(t - s)^\alpha A) - e_\alpha(-(t_0 - s)^\alpha A) \right) P f(s) \right|_{H^{\beta,q}} ds$$

$$\leq C_1 \int_{t_0}^t (t - s)^{\alpha(1-\beta)-1} |P f(s)|_q ds$$

$$+ C_1 \int_0^{t_0} ((t_0 - s)^{\alpha-1} - (t - s)^{\alpha-1})(t - s)^{-\alpha\beta} |P f(s)|_q ds$$

$$+ C_1 \int_0^{t_0-\varepsilon} (t_0 - s)^{\alpha-1} \left| \left( e_\alpha(-(t - s)^\alpha A) - e_\alpha(-(t_0 - s)^\alpha A) \right) P f(s) \right|_{H^{\beta,q}} ds$$

$$+ 2C_1 \int_{t_0-\varepsilon}^{t_0} (t_0 - s)^{\alpha(1-\beta)-1} |P f(s)|_q ds$$

$$\leq C_1 M(t) \int_{t_0}^t (t - s)^{\alpha(1-\beta)-1} s^{-\alpha(1-\beta)} ds$$

$$+ C_1 M(t) \int_0^{t_0} ((t - s)^{\alpha-1} - (t_0 - s)^{\alpha-1}) s^{-\alpha(1-\beta)} ds$$

$$+ C_1 M(t) \int_0^{t_0-\varepsilon} (t_0 - s)^{\alpha-1} \left| \left( e_\alpha(-(t - s)^\alpha A) - e_\alpha(-(t_0 - s)^\alpha A) \right) P f(s) \right|_{H^{\beta,q}} ds$$

$$+ 2C_1 M(t) \int_{t_0-\varepsilon}^{t_0} (t_0 - s)^{\alpha(1-\beta)-1} s^{-\alpha(1-\beta)} ds.$$

By the properties of the Beta function, the first two integrals and the last integral tend to 0 as $t \to t_0$ as well as $\varepsilon \to 0$. In view of Lemma 7.1, the third integral also goes to 0 as $t \to t_0$, which implies

$$|\Phi(t) - \Phi(t_0)|_{H^{\beta,q}} \to 0, \quad \text{as } t \to t_0.$$

The continuity of $\Phi(t)$ evaluated in $H^{\gamma,q}$ follows by the similar argument as above.

On the other hand, we have

$$|\Phi(t)|_{H^{\beta,q}} \leq \left| \int_0^t (t - s)^{\alpha-1} e_\alpha(-(t - s)^\alpha A) P f(s) ds \right|_{H^{\beta,q}}$$

$$\leq C_1 \int_0^t (t-s)^{\alpha(1-\beta)-1} |Pf(s)|_q ds$$

$$\leq C_1 M(t) \int_0^t (t-s)^{\alpha(1-\beta)-1} s^{-\alpha(1-\beta)} ds$$

$$= C_1 M(t) B(\alpha(1-\beta), 1 - \alpha(1-\beta)), \tag{7.7}$$

and

$$|\Phi(t)|_{H^{\gamma,q}} \leq \left| \int_0^t (t-s)^{\alpha-1} e_\alpha(-(t-s)^\alpha A) Pf(s) ds \right|_{H^{\gamma,q}}$$

$$\leq C_1 \int_0^t (t-s)^{\alpha(1-\gamma)-1} |Pf(s)|_q ds$$

$$\leq C_1 M(t) \int_0^t (t-s)^{\alpha(1-\gamma)-1} s^{-\alpha(1-\beta)} ds$$

$$= t^{-\alpha(\gamma-\beta)} C_1 M(t) B(\alpha(1-\gamma), 1 - \alpha(1-\beta)).$$

More precisely,

$$t^{\alpha(\gamma-\beta)} |\Phi(t)|_{H^{\gamma,q}} \leq C_1 M(t) B(\alpha(1-\gamma), 1 - \alpha(1-\beta)) \to 0, \text{ as } t \to 0,$$

since $M(t) \to 0$ as $t \to 0$ due to assumption (f). This ensures that $\Phi(t) \in X_\infty$ and $\|\Phi(t)\|_\infty \leq B_1 M_\infty$.

For $a \in H^{\beta,q}$. By Lemma 7.1, it is easy to see that

$$E_\alpha(-t^\alpha A)a \in C([0,\infty), H^{\beta,q}) \text{ and } E_\alpha(-t^\alpha A)a \in C((0,\infty), H^{\gamma,q}).$$

This, together with Lemma 7.6, implies that for all $t \in (0,T]$,

$$E_\alpha(-t^\alpha A)a \in X_\infty,$$

$$t^{\alpha(\gamma-\beta)} E_\alpha(-t^\alpha A)a \in C([0,\infty), H^{\gamma,q}),$$

$$\|E_\alpha(-t^\alpha A)a\|_{X_\infty} \leq C_1 |a|_{H^{\beta,q}}.$$

According to (7.5), the inequality

$$\|E_\alpha(-t^\alpha A)a + \Phi(t)\|_{X_\infty} \leq \|E_\alpha(-t^\alpha A)a\|_{X_\infty} + \|\Phi(t)\|_{X_\infty} \leq \frac{1}{4L}$$

holds, which yields that $\mathcal{F}$ has a unique fixed point.

**Claim IV.** $u(t) \to a$ in $H^{\beta,q}$ as $t \to 0$. We need to verify

$$\lim_{t \to 0} \int_0^t (t-s)^{\alpha-1} e_\alpha(-(t-s)^\alpha A) Pf(s) ds = 0,$$

$$\lim_{t \to 0} \int_0^t (t-s)^{\alpha-1} e_\alpha(-(t-s)^\alpha A) F(u(s), u(s)) ds = 0$$

in $H^{\beta,q}$. In fact, it is obvious that $\lim_{t\to0}\Phi(t)=0$ ($\lim_{t\to0}M(t)=0$) owing to (7.7). In addition,

$$\left|\int_0^t (t-s)^{\alpha-1} e_\alpha(-(t-s)^\alpha A)F(u(s),u(s))ds\right|_{H^{\beta,q}}$$

$$\leq C_1 \int_0^t (t-s)^{\alpha(1-\beta)-1}|F(u(s),u(s))|_q ds$$

$$\leq MC_1 \int_0^t (t-s)^{\alpha(1-\beta)-1}|u(s)|^2_{H^{\gamma,q}} ds$$

$$\leq MC_1 \int_0^t (t-s)^{\alpha(1-\beta)-1} s^{-2\alpha(\gamma-\beta)} ds \sup_{s\in[0,t]}\{s^{2\alpha(\gamma-\beta)}|u(s)|^2_{H^{\gamma,q}}\}$$

$$= MC_1 B(\alpha(1-\beta),1-2\alpha(\gamma-\beta)) \sup_{s\in[0,t]}\{s^{2\alpha(\gamma-\beta)}|u(s)|^2_{H^{\gamma,q}}\} \to 0, \text{ as } t\to0.$$

$\square$

### 7.2.4 Local Existence

In this subsection, we study the local mild solution of problem (7.2) in $H^{\beta,q}$ and $J_q$.

**Theorem 7.2.** Let $1<q<\infty$, $0<\beta<1$ and (f) hold. Suppose

$$\frac{n}{2q}-\frac{1}{2}<\beta. \tag{7.8}$$

Then there is a $\gamma>\max\{\beta,\frac{1}{2}\}$ such that for every $a\in H^{\beta,q}$ there exist $T_*>0$ and a unique function $u:[0,T_*]\to H^{\beta,q}$ satisfying:

(a) $u:[0,T_*]\to H^{\beta,q}$ is continuous and $u(0)=a$;
(b) $u:(0,T_*]\to H^{\gamma,q}$ is continuous and $\lim_{t\to0} t^{\alpha(\gamma-\beta)}|u(t)|_{H^{\gamma,q}}=0$;
(c) $u$ satisfies (7.4) for $t\in[0,T_*]$.

**Proof.** Let $\gamma=\frac{(1+\beta)}{2}$. Fix $a\in H^{\beta,q}$. Let $X=X[T]$ be the space of all curves $u:(0,T]\to H^{\beta,q}$ such that:

(i) $u:[0,T]\to H^{\beta,q}$ is continuous;
(ii) $u:(0,T]\to H^{\gamma,q}$ is continuous and $\lim_{t\to0} t^{\alpha(\gamma-\beta)}|u(t)|_{H^{\gamma,q}}=0$;

with its natural norm

$$\|u\|_X = \sup_{t\in[0,T]}\{t^{\alpha(\gamma-\beta)}|u(t)|_{H^{\gamma,q}}\}.$$

Similar to the proof of Theorem 7.1, it is easy to claim that $\mathcal{G}:X\times X\to X$ is continuous linear map and $\Phi(t)\in X$.

By Lemma 7.1, it is easy to see that for all $t\in(0,T]$,

$$E_\alpha(-t^\alpha A)a\in C([0,T],H^{\beta,q}),$$

$$E_\alpha(-t^\alpha A)a \in C((0,T], H^{\gamma,q}).$$

From Lemma 7.6, it follows that

$$E_\alpha(-t^\alpha A)a \in X,$$

$$t^{\alpha(\gamma-\beta)}E_\alpha(-t^\alpha A)a \in C([0,T], H^{\gamma,q}).$$

Hence, let $T_* > 0$ be sufficiently small such that

$$\|E_\alpha(-t^\alpha A)a + \Phi(t)\|_{X[T_*]} \leq \|E_\alpha(-t^\alpha A)a\|_{X[T_*]} + \|\Phi(t)\|_{X[T_*]} < \frac{1}{4L},$$

which implies that $\mathcal{F}$ has a unique fixed point due to Lemma 7.4.     □

Let $\gamma = \frac{(1+\beta)}{2}$.

**Theorem 7.3.** *Let $1 < q < \infty$, $0 < \beta < 1$ and (f) hold. Suppose that*

$$a \in H^{\beta,q} \text{ with } \frac{n}{2q} - \frac{1}{2} < \beta.$$

*Then problem (7.2) has a unique mild solution $u$ in $J_q$ for $a \in H^{\beta,q}$. Moreover, $u$ is continuous on $[0,T]$, $A^\gamma u$ is continuous in $(0,T]$ and $t^{\alpha(\gamma-\beta)}A^\gamma u(t)$ is bounded as $t \to 0$.*

**Proof. Step I.** Set

$$K(t) := \sup_{s \in (0,t]} s^{\alpha(\gamma-\beta)}|A^\gamma u(s)|_q$$

and

$$\Psi(t) := \mathcal{G}(u,u)(t) = \int_0^t (t-s)^{\alpha-1} e_\alpha(-(t-s)^\alpha A)F(u(s),u(s))ds.$$

As an immediate consequence of Claim II in Theorem 7.1, $\Psi(t)$ is continuous in $[0,T]$, $A^\gamma \Psi(t)$ exists and is continuous in $(0,T]$ with

$$|A^\gamma \Psi(t)|_q \leq MC_1 B(\alpha(1-\gamma), 1 - 2\alpha(\gamma-\beta))K^2(t)t^{-\alpha(\gamma-\beta)}. \tag{7.9}$$

We also consider the integral $\Phi(t)$. Since (f) holds, the inequality

$$|Pf(s)|_q \leq M(t)s^{\alpha(1-\beta)}$$

is satisfied with a continuous function $M(t)$. From Claim III in Theorem 7.1, we derive that $A^\gamma \Phi(t)$ is continuous in in $(0,T]$ with

$$|A^\gamma \Phi(t)|_q \leq C_1 M(t)B(\alpha(1-\gamma), 1 - \alpha(1-\beta))t^{-\alpha(\gamma-\beta)}. \tag{7.10}$$

For $|Pf(t)|_q = o(t^{-\alpha(1-\beta)})$ as $t \to 0$, we have $M(t) = 0$. Here (7.10) means $|A^\gamma \Phi(t)|_q = o(t^{-\alpha(\gamma-\beta)})$ as $t \to 0$.

We prove that $\Phi$ is continuous in $J_q$. In fact, take $0 \leq t_0 < t < T$, we have

$$|\Phi(t) - \Phi(t_0)|_q$$

$$\leq C_3 \int_{t_0}^t (t-s)^{\alpha-1}|Pf(s)|_q ds + C_3 \int_0^{t_0} ((t_0-s)^{\alpha-1} - (t-s)^{\alpha-1})|Pf(s)|_q ds$$

$$+ C_3 \int_0^{t_0-\varepsilon} (t_0 - s)^{\alpha-1} \|e_\alpha(-(t - s)^\alpha A) - e_\alpha(-(t_0 - s)^\alpha A)\| |Pf(s)|_q ds$$

$$+ 2C_3 \int_{t_0-\varepsilon}^{t_0} (t_0 - s)^{\alpha-1} |Pf(s)|_q ds$$

$$\leq C_3 M(t) \int_{t_0}^t (t - s)^{\alpha-1} s^{-\alpha(1-\beta)} ds$$

$$+ C_3 M(t) \int_0^{t_0} \left((t - s)^{\alpha-1} - (t_0 - s)^{\alpha-1}\right) s^{-\alpha(1-\beta)} ds$$

$$+ C_3 M(t) \int_0^{t_0-\varepsilon} (t_0 - s)^{\alpha-1} s^{-\alpha(1-\beta)} ds$$

$$\times \sup_{s\in[0,t-\varepsilon]} \|e_\alpha(-(t - s)^\alpha A) - e_\alpha(-(t_0 - s)^\alpha A)\|$$

$$+ 2C_3 M(t) \int_{t_0-\varepsilon}^{t_0} (t_0 - s)^{\alpha-1} s^{-\alpha(1-\beta)} ds \to 0, \text{ as } t \to t_0$$

by previous discussion.

Further, we consider the function $E_\alpha(-t^\alpha A)a$. It is obvious by Lemma 7.6 that

$$|A^\gamma E_\alpha(-t^\alpha A)a|_q \leq C_1 t^{-\alpha(\gamma-\beta)} |A^\beta a|_q = C_1 t^{-\alpha(\gamma-\beta)} |a|_{H^{\beta,q}},$$

$$\lim_{t\to 0} t^{\alpha(\gamma-\beta)} |A^\gamma E_\alpha(-t^\alpha A)a|_q = \lim_{t\to 0} t^{\alpha(\gamma-\beta)} |E_\alpha(-t^\alpha A)a|_{H^{\gamma,q}} = 0.$$

**Step II.** Now we construct the solution by the successive approximation:

$$u_0(t) = E_\alpha(-t^\alpha A)a + \Phi(t),$$

$$u_{n+1}(t) = u_0(t) + \mathcal{G}(u_n, u_n)(t), \quad n = 0, 1, 2, \ldots. \tag{7.11}$$

Making use of above results, we know that

$$K_n(t) := \sup_{s\in(0,t]} s^{\alpha(\gamma-\beta)} |A^\gamma u_n(s)|_q$$

are continuous and increasing function on $[0, T]$ with $K_n(0) = 0$. Furthermore, in virtue of (7.10) and (7.11), $K_n(t)$ fulfills the following inequality

$$K_{n+1}(t) \leq K_0(t) + MC_1 B(\alpha(1 - \gamma), 1 - 2\alpha(\gamma - \beta)) K_n^2(t). \tag{7.12}$$

For $K_0(0) = 0$, we choose a $T > 0$ such that

$$4MC_1 B(\alpha(1 - \gamma), 1 - 2\alpha(\gamma - \beta)) K_0(T) < 1. \tag{7.13}$$

Then a fundamental consideration of (7.12) ensures that the sequence $\{K_n(T)\}$ is bounded, i.e.,

$$K_n(T) \leq \rho(T), \quad n = 0, 1, 2, \ldots,$$

where

$$\rho(t) = \frac{1 - \sqrt{1 - 4MC_1 B(\alpha(1 - \gamma), 1 - 2\alpha(\gamma - \beta)) K_0(t)}}{2MC_1 B(\alpha(1 - \gamma), 1 - 2\alpha(\gamma - \beta))}.$$

Analogously, for any $t \in (0, T]$, $K_n(t) \leq \rho(t)$ holds. In the same way we note that $\rho(t) \leq 2K_0(t)$.

Let us consider the equality

$$w_{n+1}(t) = \int_0^t (t-s)^{\alpha-1} e_\alpha(-(t-s)^\alpha A)[F(u_{n+1}(s), u_{n+1}(s)) - F(u_n(s), u_n(s))]ds,$$

where $w_n = u_{n+1} - u_n$, $n = 0, 1, \ldots$, and $t \in (0, T]$. Writing

$$W_n(t) := \sup_{s \in (0,t]} s^{\alpha(\gamma-\beta)} |A^\gamma w_n(s)|_q.$$

On account of (7.6), we have

$$|F(u_{n+1}(s), u_{n+1}(s)) - F(u_n(s), u_n(s))|_q \leq M(K_{n+1}(s) + K_n(t))W_n(s)s^{-2\alpha(\gamma-\beta)},$$

which follows from Claim II in Theorem 7.1 that

$$t^{\alpha(\gamma-\beta)} |A^\gamma w_{n+1}(t)|_q \leq 2MC_1 B(\alpha(1-\gamma), 1 - \alpha(1-\beta))\rho(t)W_n(t).$$

This inequality gives

$$W_{n+1}(T) \leq 2MC_1 B(\alpha(1-\gamma), 1 - 2\alpha(\gamma-\beta))\rho(T)W_n(T)$$
$$\leq 4MC_1 B(\alpha(1-\gamma), 1 - 2\alpha(\gamma-\beta))K_0(T)W_n(T). \tag{7.14}$$

According to (7.13) and (7.14), it is easy to see that

$$\lim_{n \to 0} \frac{W_{n+1}(T)}{W_n(T)} < 4MC_1 B(\alpha(1-\gamma), 1 - 2\alpha(\gamma-\beta)) < 1,$$

thus the series $\sum_{n=0}^\infty W_n(T)$ converges. It shows that

the series $\sum_{n=0}^\infty t^{\alpha(\gamma-\beta)} A^\gamma w_n(t)$ converges uniformly for $t \in (0, T]$,

therefore, the sequence $\{t^{\alpha(\gamma-\beta)} A^\gamma u_n(t)\}$ converges uniformly in $(0, T]$. This implies that

$$\lim_{n \to \infty} u_n(t) = u(t) \in D(A^\gamma)$$

and

$$\lim_{n \to \infty} t^{\alpha(\gamma-\beta)} A^\gamma u_n(t) = t^{\alpha(\gamma-\beta)} A^\gamma u(t) \text{ uniformly,}$$

since $A^{-\gamma}$ is bounded and $A^\gamma$ is closed. Accordingly, the function

$$K(t) = \sup_{s \in (0,t]} s^{\alpha(\gamma-\beta)} |A^\gamma u(s)|_q$$

also satisfies

$$K(t) \leq \rho(t) \leq 2K_0(t), \ t \in (0, t] \tag{7.15}$$

and

$$\varsigma_n := \sup_{s \in (0,T]} s^{2\alpha(\gamma-\beta)} |F(u_n(s), u_n(s)) - F(u(s), u(s))|_q$$

$$\leq M(K_n(T) + K(T)) \sup_{s \in (0,T]} s^{\alpha(\gamma-\beta)} |A^\gamma(u_n(s) - u(s))|_q$$

$$\to 0, \text{ as } n \to \infty.$$

Finally, it remains to verify that $u$ is a mild solution of problem (7.2) in $[0, T]$. Since

$$|\mathcal{G}(u_n, u_n)(t) - \mathcal{G}(u, u)(t)|_q \leq \int_0^t (t-s)^{\alpha-1} \varsigma_n s^{-2\alpha(\gamma-\beta)} ds = t^{\alpha\beta} \varsigma_n \to 0, \ (n \to \infty),$$

we have $\mathcal{G}(u_n, u_n)(t) \to \mathcal{G}(u, u)(t)$. Taking the limits on both sides of (7.11), we derive

$$u(t) = u_0(t) + \mathcal{G}(u, u)(t). \tag{7.16}$$

Let $u(0) = a$, we find that (7.16) holds for $t \in [0, T]$ and $u \in C([0, T], J_q)$. What is more, the uniform convergence of $t^{\alpha(\gamma-\beta)} A^\gamma u_n(t)$ to $t^{\alpha(\gamma-\beta)} A^\gamma u(t)$ derives the continuity of $A^\gamma u(t)$ on $(0, T]$. From (7.15) and $K_0(0) = 0$, we get that $|A^\gamma u(t)|_q = o(t^{-\alpha(\gamma-\beta)})$ is obvious.

**Step III.** We prove that the mild solution is unique. Suppose that $u$ and $v$ are mild solutions of problem (7.2). Let $w = u - v$, we consider the equality

$$w(t) = \int_0^t (t-s)^{\alpha-1} e_\alpha(-(t-s)^\alpha A)[F(u(s), u(s)) - F(v(s), v(s))] ds.$$

Introducing the functions

$$\widetilde{K}(t) := \max \left\{ \sup_{s \in (0,t]} s^{\alpha(\gamma-\beta)} |A^\gamma u(s)|_q, \ \sup_{s \in (0,t]} s^{\alpha(\gamma-\beta)} |A^\gamma v(s)|_q \right\}.$$

By (7.6) and Lemma 7.6, we get

$$|A^\gamma w(t)|_q \leq MC_1 \widetilde{K}(t) \int_0^t (t-s)^{\alpha(1-\gamma)-1} s^{-\alpha(\gamma-\beta))} |A^\gamma w(s)|_q ds.$$

Gronwall inequality shows that $A^\gamma w(t) = 0$ for $t \in (0, T]$. This implies that $w(t) = u(t) - v(t) \equiv 0$ for $t \in [0, T]$. Therefore the mild solution is unique. □

## 7.2.5 *Regularity*

In this subsection, we consider the regularity of a solution $u$ which satisfies problem (7.2). Throughout this part we assume that:

($f_1$) $Pf(t)$ is Hölder continuous with an exponent $\vartheta \in (0, \alpha(1-\gamma))$, that is,

$$|Pf(t) - Pf(s)|_q \leq L|t - s|^\vartheta, \quad \text{for all } 0 < t, \ s \leq T.$$

**Definition 7.3.** A function $u : [0, T] \to J_q$ is called a classical solution of problem (7.2), if $u \in C([0, T], J_q)$ with ${}^C_0 D_t^\alpha u(t) \in C((0, T], J_q)$, which takes values in $D(A)$ and satisfies (7.2) for all $t \in (0, T]$.

**Lemma 7.7.** *Let* $(f_1)$ *be satisfied. If*

$$\Phi_1(t) := \int_0^t (t-s)^{\alpha-1} e_\alpha(-(t-s)^\alpha A)\big(Pf(s) - Pf(t)\big) ds, \text{ for } t \in (0,T],$$

*then* $\Phi_1(t) \in D(A)$ *and* $A\Phi_1(t) \in C^\vartheta([0,T], J_q)$.

**Proof.** For fixed $t \in (0,T]$, from Lemma 7.6 and $(f_1)$, we have

$$(t-s)^{\alpha-1}|Ae_\alpha(-(t-s)^\alpha A)\big(Pf(s) - Pf(t)\big)|_q$$
$$\leq (t-s)^{-1}|Pf(s) - Pf(t)|_q$$
$$\leq C_1 L (t-s)^{\vartheta-1} \in L^1([0,T], J_q), \tag{7.17}$$

then

$$|A\Phi_1(t)|_q \leq \int_0^t (t-s)^{\alpha-1}|Ae_\alpha(-(t-s)^\alpha A)\big(Pf(s) - Pf(t)\big)|_q ds$$
$$\leq C_1 L \int_0^t (t-s)^{\vartheta-1} ds$$
$$\leq \frac{C_1 K}{\vartheta} t^\vartheta < \infty.$$

By the closeness of $A$, we obtain $\Phi_1(t) \in D(A)$.

We need to show that $A\Phi_1(t)$ is Hölder continuous. Since

$$\frac{d}{dt}\big(t^{\alpha-1} e_\alpha(-\mu t^\alpha)\big) = t^{\alpha-2} E_{\alpha,\alpha-1}(-\mu t^\alpha),$$

then

$$\frac{d}{dt}\big(t^{\alpha-1} Ae_\alpha(-t^\alpha A)\big)$$
$$= \frac{1}{2\pi i} \int_{\Gamma_\theta} t^{\alpha-2} E_{\alpha,\alpha-1}(-\mu t^\alpha) A(\mu I + A)^{-1} d\mu$$
$$= \frac{1}{2\pi i} \int_{\Gamma_\theta} t^{\alpha-2} E_{\alpha,\alpha-1}(-\mu t^\alpha) d\mu - \frac{1}{2\pi i} \int_{\Gamma_\theta} t^{\alpha-2} \mu E_{\alpha,\alpha-1}(-\mu t^\alpha)(\mu I + A)^{-1} d\mu$$
$$= \frac{1}{2\pi i} \int_{\Gamma'_\theta} -t^{\alpha-2} E_{\alpha,\alpha-1}(\xi) \frac{1}{t^\alpha} d\xi - \frac{1}{2\pi i} \int_{\Gamma'_\theta} t^{\alpha-2} E_{\alpha,\alpha-1}(\xi) \frac{\xi}{t^\alpha}\left(-\frac{\xi}{t^\alpha} I + A\right)^{-1} \frac{1}{t^\alpha} d\xi.$$

In view of $\|(\mu I + A)^{-1}\| \leq \frac{C}{|\mu|}$, we derive that

$$\left\|\frac{d}{dt}\big(t^{\alpha-1} Ae_\alpha(-t^\alpha A)\big)\right\| \leq C_\alpha t^{-2}, \ 0 < t \leq T.$$

By the mean value theorem, for every $0 < s < t \leq T$, we have

$$\|t^{\alpha-1} Ae_\alpha(-t^\alpha A) - s^{\alpha-1} Ae_\alpha(-s^\alpha A)\|$$
$$= \left\|\int_s^t \frac{d}{d\tau}\big(\tau^{\alpha-1} Ae_\alpha(-\tau^\alpha A)\big) d\tau\right\|$$
$$\leq \int_s^t \left\|\frac{d}{d\tau}\big(\tau^{\alpha-1} Ae_\alpha(-\tau^\alpha A)\big)\right\| d\tau$$
$$\leq C_\alpha \int_s^t \tau^{-2} d\tau$$
$$= C_\alpha\big(s^{-1} - t^{-1}\big). \tag{7.18}$$

Let $h > 0$ be such that $0 < t < t + h \leq T$, then

$$
\begin{aligned}
A\Phi_1(&t+h) - A\Phi_1(t)\\
=& \int_0^t \big((t+h-s)^{\alpha-1}Ae_\alpha(-(t+h-s)^\alpha A)\\
&- (t-s)^{\alpha-1}Ae_\alpha(-(t-s)^\alpha A)\big)\big(Pf(s) - Pf(t)\big)ds\\
&+ \int_0^t (t+h-s)^{\alpha-1}Ae_\alpha(-(t+h-s)^\alpha A)\big(Pf(t) - Pf(t+h)\big)ds\\
&+ \int_t^{t+h} (t+h-s)^{\alpha-1}Ae_\alpha(-(t+h-s)^\alpha A)\big(Pf(s) - Pf(t+h)\big)ds\\
=:&\ h_1(t) + h_2(t) + h_3(t).
\end{aligned}
\tag{7.19}
$$

We estimate each of the three terms separately. For $h_1(t)$, from (7.18) and (f$_1$), we have

$$
\begin{aligned}
|h_1(t)|_q \leq& \int_0^t \|(t+h-s)^{\alpha-1}Ae_\alpha(-(t+h-s)^\alpha A)\\
&- (t-s)^{\alpha-1}Ae_\alpha(-(t-s)^\alpha A)\|\|Pf(s) - Pf(t)|_q ds\\
\leq& C_\alpha Lh \int_0^t (t+h-s)^{-1}(t-s)^{\vartheta-1}ds\\
\leq& C_\alpha Lh \int_0^t (s+h)^{-1}(t-s)^{\vartheta-1}ds\\
\leq& C_\alpha L \int_0^h \frac{h}{s+h}s^{\vartheta-1}ds + C_\alpha Lh \int_h^\infty \frac{s}{s+h}s^{\vartheta-1}ds\\
\leq& C_\alpha Lh^\vartheta.
\end{aligned}
\tag{7.20}
$$

For $h_2(t)$, we use Lemma 7.6 and (f$_1$),

$$
\begin{aligned}
|h_2(t)|_q \leq& \int_0^t (t+h-s)^{\alpha-1}|Ae_\alpha(-(t+h-s)^\alpha A)\big(Pf(t) - Pf(t+h)\big)|_q ds\\
\leq& C_1 \int_0^t (t+h-s)^{-1}|Pf(t) - Pf(t+h)|_q ds\\
\leq& C_1 Lh^\vartheta \int_0^t (t+h-s)^{-1}ds\\
=& C_1 L\big(\ln h - \ln(t+h)\big)h^\vartheta.
\end{aligned}
\tag{7.21}
$$

Furthermore, for $h_3(t)$, by Lemma 7.6 and $(f_1)$, we have

$$\begin{aligned}
|h_3(t)|_q &\leq \int_t^{t+h} (t+h-s)^{\alpha-1} |Ae_\alpha(-(t+h-s)^\alpha A)\big(Pf(s) - Pf(t+h)\big)|_q ds \\
&\leq C_1 \int_t^{t+h} (t+h-s)^{-1} |Pf(s) - Pf(t+h)|_q ds \\
&\leq C_1 L \int_t^{t+h} (t+h-s)^{\vartheta-1} ds \\
&= C_1 L \frac{h^\vartheta}{\vartheta}.
\end{aligned} \tag{7.22}$$

Combining (7.20), (7.21) with (7.22), we get that $A\Phi_1(t)$ is Hölder continuous. $\square$

**Theorem 7.4.** *Let the assumptions of Theorem 7.3 be satisfied. If $(f_1)$ holds, then for every $u \in D(A)$, the mild solution of (7.2) is a classical one.*

**Proof.** For $a \in D(A)$. Then Lemma 7.2(ii) ensures that $u(t) = E_\alpha(-t^\alpha A)a$ $(t > 0)$ is a classical solution to the following problem

$$\begin{cases} {}_0^C D_t^\alpha u = -Au, & t > 0, \\ u(0) = a. \end{cases}$$

**Step I.** We verify that

$$\Phi(t) = \int_0^t (t-s)^{\alpha-1} e_\alpha(-(t-s)^\alpha A) Pf(s) ds$$

is a classical solution to the problem

$$\begin{cases} {}_0^C D_t^\alpha u = -Au + Pf(t), & t > 0, \\ u(0) = 0. \end{cases}$$

It follows from Theorem 7.3 that $\Phi \in C([0,T], J_q)$. We rewrite $\Phi(t) = \Phi_1(t) + \Phi_2(t)$, where

$$\Phi_1(t) = \int_0^t (t-s)^{\alpha-1} e_\alpha(-(t-s)^\alpha A)\big(Pf(s) - Pf(t)\big) ds,$$

$$\Phi_2(t) = \int_0^t (t-s)^{\alpha-1} e_\alpha(-(t-s)^\alpha A) Pf(t) ds.$$

According to Lemma 7.7, we know that $\Phi_1(t) \in D(A)$. To prove the same conclusion for $\Phi_2(t)$. By Lemma 7.2(iii), we notice that

$$A\Phi_2(t) = Pf(t) - E_\alpha(-t^\alpha A) Pf(t).$$

Since $(f_1)$ holds, it follows that

$$|A\Phi_2(t)|_q \leq (1 + C_1) |Pf(t)|_q,$$

thus

$$\Phi_2(t) \in D(A) \text{ for } t \in (0,T] \text{ and } A\Phi_2(t) \in C^\nu((0,T], J_q). \tag{7.23}$$

Next, we prove $^C_0D^\alpha_t\Phi \in C((0,T],J_q)$. In view of Lemma 7.2(iv) and $\Phi(0) = 0$, we have

$$^C_0D^\alpha_t\Phi(t) = \frac{d}{dt}\left(_0D^{\alpha-1}_t\Phi(t)\right) = \frac{d}{dt}\left(E_\alpha(-t^\alpha A) * Pf\right).$$

It remains to prove that $E_\alpha(t^\alpha A) * Pf$ is continuously differentiable in $J_q$. Let $0 < h \le T - t$, one derives the following:

$$\frac{1}{h}\left(E_\alpha(-(t+h)^\alpha A) * Pf - E_\alpha(-t^\alpha A) * Pf\right)$$

$$= \int_0^t \frac{1}{h}\left(E_\alpha(-(t+h-s)^\alpha A)Pf(s) - E_\alpha(-(t-s)^\alpha A)Pf(s)\right)ds$$

$$+ \frac{1}{h}\int_t^{t+h} E_\alpha(-(t+h-s)^\alpha A)Pf(s)ds.$$

Notice that

$$\int_0^t \frac{1}{h}\left|E_\alpha(-(t+h-s)^\alpha A)Pf(s) - E_\alpha(-(t-s)^\alpha A)Pf(s)\right|_q ds$$

$$\le C_1\frac{1}{h}\int_0^t |E_\alpha(-(t+h-s)^\alpha A)Pf(s)|_q ds + C_1\frac{1}{h}\int_0^t |E_\alpha(-(t-s)^\alpha A)Pf(s)|_q ds$$

$$\le C_1 M(t)\frac{1}{h}\int_0^t (t+h-s)^{-\alpha}s^{-\alpha(1-\beta)}ds + C_1 M(t)\frac{1}{h}\int_0^t (t-s)^{-\alpha}s^{-\alpha(1-\beta)}ds$$

$$\le C_1 M(t)\frac{1}{h}\left((t+h)^{1-\alpha} + t^{1-\alpha}\right)B(1-\alpha, 1-\alpha(1-\beta)),$$

then using the dominated convergence theorem, we find

$$\lim_{h\to 0}\int_0^t \frac{1}{h}\left(E_\alpha(-(t+h-s)^\alpha A)Pf(s) - E_\alpha(-(t-s)^\alpha A)Pf(s)\right)ds$$

$$= \int_0^t (t-s)^{\alpha-1}Ae_\alpha(-(t-s)^\alpha A)Pf(s)ds$$

$$= A\Phi(t).$$

On the other hand,

$$\frac{1}{h}\int_t^{t+h} E_\alpha(-(t+h-s)^\alpha A)Pf(s)ds$$

$$= \frac{1}{h}\int_0^h E_\alpha(-s^\alpha A)Pf(t+h-s)ds$$

$$= \frac{1}{h}\int_0^h E_\alpha(-s^\alpha A)\left(Pf(t+h-s) - Pf(t-s)\right)ds$$

$$+ \frac{1}{h}\int_0^h E_\alpha(-s^\alpha A)\left(Pf(t-s) - Pf(t)\right)ds + \frac{1}{h}\int_0^h E_\alpha(-s^\alpha A)Pf(t)ds.$$

From Lemmas 7.1, 7.6 and ($f_1$), we have

$$\left|\frac{1}{h}\int_0^h E_\alpha(-s^\alpha A)\left(Pf(t+h-s) - Pf(t-s)\right)ds\right|_q \le C_1 Lh^\vartheta,$$

$$\left| \frac{1}{h} \int_0^h E_\alpha(-s^\alpha A)\big(Pf(t-s) - Pf(t)\big)ds \right|_q \le C_1 L \frac{h^\vartheta}{\vartheta + 1}.$$

Also Lemma 7.2(i) gives that $\lim_{h \to 0} \frac{1}{h} \int_0^h E_\alpha(s^\alpha A)Pf(t)ds = Pf(t)$. Hence

$$\lim_{h \to 0} \frac{1}{h} \int_t^{t+h} E_\alpha((t + h - s)^\alpha A)Pf(s)ds = Pf(t).$$

We deduce that $E_\alpha(t^\alpha A) * Pf$ is differentiable at $t_+$ and $\frac{d}{dt}\big(E_\alpha(t^\alpha A) * Pf\big)_+ = A\Phi(t) + Pf(t)$. Similarly, $E_\alpha(t^\alpha A) * Pf$ is differentiable at $t_-$ and $\frac{d}{dt}\big(E_\alpha(t^\alpha A) * Pf\big)_- = A\Phi(t) + Pf(t)$.

We show that $A\Phi = A\Phi_1 + A\Phi_2 \in C((0, T], J_q)$. In fact, it is clear that $\Phi_2(t) = Pf(t) - E_\alpha(t^\alpha A)Pf(t)$ due to Lemma 7.2(iii), which is continuous in view of Lemma 7.1. Furthermore, according to Lemma 7.7, we know that $A\Phi_1(t)$ is also continuous. Consequently, $^C_0 D_t^\alpha \Phi \in C((0, T], J_q)$.

**Step II.** Let $u$ be the mild solution of (7.2). To prove that $F(u, u) \in C^\vartheta((0, T], J_q)$, in view of (7.6), we have to verify that $A^\gamma u$ is Hölder continuous in $J_q$. Take $h > 0$ such that $0 < t < t + h$.

Denote $\varphi(t) := E_\alpha(-t^\alpha A)a$, by Lemmas 7.2(iv) and 7.6, then

$$|A^\gamma \varphi(t+h) - A^\gamma \varphi(t)|_q = \left| \int_t^{t+h} -s^{\alpha-1} A^\gamma e_\alpha(-s^\alpha A)a\, ds \right|_q$$

$$\le \int_t^{t+h} s^{\alpha-1} |A^{\gamma-\beta} e_\alpha(-s^\alpha A)A^\beta a|_q\, ds$$

$$\le C_1 \int_t^{t+h} s^{\alpha(1+\beta-\gamma)-1} ds |A^\beta a|_q$$

$$= \frac{C_1 |a|_{H^{\beta,q}}}{\alpha(1 + \beta - \gamma)}\big((t + h)^{\alpha(1+\beta-\gamma)} - t^{\alpha(1+\beta-\gamma)}\big)$$

$$\le \frac{C_1 |a|_{H^{\beta,q}}}{\alpha(1 + \beta - \gamma)} h^{\alpha(1+\beta-\gamma)}.$$

Thus, $A^\gamma \varphi \in C^\vartheta((0, T], J_q)$.

For every small $\varepsilon > 0$, take $h$ such that $\varepsilon \le t < t + h \le T$, since

$$|A^\gamma \Phi(t+h) - A^\gamma \Phi(t)|_q$$

$$\le \left| \int_t^{t+h} (t + h - s)^{\alpha-1} A^\gamma e_\alpha(-(t+h-s)^\alpha A)Pf(s)ds \right|_q$$

$$+ \left| \int_0^t A^\gamma \big((t + h - s)^{\alpha-1} e_\alpha(-(t+h-s)^\alpha A) \right.$$

$$\left. - (t - s)^{\alpha-1} e_\alpha(-(t-s)^\alpha A)\big)Pf(s)ds \right|_q$$

$$= \phi_1(t) + \phi_2(t).$$

Applying Lemma 7.6 and (f), we get

$$\phi_1(t) \le C_1 \int_t^{t+h} (t+h-s)^{\alpha(1-\gamma)-1} |Pf(s)|_q ds$$

$$\le C_1 M(t) \int_t^{t+h} (t+h-s)^{\alpha(1-\gamma)-1} s^{-\alpha(1-\beta)} ds$$

$$\le M(t) \frac{C_1}{\alpha(1-\gamma)} h^{\alpha(1-\gamma)} t^{-\alpha(1-\beta)}$$

$$\le M(t) \frac{C_1}{\alpha(1-\gamma)} h^{\alpha(1-\gamma)} \varepsilon^{-\alpha(1-\beta)}.$$

To estimate $\phi_2$, we give the inequality

$$\frac{d}{dt} \left( t^{\alpha-1} A^\gamma e_\alpha(-t^\alpha A) \right) = \frac{1}{2\pi i} \int_\Gamma \mu^\gamma t^{\alpha-2} E_{\alpha,\alpha-1}(-\mu t^\alpha)(\mu I + A)^{-1} d\mu$$

$$= \frac{1}{2\pi i} \int_{\Gamma'} -\left( -\frac{\xi}{t^\alpha} \right)^\gamma t^{\alpha-2} E_{\alpha,\alpha-1}(\xi) \left( -\frac{\xi}{t^\alpha} I + A \right)^{-1} \frac{1}{t^\alpha} d\xi,$$

this yields that $\| \frac{d}{dt} \left( t^{\alpha-1} A^\gamma e_\alpha(-t^\alpha A) \right) \| \le C_\alpha t^{\alpha(1-\gamma)-2}$. The mean value theorem shows

$$\| t^{\alpha-1} A^\gamma e_\alpha(-t^\alpha A) - s^{\alpha-1} A^\gamma e_\alpha(-s^\alpha A) \| \le \int_s^t \left\| \frac{d}{d\tau} \left( \tau^{\alpha-1} A^\gamma e_\alpha(-\tau^\alpha A) \right) \right\| d\tau$$

$$\le C_\alpha \int_s^t \tau^{\alpha(1-\gamma)-2} d\tau$$

$$= C_\alpha \left( s^{\alpha(1-\gamma)-1} - t^{\alpha(1-\gamma)-1} \right),$$

thus

$$\phi_2(t) \le \int_0^t |A^\gamma \left( (t+h-s)^{\alpha-1} e_\alpha(-(t+h-s)^\alpha A) \right.$$

$$- (t-s)^{\alpha-1} e_\alpha(-(t-s)^\alpha A)) Pf(s)|_q ds$$

$$\le \int_0^t \left( (t-s)^{\alpha(1-\gamma)-1} - (t+h-s)^{\alpha(1-\gamma)-1} \right) |Pf(s)|_q ds$$

$$\le C_\alpha M(t) \int_0^t (t-s)^{\alpha(1-\gamma)-1} s^{-\alpha(1-\beta)} ds$$

$$- C_\alpha M(t) \int_0^{t+h} (t-s+h)^{\alpha(1-\gamma)-1} s^{-\alpha(1-\beta)} ds$$

$$+ C_\alpha M(t) \int_t^{t+h} (t-s+h)^{\alpha(1-\gamma)-1} s^{-\alpha(1-\beta)} ds$$

$$\le C_\alpha M(t) \left( t^{\alpha(\beta-\gamma)} - (t+h)^{\alpha(\beta-\gamma)} \right) B(\alpha(1-\gamma), 1 - \alpha(1-\beta))$$

$$+ C_\alpha M(t) h^{\alpha(1-\gamma)} t^{-\alpha(1-\beta)}$$

$$\le C_\alpha M(t) h^{\alpha(\gamma-\beta)} [\varepsilon(\varepsilon+h)]^{\alpha(\beta-\gamma)} + C_\alpha M(t) h^{\alpha(1-\gamma)} \varepsilon^{-\alpha(1-\beta)},$$

which ensures that $A^\gamma \Phi \in C^\vartheta([\varepsilon, T], J_q)$. Therefore $A^\gamma \Phi \in C^\vartheta((0, T], J_q)$ due to arbitrary $\varepsilon$.

Recall

$$\Psi(t) = \int_0^t (t-s)^{\alpha-1} e_\alpha(-(t-s)^\alpha A) F(u(s), u(s)) ds.$$

Since $|F(u(s), u(s))|_q \le MK^2(t) s^{-2\alpha(\gamma-\beta)}$, where

$$K(t) := \sup_{s \in [0,t]} s^{\alpha(\gamma-\beta)} |u(s)|_{H^{\gamma,q}}$$

is continuous and bounded in $(0, T]$. A similar argument enable us to give the Hölder continuity of $A^\gamma \Psi$ in $C^\vartheta((0, T], J_q)$. Therefore, we have $A^\gamma u(t) = A^\gamma \varphi(t) + A^\gamma \Phi(t) + A^\gamma \Psi(t) \in C^\vartheta((0, T], J_q)$.

Since $F(u, u) \in C^\vartheta((0, T], J_q)$ is proved, according to Step II, this yields that ${}^C_0 D_t^\alpha \Psi \in C((0, T], J_q)$, $A\Psi \in C((0, T], J_q)$ and ${}^C_0 D_t^\alpha \Psi = -A\Psi + F(u, u)$. In this way we obtain that ${}^C_0 D_t^\alpha u \in C((0, T], J_q)$, $Au \in C((0, T], J_q)$ and ${}^C_0 D_t^\alpha u = -Au + F(u, u) + Pf$, we conclude that $u$ is a classical solution. $\qquad \square$

**Theorem 7.5.** *Assume that* $(f_1)$ *holds. If* $u$ *is a classical solution of* (7.2), *then* $Au \in C^\nu((0, T], J_q)$ *and* ${}^C_0 D_t^\alpha u \in C^\nu((0, T], J_q)$.

**Proof.** If $u$ is a classical solution of (7.2), then $u(t) = \varphi(t) + \Phi(t) + \Psi(t)$. It remains to show that $A\varphi \in C^{\alpha(1-\beta)}((0, T], J_q)$, it suffices to prove that $A\varphi \in C^{\alpha(1-\beta)}([\varepsilon, T], J_q)$ for every $\varepsilon > 0$. In fact, take $h$ such that $\varepsilon \le t < t + h \le T$, by Lemma 7.2(iii),

$$|A\varphi(t+h) - A\varphi(t)|_q = \left| \int_t^{t+h} -s^{\alpha-1} A^2 e_\alpha(-s^\alpha A) a ds \right|_q$$

$$\le C_1 \int_t^{t+h} s^{-\alpha(1-\beta)-1} ds |a|_{H^{\beta,q}}$$

$$= \frac{C_1 |a|_{H^{\beta,q}}}{\alpha} (t^{-\alpha(1-\beta)} - (t+h)^{-\alpha(1-\beta)})$$

$$\le \frac{C_1 |a|_{H^{\beta,q}}}{\alpha} \frac{h^{\alpha(1-\beta)}}{(\varepsilon(\varepsilon+h))^{\alpha(1-\beta)}}.$$

Similar to Lemma 7.7, we write $\Phi(t)$ as

$$\Phi(t) = \Phi_1(t) + \Phi_2(t) = \int_0^t (t-s)^{\alpha-1} e_\alpha(-(t-s)^\alpha A)(Pf(s) - Pf(t)) ds$$

$$+ \int_0^t (t-s)^{\alpha-1} e_\alpha(-(t-s)^\alpha A) Pf(t) ds,$$

for $t \in (0, T]$. It follows from Lemma 7.7 and (7.23) that $A\Phi_1(t) \in C^\nu([0, T], J_q)$ and $A\Phi_2(t) \in C^\vartheta((0, T], J_q)$, respectively.

Since $F(u, u) \in C^\vartheta((0, T], J_q)$, the result related to the function $\Psi(t)$ is proved by similar argument, which means that $A\Psi \in C^\nu((0, T], J_q)$. Therefore $Au \in C^\nu((0, T], J_q)$ and ${}^C_0 D_t^\alpha u = Au + F(u, u) + Pf \in C^\nu((0, T], J_q)$. The proof is completed. $\qquad \square$

## 7.3 Fractional Euler-Lagrange Equations

### 7.3.1 *Introduction*

In this section, we consider $a < b$ two reals, $d \in \mathbb{N}$ and the following Lagrangian functional

$$\mathfrak{L}(u) = \int_a^b L(u, {}_aD_t^\alpha u, t)dt,$$

where $L$ is a Lagrangian, i.e. a map of the form:

$$L : \mathbb{R}^d \times \mathbb{R}^d \times [a, b] \to \mathbb{R},$$

$$(x, y, t) \to L(x, y, t),$$

where ${}_aD_t^\alpha$ is the left fractional derivative of Riemann-Liouville of order $0 < \alpha < 1$ and where the variable $u$ is a function defined almost everywhere on $(a, b)$ with values in $\mathbb{R}^d$. It is well known that critical points of the functional $L$ are characterized by the solutions of the fractional Euler-Lagrange equation:

$$\frac{\partial L}{\partial x}(u, {}_aD_t^\alpha u, t) + {}_tD_b^\alpha \left( \frac{\partial L}{\partial y}(u, {}_aD_t^\alpha u, t) \right) = 0, \qquad (7.24)$$

where ${}_tD_b^\alpha$ is the right fractional derivative of Riemann-Liouville , see detailed proofs in Agrawal, 2002; Baleanu and Muslih, 2005 for example.

For any $p \geq 1, L^p := L^p((a, b), \mathbb{R}^d)$ denotes the classical Lebesgue space of $p$-integrable functions endowed with its usual norm $\| \cdot \|_{L^p}$. We denote by $| \cdot |$ the Euclidean norm of $\mathbb{R}^d$ and $C := C([a, b], \mathbb{R}^d)$ the space of continuous functions endowed with its usual norm $\| \cdot \|$. We remind that a function $f$ is an element of $AC$ if and only if $f' \in L^1$ and the following equality holds

$$\forall\, t \in [a, b], \quad f(t) = f(a) + \int_a^t f'(\xi)d\xi,$$

where $f'$ denotes the derivative of $f$. We refer to Kolmogorov, Fomine and Tihomirov, 1974 for more details concerning the absolutely continuous functions. In addition, we denote by $C_a$ (resp. $AC_a$ or $C_a^\infty$) the space of functions $f \in C$ (resp. $AC$ or $C^\infty$) such that $f(a) = 0$. In particular, $C_c^\infty \subset C_a^\infty \subset AC_a$.

**Remark 7.1.** In the whole section, an equality between functions must be understood as an equality holding for almost all $t \in (a, b)$. When it is not the case, the interval on which the equality is valid will be specified.

**Definition 7.4.** A function $u$ is said to be a weak solution of (7.24) if $u \in C$ and if $u$ satisfies (7.24) a.e. on $[a, b]$.

In the following, we will provide some properties concerning the left fractional integral operators of Riemann-Liouville. One can easily derive the analogous versions for the right ones. Proposition 7.2 is well known and one can find their proofs in the classical literature on the subject (see Lemma 2.1 in Kilbas, Srivastava and Trujillo, 2006).

**Proposition 7.2.** *For any $\alpha > 0$ and any $p \geq 1$, $_aD_t^{-\alpha}$ is linear and continuous from $L^p$ to $L^p$. Precisely, the following inequality holds:*

$$\|_aD_t^{-\alpha}f\|_{L^p} \leq \frac{(b-a)^\alpha}{\Gamma(1+\alpha)}\|f\|_{L^p}, \quad for \ f \in L^p.$$

The following classical property concerns the integration of fractional integrals. It is occasionally called fractional integration by parts:

**Proposition 7.3.** *Let $0 < \frac{1}{p} < \alpha < 1$ and $q = \frac{p}{p-1}$. Then, for any $f \in L^p$, we have*

(i) $_aD_t^{-\alpha}$ *is Hölder continuous on $[a,b]$ with exponent $\alpha - \frac{1}{p} > 0$;*

(ii) $\lim_{t \to a} {_aD_t^{-\alpha}}f(t) = 0$.

*Consequently, $_aD_t^{-\alpha}f(t)$ can be continuously extended by $0$ in $t = a$. Finally, for any $f \in L^p$, we have $_aD_t^{-\alpha}f \in C_a$. Moreover, the following inequality holds:*

$$\|_aD_t^{-\alpha}f\| \leq \frac{(b-a)^{\alpha-\frac{1}{p}}}{\Gamma(\alpha)((\alpha-1)q+1)^{\frac{1}{q}}}\|f\|_{L^p}, \quad for \ f \in L^p.$$

**Proof.** Let us note that this result is mainly proved in Section 5.2. Let $f \in L^p$. We first remind the following inequality

$$(\xi_1 - \xi_2)^q \leq \xi_1^q - \xi_2^q, \quad for \ \xi_1 \geq \xi_2 \geq 0.$$

Let us prove that $_aD_t^{-\alpha}f(t)$ is Hölder continuous on $[a,b]$. For any $a < t_1 < t_2 \leq b$, using Hölder inequality, we have

$$|_aD_t^{-\alpha}f(t_2) - {_aD_t^{-\alpha}}f(t_1)| = \frac{1}{\Gamma(\alpha)}\left|\int_a^{t_2}(t_2-\xi)^{\alpha-1}f(\xi)d\xi - \int_a^{t_1}(t_1-\xi)^{\alpha-1}f(\xi)d\xi\right|$$

$$\leq \frac{1}{\Gamma(\alpha)}\left|\int_{t_1}^{t_2}(t_2-\xi)^{\alpha-1}f(\xi)d\xi\right|$$

$$+ \frac{1}{\Gamma(\alpha)}\left|\int_a^{t_1}((t_2-\xi)^{\alpha-1} - (t_1-\xi)^{\alpha-1})f(\xi)d\xi\right|$$

$$\leq \frac{\|f\|_{L^p}}{\Gamma(\alpha)}\left(\int_{t_1}^{t_2}(t_2-\xi)^{(\alpha-1)q}d\xi\right)^{\frac{1}{q}}$$

$$+ \frac{\|f\|_{L^p}}{\Gamma(\alpha)}\left(\int_a^{t_1}((t_1-\xi)^{\alpha-1} - (t_2-\xi)^{\alpha-1})^q d\xi\right)^{\frac{1}{q}}$$

$$\leq \frac{\|f\|_{L^p}}{\Gamma(\alpha)}\left(\int_{t_1}^{t_2}(t_2-\xi)^{(\alpha-1)q}d\xi\right)^{\frac{1}{q}}$$

$$+ \frac{\|f\|_{L^p}}{\Gamma(\alpha)}\left(\int_a^{t_1}(t_1-\xi)^{(\alpha-1)q} - (t_2-\xi)^{(\alpha-1)q}d\xi\right)^{\frac{1}{q}}$$

$$\leq \frac{2\|f\|_{L^p}}{\Gamma(\alpha)((\alpha-1)q+1)^{\frac{1}{q}}}(t_2-t_1)^{\alpha-\frac{1}{p}}.$$

The proof of the first point is complete. Let us consider the second point. For any $t \in [a, b]$, we can prove in the same manner that

$$|_aD_t^{-\alpha}f(t)| \le \frac{\|f\|_{L^p}}{\Gamma(\alpha)((\alpha-1)q+1)^{\frac{1}{q}}}(t-a)^{\alpha-\frac{1}{p}}, \quad \text{as } t \to 0.$$

The proof is now complete. $\qquad\qquad\qquad\qquad\qquad\qquad\qquad\qquad\qquad\square$

In Subsection 7.3.2, we introduce an appropriate space of functions. Subsection 7.3.3 is concerned with variational structure. Subsection 7.3.4 is devoted to the existence theorem of weak solution for (7.24).

### 7.3.2 *Functional Spaces*

In order to prove the existence of a weak solution of (7.24) using a variational method, we need the introduction of an appropriate space of functions. This space has to present some properties like reflexivity, see Dacorogna, 2008.

For any $0 < \alpha < 1$ and any $p \ge 1$, we define the following space of functions

$$E_{\alpha,p} := \{u \in L^p|\ _aD_t^{\alpha}u \in L^p \text{ and } _aD_t^{-\alpha}(_aD_t^{\alpha}u) = u \text{ a.e.}\}.$$

We endow $E_{\alpha,p}$ with the following norm

$$\|\cdot\|_{\alpha,p} : E_{\alpha,p} \to \mathbb{R}^+,$$

$$u \mapsto \left(\|u\|_{L^p}^p + \|_aD_t^{\alpha}u\|_{L^p}^p\right)^{\frac{1}{p}}.$$

Let us note that

$$|\cdot|_{\alpha,p} : E_{\alpha,p} \to \mathbb{R}^+,$$

$$u \mapsto \|_aD_t^{\alpha}u\|_{L^p}$$

is an equivalent norm to $\|\cdot\|_{\alpha,p}$ for $E_{\alpha,p}$. Indeed, Proposition 7.2 leads to

$$\|u\|_{L^p} = \|_aD_t^{-\alpha}(_aD_t^{\alpha}u)\|_{L^p} \le \frac{(b-a)^{\alpha}}{\Gamma(1+\alpha)}\|_aD_t^{\alpha}u\|_{L^p}, \quad \text{for } u \in E_{\alpha,p}. \qquad (7.25)$$

The goal of this section is to prove the following proposition.

**Proposition 7.4.** *Assuming* $0 < \frac{1}{p} < \alpha < 1$, $E_{\alpha,p}$ *is a reflexive separable Banach space and the compact embedding* $E_{\alpha,p} \hookrightarrow C_a$ *holds.*

**Proof.** Consider that

$$0 < \frac{1}{p} < \alpha < 1 \quad \text{and } q = \frac{p}{p-1}.$$

Now, we divide the proof into several steps.

**Claim I.** $E_{\alpha,p}$ is a reflexive separable Banach space.

Let us consider $(L^p)^2$ the set $L^p \times L^p$ endowed with the norm $\|(u,v)\|_{(L_p)^2} = (\|u\|_{L^p}^p + \|v\|_{L^p}^p)^{\frac{1}{p}}$. Since $p > 1$, $(L^p, \|\cdot\|_{L^p})$ is a reflexive separable Banach space and therefore, $((L^p)^2, \|\cdot\|_{(L^p)^2})$ is also a reflexive separable Banach space. We define $\Omega := \{(u, {}_aD_t^\alpha u) : u \in E_{\alpha,p}\}$. Let us prove that $\Omega$ is a closed subspace of $((L^p)^2, \|\cdot\|_{(L^p)^2})$. Let $(u_n, v_n)_{n\in\mathbb{N}} \subset \Omega$ such that

$$(u_n, v_n) \xrightarrow{(L^p)^2} (u, v).$$

Then, we prove that $(u, v) \in \Omega$. For any $n \in \mathbb{N}$, $(u_n, v_n) \in \Omega$. Thus, $u_n \in E_{\alpha,p}$ and $v_n = {}_aD_t^\alpha u_n$. Consequently, we have

$$u_n \xrightarrow{L^p} u \quad \text{and} \quad {}_aD_t^\alpha u_n \xrightarrow{L^p} v.$$

For any $n \in \mathbb{N}$, since $u_n \in E_{\alpha,p}$ and ${}_aD_t^{-\alpha}$ is continuous from $L^p$ to $L^p$, we have

$$u_n = {}_aD_t^{-\alpha}({}_aD_t^\alpha u_n) \xrightarrow{L^p} {}_aD_t^{-\alpha}v.$$

Thus, $u = {}_aD_t^{-\alpha}v$, ${}_aD_t^\alpha u = {}_aD_t^\alpha({}_aD_t^{-\alpha}v) = v \in L^p$ and ${}_aD_t^{-\alpha}({}_aD_t^\alpha u) = {}_aD_t^{-\alpha}v = u$. Hence, $u \in E_{\alpha,p}$ and $(u, v) = (u, {}_aD_t^\alpha u) \in \Omega$. In conclusion, $\Omega$ is a closed subspace of $((L^p)^2, \|\cdot\|_{(L^p)^2})$ and then $\Omega$ is a reflexive separable Banach space. Finally, defining the following operator

$$A : E_{\alpha,p} \to \Omega,$$
$$u \mapsto (u, {}_aD_t^\alpha u),$$

we prove that $E_{\alpha,p}$ is isometric isomorphic to $\Omega$. This completes the proof of Step I.

**Claim II.** The continuous embedding $E_{\alpha,p} \hookrightarrow C_a$.

Let $u \in E_{\alpha,p}$ and then ${}_aD_t^\alpha u \in L^p$. Since $0 < \frac{1}{p} < \alpha < 1$, Proposition 7.3 leads to ${}_aD_t^{-\alpha}({}_aD_t^\alpha u) \in C_a$. Furthermore, $u = {}_aD_t^{-\alpha}({}_aD_t^\alpha u)$ and consequently, $u$ can be identified to its continuous representative. Finally, Proposition 7.3 also gives

$$\|u\| = \|{}_aD_t^{-\alpha}({}_aD_t^\alpha u)\| \leq \frac{(b-a)^{\alpha-\frac{1}{p}}}{\Gamma(\alpha)((\alpha-1)q+1)^{\frac{1}{q}}} |u|_{\alpha,p}, \quad \text{for } u \in E_{\alpha,p}.$$

Since $\|\cdot\|_{\alpha,p}$ and $|\cdot|_{\alpha,p}$ are equivalent norms, the proof of Step II is complete.

**Claim III.** The compact embedding $E_{\alpha,p} \hookrightarrow\hookrightarrow C_a$.

Since $E_{\alpha,p}$ is a reflexive Banach space, we only have to prove that

$$\forall (u_n)_{n\in\mathbb{N}} \subset E_{\alpha,p} \quad \text{such that} \quad u_n \xrightarrow{E_{\alpha,p}} u, \quad \text{then} \quad u_n \xrightarrow{C} u.$$

Let $(u_n)_{n\in\mathbb{N}} \subset E_{\alpha,p}$ such that

$$u_n \xrightarrow{E_{\alpha,p}} u.$$

Since $E_{\alpha,p} \hookrightarrow C_a$, we have

$$u_n \xrightarrow{C} u.$$

Since $(u_n)_{n\in\mathbb{N}}$ converges weakly in $E_{\alpha,p}$, $(u_n)_{n\in\mathbb{N}}$ is bounded in $E_{\alpha,p}$. Consequently, $({}_aD_t^\alpha u_n)_{n\in\mathbb{N}}$ is bounded in $L_p$ by a constant $M \geq 0$. Let us prove that

$(u_n)_{n \in \mathbb{N}} \subset C_a$ is uniformly Lipschitzian on $[a, b]$. According to the proof of Proposition 7.3, for every $n \in \mathbb{N}$, and $a \leq t_1 < t_2 \leq b$, we have,

$$
\begin{aligned}
|u_n(t_2) - u_n(t_1)| &\leq |_aD_t^{-\alpha}(_aD_t^\alpha u_n(t_2)) - {_aD_t^{-\alpha}}(_aD_t^\alpha u_n(t_1))| \\
&\leq \frac{2\|_aD_t^\alpha u_n\|_{L^p}}{\Gamma(\alpha)((\alpha-1)q+1)^{\frac{1}{p}}} (t_2 - t_1)^{\alpha - \frac{1}{p}} \\
&\leq \frac{2M}{\Gamma(\alpha)((\alpha-1)q+1)^{\frac{1}{p}}} (t_2 - t_1)^{\alpha - \frac{1}{p}}.
\end{aligned}
$$

Hence, from Arzela-Ascoli theorem, $(u_n)_{n \in \mathbb{N}}$ is relatively compact in $C$. Consequently, there exists a subsequence of $(u_n)_{n \in \mathbb{N}}$ converging strongly in $C$ and the limit is $u$ by uniqueness of the weak limit.

Now, let us prove by contradiction that the whole sequence $(u_n)_{n \in \mathbb{N}}$ converges strongly to $u$ in $C$. If not, there exist $\varepsilon > 0$ and a subsequence $(u_{n_k})_{k \in \mathbb{N}}$ such that

$$
\|u_{n_k} - u\| > \varepsilon > 0, \quad \text{for } k \in \mathbb{N}. \tag{7.26}
$$

Nevertheless, since $(u_{n_k})_{k \in \mathbb{N}}$ is a subsequence of $(u_n)_{n \in \mathbb{N}}$, then it satisfies

$$
u_{n_k} \overset{E_{\alpha,p}}{\rightharpoonup} u.
$$

In the same way (using Arzela-Ascoli theorem), we can construct a subsequence of $(u_{n_k})_{k \in \mathbb{N}}$ converging strongly to $u$ in $C$ which is a contradiction to (7.26). The proof of Step III is now complete. $\qquad \square$

Let us remind the following property

$$
_aD_t^{-\alpha}\varphi \in C_a^\infty, \quad \text{for } \varphi \in C_c^\infty.
$$

From this result, we get the following results.

**Proposition 7.5.** $C_a^\infty$ *is dense in* $E_{\alpha,p}$.

**Proof.** Indeed, let us first prove that $C_a^\infty \subset E_{\alpha,p}$. Let $u \in C_a^\infty \subset L^p$. Since $u \in AC_a$ and $u' \in L^p$, we have $_aD_t^\alpha u = {_aD_t^{\alpha-1}}u' \in L^p$. Since $u \in AC$, we also have $_aD_t^{-\alpha}(_aD_t^\alpha u) = u$. Finally, $u \in E_{\alpha,p}$. Now, let us prove that $C_a^\infty$ is dense in $E_{\alpha,p}$. Let $u \in E_{\alpha,p}$, then $_aD_t^\alpha u \in L^p$. Consequently, there exists $(v_n)_{n \in \mathbb{N}} \subset C_c^\infty$ such that

$$
v_n \xrightarrow{L^p} {_aD_t^\alpha}u \quad \text{and then} \quad _aD_t^{-\alpha}v_n \xrightarrow{L^p} {_aD_t^{-\alpha}}(_aD_t^\alpha u) = u,
$$

since $_aD_t^{-\alpha}$ is continuous from $L^p$ to $L^p$. Defining $u_n := {_aD_t^{-\alpha}}v_n \in C_a^\infty$ for any $n \in \mathbb{N}$, we obtain

$$
u_n \xrightarrow{L^p} u \quad \text{and} \quad _aD_t^\alpha u_n = {_aD_t^\alpha}(_aD_t^{-\alpha}v_n) = v_n \xrightarrow{L^p} {_aD_t^\alpha}u.
$$

Finally, $(u_n)_{n \in \mathbb{N}} \subset C_a^\infty$ and converges to $u$ in $E_{\alpha,p}$. The proof is completed. $\qquad \square$

**Proposition 7.6.** *If* $\frac{1}{p} < \min(\alpha, 1-\alpha)$, *then* $E_{\alpha,p} = \{u \in L^p : {_aD_t^\alpha}u \in L^p\}$.

**Proof.** Indeed, let $u \in L^p$ satisfying $_aD_t^\alpha u \in L^p$ and let us prove that $_aD_t^{-\alpha}(_aD_t^\alpha u) = u$. Let $\varphi \in C_c^\infty \subset L^1$. Since $_aD_t^\alpha u \in L^p$, Proposition 1.10 leads to

$$\int_a^b {_aD_t^{-\alpha}}(_aD_t^\alpha u) \cdot \varphi dt = \int_a^b {_aD_t^\alpha u} \cdot {_tD_b^{-\alpha}}\varphi dt = \int_a^b \frac{d}{dt}(_aD_t^{\alpha-1}u) \cdot {_tD_b^{-\alpha}}\varphi dt.$$

Then, an integration by parts gives

$$\int_a^b {_aD_t^{-\alpha}}(_aD_t^\alpha u) \cdot \varphi dt = \int_a^b {_aD_t^{\alpha-1}u} \cdot {_tD_b^{1-\alpha}}u dt.$$

Indeed, $_tD_b^{-\alpha}\varphi(b) = 0$ since $\varphi \in C_c^\infty$ and $_aD_t^{\alpha-1}u(a) = 0$ since $u \in L^p$ and $\frac{1}{p} < 1-\alpha$. Finally, using Proposition 1.10 again, we obtain

$$\int_a^b {_aD_t^{-\alpha}}(_aD_t^\alpha u) \cdot \varphi dt = \int_a^b u \cdot {_tD_b^{\alpha-1}}(_tD_b^{1-\alpha}\varphi) dt = \int_a^b u \cdot \varphi dt,$$

this completes the proof. $\qquad\qquad\qquad\qquad\qquad\qquad\qquad\qquad\qquad\quad\square$

**Remark 7.2.** In the Proposition 7.6, let us note that such a definition of $E_{\alpha,p}$ could lead us to name it fractional Sobolev space and to denote it by $W^{\alpha,p}$. Nevertheless, these notions and notations are already used, see Brezis, 2011.

### 7.3.3    *Variational Structure*

In this subsection, we assume that Lagrangian L is of class $C^1$ and we define the Lagrangian functional $\mathfrak{L}$ on $E_{\alpha,p}$ (with $0 < \frac{1}{p} < \alpha < 1$). Precisely, we define

$$\mathfrak{L} : E_{\alpha,p} \to \mathbb{R},$$

$$u \mapsto \int_a^b L(u, {_aD_t^\alpha}u, t)dt.$$

$\mathfrak{L}$ is said to be Gâteaux differentiable in $u \in E_{\alpha,p}$ if the map

$$D\mathfrak{L}(u) : E_{\alpha,p} \to \mathbb{R},$$

$$v \mapsto D\mathfrak{L}(u)(v) := \lim_{h \to 0} \frac{\mathfrak{L}(u + hv) - \mathfrak{L}(u)}{h}$$

is well-defined for any $v \in E_{\alpha,p}$ and if it is linear and continuous. A critical point $u \in E_{\alpha,p}$ of $\mathfrak{L}$ is defined by $D\mathfrak{L}(u) = 0$.

We introduce the following hypotheses:

**(H1)** there exist $0 \leq d_1 \leq p$ and $r_1, s_1 \in C(\mathbb{R}^d \times [a,b], \mathbb{R}^+)$ such that

$$|L(x,y,t) - L(x,0,t)| \leq r_1(x,t)\|y\|^{d_1} + s_1(x,t), \quad \text{for } (x,y,t) \in \mathbb{R}^d \times \mathbb{R}^d \times [a,b];$$

**(H2)** there exist $0 \leq d_2 \leq p$ and $r_2, s_2 \in C(\mathbb{R}^d \times [a,b], \mathbb{R}^+)$ such that

$$\left\|\frac{\partial L}{\partial x}(x,y,t)\right\| \leq r_2(x,t)\|y\|^{d_2} + s_2(x,t), \quad \text{for } (x,y,t) \in \mathbb{R}^d \times \mathbb{R}^d \times [a,b];$$

**(H3)** there exist $0 \leq d_3 \leq p-1$ and $r_3, s_3 \in C(\mathbb{R}^d \times [a,b], \mathbb{R}^+)$ such that
$$\left\| \frac{\partial L}{\partial y}(x,y,t) \right\| \leq r_3(x,t)\|y\|^{d_3} + s_3(x,t), \quad \text{for } (x,y,t) \in \mathbb{R}^d \times \mathbb{R}^d \times [a,b];$$

**(H4)** *coercivity condition:* there exist $\gamma > 0, 1 \leq d_4 < p, c_1 \in C(\mathbb{R}^d \times [a,b], [\gamma, \infty))$, $c_2, c_3 \in C([a,b], \mathbb{R})$ such that
$$\forall (x,y,t) \in \mathbb{R}^d \times \mathbb{R}^d \times [a,b], \quad L(x,y,t) \geq c_1(x,t)\|y\|^p + c_2(t)\|x\|^{d_4} + c_3(t);$$

**(H5)** *convexity condition:*
$$\forall t \in [a,b], \quad L(\cdot, \cdot, t) \text{ is convex.}$$

Hypotheses denoted by (H1)-(H3) are usually called regularity hypotheses (see Cesari, 1983; Dacorogna, 2008).

Let us prove the following results.

**Lemma 7.8.** *The following implications hold:*

**(i)** $L$ *satisfies* $(H1) \Rightarrow$ *for any* $u \in E_{\alpha,p}, L(u, {}_aD_t^\alpha u, t) \in L^1$ *and then* $\mathfrak{L}(u)$ *exists in* $\mathbb{R}$;

**(ii)** $L$ *satisfies* $(H2) \Rightarrow$ *for any* $u \in E_{\alpha,p}, \partial L/\partial x(u, {}_aD_t^\alpha u, t) \in L^1$;

**(iii)** $L$ *satisfies* $(H3) \Rightarrow$ *for any* $u \in E_{\alpha,p}, \partial L/\partial y(u, {}_aD_t^\alpha u, t) \in L^q$, *where* $q = \frac{p}{p-1}$.

**Proof.** Let us assume that $\mathfrak{L}$ satisfies (H1) and let $u \in E_{\alpha,p} \subset C_a$. Then, $\|{}_aD_t^\alpha u\|^{d_1} \in L^{p/d_1} \subset L^1$ and the three maps $t \to r_1(u(t),t)$, $s_1(u(t),t)$, $|L(u(t),0,t)| \in C([a,b], \mathbb{R}^+) \subset L^\infty \subset L^1$. hypothesis (H1) implies for almost all $t \in [a,b]$
$$|L(u(t), {}_aD_t^\alpha u(t), t)| \leq r_1(u(t),t)\|{}_aD_t^\alpha u(t)\|^{d_1} + s_1(u(t),t) + |L(u(t),0,t)|.$$

Hence, $L(u, {}_aD_t^\alpha u(t), t) \in L^1$ and then $L(u)$ exists in $\mathbb{R}$. We proceed in the same manner in order to prove the second point of Lemma 7.8. Now, assuming that $L$ satisfies (H3), we have $\|{}_aD_t^\alpha u\|^{d_3} \in L^{p/d_3} \subset L^q$ for any $u \in E_{\alpha,p}$. An analogous argument gives the third point of Lemma 7.8. This completes the proof. $\square$

**Lemma 7.9.** *Assuming that $L$ satisfies hypotheses (H1)-(H3), $\mathfrak{L}$ is Gâteaux differentiable in any $u \in E_{\alpha,p}$ and*
$$D\mathfrak{L}(u)(v) = \int_a^b \left( \frac{\partial L}{\partial x}(u, {}_aD_t^\alpha u, t) \cdot v + \frac{\partial L}{\partial y}(u, {}_aD_t^\alpha u, t) \cdot {}_aD_t^\alpha v \right) dt, \quad \text{for } u, v \in E_{\alpha,p}.$$

**Proof.** Let $u, v \in E_{\alpha,p} \subset C_a$. Let $\psi_{u,v}$ defined for any $h \in [-1,1]$ and for almost all $t \in [a,b]$ by
$$\psi_{u,v}(t,h) := L(u(t) + hv(t), {}_aD_t^\alpha u(t) + h {}_aD_t^\alpha v(t), t).$$

Then, we define the following mapping
$$\phi_{u,v} : [-1,1] \to \mathbb{R},$$
$$h \mapsto \int_a^b L(u + hv(t), {}_aD_t^\alpha u + h {}_aD_t^\alpha v, t)dt = \int_a^b \psi_{u,v}(t,h)dt.$$

Our aim is to prove that the following term

$$D\mathcal{L}(u)(v) = \lim_{h \to 0} \frac{\mathcal{L}(u + hv) - \mathcal{L}(u)}{h} = \lim_{h \to 0} \frac{\phi_{u,v}(h) - \phi_{u,v}(0)}{h} = \phi_{u,v}'(0)$$

exists in $\mathbb{R}$. In order to differentiate $\phi_{u,v}$, we use the theorem of differentiation under the integral sign. Indeed, we have for almost all $t \in [a, b]$, $\psi_{u,v}(t, \cdot)$ is differentiable on $[-1, 1]$ with

$$\frac{\partial \psi_{u,v}}{\partial h}(t, h) = \frac{\partial L}{\partial x}(u(t) + hv(t), {}_aD_t^\alpha u(t) + h\,{}_aD_t^\alpha v(t), t) \cdot v(t)$$

$$+ \frac{\partial L}{\partial y}(u(t) + hv(t), {}_aD_t^\alpha u(t) + h\,{}_aD_t^\alpha v(t), t) \cdot {}_aD_t^\alpha v(t).$$

Then, from hypotheses (H2) and (H3), we have for any $h \in [-1, 1]$ and for almost all $t \in [a, b]$

$$\left| \frac{\partial \psi_{u,v}}{\partial h}(t, h) \right|$$

$$\leq \left[ r_2(u(t) + hv(t), t) \|{}_aD_t^\alpha u(t) + h\,{}_aD_t^\alpha v(t)\|^{d_2} + s_2(u(t) + hv(t), t) \right] \|v(t)\|$$

$$+ \left[ r_3(u(t) + hv(t), t) \|{}_aD_t^\alpha u(t) + h\,{}_aD_t^\alpha v(t)\|^{d_3} + s_3(u(t) + hv(t), t) \right] \|{}_aD_t^\alpha v(t)\|.$$

We define

$$r_{2,0} := \max_{(t,h) \in [a,b] \times [-1,1]} r_2(u(t) + hv(t), t)$$

and we define similarly $s_{2,0}, r_{3,0}, s_{3,0}$. Finally, it holds

$$\left| \frac{\partial \psi_{u,v}}{\partial h}(t, h) \right| \leq 2^{d_2} r_{2,0} \underbrace{(\|{}_aD_t^\alpha u(t)\|^{d_2} + \|{}_aD_t^\alpha v(t)\|^{d_2})}_{\in L^{p/d_2} \subset L^1} \underbrace{\|v(t)\|}_{\in C_a \subset L^\infty} + s_{2,0} \underbrace{\|v(t)\|}_{\in C_a \subset L^1}$$

$$+ 2^{d_3} r_{3,0} \underbrace{(\|{}_aD_t^\alpha u(t)\|^{d_3} + \|{}_aD_t^\alpha v(t)\|^{d_3})}_{\in L^{p/d_3} \subset L^q} \underbrace{\|{}_aD_t^\alpha v(t)\|}_{\in L^p} + s_{3,0} \underbrace{\|{}_aD_t^\alpha v(t)\|}_{\in L^p \subset L^1}.$$

The right term is then a $L^1$ function independent of $h$. Consequently, applying the theorem of differentiation under the integral sign, $\phi_{u,v}$ is differentiable with

$$\phi_{u,v}'(h) = \int_a^b \frac{\partial \psi_{u,v}}{h}(t, h)dt, \quad \text{for } h \in [-1, 1].$$

Hence

$$D\mathcal{L}(u)(v) = \phi_{u,v}'(0) = \int_a^b \frac{\partial \psi_{u,v}}{h}(t, 0)dt$$

$$= \int_a^b \left( \frac{\partial L}{\partial x}(u, {}_aD_t^\alpha, t)\, v + \frac{\partial L}{\partial y}(u, {}_aD_t^\alpha u, t)\, {}_aD_t^\alpha v \right) dt.$$

From Lemma 7.8, it holds

$$\frac{\partial L}{\partial x}(u, {}_aD_t^\alpha u, t) \in L^1 \quad \text{and} \quad \frac{\partial L}{\partial y}(u, {}_aD_t^\alpha u, t) \in L^q.$$

Since $v \in C_a \subset L^\infty$ and $_aD_t^\alpha \in L^p$, $D\mathfrak{L}(u)(v)$ exists in $\mathbb{R}$. Moreover, we have

$$|D\mathfrak{L}(u)(v)| \le \left\| \frac{\partial L}{\partial x}(u, {}_aD_t^\alpha u, t) \right\|_{L^1} \|v\| + \left\| \frac{\partial L}{\partial y}(u, {}_aD_t^\alpha u, t) \right\|_{L^q} \|_aD_t^\alpha v\|_{L^p}$$

$$\le \left( \frac{(b-a)^{\alpha - \frac{1}{p}}}{\Gamma(\alpha)((\alpha-1)q+1)^{\frac{1}{q}}} \left\| \frac{\partial L}{\partial x}(u, {}_aD_t^\alpha u, t) \right\|_{L^1} + \left\| \frac{\partial L}{\partial y}(u, {}_aD_t^\alpha u, t) \right\|_{L^q} \right) |v|_{\alpha,p}.$$

Consequently, $D\mathfrak{L}(u)$ is linear and continuous from $E_{\alpha,p}$ to $\mathbb{R}$. The proof is completed. $\qquad\square$

### 7.3.4 Existence of Weak Solution

In this subsection, we will present the existence theorem of weak solution for (7.24). We firstly give two preliminary theorems.

**Theorem 7.6.** *Assume that $L$ satisfies hypotheses (H1)-(H3). If $u$ is a critical point of $\mathfrak{L}$, $u$ is a weak solution of (7.24).*

**Proof.** Let $u$ be a critical point of $\mathfrak{L}$. Then, we have in particular

$$D\mathfrak{L}(u)(v) = \int_a^b \left( \frac{\partial L}{\partial x}(u, {}_aD_t^\alpha u, t) \cdot v + \frac{\partial L}{\partial y}(u, {}_aD_t^\alpha u, t) \cdot {}_aD_t^\alpha v \right) dt = 0, \quad \text{for } v \in C_c^\infty.$$

For any $v \in C_c^\infty \subset AC_a$, $_aD_t^\alpha v = {}_aD_t^{\alpha-1}v' \in C_a^\infty$. Since $\partial L/\partial y(u, {}_aD_t^\alpha u, t) \in L^q$, Proposition 1.10 gives

$$\int_a^b \left[ \frac{\partial L}{\partial x}(u, {}_aD_t^\alpha u, t) \cdot v + {}_tD_b^{\alpha-1}\left( \frac{\partial L}{\partial y}(u, {}_aD_t^\alpha u, t) \right) \cdot v' \right] dt = 0, \quad \text{for } v \in C_c^\infty.$$

Finally, we define

$$w_u(t) = \int_a^t \frac{\partial L}{\partial x}(u, {}_aD_t^\alpha u, t)dt, \quad \text{for } t \in [a, b].$$

Since $\partial L/\partial x(u, {}_aD_t^\alpha, t) \in L^1$, $w_u \in AC_a$ and $w_u' = \partial L/\partial x(u, {}_aD_t^\alpha u, t)$. Then, an integration by parts leads to

$$\int_a^b \left( {}_tD_b^{\alpha-1}\left( \frac{\partial L}{\partial y}(u, {}_aD_t^\alpha u, t) \right) - w_u \right) v' dt = 0, \quad \text{for } v \in C_c^\infty.$$

Consequently, there exists a constant $C \in \mathbb{R}^d$ such that

$$_tD_b^{\alpha-1}\left( \frac{\partial L}{\partial y}(u, {}_aD_t^\alpha u, t) \right) = C + w_u \in AC.$$

By differentiation, we obtain

$$-_tD_b^\alpha\left( \frac{\partial L}{\partial y}(u, {}_aD_t^\alpha u, t) \right) = \frac{\partial L}{\partial x}(u, {}_aD_t^\alpha u, t),$$

and then $u \in E_{\alpha,p} \subset C$ satisfies (7.24) a.e. on $[a, b]$. The proof is completed. $\qquad\square$

As usual in a variational method, in order to prove the existence of a global minimizer of a functional, coercivity and convexity hypotheses need to be added on the Lagrangian. We have already define hypotheses (H4) (coercivity) and (H5) (convexity). Next, we introduce two different convexity hypotheses (H5)′ and (H5)″:

**(H5)′** $\forall\ (x,t) \in \mathbb{R}^d \times [a,b]$, $L(x,\cdot,t)$ is convex and $(L(\cdot,y,t))_{(y,t)\in\mathbb{R}^d\times[a,b]}$ is uniformly equicontinuous on $\mathbb{R}^d$, i.e.,

$$\forall\ \varepsilon > 0,\ \exists\ \delta > 0,\ \forall\ (x_1,x_2) \in (\mathbb{R}^d)^2,\ \|x_2 - x_1\| < \delta$$
$$\Rightarrow \forall\ (y,t) \in \mathbb{R}^d \times [a,b],\ |L(x_2,y,t) - L(x_1,y,t)| < \varepsilon.$$

**(H5)″** $\forall\ (x,t) \in \mathbb{R}^d \times [a,b]$, $L(x,\cdot,t)$ is convex.

Let us note that hypotheses (H5) and (H5)′ are independent. hypothesis (H5)″ is the weakest. Nevertheless, in this case, the detailed proof of Theorem 7.7 is more complicated. Consequently, in the case of hypothesis (H5)″, we do not develop the proof and we use a strong result proved in Dacorogna, 2008. Let us prove the following preliminary result.

**Lemma 7.10.** *Assume that $L$ satisfies hypothesis (H4). Then, $\mathfrak{L}$ is coercive in the sense that*

$$\lim_{\|u\|_{\alpha,p}\to+\infty} \mathfrak{L}(u) = +\infty.$$

**Proof.** Let $u \in E_{\alpha,p}$, we have

$$\mathfrak{L}(u) = \int_a^b L(u, {_aD_t^\alpha}u, t)dt \geq \int_a^b c_1(u,t)\|_aD_t^\alpha u\|^p + c_2(t)\|u\|^{d_4} + c_3(t)dt.$$

Equation (7.25) implies that

$$\|u\|_{L^{d_4}}^{d_4} \leq (b-a)^{1-\frac{d_4}{p}}\|u\|_{L^p}^{d_4} \leq \frac{(b-a)^{\alpha+1-\frac{d_4}{p}}}{\Gamma(\alpha+1)}\|_aD_t^\alpha u\|_{L^p}^{d_4} = \frac{(b-a)^{\alpha+1-\frac{d_4}{p}}}{\Gamma(\alpha+1)}|u|_{\alpha,p}^{d_4}.$$

Finally, we conclude that

$$\mathfrak{L}(u) \geq \gamma\|_aD_t^\alpha u\|_{L^p}^p - \|c_2\|\|u\|_{L^{d_4}}^{d_4} - (b-a)\|c_3\|$$

$$\geq \gamma|u|_{\alpha,p}^p - \frac{\|c_2\|(b-a)^{\alpha+1-\frac{d_4}{p}}}{\Gamma(\alpha+1)}|u|_{\alpha,p}^{d_4} - (b-a)\|c_3\|, \quad \text{for } u \in E_{\alpha,p}.$$

Since $d_4 < p$ and the norms $|\cdot|_{\alpha,p}$ and $\|\cdot\|_{\alpha,p}$ are equivalent, the proof is completed. $\qquad\square$

**Theorem 7.7.** *Assume that $L$ satisfies hypotheses (H1)-(H4) and one of hypotheses (H5), (H5)′ or (H5)″. Then, $\mathfrak{L}$ admits a global minimizer.*

**Proof.** Let $(u_n)_{n\in\mathbb{N}}$ be a sequence in $E_{\alpha,p}$ satisfying

$$\mathfrak{L}(u_n) \to \inf_{v\in E_{\alpha,p}} \mathfrak{L}(v) =: K.$$

Since $L$ satisfies hypothesis (H1), $\mathfrak{L}(u) \in \mathbb{R}$ for any $u \in E_{\alpha,p}$. Hence, $K < +\infty$. Let us prove by contradiction that $(u_n)_{n \in \mathbb{N}}$ is bounded in $E_{\alpha,p}$. In the negative case, we can construct a subsequence $(u_{n_k})_{k \in \mathbb{N}}$ satisfying $\|u_{n_k}\|_{\alpha,p} \to \infty$. Since $L$ satisfies hypothesis (H4), Lemma 7.10 gives:

$$K = \lim_{k \in \mathbb{N}} \mathfrak{L}(u_{n_k}) = +\infty,$$

which is a contradiction. Hence, $(u_n)_{n \in \mathbb{N}}$ is bounded in $E_{\alpha,p}$. Since $E_{\alpha,p}$ is reflexive, there exists a subsequence still denoted by $(u_n)_{n \in \mathbb{N}}$ converging weakly in $E_{\alpha,p}$ to an element denoted by $u \in E_{\alpha,p}$. Let us prove that u is a global minimizer of $\mathfrak{L}$. Since

$$u_n \xrightarrow{E_{\alpha,p}} u \quad \text{and} \quad E_{\alpha,p} \hookrightarrow C_a,$$

we have

$$u_n \xrightarrow{C} u \quad \text{and} \quad {_aD_t^\alpha} u_n \xrightarrow{L^p} {_aD_t^\alpha} u. \tag{7.27}$$

Case $L$ satisfies (H5): by convexity, it holds for any $n \in \mathbb{N}$

$$\mathfrak{L}(u_n) = \int_a^b L(u_n, {_aD_t^\alpha} u_n, t) dt$$

$$\geq \int_a^b L(u, {_aD_t^\alpha} u, t) dt + \int_a^b \frac{\partial L}{\partial x}(u, {_aD_t^\alpha} u, t) \, (u_n - u) dt$$

$$+ \int_a^b \frac{\partial L}{\partial y}(u, {_aD_t^\alpha} u, t) \, ({_aD_t^\alpha} u_n - {_aD_t^\alpha} u) dt.$$

Since $L$ satisfies hypotheses (H2) and (H3), $\partial L/\partial x(u, {_aD_t^\alpha} u, t) \in L^1$ and $\partial L/\partial y(u, {_aD_t^\alpha} u, t) \in L^q$. Consequently, using (7.27) and making $n$ tend to $+\infty$, we obtain

$$K = \inf_{v \in E_{\alpha,p}} \mathfrak{L}(v) \geq \int_a^b L(u, {_aD_t^\alpha} u, t) dt = \mathfrak{L}(u).$$

Consequently, $u$ is a global minimizer of $\mathfrak{L}$.

Case $L$ satisfies (H5)$'$: let $\varepsilon > 0$. Since $(u_n)_{n \in \mathbb{N}}$ converges strongly in $C$ to $u$, we have

$$\exists\, N \in \mathbb{N}, \ \forall\, n \geq N, \ \|u_n - u\| < \delta,$$

where $\delta$ is given in the definition of (H5)$'$. In consequence, it holds a.e. on [a,b]

$$|L(u_n(t), {_aD_t^\alpha} u_n(t), t) - L(u(t), {_aD_t^\alpha} u_n(t), t)| < \varepsilon, \quad \text{for } n \geq N. \tag{7.28}$$

Moreover, for any $n \geq N$, we have

$$\mathfrak{L}(u_n) = \int_a^b L(u, {_aD_t^\alpha} u, t) dt + \int_a^b \left( L(u_n, {_aD_t^\alpha} u_n, t) - L(u, {_aD_t^\alpha} u_n, t) \right) dt$$

$$+ \int_a^b \left( L(u, {_aD_t^\alpha} u_n, t) - L(u, {_aD_t^\alpha} u, t) \right) dt.$$

Then, for any $n \geq N$, it holds by convexity

$$\mathfrak{L}(u_n) \geq \int_a^b L(u, {}_aD_t^\alpha u, t)dt - \int_a^b |L(u_n, {}_aD_t^\alpha u_n, t) - L(u, {}_aD_t^\alpha u_n, t)|dt$$

$$+ \int_a^b \frac{\partial L}{\partial y}(u, {}_aD_t^\alpha u, t) \left({}_aD_t^\alpha u_n - {}_aD_t^\alpha u\right)dt.$$

And, using equation (7.28), we obtain for any $n \geq N$

$$\mathfrak{L}(u_n) \geq \int_a^b L(u, {}_aD_t^\alpha u, t)dt - \varepsilon(b-a) + \int_a^b \frac{\partial L}{\partial y}(u, {}_aD_t^\alpha u, t) \left({}_aD_t^\alpha u_n - {}_aD_t^\alpha u\right)dt.$$

We remind that $\partial L/\partial y(u, {}_aD_t^\alpha u, t) \in L^q$ since $L$ satisfies (H3). Since $({}_aD_t^\alpha u_n)_{n \in \mathbb{N}}$ converges weakly in $L^p$ to ${}_aD_t^\alpha u$ we obtain by making $n$ tend to $+\infty$ and then by making $\varepsilon$ tend to 0

$$K = \inf_{v \in E_{\alpha,p}} \mathfrak{L}(v) \geq \int_a^b L(u, {}_aD_t^\alpha u, t)dt = \mathfrak{L}(u).$$

Consequently, $u$ is a global minimizer of $\mathfrak{L}$.

Case $L$ satisfies (H5)″: we refer to Theorem 3.23 in Bacorogna, 2008. □

Finally, we give the existence theorem of weak solution for (7.24).

**Theorem 7.8.** *Let $L$ be a Lagrangian of class $C^1$ and $0 < \frac{1}{p} < \alpha < 1$. If $L$ satisfies the hypotheses denoted by (H1)-(H5). Then (7.24) admits a weak solution.*

Combining Theorem 7.6 and Theorem 4.7, the proof of Theorem 4.1 is obvious.

Let us consider some examples of Lagrangian $L$ satisfying hypotheses of Theorem 4.1. Consequently, the fractional Euler-Lagrange equation (7.24) associated admits a weak solution $u \in E_{\alpha,p}$.

**Example 7.1.** The most classical example is the Dirichlet integral, i.e. the Lagrangian functional associated to the Lagrangian $L$ given by

$$L(x, y, t) = \frac{1}{2}\|y\|^2.$$

In this case, $L$ satisfies hypotheses (H1)-(H5) for $p = 2$. Hence, the fractional Euler-Lagrange equation (7.24) associated admits a weak solution in $E_{\alpha,p}$ for $\frac{1}{2} < \alpha < 1$.

In a more general case, the following Lagrangian $L$

$$L(x, y, t) = \frac{1}{p}\|y\|^p + a(x, t),$$

where $p > 1$ and $a \in C^1(\mathbb{R}^d \times [a, b], \mathbb{R}^+)$, satisfies hypotheses (H1)-(H4) and (H5)″. Consequently, the fractional Euler-Lagrange equation (7.24) associated to $L$ admits a weak solution in $E_{\alpha,p}$ for any $\frac{1}{p} < \alpha < 1$. Let us note that if for any $t \in [a, b]$, $a(\cdot, t)$ is convex, then $L$ satisfies hypothesis (H5).

In the unidimensional case $d = 1$, let us take a Lagrangian with a second term linear in its first variable, i.e.

$$L(x, y, t) = \frac{1}{p}|y|^p + f(t)x,$$

where $p > 1$ and $f \in C^1([a, b], \mathbb{R})$. Then, $L$ satisfies hypotheses (H1)-(H5). Then, the fractional Euler-Lagrange equation (7.24) associated admits a weak solution in $E_{\alpha,p}$ for any $\frac{1}{p} < \alpha < 1$.

Theorem 7.8 is a result based on strong conditions on Lagrangian $L$. Consequently, some Lagrangian do not satisfy all hypotheses of Theorem 7.8. We can cite Bolza's example in dimension $d = 1$ given by

$$L(x, y, t) = (y^2 - 1)^2 + x^4.$$

$L$ does not satisfy hypothesis (H4) neither hypothesis (H5)″. Nevertheless, as usual with variational methods, the conditions of regularity, coercivity and/or convexity can often be replaced by weaker assumptions specific to the studied problem. As an example, we can cite Ammi and Torres, 2008 and references therein about higher-order integrals of the calculus of variations. Indeed, in this subsection, it is proved that calculus of variations is still valid with weaker regularity assumptions.

## 7.4  Fractional Diffusion Equations

### 7.4.1  *Introduction*

In this section, as a first problem, we study the existence and regularity of solutions to the following initial value problem

$$\begin{cases} {}^C_0 D^\alpha_t u(x, t) + \mathcal{A}u(x, t) = f(x, t) & \text{in } \mathbb{R}^n \times (0, b), \\ u(x, 0) = u_0(x), & \text{in } \mathbb{R}^n, \end{cases} \tag{7.29}$$

where ${}^C_0 D^\alpha_t$ is the Caputo fractional derivative of order $\alpha \in (0, 1)$ with respect to $t$ (see Kilbas, Srivastava and Trujillo, 2006), $f \in \mathcal{F}^{q,\sigma}((0, b], H^{-1}(\mathbb{R}^n))$, $\mathcal{F}^{q,\sigma}$ is a space of weighted Hölder continuous functions to be defined later, $0 < \sigma < q \leq 1$, $u_0$ is a given initial data for $u$,

$$\mathcal{A}u(x, t) = -\sum_{i=1}^n \frac{\partial}{\partial x_i}\left(\sum_{j=1}^n a_{ij}(x)\frac{\partial}{\partial x_j}u(x, t)\right) + b(x)u(x, t), \tag{7.30}$$

the real-valued functions $a_{ij}$ satisfy

$$a_{ij} \in L^\infty(\mathbb{R}^n), \quad 1 \leq i, j \leq n, \tag{7.31}$$

$$C_0 \sum_{i=1}^n \xi_i^2 \leq \sum_{i,j=1}^n a_{ij}(x)\xi_i\xi_j, \quad \text{a.e. } x \in \mathbb{R}^n, \; \xi \in \mathbb{R}^n, \tag{7.32}$$

with some constant $C_0 > 0$, and $b(x)$ is a real-valued function satisfying

$$b \in L^\infty(\mathbb{R}^n) \text{ and } b(x) \geq B_0 > 0, \text{ a.e. } x \in \mathbb{R}^n. \tag{7.33}$$

The second problem deals with the existence and regularity of solutions for the Dirichlet type initial-boundary value problem

$$\begin{cases} {}^{C}_{0}D^{\alpha}_{t}u(x,t) + \mathcal{A}u(x,t) = f(x,t), & \text{in } \Omega \times (0,b), \\ u = 0, & \text{on } \partial\Omega \times (0,b), \\ u(x,0) = u_0(x), & \text{in } \Omega, \end{cases} \tag{7.34}$$

where $\Omega \subset \mathbb{R}^n$, $f \in \mathcal{F}^{q,\sigma}((0,b], H^{-1}(\Omega))$, $0 < \sigma < q \leq 1$, $u_0$ is a given initial data for $u$, $\mathcal{A}$ is the same as (7.30), the real-valued functions $a_{ij}$ satisfy (7.31)-(7.32) on $\Omega \subset \mathbb{R}^n$, and real-valued function $b(x)$ satisfies (7.33).

Thirdly, we investigate the existence and regularity of solutions to an initial-boundary value problem of Neumann-type

$$\begin{cases} {}^{C}_{0}D^{\alpha}_{t}u(x,t) + \mathcal{A}u(x,t) = f(x,t) & \text{in } \Omega \times (0,b), \\ \displaystyle\sum_{i,j=1}^{n} \nu_i(x)a_{ij}(x)\frac{\partial}{\partial x_j}u(x,t) = 0, & \text{on } \partial\Omega \times (0,b), \\ u(x,0) = u_0(x), & \text{in } \Omega, \end{cases} \tag{7.35}$$

where $\Omega \subset \mathbb{R}^n$, $f \in \mathcal{F}^{q,\sigma}((0,b], H^1(\Omega)^*)$, $0 < \sigma < q \leq 1$, $u_0$ is a given initial data for $u$, $\mathcal{A}$ is the same as (7.30), real-valued functions $a_{ij}$ satisfy (7.31)-(7.32) on $\Omega \subset \mathbb{R}^n$, and real-valued function $b(x)$ satisfies (7.33).

In the mathematical modeling of physical phenomena, fractional differential equations are found to be better tools than their corresponding integer-order counterparts, for example, the description of anomalous diffusion via such equations leads to more informative and interesting model, see Metzler and Klafter, 2000a. It has been due to the nonlocal nature of fractional operators which can take into account the hereditary characteristics of the phenomena and processes involved in the modeling of real-world problems. It has been observed that the mean square displacement of a diffusive material is proportional to $t$ when $t \to \infty$ in normal diffusion (integer-order diffusion), but this proportionality becomes of the order $t^\alpha$ for $t \to \infty$ in case of anomalous diffusion. This aspect clearly indicates the importance and adaptation of fractional-order operators in the mathematical modeling of scientific and technical problems. For instance, in the study of fractional advection-dispersion equation and fractional Fokker-Planck equation (Benson, Wheatcraft and Meerschaert, 2000a,b), a laboratory tracer test has indicated that fractional differential equations approximate Lévy motion better than integer-order equations. For more details, see Bajlekova, 2001; Chaves, 1998; Clement and Zacher, 2008; Desch and Londen, 2013; Magdziarz and Zorawik, 2015; Caffarelli and Vasseur, 2010; Zheng, Liu, Turner *et al.*, 2015; Liu and Li, 2015.

In Luchko, 2009b, the author proved a maximum principle for the generalized time-fractional diffusion equation on an open bounded domain (like in (7.34)) and then applied it to show the existence of at most one classical solution for an initial-boundary value problem involving this equation, while the uniqueness and existence results for the same equation were obtained in Luchko, 2010. The stability properties and uniqueness of solutions for the homogeneous generalized time-fractional diffusion equation ((7.34) with $f = 0$) were discussed in Sakamoto and Yamamoto, 2011. Some recent results on the topic can be found in Alsaedi, Ahmad and Kirane, 2015; Zhou, Zhang and Shen, 2013; Zhou, Jiao and J. Pečarić, 2013; Zhou, Vijayakumar and Murugesu, 2015.

It is natural to investigate a usual initial value problem or initial-boundary value problem form the physical point of view. The initial condition $u(x, 0) = u_0(x)$ determines the necessity to use Caputo fractional derivative. In case we take Riemann-Liouville fractional derivative , the proper initial data will be the limiting value of the fractional integral of solution of order $1 - \alpha$ as $t \to 0$, but it will not be the limiting value of the solution itself (see Kilbas, Srivastava and Trujillo, 2006). A classical solution for the initial value problem of fractional diffusion equation has been constructed and studied in Eidelman and Kochubei, 2004, when the bounded function $f$ is jointly continuous in $(x, t)$ and locally Hölder continuous in $x$, and the coefficients of $\mathcal{A}$ are bounded Hölder continuous. In Zacher, 2013, the author obtained an interior Hölder estimate for a bounded weak solution to an initial value problem and the $L^\infty$ bound of the solution. In Luchko, 2011, a maximum principle for linear diffusion equation with multiple fractional time derivatives was established and a generalized solution with $f = 0$ was constructed. In Beckers and Yamamoto, 2013, some basic properties such as existence and regularity of solution to an initial-boundary value problem for linear diffusion equation with multiple fractional time derivatives were discussed.

There are also several papers devoted to abstract form of the equations considered in this section, see El-Borai, 2002; Wang, Chen and Xiao, 2012; Ponce, 2013, where the existence of classical solution is obtained under Hölder condition.

Assuming $f$ to be weighted Hölder continuous , which is weaker than Hölder continuous, we show the existence of classical solutions to the abstract form of (7.29), (7.34) and (7.35). This conclusion extends and partly improves the results in El-Borai, 2002; Wang, Chen and Xiao, 2012. For more results involving weighted Hölder continuous functions, we refer the reader to the works Ashyralyev and Tetikoglu, 2015; Han, 2008; Mikulevicius, Pragarauskas and Sonnadara, 2007; Yagi, 2010 and the references therein. Furthermore, if $f$ is assumed to be weighted Hölder continuous, then ${}^C_0 D^\alpha_t u$ and $\mathcal{A}u$ also belong to the function space for $f$. In this scenario, the regularity of solutions is termed as maximal regularity. On the other hand, if $f$ possesses spatial regularity, then the classical solutions of (7.29), (7.34) and (7.35) do have the same regularity.

This section is organized as follows. Subsection 7.4.2 provides the definitions and preliminary results to be used in the sequel. In Subsection 7.4.3, the existence and regularity results for (7.29), (7.34) and (7.35) are investigated. We also derive some new estimations for regularity.

### 7.4.2   *Preliminaries*

Let $X$ be a Banach space equipped with the norm $|\cdot|$. The symbol $H^1(\Omega)$ refers to the usual Sobolev space (see Adams, 1975), and the space $\overset{\circ}{H}{}^1(\Omega)$ represents the closure of $C_0^\infty(\Omega)$ in the space $H^1(\Omega)$. Let the dual space of $\overset{\circ}{H}{}^1(\Omega)$ be denoted by $H^{-1}(\Omega)$, that is, $H^{-1}(\Omega) = \{\overset{\circ}{H}{}^1(\Omega)\}^*$. Define

$$C^\alpha((0,b],X) := \{F : (0,b] \to X \mid F \text{ and } {}^C_0 D^\alpha_t F \text{ are continuous on } (0,b]\}.$$

By $\mathcal{B}([0,b],X)$ we denote the space of uniformly bounded functions on $[0,b]$ (not necessarily smooth or measurable). This space is a Banach space with the supremum norm

$$\|F\|_\mathcal{B} = \sup_{0\leq t\leq b} |F(t)|.$$

For $\eta > 0$, the space $\mathcal{B}^{-\eta}((0,b],X)$ represents the space of $X$-valued functions which are uniformly bounded on $(a,b]$ with weight function $t^\eta$, namely, $F \in \mathcal{B}^{-\eta}((0,b],X)$ if and only if $t^\eta F \in \mathcal{B}((0,b],X)$. We equip this space with the norm

$$\|F\|_{\mathcal{B}^{-\eta}} = \|t^\eta F\|_\mathcal{B}.$$

The space of Hölder continuous functions on $[0,b]$ with exponent $\sigma(0 < \sigma < q \leq 1)$ is denoted by $\mathcal{F}^\sigma([0,b],X)$ which is equipped with the norm

$$\|F\|_{\mathcal{F}^\sigma} = \sup_{0\leq t\leq b} |F(t)| + \sup_{0\leq s<t\leq b} \frac{|F(t)-F(s)|}{(t-s)^\sigma}.$$

Let $\mathcal{F}^{q,\sigma}((0,b],X)$ denote the weighted Hölder continuous function space of continuous functions on $(0,b]$ $([0,b])$ for $0 < q < 1$ $(q=1)$ with the following properties:

(i)  When $0 < q < 1$, $\lim_{t\to 0} t^{1-q}F(t)$ exists;

(ii) $F$ is Hölder continuous with the exponent $\sigma$ and with the weight $s^{1-q+\sigma}$, that is,

$$\sup_{0\leq s<t\leq b} \frac{s^{1-q+\sigma}|F(t)-F(s)|}{(t-s)^\sigma} < \infty; \tag{7.36}$$

(iii)

$$\lim_{t\to 0} \omega_F(t) = \lim_{t\to 0} \sup_{0\leq s<t} \frac{s^{1-q+\sigma}|F(t)-F(s)|}{(t-s)^\sigma} \to 0. \tag{7.37}$$

Then the space $\mathcal{F}^{q,\sigma}((0,b],X)$ is a Banach space with the norm

$$\|F\|_{\mathcal{F}^{q,\sigma}} = \sup_{0\leq t\leq b} t^{1-q}|F(t)| + \sup_{0\leq s<t\leq b} \frac{s^{1-q+\sigma}|F(t)-F(s)|}{(t-s)^\sigma}. \tag{7.38}$$

For $0 < \sigma < \sigma' < q \leq 1$, we have $\mathcal{F}^{q,\sigma'}((0,b],X) \subset \mathcal{F}^{q,\sigma}((0,b],X)$ with continuous embedding, see Yagi, 2010.

The relationship between Hölder continuous function and weighted Hölder continuous function is shown in the following result.

**Remark 7.3.** (Yagi, 2010) For $0 < \sigma < q < 1$,

$$F(t) = t^{q-1}G(t), \quad G(t) \in \mathcal{F}^\sigma([0,b], X), G(0) = 0,$$

belongs to $\mathcal{F}^{q,\sigma}((0,b], X)$. For $0 < \sigma < q = 1$, $F \in F^\sigma([0,b], X)$ belongs to $\mathcal{F}^{q,\sigma}((0,b], X)$.

Next, let us handle (7.29) in the space $X = H^{-1}(\mathbb{R}^n)$. Then problem (7.29)) can be formulated in an abstract form as

$$\begin{cases} {}^C_0 D^\alpha_t U(t) + AU(t) = F(t), & 0 < t \le b, \\ U(0) = U_0, \end{cases} \tag{7.39}$$

where $A$ is a sectorial operator in $X$ with angle $0 \le \omega_A < \frac{\pi}{2}$. Indeed, by Theorem 2.2 of Yagi, 2010, we deduce that $\mathcal{A}$ is a sectorial operator in $H^{-1}(\mathbb{R}^n)$ with domain $D(\mathcal{A}) = H^1(\mathbb{R}^n)$ and with angle $\omega_A < \frac{\pi}{2}$, and $-\mathcal{A}$ generates an analytic semigroup on $H^{-1}(\mathbb{R}^n)$.

Then we consider (7.34) in the space $X = H^{-1}(\Omega) = \{\mathring{H}^1(\Omega)\}^*$, which can be formulated as (7.39) in $H^{-1}(\Omega)$. By Theorem 2.3 of Yagi, 2010, we deduce that $\mathcal{A}$ is a sectorial operator in $H^{-1}(\Omega)$ with domain $D(\mathcal{A}) = \mathring{H}^1(\Omega)$ and with angle $\omega_A < \frac{\pi}{2}$. In this case, $-\mathcal{A}$ generates an analytic semigroup on $H^{-1}(\Omega)$.

Finally, let us handle (7.35) in the space $X = H^1(\Omega)^*$, which can be formulated as (7.39) in $H^1(\Omega)^*$. By Theorem 2.4 of Yagi, 2010, we deduce that $\mathcal{A}$ is a sectorial operator in $H^1(\Omega)^*$ with domain $D(\mathcal{A}) = H^1(\Omega)$, the angle $\omega_A < \frac{\pi}{2}$, and $-\mathcal{A}$ generates an analytic semigroup on $H^1(\Omega)^*$. As argued in Pazy, 1983 and Yagi, 2010, we define

$$\sigma(A) \subset \Sigma_\omega = \{\lambda \in \mathbb{C} : |\arg \lambda| < \omega\}, \quad \omega_A < \omega < \frac{\pi}{2}, \tag{7.40}$$

and

$$|(\lambda I - A)^{-1}| \le \frac{M_\omega}{|\lambda|}, \quad \lambda \notin \Sigma_\omega. \tag{7.41}$$

Then $0 \in \sigma(A)$ and $-A$ generates an analytic semigroup $\{T(t)\}_{t \ge 0}$ satisfying

$$\|T(t)\|_{B(X)} \le M_\omega, \quad t \ge 0, \tag{7.42}$$

$$\|A^\eta T(t)\|_{B(X)} \le C_\eta t^{-\eta}, \quad t > 0, \eta > 0. \tag{7.43}$$

For $\eta > 0$ and $U \in D(A^\eta)$, it is clear that $t^\eta T(t)A^\eta U \to 0$ in $X$ as $t \to 0$. However, for any $F \in X$, the convergence is obtained by the uniform boundedness of $\|t^\eta A^\eta T(t)\|_{B(X)}$ due to (7.43) and the denseness of $D(A^\eta)$ in $X$. That is, for $\eta > 0$ and $t \to 0$,

$$t^\eta A^\eta T(t) \text{ converges to } 0 \text{ strongly on } X. \tag{7.44}$$

Let $0 < \eta \le 1$. Then

$$\|(T(t)-I)A^{-\eta}\|_{B(X)} \le \left\| \int_0^t A^{1-\eta}T(t)d\tau \right\|_{B(X)} \le C\int_0^t \tau^{\eta-1}d\tau \le Ct^\eta, \quad 0 < t < \infty. \tag{7.45}$$

Let $0 < t_1 \le t_2$ and $x \in X$. By the analyticity of $T(t)$ and (7.42) we obtain

$$T(t_2)x - T(t_1)x = \int_{t_1}^{t_2} AT(s)x\,ds = \int_{t_1}^{t_2} T(s - t_1)AT(t_1)\,ds,$$

which implies that

$$|T(t_2)x - T(t_1)x| \le M_\omega(t_2 - t_1)\|AT(t_1)\|_{B(X)}|x|. \tag{7.46}$$

$A^\eta$ $(\eta > 0)$ is the fractional powers of $A$, and it is a densely defined, closed linear operator of $X$. Moreover,

$$\lim_{h \to 0} \frac{1}{h} \int_0^h T(s)x\,ds = x, \quad \text{for } x \in X. \tag{7.47}$$

**Definition 7.5.** A function $U : [0, b] \to X$ is called a (classical) solution of (7.39) if $u$ is continuous on $[0, b]$, ${}_0^C D_t^\alpha u$ exists and is continuous on $(0, b]$, $u(t) \in D(A)$ for $t \in (0, b]$ and (7.39) is satisfied on $[0, b]$.

As in Zhou and Jiao, 2010a and Wang, Chen and Xiao, 2012, let $U : [0, b] \to X$ be a solution of (7.39). Then

$$U(t) = S_\alpha(t)u_0 + \int_0^t (t - s)^{\alpha-1} P_\alpha(t - s)F(s)\,ds, \tag{7.48}$$

where

$$S_\alpha(t) = \int_0^\infty M_\alpha(\theta)T(t^\alpha\theta)\,d\theta, \quad \Psi(t) = \alpha \int_0^\infty \theta M_\alpha(\theta)T(t^\alpha\theta)\,d\theta, \tag{7.49}$$

$M_\alpha$ is the Wright function (see Definition 1.9).

By a similar argument as in Wang, Chen and Xiao, 2012, and Wang and Zhou, 2011a, we have that $\{S_\alpha(t)\}_{t \ge 0}$ and $\{P_\alpha(t)\}_{t \ge 0}$ are strongly continuous with $S_\alpha(t)$ and $P_\alpha(t)$ continuous in the uniform operator topology for $t > 0$. Moreover, for $t > 0$, $x \in X$,

$$S_\alpha'(t)x = -t^{\alpha-1}AP_\alpha(t)x, \tag{7.50}$$

$$S_\alpha(t) = J_t^{1-\alpha}(t^{\alpha-1}P_\alpha(t)), \tag{7.51}$$

$$\|S_\alpha(t)\|_{B(X)} \le M_\omega, \quad \|P_\alpha(t)\|_{B(X)} \le \frac{M_\omega}{\Gamma(\alpha)}. \tag{7.52}$$

By (7.50), (7.51) and the closure property of $A$, we infer that

$${}_0^C D_t^\alpha S_\alpha(t)x = {}_0 D_t^{\alpha-1} S_\alpha'(t)x = -AS_\alpha(t)x, \quad \text{for } t > 0, \ x \in X. \tag{7.53}$$

By virtue of (7.43), for $0 < \eta < 1$ and $0 < t < \infty$, we get

$$\|A^\eta S_\alpha(t)\|_{B(X)} = \left\|\int_0^\infty M_\alpha(\theta)A^\eta T(t^\alpha\theta)\,d\theta\right\|_{B(X)} \tag{7.54}$$

$$\le Ct^{-\alpha\eta}, \tag{7.55}$$

and for $0 < \eta < 2$ and $0 < t < \infty$, we have

$$\|A^\eta P_\alpha(t)\|_{B(X)} = \alpha \left\| \int_0^\infty \theta M_\alpha(\theta) A^\eta T(t^\alpha \theta) d\theta \right\|_{B(X)}$$

$$\leq C t^{-\alpha \eta}. \tag{7.56}$$

Using an argument similar to (7.44), for $0 < \eta < 1$ and $t \to 0$, we find that

$$t^{\alpha \eta} A^\eta S_\alpha(t) \text{ converges to 0 strongly on } X. \tag{7.57}$$

Similarly, for $0 < \eta < 1$, as $t \to 0$,

$$t^{-\alpha \eta} \left( S_\alpha(t) - I \right) A^{-\eta} \text{ converges to 0 strongly on } X. \tag{7.58}$$

In fact, (7.50) and (7.54) imply that

$$\|(S_\alpha(t) - I)A^{-\eta}\|_{B(X)} \leq \int_0^t s^{\alpha - 1} \|A^{1-\eta} P_\alpha(s)\|_{B(X)} ds$$

$$\leq C \int_0^t s^{\alpha - 1 - \alpha(1-\eta)} ds \leq C t^{\alpha \eta}, \quad 0 < t < \infty. \tag{7.59}$$

Then, if $U \in D(A^{1-\eta})$, then $t^{-\alpha\eta}(S_\alpha(t) - I)A^{-\eta}U = t^{-\alpha\eta}(S_\alpha(t) - I)A^{-1}A^{1-\eta}U \to 0$ in $X$ in the limit $t \to 0$ by (7.59). Furthermore, for any $F \in X$, the convergence follows from the uniform boundedness of $\|t^{-\alpha\eta}(S_\alpha(t) - I)A^{-\eta}\|_{B(X)}$ due to (7.59) and the denseness of $D(A^{1-\eta})$ in $X$.

### 7.4.3 *Existence and Regularity*

**Theorem 7.9.** *Let $0 < \sigma < q \leq 1$ and $U_0 \in D(A^q)$. Then, for any $0 < b < \infty$,*

$$AS_\alpha(t)U_0 \in \mathcal{F}^{q,\sigma}((0, b], X), \tag{7.60}$$

*with*

$$\|AS_\alpha(t)U_0\|_{\mathcal{F}^{q,\sigma}((0,b],X)} \leq C|A^q U_0|. \tag{7.61}$$

**Proof.** In view of the fact

$$t^{1-q} AS_\alpha(t)U_0 = t^{(1-\alpha)(1-q)} t^{\alpha(1-q)} A^{1-q} S_\alpha(t) A^q U_0,$$

(7.57) yields that

$$\lim_{t \to 0} t^{1-q} AS_\alpha(t)U_0 = 0, \quad 0 < q < 1.$$

Moreover, (7.45) implies that

$$|A(T(t^\alpha \theta) - T(s^\alpha \theta))U_0|$$
$$= s^{-\alpha(1-q+\frac{\nu}{\alpha})} \theta^{-1+q-\frac{\nu}{\alpha}}$$
$$\times \left| \left( T\left(t^\alpha \theta - s^\alpha \theta\right) - I \right) A^{-\frac{\nu}{\alpha}} (s^\alpha \theta)^{1-q+\frac{\nu}{\alpha}} A^{1-q+\frac{\nu}{\alpha}} T(s^\alpha \theta) A^q U_0 \right|$$
$$\leq C s^{-\alpha(1-q+\frac{\nu}{\alpha})} (t-s)^\nu \theta^{q-1} \left| (s^\alpha \theta)^{1-q+\frac{\nu}{\alpha}} A^{1-q+\frac{\nu}{\alpha}} T(s^\alpha \theta) A^q U_0 \right|. \tag{7.62}$$

Select a constant $\nu$ satisfying $\sigma < \nu < \min\{\alpha q, (1-\alpha)(1-q)+\sigma\}$. Then $(1-\alpha)(1-q)+\sigma-\nu > 0$, $\nu-\sigma > 0$, $0 < \nu_2 := 1-q+\frac{\nu}{\alpha} < 1$. In consequence, (7.62) implies that

$$\frac{s^{1-q+\sigma}\left|A\big(S_\alpha(t)-S_\alpha(s)\big)U_0\right|}{(t-s)^\sigma}$$

$$\leq Cs^{(1-\alpha)(1-q)+\sigma-\nu}(t-s)^{\nu-\sigma}$$

$$\times \int_0^\infty M_\alpha(\theta)\theta^{q-1}\left|(s^\alpha\theta)^{1-q+\frac{\nu}{\alpha}}A^{1-q+\frac{\nu}{\alpha}}T(s^\alpha\theta)A^qU_0\right|d\theta$$

$$\leq Cb^{\nu_1}\int_0^\infty M_\alpha(\theta)\theta^{q-1}\left|(s^\alpha\theta)^{\nu_2}A^{\nu_2}T(s^\alpha\theta)A^qU_0\right|d\theta, \tag{7.63}$$

where $0 \leq s < t \leq b$, $\nu_1 = (1-\alpha)(1-q)$.

We deduce from (7.43), (7.44), (7.63) and Lebesgue dominated convergence theorem that

$$\frac{s^{1-q+\sigma}\left|A\big(S_\alpha(t)-S_\alpha(s)\big)U_0\right|}{(t-s)^\sigma} \leq C\left|A^qU_0\right|,$$

$$\lim_{t\to 0}\sup_{0\leq s<t}\frac{s^{1-q+\sigma}\left|A\big(S_\alpha(t)-S_\alpha(s)\big)U_0\right|}{(t-s)^\sigma} = 0.$$

Thus, (7.60) and (7.61) are established. $\qquad\square$

**Theorem 7.10.** *Suppose that $F \in \mathcal{F}^{q,\sigma}((0,b], X)$ with $0 < \max\{\sigma, 1-\alpha\} < q \leq 1$, and let $U_0 \in X$. Then (7.39) has a unique solution $U$ in the function space*

$$U \in C([0,b], X) \cap C((0,b], D(A)) \cap C^\alpha((0,b], X) \tag{7.64}$$

*with the estimate*

$$|U(t)| + t^\alpha\left|{}^C_0D^\alpha_t U(t)\right| + t^\alpha|AU(t)| \leq C\left(|U_0| + t^{\alpha+q-1}\|F\|_{\mathcal{F}^{q,\sigma}}\right), \quad t \in (0,b]. \tag{7.65}$$

**Proof.** Let $U_1 = S_\alpha(t)U_0$ $(t \geq 0)$. We deduce from (7.53) that $U_1(t)$ is a solution of the following problem

$$\begin{cases} {}^C_0D^\alpha_t U(t) + AU(t) = 0, & 0 < t \leq b, \\ U(0) = U_0. \end{cases} \tag{7.66}$$

Owing to (7.48), it is easy to see that $U_1(t)$ is the only solution to problem (7.66). Set

$$W(t) = \int_0^t (t-s)^{\alpha-1}P_\alpha(t-s)F(s)ds, \quad 0 < t \leq b.$$

Then from (7.50), (7.56) and the assumption on $F$, we obtain

$$|AW(t)| \leq \int_0^t (t-s)^{\alpha-1}\|AP_\alpha(t-s)\|_{B(X)}|F(s) - F(t)|ds + \|I - S_\alpha(t)\|_{B(X)}|F(t)|$$

$$\leq C\int_0^t (t-s)^{\sigma-1}s^{q-1-\sigma}ds\|F\|_{\mathcal{F}^{q,\sigma}} + (1+M_\omega)\|F\|_{\mathcal{F}^{q,\sigma}}t^{q-1}$$

$$\leq CB(\sigma, q-\sigma)t^{q-1}\|F\|_{\mathcal{F}^{q,\sigma}} + (1+M_\omega)t^{q-1}\|F\|_{\mathcal{F}^{q,\sigma}}. \tag{7.67}$$

Thus, $W(t) \in D(A)$ for $\epsilon \leq t \leq b$ ($0 < \epsilon < b$ and $\epsilon$ is arbitrary).

Next, we show that $_0^C D_t^\alpha W(t) \in C((0,b],X)$. In view of (7.51) and $W(0) = 0$, we have that

$$_0^C D_t^\alpha W(t) = \frac{d}{dt} {_0}D_t^{\alpha-1} W(t) = \frac{d}{dt}\left({_0}D_t^{\alpha-1}(t^{\alpha-1} P_\alpha(t)) * F\right) = \frac{d}{dt}(S_\alpha * F). \quad (7.68)$$

It suffices to show that $Z(t) := (S_\alpha * F) \in C^1((0,b],X)$. By a direct calculation we get

$$\frac{Z(t+h) - Z(t)}{h} = \int_0^t \frac{S_\alpha(t+h-s) - S_\alpha(t-s)}{h} F(s)ds$$
$$+ \frac{1}{h}\int_t^{t+h} S_\alpha(t+h-s)F(s)ds. \quad (7.69)$$

Since (7.56) implies

$$|(t-s)^{\alpha-1} A P_\alpha(t-s)F(s)| \leq (t-s)^{-1} s^{q-1} \|F\|_{F^{q,\sigma}} \in L^1([\epsilon,t),\mathbb{R}),$$

therefore it follows from (7.50) together with Lebesgue dominated convergence theorem that

$$\lim_{h \to 0} \int_\epsilon^t \frac{S_\alpha(t+h-s) - S_\alpha(t-s)}{h} F(s)ds = \int_\epsilon^t (t-s)^{\alpha-1}(-A)P_\alpha(t-s)F(s)ds. \quad (7.70)$$

Letting $\epsilon \to 0$, we get

$$\lim_{h \to 0} \int_0^t \frac{S_\alpha(t+h-s) - S_\alpha(t-s)}{h} F(s)ds = -AW(t).$$

Furthermore, we deduce that

$$\frac{1}{h}\int_t^{t+h} S_\alpha(t+h-s)F(s)ds = \frac{1}{h}\int_0^h S_\alpha(s)\big(F(t+h-s) - F(t)\big)ds$$
$$+ \frac{1}{h}\int_0^h S_\alpha(s)F(t)ds$$
$$=: K_1 + K_2.$$

From (7.52) and the assumption on $F$, we infer that

$$|K_1| \leq \frac{C}{h}\int_0^h t^{q-1-\sigma}(h-s)^\sigma ds \leq C t^{q-1-\sigma} h^\sigma.$$

Equations (7.47) and (7.49) imply that $\lim_{h\to 0} K_2 = F(t)$ for $0 \leq t \leq b$. Therefore by (7.47), for $\epsilon \leq t \leq b$, we have

$$\frac{1}{h}\int_t^{t+h} S_\alpha(t+h-s)F(s)ds \to F(t) \quad \text{as } h \to 0^+. \quad (7.71)$$

Using (7.69), (7.70), and (7.71), we deduce that $z$ is differentiable from the right at $t$ and $z'_+(t) + AW(t) = F(t)$ for $t \in [\epsilon,b]$. Similarly, $z$ is differentiable from the left at $t$ and $z'_-(t) + AW(t) = F(t)$ for $t \in [\epsilon,b]$.

In the following, we show that $AW(t) \in C((0, b], X)$. Using (7.50), we obtain

$$AW(t) = \int_0^t (t-s)^{\alpha-1} AP_\alpha(t-s)\big(F(s) - F(t)\big)ds$$

$$+ \int_0^t (t-s)^{\alpha-1} AP_\alpha(t-s)F(t)ds$$

$$= \int_0^t (t-s)^{\alpha-1} AP_\alpha(t-s)\big(F(s) - F(t)\big)ds + \big(I - S_\alpha(t)\big)F(t)$$

$$=: Y_1 + Y_2. \tag{7.72}$$

Again, by the assumption on $F$ and the uniform continuity of $S_\alpha(t)$ for $t > 0$, one can get that $Y_2(t)$ is continuous for $\epsilon \le t \le b$. On the other hand,

$$Y_1(t+h) - Y_1(t)$$

$$= \int_0^t \big((t+h-s)^{\alpha-1} AP_\alpha(t+h-s) - (t-s)^{\alpha-1} AP_\alpha(t-s)\big)\big(F(s) - F(t)\big)ds$$

$$+ \int_0^t (t+h-s)^{\alpha-1} AP_\alpha(t+h-s)\big(F(t) - F(t+h)\big)ds$$

$$+ \int_t^{t+h} (t+h-s)^{\alpha-1} AP_\alpha(t+h-s)\big(F(s) - F(t+h)\big)ds$$

$$=: S_1(t) + S_2(t) + S_3(t). \tag{7.73}$$

From (7.56) and the assumption on $F$, we have

$$\big|(t+h-s)^{\alpha-1} AP_\alpha(t+h-s)\big(F(s) - F(t)\big)\big| \le C(t-s)^{\sigma-1}s^{q-1-\sigma} \in L^1((0,t), \mathbb{R}).$$

It is immediate from the strong continuity of $\{P_\alpha(t)\}_{t\ge0}$ and Lebesgue dominated convergence theorem that

$$\lim_{h\to 0^+} \int_0^t (t+h-s)^{\alpha-1} AP_\alpha(t+h-s)\big(F(s) - F(t)\big)ds$$

$$= \int_0^t (t-s)^{\alpha-1} AP_\alpha(t-s)\big(F(s) - F(t)\big)ds,$$

which implies that for $0 < t \le b$, $S_1(t) \to 0$ as $h \to 0^+$. Furthermore, it follows from (7.50) that

$$S_2(t) = \big(S_\alpha(h) - S_\alpha(t+h)\big)\big(F(t) - F(t+h)\big).$$

Thus, in view of (7.52) and the assumption on $F$, we have that $S_2(t)$ is continuous for $0 < t \le b$. Using (7.56) and the assumption on $F$, we obtain

$$|S_3(t)| \le C \int_t^{t+h} (t+h-s)^{\sigma-1} s^{q-\sigma-1} ds \sup_{0<s<t+h} \frac{s^{1-q+\sigma}|F(t+h) - F(s)|}{(t+h-s)^\sigma}$$

$$\le Ch^\sigma t^{q-\sigma-1} \omega_F(t+h),$$

which implies that $\lim_{h \to 0^+} S_3(t) = 0$ for $\epsilon < t \le b$. Thus for $\epsilon \le t \le b$, we have $Y_1(t - h) - Y_1(t) \to 0$ as $h \to 0^+$. By a similar argument, for $\epsilon \le t \le b$, $Y_1(t - h) - Y_1(h) \to 0$ as $h \to 0^-$. Thus, $AW \in C([\epsilon, b], X)$, which implies that $Z' \in C([\epsilon, b], X)$ by the assumption on $F$. Then $^C_0D^\alpha_t W \in C([\epsilon, b], X)$ by (7.68). Since $\epsilon$ is arbitrary, we find that $AW \in C((0, b], X)$ and $^C_0D^\alpha_t W \in C((0, b], X)$.

Set $U = U_1 + W$. By the foregoing arguments, it is easy to find that $U$ is continuous on $(0, b]$. Furthermore, in view of (7.52) and the assumption on $F$, we obtain

$$|U(t) - U(0)| \le |(S_\alpha(t) - I)U_0| + \int_0^t (t - s)^{\alpha-1} \|P_\alpha(t - s)\|_{B(X)} |F(s)| ds$$

$$\le |(S_\alpha(t) - I)U_0| + C \int_0^\infty (t - s)^{\alpha-1} s^{q-1} ds$$

$$\le |(S_\alpha(t) - I)U_0| + Ct^{\alpha-1+q} B(\alpha, q).$$

This estimate, together with the strong continuity of $\{S_\alpha(t)\}_{t \ge 0}$, implies that $U$ is continuous at $t = 0$. Therefore, $U$ is a solution of (7.39), and it is unique by (7.48). It is easy to find that (7.64) holds. The estimate (7.65) can also be obtained by means of (7.52), (7.54) and (7.67). This completes the proof. $\square$

**Corollary 7.1.** *If $q = 1$, then $\mathcal{F}^{q,\sigma}((0, b], X) = \mathcal{F}^\sigma((0, b], X)$. That is, $F$ is Hölder continuous. Thus Theorem 7.10 extends and improves the part of results in El-Borai, 2002; Wang, Chen and Xiao, 2012, in which $F$ is Hölder continuous and $U_0 \in D(A)$. In Theorem 7.10, $U_0 \in X$ due to the analyticity of $\{T(t)\}_{t \ge 0}$, which is also a differentiable semigroup.*

**Theorem 7.11.** *Suppose that $F \in \mathcal{F}^{q,\sigma}((0, b], X)$ with $0 < \max\{\sigma, 1 - \alpha\} < q \le 1$ and $\sigma < \alpha < 1$. If $U_0 \in D(A^q)$, then (7.39) has a solution $U$ with the following regularity:*

$$A^{q_1} U \in C([0, b], X), \quad 0 < q_1 \le \frac{\alpha + q - 1}{\alpha}, \tag{7.74}$$

$$^C_0D^\alpha_t U, \; AU \in \mathcal{F}^{q,\sigma}((0, b], X), \tag{7.75}$$

*and*

$$\|A^{q_1} U\|_C \le C(|A^q U_0| + \|F\|_{F^{q,\sigma}}), \tag{7.76}$$

$$\|^C_0D^\alpha_t U\|_{F^{q,\sigma}} + \|AU\|_{F^{q,\sigma}} \le C(|A^q U_0| + \|F\|_{F^{q,\sigma}}). \tag{7.77}$$

**Proof.** The existence of solution follows from Theorem 7.10. If $U$ is a solution of (7.39), we already know that $A^{q_1} U(t)$ is continuous for $t > 0$. Next we show that $A^{q_1} U(t)$ is continuous at $t = 0$. In view of (7.48) and (7.50), we can write

$$A^{q_1}(U(t) - U_0) = (S_\alpha(t) - I)A^{q_1} U_0 + \int_0^t (t - s)^{\alpha-1} A^{q_1} P_\alpha(t - s)(F(s) - F(t)) ds$$

$$+ (I - S_\alpha(t)(t)) A^{q_1-1} F(t)$$

$$=: J_1(t) + J_2(t) + J_3(t).$$

Due to the strong continuity of $\{S_\alpha(t)\}_{t\geq 0}$, $\lim_{t\to 0} J_1 = 0$. We also find that

$$|J_2(t)| \leq C \int_0^t (t-s)^{\alpha-1-\alpha q_1+\sigma} s^{q-\sigma-1} ds \sup_{0\leq s<t} \frac{s^{1-q+\sigma}|F(t)-F(s)|}{(t-s)^\sigma}$$

$$\leq C b^{\alpha-1-\alpha q_1+q} B(\alpha - \alpha q_1 + \sigma, q - \sigma) \omega_F(t).$$

In view of the assumption on $F$, we have $\lim_{t\to 0} J_2 = 0$. Using (7.58) and the assumption on $F$, we get $\lim_{t\to 0} J_3 = 0$, where

$$J_3 = -t^{q-1-\alpha(q_1-1)} t^{\alpha(q_1-1)} (S_\alpha(t)(t) - I) A^{q_1-1} t^{1-q} F(t).$$

The estimate (7.76) can easily be obtained at the same time.

Let us now show (7.75). We also know that

$$AU(t) = AS_\alpha(t)U_0 + \int_0^t (t-s)^{\alpha-1} AP_\alpha(t-s)(F(s)-F(t)) ds$$

$$+ (I - S_\alpha(t)(t)) F(t)$$

$$= I_1(t) + I_2(t) + I_3(t). \tag{7.78}$$

By Theorem 7.9, we get $I_1 \in \mathcal{F}^{q,\sigma}((0,b], X)$. So, it suffices to prove that $I_i \in \mathcal{F}^{q,\sigma}((0,b], X)$ for $i = 2, 3$.

Indeed,

$$|I_2(t)| \leq C \int_0^t (t-s)^{\sigma-1} s^{q-\sigma-1} ds \sup_{0\leq s<t} \frac{s^{1-q+\sigma}|F(t)-F(s)|}{(t-s)^\sigma}$$

$$\leq C t^{q-1} B(\sigma, q-\sigma) \omega_F(t),$$

From the assumption on $F$, we have $\lim_{t\to 0} t^{1-q} I_2 = 0$.

Similar to (7.73), for $0 \leq s < t \leq b$, one can write

$$I_2(t) - I_2(s)$$

$$= \int_s^t (t-\tau)^{\alpha-1} AP_\alpha(t-\tau)(F(\tau)-F(t)) d\tau$$

$$+ \int_0^s (t-\tau)^{\alpha-1} AP_\alpha(t-\tau)(F(s)-F(t)) d\tau$$

$$+ \int_0^s ((t-\tau)^{\alpha-1} AP_\alpha(t-\tau) - (s-\tau)^{\alpha-1} AP_\alpha(s-\tau)) (F(\tau)-F(s)) d\tau$$

$$:= I_{21}(t,s) + I_{22}(t,s) + I_{23}(t,s).$$

From (7.56), it follows that

$$|I_{21}(t,s)| \leq \int_s^t (t-\tau)^{\sigma-1} \tau^{q-1-\sigma} d\tau \sup_{0<\tau<t} \frac{\tau^{1-q+\sigma}|F(t)-F(\tau)|}{(t-\tau)^\sigma}$$

$$\leq C(t-s)^\sigma s^{q-\sigma-1} \omega_F(t).$$

On the other hand, (7.52) implies that

$$|I_{22}(t,s)| = \|S_\alpha(t-s) - S_\alpha(t)\|_{B(X)} |F(s)-F(t)|$$

$$\leq C(t-s)^\sigma s^{q-\sigma-1} \omega_F(t).$$

Further, using (7.43), we have

$$\left\| (t-\tau)^{\alpha-1} AT\left((t-\tau)^\alpha \theta\right) - (s-\tau)^{\alpha-1} AT\left((s-\tau)^\alpha \theta\right) \right\|_{B(X)}$$

$$\leq (t-\tau)^{\alpha-1} \left\| A\left(T(t-\tau)^\alpha \theta - T(s-\tau)^\alpha \theta\right) \right\|_{B(X)}$$

$$+ \left((s-\tau)^{\alpha-1} - (t-\tau)^{\alpha-1}\right) \left\| AT((s-\tau)^\alpha \theta) \right\|_{B(X)}$$

$$\leq \alpha\theta(t-\tau)^{\alpha-1} \int_{s-\tau}^{t-\tau} A^2 T(\rho^\alpha \theta) \rho^{\alpha-1} d\rho$$

$$+ C\left((s-\tau)^{\alpha-1} - (t-\tau)^{\alpha-1}\right)(s-\tau)^{-\alpha}\theta^{-1}$$

$$\leq C\theta^{-1}\left((s-\tau)^{-1} - (t-\tau)^{-1}\right),$$

for $0 < \tau < s < t$. Then we get

$$|I_{23}(t,s)| \leq C \int_0^s \left((s-\tau)^{-1} - (t-\tau)^{-1}\right)|F(\tau) - F(s)| d\tau$$

$$\leq C\omega_F(s)(t-s) \int_0^s (s-\tau)^{\sigma-1}(t-\tau)^{-1}\tau^{q-1-\sigma} d\tau$$

$$= C\omega_F(s)(t-s) \int_0^s (t-s+\tau)^{-1}\tau^{\sigma-1}(s-\tau)^{q-1-\sigma} d\tau.$$

Since

$$(t-s) \int_{\frac{s}{2}}^s (t-s+\tau)^{-1}\tau^{\sigma-1}(s-\tau)^{q-1-\sigma} d\tau$$

$$\leq 2(t-s)^\sigma s^{-1} \int_{\frac{s}{2}}^s (t-s)^{1-\sigma}\tau^\sigma (t-s+\tau)^{-1}(s-\tau)^{q-1-\sigma} d\tau$$

$$\leq 2(t-s)^\sigma s^{-1} \int_0^s (s-\tau)^{q-1-\sigma} d\tau$$

$$= \frac{2}{q-\sigma}(t-s)^\sigma s^{q-1-\sigma},$$

and

$$(t-s) \int_0^{\frac{s}{2}} (t-s+\tau)^{-1}\tau^{\sigma-1}(s-\tau)^{q-1-\sigma} d\tau$$

$$\leq (t-s)2^{1-q+\sigma} s^{q-1-\sigma} \int_0^\infty (t-s+\tau)^{-1}\tau^{\sigma-1} d\tau$$

$$= \frac{2^{1-q+\sigma}\pi}{\sin\sigma\pi}(t-s)^\sigma s^{q-1-\sigma},$$

we get

$$|I_{23}(t,s)| \leq C\omega_F(s)(t-s)^\sigma s^{q-1-\sigma},$$

for $0 \leq s < t \leq b$. Therefore, $I_2 \in \mathcal{F}^{q,\sigma}((0,b], X)$.

Next we show $I_3 \in \mathcal{F}^{q,\sigma}((0,b], X)$. The strong continuity of $\{S_\alpha(t)\}_{t\geq 0}$, $t^{1-q}I_3(t) = (sI - S_\alpha(t))(F(\tau) - F(s))t^{1-q}F(t)$ implies that $\lim_{t\to 0} t^{1-q}I_3(t) = 0$. On the other hand,

$$I_3(t) - I_3(s) = (I - S_\alpha(t))(F(t) - F(s)) - (S_\alpha(t) - S_\alpha(s))F(s).$$

Indeed, one can deduce from (7.52) and the assumption on $F$ that

$$\left|(I - S_\alpha(t))(F(t) - F(s))\right| \leq C\omega_F(t)(t - s)^\sigma s^{q-\sigma-1}.$$

Similarly, it follows from (7.45) that

$$\left|(T(t^\alpha\theta) - T(s^\alpha\theta))F(s)\right|$$
$$\leq (s^\alpha\theta)^{-\frac{\sigma}{\alpha}}s^{q-1}\left\|(T(t^\alpha\theta - s^\alpha\theta) - I)A^{-\frac{\sigma}{\alpha}}\right\|_{B(X)}\left|(s^\alpha\theta)^{\frac{\sigma}{\alpha}}A^{\frac{\sigma}{\alpha}}T(s^\alpha\theta)s^{1-q}F(s)\right|$$
$$\leq Cs^{q-1-\sigma}(t - s)^\sigma\left|(s^\alpha\theta)^{\frac{\sigma}{\alpha}}A^{\frac{\sigma}{\alpha}}T(s^\alpha\theta)s^{1-q}F(s)\right|.$$

Hence, by (7.43), (7.44) and Lebesgue dominated convergence theorem, we get

$$\lim_{t\to 0}\sup_{0\leq s<t}\int_0^\infty M_\alpha(\theta)\left|(s^\alpha\theta)^{\frac{\sigma}{\alpha}}A^{\frac{\sigma}{\alpha}}T(s^\alpha\theta)s^{1-q}F(s)\right|d\theta = 0.$$

In this way, $I_3 \in \mathcal{F}^{q,\sigma}((0,b], X)$.

Thus $AU \in \mathcal{F}^{q,\sigma}((0,b], X)$. In view of ${}_0^C D_t^\alpha U(t) + AU(t) = F(t)$, we obtain (7.77). This concludes the proof. $\qquad\square$

**Theorem 7.12.** *Let* $0 < \max\{\sigma, 1 - \alpha\} < q \leq 1$, $0 < \rho < \min\{\frac{\sigma}{\alpha}, q\}$, $F \in \mathcal{F}^{q,\sigma}((0,b], X)$, $F(t) \in D(A^\rho)$ *for all* $0 < t \leq b$ *and*

$$A^\rho F \in \mathcal{B}^{q-1-\alpha\rho}((0,b], X),$$

*with* $U_0 \in D(A^q)$. *Then* (7.35) *has a unique solution* $U$ *with* $A^\rho {}_0^C D_t^\alpha U$, $A^{1+\rho}U \in \mathcal{B}^{q-1-\alpha\rho}((0,b], X)$ *and*

$$\left\|A^\rho {}_0^C D_t^\alpha U\right\|_{\mathcal{B}^{q-1-\alpha\rho}} + \left\|A^{1+\rho}U\right\|_{\mathcal{B}^{q-1-\alpha\rho}} \leq C\left(|A^q U_0| + \|F\|_{\mathcal{F}^{q,\sigma}} + \|A^\rho F\|_{\mathcal{B}^{q-1-\alpha\rho}}\right).$$

**Proof.** By (7.52), (7.54), (7.56), (7.78) and the assumption on $F$, we deduce that

$$\left|t^{1+\alpha\rho-q}A^{1+\rho}U(t)\right| \leq \left\|t^{1+\alpha\rho-q}A^{1+\rho-q}\Phi(t)\right\|\left|A^q U_0\right|$$

$$+ CB(\sigma - \alpha\rho, q - \sigma)\sup_{0\leq s<t\leq b}\frac{s^{1-q+\sigma}|F(t) - F(s)|}{(t - s)^\sigma}$$

$$+ (1 + M_\omega)\left|t^{1+\alpha\rho-q}A^\rho F(t)\right|$$

$$\leq C|A^q U_0| + C\|F\|_{\mathcal{F}^{q,\sigma}} + C\|A^\rho F\|_{\mathcal{B}^{q-1-\alpha\rho}},$$

which completes the proof. $\qquad\square$

As application of 7.10, we can obtain the following results.

**Theorem 7.13.** *Suppose that* $f \in \mathcal{F}^{q,\sigma}((0,b], H^{-1}(\mathbb{R}^n))$ *with* $0 < \max\{\sigma, 1 - \alpha\} < q \leq 1$, *and let* $u_0 \in H^{-1}(\mathbb{R}^n)$. *Then* (7.29) *has a unique solution* $u$ *satisfying*

$$u \in C([0,b], H^{-1}(\mathbb{R}^n)) \cap C((0,b], D(\mathcal{A})) \cap C^\alpha((0,b], H^{-1}(\mathbb{R}^n))$$

*with the estimate*

$$|u(t)|_{H^{-1}} + t^\alpha|{}_0^C D_t^\alpha u(t)|_{H^{-1}} + t^\alpha|Au(t)|_{H^{-1}}$$
$$\leq C\left(|u_0|_{H^{-1}} + t^{\alpha+q-1}\|f\|_{\mathcal{F}^{q,\sigma}}\right), \quad t \in (0,b].$$

**Theorem 7.14.** *Suppose that* $f \in \mathcal{F}^{q,\sigma}((0,b], H^{-1}(\Omega))$ *with* $0 < \max\{\sigma, 1-\alpha\} < q \leq 1$, *and let* $u_0 \in H^{-1}(\Omega)$. *Then* (7.34) *has a unique solution* $u$ *satisfying*

$$u \in C([0,b], H^{-1}(\Omega)) \cap C((0,b], D(\mathcal{A})) \cap C^{\alpha}((0,b], H^{-1}(\Omega))$$

*with the estimate*

$$|u(t)|_{H^{-1}} + t^{\alpha}|{}_0^C D_t^{\alpha} u(t)|_{H^{-1}} + t^{\alpha}|\mathcal{A}u(t)|_{H^{-1}}$$
$$\leq C\left(|u_0|_{H^{-1}} + t^{\alpha+q-1}\|f\|_{\mathcal{F}^{q,\sigma}}\right), \quad t \in (0,b].$$

**Theorem 7.15.** *Suppose that* $f \in \mathcal{F}^{q,\sigma}((0,b], H^1(\Omega^*))$ *with* $0 < \max\{\sigma, 1-\alpha\} < q \leq 1$, *and let* $u_0 \in H^1(\Omega)^*$. *Then* (7.35) *has a unique solution* $u$ *satisfying*

$$u \in C([0,b], H^1(\Omega^*)) \cap C((0,b], D(\mathcal{A})) \cap C^{\alpha}((0,b], H^1(\Omega^*))$$

*with the estimate*

$$|u(t)|_{H^{1*}} + t^{\alpha}|{}_0^C D_t^{\alpha} u(t)|_{H^{1*}} + t^{\alpha}|\mathcal{A}u(t)|_{H^{1*}}$$
$$\leq C\left(|u_0|_{H^{1*}} + t^{\alpha+q-1}\|f\|_{\mathcal{F}^{q,\sigma}}\right), \quad t \in (0,b].$$

Concerning the application of Theorem 7.11, we can get the following theorems.

**Theorem 7.16.** *For any* $f \in \mathcal{F}^{q,\sigma}((0,b], H^{-1}(\mathbb{R}^n))$, $0 < \max\{\sigma, 1-\alpha\} < q \leq 1$, $\sigma < \alpha < 1$, $0 < q_1 \leq 1 + \frac{q-1}{\alpha}$, $u_0 \in D(\mathcal{A}^q)$, (7.29) *has a unique solution* $u$ *in the function space:*

$$\begin{cases} u \in C((0,b], H^1(\mathbb{R}^n)) \cap C([0,b], D(\mathcal{A}^{q_1})) \cap C^{\alpha}((0,b], H^{-1}(\mathbb{R}^n)), \\ {}_0^C D_t^{\alpha} u, \mathcal{A}u \in \mathcal{F}^{q,\sigma}((0,b], H^{-1}(\mathbb{R}^n)), \end{cases}$$

*with the estimates*

$$\|\mathcal{A}^{q_1} u\|_C + \|{}_0^C D_t^{\alpha} u\|_{\mathcal{F}^{q,\sigma}} + \|\mathcal{A}u\|_{\mathcal{F}^{q,\sigma}} \leq C\left(|\mathcal{A}^q u_0|_{H^{-1}} + \|f\|_{\mathcal{F}^{q,\sigma}}\right).$$

**Theorem 7.17.** *For any* $f \in \mathcal{F}^{q,\sigma}((0,b], H^{-1}(\Omega))$, $0 < \max\{\sigma, 1-\alpha\} < q \leq 1$, $\sigma < \alpha < 1$, $0 < q_1 \leq 1 + \frac{q-1}{\alpha}$, $u_0 \in D(\mathcal{A}^q)$, (7.34) *has a unique solution* $u$ *in the function space:*

$$\begin{cases} u \in C((0,b], \mathring{H}^1(\Omega)) \cap C([0,b], D(\mathcal{A}^{q_1})) \cap C^{\alpha}((0,b], H^{-1}(\Omega)), \\ {}_0^C D_t^{\alpha} u, \mathcal{A}u \in \mathcal{F}^{q,\sigma}((0,b], H^{-1}(\Omega)), \end{cases}$$

*with the estimates*

$$\|\mathcal{A}^{q_1} u\|_C + \|{}_0^C D_t^{\alpha} u\|_{\mathcal{F}^{q,\sigma}} + \|\mathcal{A}u\|_{\mathcal{F}^{q,\sigma}} \leq C\left(|\mathcal{A}^q u_0|_{H^{-1}} + \|f\|_{\mathcal{F}^{q,\sigma}}\right).$$

**Theorem 7.18.** *For any* $f \in \mathcal{F}^{q,\sigma}((0,b], H^1(\Omega)^*)$, $0 < \max\{\sigma, 1-\alpha\} < q \leq 1$, $\sigma < \alpha < 1$, $0 < q_1 \leq 1 + \frac{q-1}{\alpha}$, $u_0 \in D(\mathcal{A}^q)$, (7.35) *has a unique solution* $u$ *in the function space:*

$$\begin{cases} u \in C((0,b], H^1(\Omega)) \cap C([0,b], D(\mathcal{A}^{q_1})) \cap C^{\alpha}((0,b], H^1(\Omega)^*), \\ {}_0^C D_t^{\alpha} u, \mathcal{A}u \in \mathcal{F}^{q,\sigma}((0,b], H^1(\Omega)^*), \end{cases}$$

*with the estimates*

$$\|\mathcal{A}^{q_1} u\|_C + \|{}_0^C D_t^{\alpha} u\|_{\mathcal{F}^{q,\sigma}} + \|\mathcal{A}u\|_{\mathcal{F}^{q,\sigma}} \leq C\left(|\mathcal{A}^{q_1} u_0|_{H^{1*}} + \|f\|_{\mathcal{F}^{q,\sigma}}\right).$$

By a direct consequence of Theorem 7.12, the following results hold.

**Theorem 7.19.** *Let* $0 < \max\{\sigma, 1 - \alpha\} < q \leq 1$, $f \in \mathcal{F}^{q,\sigma}((0,b], H^{-1}(\mathbb{R}^n))$, $f(t) \in \mathcal{D}(\mathcal{A}^\rho)$ *for all* $0 < t \leq b$,

$$\mathcal{A}^\rho f \in \mathcal{B}^{q-1-\alpha\rho}((0,b], H^{-1}(\mathbb{R}^n)), \quad 0 < \rho < \min\left\{\frac{\sigma}{\alpha}, q\right\},$$

*and let* $u_0 \in \mathcal{D}(\mathcal{A}^q)$. *Then* (7.29) *has a unique solution* $u$ *with* $\mathcal{A}^\rho {}_0^C D_t^\alpha u$, $\mathcal{A}^{1+\rho} u \in \mathcal{B}^{q-1-\alpha\rho}((0,b], H^{-1}(\mathbb{R}^n))$ *and*

$$\left\|\mathcal{A}^\rho {}_0^C D_t^\alpha u\right\|_{\mathcal{B}^{q-1-\alpha\rho}} + \left\|\mathcal{A}^{1+\rho} u\right\|_{\mathcal{B}^{q-1-\alpha\rho}} \leq C\left(\left\|\mathcal{A}^q u_0\right\|_{H^{-1}} + \|f\|_{\mathcal{F}^{q,\sigma}} + \left\|\mathcal{A}^\rho f\right\|_{\mathcal{B}^{q-1-\alpha\rho}}\right).$$

**Theorem 7.20.** *Let* $0 < \max\{\sigma, 1 - \alpha\} < q \leq 1$, $f \in \mathcal{F}^{q,\sigma}((0,b], H^{-1}(\Omega))$, $f(t) \in \mathcal{D}(\mathcal{A}^\rho)$ *for all* $0 < t \leq b$,

$$\mathcal{A}^\rho f \in \mathcal{B}^{q-1-\alpha\rho}((0,b], H^{-1}(\Omega)), \quad 0 < \rho < \min\left\{\frac{\sigma}{\alpha}, q\right\},$$

*and let* $u_0 \in \mathcal{D}(\mathcal{A}^q)$. *Then* (7.34) *has a unique solution* $u$ *with* $\mathcal{A}^\rho {}_0^C D_t^\alpha u$, $\mathcal{A}^{1+\rho} u \in \mathcal{B}^{q-1-\alpha\rho}((0,b], H^{-1}(\Omega))$ *and*

$$\left\|\mathcal{A}^\rho {}_0^C D_t^\alpha u\right\|_{\mathcal{B}^{q-1-\alpha\rho}} + \left\|\mathcal{A}^{1+\rho} u\right\|_{\mathcal{B}^{q-1-\alpha\rho}} \leq C\left(\left\|\mathcal{A}^q u_0\right\|_{H^{-1}} + \|f\|_{\mathcal{F}^{q,\sigma}} + \left\|\mathcal{A}^\rho f\right\|_{\mathcal{B}^{q-1-\alpha\rho}}\right).$$

**Theorem 7.21.** *Let* $0 < \max\{\sigma, 1 - \alpha\} < q \leq 1$, $f \in \mathcal{F}^{q,\sigma}((0,b], H^1(\Omega)^*)$, $f(t) \in \mathcal{D}(\mathcal{A}^\rho)$ *for all* $0 < t \leq b$,

$$\mathcal{A}^\rho f \in \mathcal{B}^{q-1-\alpha\rho}((0,b], H^1(\Omega)^*), \quad 0 < \rho < \min\left\{\frac{\sigma}{\alpha}, q\right\},$$

*and let* $u_0 \in \mathcal{D}(\mathcal{A}^q)$. *Then* (7.35) *has a unique solution* $u$ *with* $\mathcal{A}^\rho {}_0^C D_t^\alpha u$, $\mathcal{A}^{1+\rho} u \in \mathcal{B}^{q-1-\alpha\rho}((0,b], H^1(\Omega)^*)$ *and*

$$\left\|\mathcal{A}^\rho {}_0^C D_t^\alpha u\right\|_{\mathcal{B}^{q-1-\alpha\rho}} + \left\|\mathcal{A}^{1+\rho} u\right\|_{\mathcal{B}^{q-1-\alpha\rho}} \leq C\left(\left\|\mathcal{A}^q u_0\right\|_{H^{1*}} + \|f\|_{\mathcal{F}^{q,\sigma}} + \left\|\mathcal{A}^\rho f\right\|_{\mathcal{B}^{q-1-\alpha\rho}}\right).$$

## 7.5 Fractional Schrödinger Equations

### 7.5.1 *Introduction*

Schrödinger equations have received a great deal of interest from the mathematicians in the past twenty years or so, due in particular to their applications to optics. Indeed, simplified versions or limits of Zakharov's system lead to certain Schrödinger equations. Schrödinger equations also arise in quantum field theory, and in particular in the Hartree-Fock theory. From the mathematical point of view, Schrödinger equations appears a delicate problem, since it possesses a mixture of the properties of parabolic and hyperbolic equations. Indeed, it is almost reversible, it has conservation laws and also some dispersive properties like the Kelin-Gordon equation, but it has an infinite speed of propagation. On the other hand, Schrödinger equations has a kind of smoothing effect shared by parabolic problems but the time-reversibility it from generating an analytic semigroup. For more details, one can see the monographs Cazenave, 2003; Sulem and Sulem, 1999 and the papers Buslaev and Sulem, 2013; Cazenave, 1983; Cazenave and Lions, 1982; Cuccagna, 2001; Eid,

Muslih and Baleanu *et al.*, 2009; Fibich, 2011; Floer and Weinstein, 1986; Guo and Wu, 1995; Tsai, 1995; Wang, 2008 and etc.

In this section we study the initial value problem of the following fractional Schrödinger equations with potential

$$\begin{cases} \frac{1}{i} \, {}_0^C D_t^\alpha x(t,y) - \Delta x(t,y) + kV(y)x(t,y) = 0, & \alpha \in (0,1), \ y \in \Omega, \ t \in (0,T], \\ x(0,y) = x_0(y), & y \in \Omega, \end{cases}$$
(7.79)

where ${}_0^C D_t^\alpha$ is Caputo fractional derivative of order $\alpha$ in time $t$, $\Omega \subseteq \mathbb{R}^2$ is a bounded domain with a smooth boundary $\partial\Omega$, $\Delta$ denotes the Laplace operator in $\mathbb{R}^2$, $x$ is a complex valued function in $[0,T] \times \mathbb{R}^2$, $k := \max_{t \in [0,T]} |\chi(t)|$ with $\chi \in C(J,\mathbb{R})$, is a positive constant, and the function $V$ is called potential.

In Subsection 7.5.2, we give the preliminaries of linear Schrödinger equations and introduce a suitable concept on a mild solution for problem (7.79). In Subsection 7.5.3, existence and uniqueness of mild solutions are presented.

### 7.5.2   *Preliminaries*

We recall the following initial value problem for linear Schrödinger equations

$$\begin{cases} \frac{1}{i} \frac{\partial}{\partial t} x(t,y) - \Delta x(t,y) = 0, & y \in \Omega, \ t \in (0,T], \\ x(0,y) = x_0(y), & y \in \Omega, \end{cases}$$
(7.80)

where $\Omega \subseteq \mathbb{R}^2$ is a bounded domain with a smooth boundary $\partial\Omega$, $\Delta$ denotes the Laplace operator in $\mathbb{R}^2$, and $x$ is a complex valued function which defined in $[0,T] \times \mathbb{R}^2$.

Take $X = L^2(\Omega)$, $D(A) = H^2(\Omega) \cap H_0^1(\Omega)$, $x \in D(A)$, define $Ax = i\Delta x$. By virtue of the well-known Hille-Yosida theorem, it is obvious that $A$ is the infinitesimal generator of a strongly continuous group $\{S(t), -\infty < t < \infty\}$ in $X$. Moreover, $\{S(t), -\infty < t < \infty\}$ can be given by

$$(S(t)x)(y) = \frac{1}{4\pi it} \int_\Omega e^{\frac{i|y-z|^2}{4t}} x(z)dz.$$
(7.81)

In order to derive the expression (7.81), we define

$$\bar{x}(t,y) = \begin{cases} x(t,y), & y \in \Omega, \\ 0, & y \in \mathbb{R}^2 \setminus \Omega, \end{cases}$$
(7.82)

and

$$\bar{x}_0(y) = \begin{cases} x_0(y), & y \in \Omega, \\ 0, & y \in \mathbb{R}^2 \setminus \Omega. \end{cases}$$
(7.83)

Then equation (7.80) can be rewritten as

$$\begin{cases} \frac{\partial}{\partial t} \bar{x}(t,y) = i\Delta \bar{x}(t,y), & y \in \mathbb{R}^2, \ t \in (0,T], \\ \bar{x}(0,y) = \bar{x}_0(y), & y \in \mathbb{R}^2, \ \bar{x}_0 \in H^2(\mathbb{R}^2). \end{cases}$$
(7.84)

Applying Fourier transform to equation (7.84), we obtain

$$\begin{cases} \dfrac{d}{dt}\tilde{x}(t,z) = -iz^2\tilde{x}(t,z), & z \in \mathbb{R}^2, \ t \in (0,T], \\ \tilde{x}(0,z) = \tilde{x}_0(z), & z \in \mathbb{R}^2. \end{cases} \tag{7.85}$$

Thus, the classical solution of equation (7.85) can be given by

$$\tilde{x}(t,z) = e^{-i|z|^2 t}\tilde{x}_0(z).$$

Thus,

$$\begin{aligned} \bar{x}(t,y) &= \frac{1}{4\pi^2}\int_{\mathbb{R}^2}\int_{\mathbb{R}^2} e^{(i(y-\tau)z - i|z|^2 t)}\bar{x}_0(\tau)d\tau dz \\ &= \frac{1}{4\pi it}\int_{\mathbb{R}^2} e^{\frac{i|y-\xi|^2}{4t}}\bar{x}_0(\xi)d\xi. \end{aligned}$$

It comes from the inverse of Fourier transform that

$$x(t,y) = \frac{1}{4\pi it}\int_{\Omega} e^{\frac{i|y-z|^2}{4t}}\bar{x}_0(z)dz. \tag{7.86}$$

On the other hand, equation (7.84) has a unique classical solution

$$\bar{x}(t,y) = S(t)\bar{x}_0(y), \quad y \in \mathbb{R}^2.$$

Keeping in mind of (7.82) and (7.83), we have

$$\bar{x}(t,y) = S(t)\bar{x}_0(y), \quad y \in \Omega. \tag{7.87}$$

Combined (7.86) and (7.87), we obtain

$$(S(t)x_0)(y) = \frac{1}{4\pi it}\int_{\Omega} e^{\frac{i|y-z|^2}{4t}}x_0(z)dz.$$

By Lemma 1.1 in Pazy, 1983, we have the estimation immediately.

**Lemma 7.11.** *Let* $\{S(t), t \geq 0\}$ *be the* $C_0$*-semigroup given by* (7.81)*. Then* $S(\cdot)$ *can be extended in a unique way to a bounded operator from* $L^2(\Omega)$ *into* $L^2(\Omega)$ *and*

$$\|S(t)x\|_{L^2\Omega} \leq \|x\|_{L^2\Omega}.$$

To achieve our aim, we adopt the idea of Zhou and Jiao, 2010a,b and introduce the following two characteristic solution operators:

$$\begin{aligned} (\mathscr{T}(t)x)(y) &:= \int_0^{\infty} M_\alpha(\theta)(S(t^\alpha\theta)x)(y)d\theta \\ &= \int_0^{\infty} M_\alpha(\theta)\left(\frac{1}{4\pi it^\alpha\theta}\int_{\Omega} e^{\frac{i|y-z|^2}{4t^\alpha\theta}}x(z)dz\right)d\theta, \end{aligned} \tag{7.88}$$

and

$$\begin{aligned} (\mathscr{S}(t)x)(y) &:= \alpha\int_0^{\infty} \theta M_\alpha(\theta)(S(t^\alpha\theta)x)(y)d\theta \\ &= \alpha\int_0^{\infty} \theta M_\alpha(\theta)\left(\frac{1}{4\pi it^\alpha\theta}\int_{\Omega} e^{\frac{i|y-z|^2}{4t^\alpha\theta}}x(z)dz\right)d\theta, \end{aligned} \tag{7.89}$$

where $M_\alpha$ is the Wright function (see Definition 1.9).

Then, we can introduce the following definition for our problem.

**Definition 7.6.** By a mild solution of the system (7.79), we mean that a continuous function $x : J \to L^2(\Omega)$ which satisfies

$$
x(t,y) = \int_0^\infty M_\alpha(\theta)(S(t^\alpha\theta)x_0)(y)d\theta
$$

$$
+ \int_0^t (t-s)^{\alpha-1} \left( \alpha \int_0^\infty \theta M_\alpha(\theta) S((t-s)^\alpha\theta)(kVx)(s)(y)d\theta \right) ds
$$

$$
= \int_0^\infty M_\alpha(\theta) \left( \frac{1}{4\pi i t^\alpha\theta} \int_\Omega e^{\frac{i|y-z|^2}{4t^\alpha\theta}} x_0(z)dz \right) d\theta
$$

$$
+ \int_0^t (t-s)^{\alpha-1} \left[ \alpha \int_0^\infty \theta M_\alpha(\theta) \left( \frac{1}{4\pi i(t-s)^\alpha\theta} \right. \right.
$$

$$
\left. \left. \times \int_\Omega e^{\frac{i|y-z|^2}{4(t-s)^\alpha\theta}} kV(z)x(s,z)dz \right) d\theta \right] ds, \tag{7.90}
$$

for $t \in J$.

**Remark 7.4.** Keeping in mind of that $\int_0^\infty M_\alpha(\theta)d\theta$ and $\int_0^\infty \theta M_\alpha(\theta)d\theta$ are absolutely convergence respectively, and $\lim_{\theta\to\infty} M_\alpha(\theta) = 0$ and $\lim_{\theta\to\infty} \theta M_\alpha(\theta) = 0$, there exist two constants $M_1 > 0$ and $M_2 > 0$ respectively such that

$$
\int_0^\infty M_\alpha^2(\theta)d\theta \le M_1, \quad \int_0^\infty \theta^2 M_\alpha^2(\theta)d\theta \le M_2.
$$

The following properties of $\{\mathscr{T}(t), t \ge 0\}$ and $\{\mathscr{S}(t), t \ge 0\}$ are widely used in the sequel.

**Lemma 7.12.** *Let $\{\mathscr{T}(t), t \ge 0\}$ and $\{\mathscr{S}(t), t \ge 0\}$ be two solution operators defined by (7.88) and (7.89) respectively. Then $\mathscr{T}(t)$ and $\mathscr{S}(t)$ can be extended in a unique way to the bounded operators from $L^2(\Omega)$ into $L^2(\Omega)$ and*

$$
\|\mathscr{T}(t)x\|_{L^2\Omega} \le \sqrt{M_1}\|x\|_{L^2\Omega}, \quad \|\mathscr{S}(t)x\|_{L^2\Omega} \le \alpha\sqrt{M_2}\|x\|_{L^2\Omega}.
$$

**Proof.** For any $x \in L^2(\Omega)$, keeping in mind of Lemma 7.11, we obtain

$$
\|\mathscr{T}(t)x\|_{L^2\Omega}^2 = \int_\Omega \left| \int_0^\infty M_\alpha(\theta) \left( \frac{1}{4\pi i t^\alpha\theta} \int_\Omega e^{\frac{i|y-z|^2}{4t^\alpha\theta}} x(z)dz \right) d\theta \right|^2 dy
$$

$$
\le \int_0^\infty M_\alpha^2(\theta)d\theta \cdot \int_\Omega \int_0^\infty \left| \left( \frac{1}{4\pi i t^\alpha\theta} \int_\Omega e^{\frac{i|y-z|^2}{4t^\alpha\theta}} x(z)dz \right) \right|^2 d\theta dy
$$

$$
\le M_1 \|S(t^\alpha)x\|_{L^2\Omega}^2
$$

$$
\le M_1 \|x\|_{L^2\Omega}^2.
$$

Thus, one can obtain the first estimation immediately. Using the similar method, one can derive the second estimation. $\qquad\square$

**Lemma 7.13.** *Operators $\{\mathscr{T}(t), t \ge 0\}$ and $\{\mathscr{S}(t), t \ge 0\}$ are strongly continuous, which means that for all $x \in L^2(\Omega)$ and $0 \le t' < t'' \le T$, we have*

$$
\|\mathscr{T}(t'')x - \mathscr{T}(t')x\|_{L^2\Omega} \to 0 \quad \text{and} \quad \|\mathscr{S}(t'')x - \mathscr{S}(t')x\|_{L^2\Omega} \to 0, \quad \text{as } t' \to t''.
$$

**Proof.** For any $x \in L^2(\Omega)$ and $0 \le t' < t'' \le T$, we get that

$$\|\mathscr{T}(t'')x - \mathscr{T}(t')x\|_{L^2\Omega}^2 = \int_\Omega \left| \int_0^\infty M_\alpha(\theta)[S((t'')^\alpha\theta) - S((t')^\alpha\theta)]x(y)d\theta \right|^2 dy$$

$$\le M_1 \|[S((t'')^q\theta - (t')^q\theta) - I]x\|_{L^2\Omega}^2.$$

According to the strongly continuity of $\{S(t), t \ge 0\}$, we know that $\|\mathscr{T}(t'')x - \mathscr{T}(t')x\|_{L^2\Omega}$ tends to zero as $t'' - t' \to 0$, which means that $\{\mathscr{T}(t), t \ge 0\}$ is strongly continuous. Using the similar method, we can also obtain that $\{\mathscr{S}(t), t \ge 0\}$ is also strongly continuous. $\qquad\square$

### 7.5.3 Existence and Uniqueness

In this subsection, we study the existence and uniqueness of mild solutions for system (7.79).

**Lemma 7.14.** *Let* $\Omega$ *be a measurable subset of* $\mathbb{R}^2$, $k = \max_{t \in [0,T]} |\chi(t)|$ *and* $V \in H^2(\Omega)$. *Then, we have*

$$\|kVx\|_{L^2\Omega} \le k\|V\|_{L^\infty\Omega}\|x\|_{L^2\Omega}. \tag{7.91}$$

**Proof.** Since $H^2(\Omega)$ can be continuous embedded in the space $C^0(\Omega) = \{V \in C(\Omega) : V \in L^\infty(\Omega)\}$ and $V \in L^\infty(\Omega)$, for arbitrary $\varepsilon > 0$ there exists $\Omega_\varepsilon \subset \Omega$, such that $\alpha(\Omega_\varepsilon) = 0$ and

$$\sup_{\Omega \backslash \Omega_\varepsilon} |V(y)| < \|V\|_{L^\infty\Omega} + \varepsilon.$$

Thus,

$$\|kVx\|_{L^2\Omega}^2 = \int_{\Omega_\varepsilon} |kV(y)x(y)|^2 dy + \int_{\Omega \backslash \Omega_\varepsilon} |kV(y)x(y)|^2 dy$$

$$= k \int_{\Omega \backslash \Omega_\varepsilon} |V(y)x(y)|^2 dy$$

$$\le k\left(\|V\|_{L^\infty\Omega} + \varepsilon\right)^2 \int_\Omega |x(y)|^2 dy$$

$$\le k\left(\|V\|_{L^\infty\Omega} + \varepsilon\right)^2 \|x\|_{L^2\Omega}^2.$$

Let $\varepsilon \to 0$ and taking the limit in the above inequality, one can obtain the inequality (7.91) immediately. $\qquad\square$

In order to discuss the existence of mild solutions for system (7.79), we need the following important priori estimate.

**Lemma 7.15.** *Let* $V \in H^2(\Omega)$. *Suppose system (7.79) has a mild solution on* $[0, T]$, *then there exists a constant* $\rho > 0$ *such that*

$$\|x(t)\|_{L^2\Omega} \le \rho \text{ for all } t \in [0, T].$$

**Proof.** If $x$ is a mild solution $x$ of system (7.79) on $[0, T]$, then $x$ satisfies (7.90). Keeping in mind of Lemma 7.12 and Lemma 7.14, we have

$$\|x(t)\|^2_{L^2\Omega} \leq \int_\Omega \left| \int_0^\infty M_\alpha(\theta) \left( \frac{1}{4\pi i t^\alpha \theta} \int_\Omega e^{\frac{i|y-z|^2}{4t^\alpha \theta}} x_0(z)dz \right) d\theta \right|^2 dy$$

$$+ \int_\Omega \left| \int_0^t (t-s)^{\alpha-1} \left[ \alpha \int_0^\infty \theta M_\alpha(\theta) \left( \frac{1}{4\pi i (t-s)^\alpha \theta} \right. \right. \right.$$

$$\left. \left. \left. \times \int_\Omega e^{\frac{i|y-z|^2}{4(t-s)^\alpha \theta}} k V(z) x(s,z)dz \right) d\theta \right] ds \right|^2 dy$$

$$\leq M_1 \|x_0\|^2_{L^2\Omega} + \alpha^2 M_2 k^2 \|V\|^2_{L^\infty\Omega} \int_0^t (t-s)^{\alpha-1} \|x(s)\|^2_{L^2\Omega} ds.$$

Choose $\tilde{a}(t) = M_1 \|x_0\|^2_{L^2\Omega}$, and $\tilde{g}(t) = \alpha^2 M_2 k^2 \|V\|^2_{L^\infty\Omega}$ in Theorem 5.2 of Chapter 5, by Theorem 5.2 and Remark 5.1, there exists a constant $M_{\alpha^2 M_2 k^2 \|V\|^2_{L^\infty\Omega}} > 0$ such that

$$\|x(t)\|^2_{L^2\Omega} \leq M_1 \|x_0\|^2_{L^2\Omega} M_{\alpha^2 M_2 k^2 \|V\|^2_{L^\infty\Omega}} =: \rho^2, \quad \text{for all } t \in [0, T].$$

The proof is completed. $\qquad\square$

**Theorem 7.22.** *Let* $V \in H^2(\Omega)$. *System* (7.79) *has a unique mild solution* $x \in C([0, T], L^2(\Omega))$.

**Proof.** Fixed $x_0 \in L^2(\Omega)$, define

$$\mathcal{B}(x_0, 1) = \{x \in C([0, T_1], L^2(\Omega)) : \|x(t) - x_0\|_{L^2_\Omega} \leq 1, \ t \in [0, T_1]\},$$

where $T_1$ will be chosen latter. It is obvious that $\mathcal{B}(x_0, 1)$ is a closed and convex subset of $C([0, T_1], L^2(\Omega))$.

Define an operator $P : \mathcal{B}(x_0, 1) \to \mathcal{B}(x_0, 1)$ as follows:

$$(Px)(t, y) = \int_0^\infty M_\alpha(\theta) \left( \frac{1}{4\pi i t^\alpha \theta} \int_\Omega e^{\frac{i|y-z|^2}{4t^\alpha \theta}} x_0(z)dz \right) d\theta$$

$$+ \int_0^t (t-s)^{\alpha-1} \left[ \alpha \int_0^\infty \theta M_\alpha(\theta) \left( \frac{1}{4\pi i (t-s)^\alpha \theta} \right. \right.$$

$$\left. \left. \times \int_\Omega e^{\frac{i|y-z|^2}{4(t-s)^\alpha \theta}} k V(z) x(s,z)dz \right) d\theta \right] ds.$$

It is easy to see that $P$ is well defined.

By Lemma 7.12, for all $x \in C([0, T_1], L^2(\Omega))$,

$$\|(Px)(t) - x_0\|_{L^2\Omega}$$

$$\leq \int_\Omega \left| \int_0^\infty \xi_q(\theta) \left( \frac{1}{4\pi i t^\alpha \theta} \int_\Omega e^{\frac{i|y-z|^2}{4t^\alpha \theta}} x_0(z)dz \right) d\theta - x_0(y) \right|^2 dy$$

$$+ \int_\Omega \left| \int_0^t (t-s)^{\alpha-1} \left[ \alpha \int_0^\infty \theta M_\alpha(\theta) \left( \frac{1}{4\pi i (t-s)^\alpha \theta} \right. \right. \right.$$

$$\left. \left. \left. \times \int_\Omega e^{\frac{i|y-z|^2}{4(t-s)^\alpha \theta}} k V(z) x(s,z)dz \right) d\theta \right] ds \right|^2 dy$$

$$\leq \|\mathcal{T}(t)x_0 - x_0\|_{L^2\Omega} + \alpha\sqrt{M_2}\int_0^t (t-s)^{\alpha-1}\|(kVx)(s)\|_{L^2\Omega}ds$$

$$\leq \|\mathcal{T}(t)x_0 - x_0\|_{L^2\Omega}$$

$$+\alpha\sqrt{M_2}k\|V\|_{L^\infty\Omega}\int_0^t (t-s)^{\alpha-1}(1 + \|x_0\|_{L^2\Omega})ds$$

$$= \|\mathcal{T}(t)x_0 - x_0\|_{L^2\Omega} + t^\alpha\sqrt{M_2}k\|V\|_{L^\infty\Omega}(1 + \|x_0\|_{L^2\Omega}). \tag{7.92}$$

By Lemma 7.13, $\{\mathcal{T}(t): t \geq 0\}$ is a strongly continuous operator in $L^2(\Omega)$. Thus, there exists a $\bar{t} > 0$ such that

$$\|\mathcal{T}(t)x_0 - x_0\|_{L^2\Omega} \leq \frac{1}{2}, \quad t \leq \bar{t}.$$

Let

$$T_{11} = \min\left\{\bar{t}, \left(\frac{1}{2\sqrt{M_2}k\|V\|_{L^\infty\Omega}(1 + \|x_0\|_{L^2\Omega})}\right)^{\frac{1}{\alpha}}\right\},$$

then for all $t \leq T_{11}$, it comes from (7.92) that

$$\|(Px)(t) - x_0\|_{L^2\Omega} \leq 1.$$

Let $x_1, x_2 \in \mathcal{B}(x_0, 1)$, we have

$$\|(Px_1)(t) - (Px_2)(t)\|_{L^2\Omega}$$

$$\leq \int_\Omega \left| \int_0^t (t-s)^{\alpha-1}\left(\alpha\int_0^\infty \theta M_\alpha(\theta)\left(\frac{1}{4\pi i(t-s)^\alpha\theta}\right.\right.$$

$$\times \left.\left.\int_\Omega e^{\frac{i|y-z|^2}{4(t-s)^\alpha\theta}}kV(z)[x_1(s,z) - x_2(s,z)]dz\right)d\theta\right)ds\right|^2 dy$$

$$\leq \alpha\sqrt{M_2}k\|V\|_{L^\infty\Omega}\int_0^t (t-s)^{\alpha-1}\|x_1(s) - x_2(s)\|_{L^2\Omega}ds$$

$$\leq t^\alpha\sqrt{M_2}k\|V\|_{L^\infty\Omega}\|x_1 - x_2\|_{C([0,T_1],L^2(\Omega))}.$$

Let

$$T_{12} = \frac{1}{2}\left(\frac{1}{2\sqrt{M_2}k\|V\|_{L^\infty\Omega}}\right)^{\frac{1}{\alpha}}, \quad T_1 = \min\{T_{11}, T_{12}\},$$

then $P$ is a contraction map on $\mathcal{B}(x_0, 1)$. It follows from Banach contraction mapping principle that $P$ has a unique fixed point $x \in \mathcal{B}(x_0, 1)$, and $x$ is the unique mild solution of system (7.79) on $[0, T_1]$. $\square$

## 7.6 Notes and Remarks

The results in Section 7.2 are adopted from Zhou and Peng, 2016. The results in Section 7.3 are taken from Bourdin, 2013. The material in Section 7.4 are adopted Mu, Ahmad and Huang, 2016. The results of Section 7.5 are from Wang, Zhou and Wei, 2012d.

# Bibliography

N. Abada, M. Benchohra and H. Hammouche, Existence and controllability results for non-densely defined impulsive semilinear functional differential inclusions, J. Differential Equations, 246(2009), 3834-3863.

S. Abbas, M. Benchohra and Y. Zhou, Darboux problem for fractional order neutral functional partial hyperbolic differential equations, Int. J. Dyn. Syst. Diff. Equs., 2(2009), 301-312.

S. Abbas, M. Benchohra and Y. Zhou, Fractional order partial functional differential inclusions with infinite delay, Proc. A. Razmadze Math. Inst., 154(2010), 1-19.

S. Abbas, M. Benchohra and Y. Zhou, Fractional order partial hyperbolic functional differential equations with state-dependent delay, Int. J. Dyn. Syst. Diff. Equs., 3(2011), 459-490.

R. A. Adams, Sobolev Spaces, Academic Press, New York, (1999).

E. E. Adams and L. W. Gelhar, Field study of dispersion in a heterogeneous aquifer 2: Spatial moments analysis, Water Resources Res., 28(1992), 3293-3307.

R. P. Agarwal and B. Ahmad, Existence of solutions for impulsive anti-periodic boundary value problems of fractional semilinear evolution equations, Dyn. Contin. Discrete Impuls. Syst. Ser. A: Math. Anal., 18(2011), 457-470.

R. P. Agarwal, M. Benchohra and S. Hamani, A survey on existence results for boundary value problems of nonlinear fractional differential equations, Acta Appl. Math., 109(2010), 973-1033.

R. P. Agarwal, Y. Zhou and Y. He, Existence of fractional neutral functional differential equations with bounded delay, Comput. Math. Appl., 59(2010), 1095-1100.

R. P. Agarwal, Y. Zhou, J. R. Wang et al., Fractional functional differential equations with causal operators in Banach spaces, Math. Comput. Modelling, 54(2011), 1440-1452.

O. P. Agrawal, Formulation of Euler-Lagrange equations for fractional variational problems, J. Math. Anal. Appl., 272(2002), 368-379.

O. P. Agrawal, Generalized variational problems and Euler-Lagrange equations, Comput. Math. Appl., 59(2010), 1852-1864.

R. P. Agarwal, M. Belmekki and M. Benchohra, A survey on semilinear differential equations and inclusions involving Riemann-Liouville fractional derivative, Adv. Difference Equ., 2009(2009), Article ID 981728, 47 pages.

O. P. Agrawal, J. T. Machado and J. Sabatier, Fractional Derivatives and their Application: Nonlinear Dynamics, Springer-Verlag, Berlin, (2004).

B. Ahmad and P. Eloe, A nonlocal boundary value problem for a nonlinear fractional differential equation with two indices, Commun. Appl. Nonlinear Anal., 17(2010), 69-80.

B. Ahmad and J. J. Nieto, Existence results for a coupled system of nonlinear fractional differential equations with three-point boundary conditions, Comput. Math. Appl., 58(2009), 1838-1843.

B. Ahmad and J. J. Nieto, Existence of solutions for anti-periodic boundary value problems involving fractional differential equations via Leray-Schauder degree theory, Topol. Methods Nonlinear Anal., 35(2010), 295-304.

B. Ahmad, J. J. Nieto, A. Alsaedi *et al.*, A study of nonlinear Langevin equation involving two fractional orders in different intervals, Nonlinear Anal.: RWA, 13(2012), 599-606.

B. Ahmad and S. Sivasundaram, Existence results for nonlinear impulsive hybrid boundary value problems involving fractional differential equations, Nonlinear Anal.: HS, 3(2009), 251-258.

B. Ahmad and S. Sivasundaram, Existence of solutions for impulsive integral boundary value problems of fractional order, Nonlinear Anal.: HS, 4(2010), 134-141.

B. Ahmad and G. Wang, Impulsive anti-periodic boundary value problem for nonlinear differential equations of fractional order, Comput. Math. Appl., 59(2010), 1341-1349.

N. U. Ahmed, Existence of optimal controls for a general class of impulsive systems on Banach space, SIAM J. Control Optimal, 42(2003), 669-685.

N. U. Ahmed, Optimal feedback control for impulsive systems on the space of finitely additive measures, Publ. Math. Debrecen, 70(2007), 371-393.

M. U. Akhmet, On the smoothness of solutions of impulsive autonomous systems, Nonlinear Anal.: TMA, 60(2005), 311-324.

R. Almeida, A. B. Malinowska and D. F. M. Torres, A fractional calculus of variations for multiple integrals with application to vibrating string, J. Math. Phys., 51(2010), 033503.

A. Alsaedi, B. Ahmad and M. Kirane, Maximum principle for certain generalized time and space fractional diffusion equations, Quart. Appl. Math., 73(2015), 163-175.

A. Ambrosetti and P. Rabinowitz, Dual variational methods in critical points theory and applications, J. Funct. Anal., 14(1973), 349-381.

A. Ambrosetti and V. C. Zelati, Multiple homoclinic orbits for a class of conservative systems, Rend. Sem. Mat. Univ. Padova, 89(1993), 177-194.

M. R. S. Ammi and D. F. M. Torres, Regularity of solutions to higher-order integrals of the calculus of variations, Internat. J. Systems Sci., 39(2008), 889-895.

C. Amrouche, A. Rejaiba, $L^p$-theory for Stokes and Navier-Stokes equations with Navier boundary condition, J. Differential Equations, 256(2014), 1515-1547.

V. V. Anh and N. N. Leonenko, Spectral analysis of fractional kinetic equations with random data, J. Statist. Phys., 104(2001), 1349-1387.

J. M. Arrieta, Rates of eigenvalues on a Dumbbell domain. Simple eigenvalue case, Tran. Amer. Math. Soc., 347(1995), 3503-3531.

J. M. Arrieta, A. Carvalho and G. Lozada-Cruz, Dynamics in dumbbell domains I. Continuity of the set of equilibria, J. Differential Equations, 231(2006), 551-597.

J. M. Arrieta, A. Carvalho and G. Lozada-Cruz, Dynamics in dumbbell domains II. The limiting problem, J. Differential Equations, 247(2009a), 174-202.

J. M. Arrieta, A. Carvalho and G. Lozada-Cruz, Dynamics in dumbbell domains III. Continuity of attractors, J. Differential Equations, 247(2009b), 225-259.

A. Ashyralyev and F. Tetikoglu, A note on fractional spaces generated by the positive operator with periodic conditions and applications, Bound. Value Probl., 1(2015), 1-17.

T. Atanackovic and B. Stankovic, On a class of differential equations with left and right fractional derivatives, ZAMM., 87(2007), 537-539.

C. Atkinson and A. Osseiran, Rational solutions for the time-fractional diffusion equation, SIAM J . Appl. Math., 71(2011), 92-106.

D. Averna and G. Bonanno, A mountain pass theorem for a suitable class of functions, Rocky Mountain J. Math., 39(2009), 707-727.

O. Baghani, On fractional Langevin equation involving two fractional orders, Commun. Nonlinear Sci. Numer. Simul., (2016), http://dx.doi:10.1016/j.cnsns.2016.05.023

E. G. Bajlekova, Fractional evolution equations in Banach spaces, Thesis (Dr.) Technische Universiteit Eindhoven (The Netherlands), ProQuest LLC, Ann Arbor, MI, 2001, 113 pp.

Z. Bai, X. Dong and C. Yin, Existence results for impulsive nonlinear fractional differential equation with mixed boundary conditions, Bound. Value Probl., 2016(2016), 1-11.

Z. Bai and H. Lu, Positive solutions for boundary value problem of nonlinear fractional differential equation, J. Math. Anal. Appl., 311(2005), 495-505.

D. D. Bainov and P. S. Simeonov, Impulsive differential equations: periodic solutions and applications, New York, Longman Scientific and Technical Group. Limited, (1993).

K. Balachandran and S. Kiruthika, Existence of solutions of abstract fractional impulsive semilinear evolution equations, E. J. Qualitative Theory Diff. Equ., 4(2010), 1-12.

K. Balachandran, S. Kiruthika and J. J. Trujillo, Existence results for fractional impulsive integrodifferential equations in Banach spaces, Commun. Nonlinear Sci. Numer. Simul., 16(2011), 1970-1977.

K. Balachandran, Y. Zhou and J. Kokila, Relative controllability of fractional dynamical systems with delays in control, Commun. Nonlinear Sci. Numer. Simul., 17(2012a), 3508-3520.

K. Balachandran, Y. Zhou and J. Kokila, Relative controllability of fractional dynamical systems with distributed delays in control, Comput. Math. Appl., 64(2012b), 3201-3209.

D. Baleanu, J. A. T. Machado and A. C.-J. Luo, Fractional Dynamics and Control, Springer Science and Business Media, (2011).

D. Baleanu and S. I. Muslih, Lagrangian formulation of classical fields within Riemann-Liouville fractional derivatives, Phys. Scripta, 72(2-3)(2005), 119-121.

D. Baleanu and J. Trujillo, On exact solutions of a class of fractional Euler-Lagrange equations, Nonlinear Dynam., 52(2008), 331-335.

D. Baleanu, J. A. T. Machado and Z. B. Guvenc (Eds.), New Trends in Nanotechnology and Fractional Calculus Applications, Springer, (2010).

Z. Bai, On positive solutions of a nonlocal fractional boundary value problem, Nonlinear Anal.: TMA, 72(2010), 916-924.

J. Banaś and K. Goebel, Measure of Noncompactness in Banach Spaces, Marcel Dekker, Inc., New York, (1980).

T. Bartsch, Infinitely many solutions of a symmetric Dirchlet problem, Nonlinear Anal.: TMA, 20(1993), 1205-1216.

T. Bartsch and M. Willem, On an elliptic equation with concave and convex nonlinearities, Proc. Amer. Math. Soc., 123(1995), 3555-3561.

F. Battelli and M. Fečkan, Chaos in singular impulsive O.D.E., Nonlinear Anal.: TMA, 28(1997), 655-671.

A. Bazzani, G. Bassi and G. Turchetti, Diffusion and memory effects for stochastic processes and fractional Langevin equations, Physica A: Stat. Mech. Appl., 324(2003), 530-550.

S. Beckers and M. Yamamoto, Regularity and unique existence of solution to linear diffusion equation with multiple time-fractional derivatives, Control and Optimization with PDE Constraints, K. Bredies, C. Clason, K. Kunisch and G. von Winckel (Eds.), Basel: Birkhäuser, (2013).

M. Belmekki and M. Benchohra, Existence results for fractional order semilinear functional differential equations with nondense domain, Nonlinear Anal.: TMA, 72(2010), 925-932.

M. Benchohra, S. Hamani and S. K. Ntouyas, Boundary value problems for differential equations with fractional order and nonlocal conditions, Nonlinear Anal.: TMA, 71(2009), 2391-2396.

M. Benchohra, J. Henderson and S. K. Ntouyas, Impulsive differential equations and inclusions, Hindawi Publishing Corporation, vol. 2, New York, (2006).

M. Benchohra, J. Henderson, S. K. Ntouyas *et al.*, Existence results for fractional order functional differential equations with infinite delay, J. Math. Anal. Appl., 338(2008), 1340-1350.

M. Benchohra and D. Seba, Impulsive fractional differential equations in Banach spaces, E. J. Qualitative Theory Diff. Equ., Spec. Ed. I, 8(2009), 1-14.

D. A. Benson, R. Schumer, M. M. Meerschaert *et al.*, Fractional dispersion, Lévy motion, and the MADE tracer test, Transp. Porous Media, 42(2001), 211-240.

D. A. Benson, S. W. Wheatcraft and M. M. Meerschaert, Application of a fractional advection-dispersion equation, Water Resour. Res., 36(2000a), 1403-1412.

D. A. Benson, S. W Wheatcraft and M. M. Meerschaert, The fractional-order governing equation of Lévy motion, Water Resour. Res., 36(2000b) 1413-1423.

D. A. Benson, S. W. Wheatcraft and M. M. Meerschaert, Application of a fractional advection-dispersion equation, Water Resources Res., 36(2000a), 1403-1412.

D. A. Benson, S. W. Wheatcraft and M. M. Meerschaert, The fractional-order governing equation of Lévy motion, Water Resource Res., 36(2000b), 1413-1423.

B. Berkowitz, H. Scher and S. E. Silliman, Anomalous transport in laboratory-scale heterogeneous porous media, Water Resource Res., 36(2000), 149-158.

G. Bonanno and P. Candito, Non-differentiable functionals and applications to elliptic problems with discontinuous nonlinearities, J. Differential Equations, 244(2008), 3031-3059.

B. Bonilla, M. Rivero and J. J. Trujillo, On systems of linear fractional differential equations with constant coefficients, Appl. Math. Comput., 187(2007), 68-78.

D. Bothe, Multivalued perturbation of m-accretive differential inclusions, Israel J. Math., 108(1998), 109-138.

A. Boucherif and R. Precup, On the nonlocal initial value problem for first order differential equations, Fixed Point Theory, 4(2003), 205-212.

A. Boucherif and R. Precup, Semilinear evolution equations with nonlocal initial conditions, Dyn. Sys. Appl., 16(2007), 507-516.

L. Bourdin, Existence of a weak solution for fractional Euler-Lagrange equations, J. Math. Anal. Appl., 399(2013), 239-251.

L. Bourdin, J. Cresson and I. Greff, A continuous/discrete fractional Noether's theorem, Commun. Nonlinear Sci. Numer. Simul., 18(2013), 878-887.

H. Brezis, Functional Analysis, Sobolev Spaces and Partial Differential Equations, Springer, New York, (2011).

D. Bugajewska and P. Kasprzak, On the existence, uniqueness and topological structure of solution sets to a certain fractional differential equation, Comput. Math. Appl., 59(2009), 1108-1116.

S. Bushnaq, S. Momani and Y. Zhou, A reproducing kernel Hilbert space method for solving integro-differential equations of fractional order, J. Optim. Theory Appl., 156(2013), 96-105.

V. S. Buslaev and C. Sulem, On asymptotic stability of solitary waves for nonlinear Schrödinger equations, Ann. Inst. H. Poincaré Anal. Non Linéaire, 20(2003), 419-475.

L. Byszewski, Theorems about existence and uniqueness of solutions of a semi-linear evolution nonlocal Cauchy problem, J. Math. Anal. Appl., 162(1991), 494-505.

L. Byszewski and V. Lakshmikantham, Theorem about the existence and uniqueness of a solution of a nonlocal abstract Cauchy problem in a Banach space, Appl. Anal., 40(1991), 11-19.

L. Caffarelli and A. Vasseur, Drift diffusion equations with fractional diffusion and the quasi-geostrophic equation, Ann. of Math., 171(2010), 1903-1930.

M. Cannone, Ondelettes, Paraproduits et Navier-Stokes, Diderot Editeur, (1995).

J. Cao and H. Chen, Some results on impulsive boundary value problem for fractional differential inclusions, Electron. J. Qual. Theory Differ. Equ., 11(2010), 1-24.

J. Cao and H. Chen, Impulsive fractional differential equations with nonlinear boundary conditions, Math. Comput. Model., 55(2012), 303-311.

B. A. Carreras, V. E. Lynch and G. M. Zaslavsky, Anomalous diffusion and exit time distribution of particle tracers in plasma turbulence models, Phys. Plasmas, 8(2001), 5096-5103.

A. N. Carvalho, T. Dlotko and M. J. D. Nascimento, Nonautonomous semilinear evolution equations with almost sectorial operators, J. Evol. Equ., 8(2008), 631-659.

T. Cazenave, Stable solutions of the logarithmic Schrödinger equation, Nonlinear Anal.: TMA, 7(1983), 1127-1140.

T. Cazenave, Semilinear Schrödinger equations, Courant Lect. Notes Math., 10. New York University, Courant Institute of Mathematical Sciences, New York; American Mathematical Society, Providence, RI, (2003).

T. Cazenave and A. Haraux, An Introduction to Semilinear Evolution Equations, Clarendon Press, Oxford, (1998).

T. Cazenave and P.-L. Lions, Orbital stability of standing waves for some nonlinear Schrödinger equations, Comm. Math. Phys., 85(1982), 549-561.

G. Cerami, An existence criterion for the critical points on unbounded manifolds, Istit. Lombardo Accad. Sci. Lett. Rend. A, 112(1978), 332-336 (in Italian).

L. Cesari, Optimization-Theory and Applications, in: Applications of Mathematics (New York), vol. 17, Problems with ordinary differential equations, Springer-Verlag, New York, (1983).

Y. K. Chang and J. J. Nieto, Some new existence results for fractional differential inclusions with boundary conditions, Math. Comput. Modelling, 49(2009), 605-609.

A. S. Chaves, A fractional diffusion equation to describe Lévy flights, Phys. Lett. A, 239(1998), 13-16.

A. V. Chechkin, R. Gorenflo and I. M. Sokolov, Retarding subdiffusion and accelerating superdiffusion governed by distributed order fractional diffusion equations, Phys. Rev. E (3), 66(2002), 1-7.

A. V. Chechkin, R. Gorenflo, I. M. Sokolov *et al.*, Distributed order time fractional diffusion equation, Fract. Calc. Appl. Anal., 6(2003), 259-279.

J. Y. Chemin and I. Gallagher, Large, global solutions to the Navier-Stokes equations, slowly varying in one direction, Trans. Amer. Math. Soc., 362(2010), 2859-2873.

J. Y. Chemin, I. Gallagher and M. Paicu, Global regularity for some classes of large solutions to the Navier-Stokes equations, Ann. of Math., 173(2011), 983-1012.

D. H. Chen, R. N. Wang and Y. Zhou, Nonlinear evolution inclusions: Topological characterizations of solution sets and applications, J. Funct. Anal., 265(2013), 2039-2073.

F. Chen, X. Luo and Y. Zhou, Existence results for nonlinear fractional difference equation, Adv. Difference Equ., (2011), Article ID 713201, 12 pages.

F. Chen, J. J. Neito and Y. Zhou, Global attractivity for nonlinear fractional differential equations, Nonlinear Anal.: RWA, 13(2012), 287-298.

J. Chen and X. H. Tang, Existence and multiplicity of solutions for some fractional boundary value problem via critical point theory, Abstr. Appl. Anal., (2012), Article ID 648635, 21 pages.

J. Chen and X. H. Tang, Infinitely many solutions for a class of fractional boundary value problem, Bull. Malays. Math. Sci. Soc., 36(2013), 1083-1097

F. Chen and Y. Zhou, Attractivity of fractional functional differential equation, Comput. Math. Appl., 62(2011), 1359-1369.

H. J. Choe, Boundary regularity of suitable weak solution for the Navier-Stokes equations, J. Funct. Anal., 268(2015), 2171-2187.

M. Choulli, Une Introduction aux Problems Inverses Elliptiques et Paraboliques. Springer-Verlag, (2009).

P. Clement and R. Zacher, Global smooth solutions to a fourth-order quasilinear fractional evolution equation. Functional analysis and evolution equations, 131-146, Birkhäuser, Basel, (2008).

F. Comte, Opérateurs fractionnaires en économétrie et en finance, Prépublication MAP5, (2001).

C. Corduneanu, Functional Equations with Causal Operators, Taylor Francis, London, (2002).

J.-N. Corvellec, V. V. Motreanu and C. Saccon, Doubly resonant semilinear elliptic problems via nonsmooth critical point theory, J. Differential Equations, 248(2010), 2064-2091.

M. Cowling, I. Doust, A. McIntosh *et al.*, Banach space operators with a bounded $H^\infty$ calculus, J. Austral. Math. Soc. Set. A 60 (1996), 51-89.

J. Cresson, Inverse problem of fractional calculus of variations for partial differential equations, Commun. Nonlinear Sci. Numer. Simul., 15(2010), 987-996.

J. Cresson, I. Greff and P. Inizan, Lagrangian for the convection-diffusion equation, Math. Methods Appl. Sci., (2011).

J. Cresson and P. Inizan, Variational formulations of differential equations and asymmetric fractional embedding, J. Math. Anal. Appl., 385(2012), 975-997.

S. Cuccagna, Stabilization of solutions to nonlinear Schrödinger equations, Comm. Pure Appl. Math., 54(2001), 1110-1145.

B. Dacorogna, Direct Methods in the Calculus of Variations (2nd ed.), in: Applied Mathematical Sciences, vol. 78, Springer, New York, (2008).

V. Daftardar-Gejji and S. Bhalekar, Boundary value problems for multi-term fractional differential equations, J. Math. Anal. Appl., 345(2008), 754-765.

V. Daftardar-Gejji and H. Jafari, Analysis of a system of nonautonomous fractional differential equations involving Caputo derivatives, J. Math. Anal. Appl., 328(2007), 1026-1033.

E. N. Dancer and D. Daners, Domain perturbation of elliptic equations subject to Robin boundary conditions, J. Differential Equations, 74(1997), 86-132.

R. Danchin, Global existence in critical spaces for compressible Navier-Stokes equations, Invent. Math., 141(2000), 579-614.

M. A. Darwish, On monotonic solutions of an integral equation of Abel type, Mathematica Bohemica, 133(2008), 407-420.

M. A. Darwish, J. Henderson and S. K. Ntouyas, Fractional order semilinear mixed type functional differential equations and inclusions, Nonlinear Stud., 16(2009), 197-219.

M. F. de Almeida and L. C. F. Ferreira, On the Navier-Stokes equations in the half-space with initial and boundary rough data in Morrey spaces, J. Differential Equations, 254(2013), 1548-1570.

P. M. De Carvalho-Neto and P. Gabriela, Mild solutions to the time fractional Navier-Stokes equations in $R^N$, J. Differential Equations, 259(2015), 2948-2980.

K. Deimling, Nonlinear Functional Analysis, Springer-Verlag, (1985).

R. deLaubenfels, Existence Families, Functional Calculi and Evolution Equations, Springer-Verlag, Berlin, (1994).

D. Delbosco and L. Rodino, Existence and uniqueness for a nonlinear fractional differential equation, J. Math. Anal. Appl., 204(1996), 609-625.

J. Deng and S. Wang, Existence of solutions of nonlocal Cauchy problem for some fractional abstract differential equation, Appl. Math. Lett., 55(2016), 42-48.

G. Desch and S. O. Londen, Evolutionary equations driven by fractional Brownian motion, Stoch. Partial Differ. Equ. Anal. Comput., 1(2013), 424-454.

J. V. Devi and V. Lakshmikantham, Nonsmooth analysis and fractional differential equations, Nonlinear Anal.: TMA, 70(2009), 4151-4157.

B. C. Dhage, Existence of extremal solutions for discontinuous functional integral equations, Appl. Math. Lett., 19(2006), 881-886.

T. Diagana, G. Mophou and G. N'guérékata, On the existence of mild solutions to some semilinear fractional integro-differential equations, Electron. J. Qual. Theory Differ. Equ., 58(2010), 1-17.

J. Diblík and Z. Svoboda, Positive solutions of $p$-type retarded functional differential equations, Nonlinear Anal.: TMA, 63(2005), 813-821.

K. Diethelm, Analysis of fractional differential equations, J. Math. Anal. Appl., 265(2002), 229-248.

K. Diethelm, The Analysis of Fractional Differential Equations, Lecture Notes in Mathematics, Springer, (2010).

K. Diethelm and Y. Luchko, Numerical solution of linear multi-term initial value problems of fractional order. J. Comput. Anal. Appl., 6(2004), 243-263.

J. Dieudonne, Deux examples dequations differentielles, Acta. Sci. Math., 12B(1950), 38-40.

X. W. Dong, J. R. Wang and Y. Zhou, On nonlocal problems for fractional differential equations in Banach spaces, Opuscula Mathematica, 31(2011), 341-357.

M. M. Dzhrbashyan and A. B. Nersessyan, Fractional derivatives and Cauchy problem for differential equations of fractional order, Izv. AN Arm. SSR. Mat., 3(1968), 3-29 (in Russian).

R. Eid, S. I. Muslih and D. Baleanu et al., On fractional Schrödinger equation in $\alpha$-dimensional fractional space, Nonlinear Anal.: RWA, 10(2009), 1299-1304.

S. D. Eidelman and A. N. Kochubei, Cauchy problem for fractional diffusion equations, J. Differential Equations, 199(2004), 211-255.

M. M. El-Borai, Some probability densities and fundamental solutions of fractional evolution equations, Chaos, Sol. Frac., 14(2002), 433-440.

M. M. El-Borai, The fundamental solutions for fractional evolution equations of parabolic type, J. Appl. Math. Stoch. Anal., 3(2004), 197-211.

A. M. A. El-Sayed, Nonlinear functional differential equations of arbitrary orders, Nonlinear Anal.: TMA, 33(1998), 181-186.

A. M. A. El-Sayed, W. G. El-Sayed and O. L. Moustafa, On some fractional functional equations, Pure Math. Appl., 6(1995), 321-332.

M. El-Shahed and A. Salem, On the generalized Navier-Stokes equations, Appl. Math. Comput., 156(1)(2004), 287-293.

V. J. Ervin and J. P. Roop, Variational formulation for the stationary fractional advection dispersion equation, Numer. Meth. Part. Diff. Equs., 22(2006), 558-576.

K. S. Fa, Generalized Langevin equation with fractional derivative and long-time correlation function, Phys. Rev., 73(2006), 061104.

K. S. Fa, Fractional Langevin equation and Riemann-Liouville fractional derivative, The Eur. Phys. J., 24(2007), 139-143.

M. M. Fall and T. Weth, Nonexistence results for a class of fractional elliptic boundary value problems, J. Funct. Anal., 263(2012), 2205-2227.

Z. Fan, Impulsive problems for semilinear differential equations with nonlocal conditions, Nonlinear Anal.: TMA, 72(2010), 1104-1109.

Z. Fan and G. Li, Existence results for semilinear differential equations with nonlocal and impulsive conditions, J. Funct. Anal., 258(2010), 1709-1727.

M. Fečkan, Topological Degree Approach to Bifurcation Problems, Springer, (2008).

M. Fečkan, J. R. Wang and Y. Zhou, Controllability of fractional functional evolution equations of Sobolev type via characteristic solution operators, J. Optim. Theory Appl., 156(1)(2013), 79-95.

M. Fečkan, Y. Zhou and J. R. Wang, On the concept and existence of solutions for impulsive fractional differential equations, Commun. Nonlinear Sci. Numer. Simul., 17(2012), 3050-3060.

G. Fibich, Singular solutions of the subcritical nonlinear Schrödinger equation, Physica D: Nonlinear Phenomena, 240(2011), 1119-1122.

G. J. Fix and J. P. Roop, Least squares finite-element solution of a fractional order two-point boundary value problem, Comput. Math. Appl., 48(2004), 1017-1033.

A. Floer and A. Weinstein, Nonspreading wave packets for the cubic Schrödinger equation with a bounded potential, J. Funct. Anal., 69(1986), 397-408.

G. S. F. Frederico and D. F. M. Torres, A formulation of Noether's theorem for fractional problems of the calculus of variations, J. Math. Anal. Appl., 334(2007), 834-846.

R. R. Gadyl'shin, On the eigenvalues of a dumb-bell with a thin handle, Izv. Math., 69(2005), 265-329.

G. P. Galdi, An Introduction to the Mathematical Theory of the Navier-Stokes Equations: Nonlinear Steady Problems, Springer Tracts in Natural Philosophy, New York, (1998).

Z. Z. Ganji, D. D. Ganji, A. Ganji *et al.*, Analytical solution of time-fractional Navier-Stokes equation in polar coordinate by homotopy perturbation method, Numer. Methods Partial Diff. Equ., 26(2010), 117-124.

Y. Giga and S. Yoshikazu, Abstract $L^p$ estimates for the Cauchy problem with applications to the Navier-Stokes equations in exterior domains, J. Funct. Anal., 102(1991), 72-94.

M. Giona, S. Gerbelli and H. E. Roman, Fractional diffusion equation and relaxation in complex viscoelastic materials, Physica A, 191(1992), 449-453.

C. Godbillon, Géométrie Différentielle et Mécanique Analytique, Hermann, Paris, (1969).

R. Gorenflo, Y. Luchko and P. P. Zabrejko, On solvability of linear fractional differential equations in Banach spaces, Frac. Calc. Appl. Anal., 2(1999), 163-176.

D. J. Guo, Existence of positive solutions for $n$th-order nonlinear impulsive singular integro-differential equations in Banach spaces, Nonlinear Anal.: TMA, 68(2008), 2727-2740.

D. J. Guo, V. Lakshmikantham and X. Z. Liu, Nonlinear Integral Equations in Abstract Spaces, Kluwer Academic, Dordrecht, (1996).

B. L. Guo and Y. P. Wu, Orbital stability of solitary waves for the nonlinear derivative Schrödinger equation, J. Differential Equations, 123(1995), 35-55.

J. K. Hale, Theory of Functional Differential Equations, Springer-Verlag, New York, (1977).

J. K. Hale and J. Kato, Phase space for retarded equations with infinite delay, Funkcial. Ekvac., 21(1978), 11-41.

B. Han, Refinable functions and cascade algorithms in weighted spaces with Hölder continuous masks, SIAM J. Math. Anal., 40(2008), 70-102.

A. Hanyga, Multidimensional solutions of time-fractional diffusion-wave equations, Proc. R. Soc. Lond. Ser. A: Math. Phys. Eng. Sci., 458(2002), 933-957.

D. Haroske and H. Triebel, Distributions, Sobolev Spaces, Elliptic Equations, European Mathematical Society, Switzerland, (2008).

Y. Hatno and N. Hatano, Dispersive transport of ions in column experiments: an explanation of long-tailed profiles, Water Resources Res., 34(1980), 1027-1033.

N. Hayashi, E. I. Kaikina and P. I. Naumkin, Asymptotics for fractional nonlinear heat equations, J. London Math. Soc., 72(2005), 663-688.

H. Heck, H. Kim and H. Kozono, Weak solutions of the stationary Navier-Stokes equations for a viscous incompressible fluid past an obstacle, Math. Ann., 356(2013), 653-681.

S. Heikkilä and V. Lakshmikantham, Monotone Iterative for Discontinuous Nonlinear Differential Equations, Monographs and Textbooks in Pure and Applied Mathematics, 181. Marcel Dekker, Inc., New York, (1994).

H.-P. Heinz, On the behaviour of measure of noncompactness with respect to differentiation and integration of vector-valued functions, Nonlinear Anal.: TMA, 7(1983), 1351-1371.

D. Henry, Geometric Theory of Semilinear Parabolic Equations, LNM 840, Springer-Verlag, Berlin, Heidelberg, New York, (1981).

E. Hernández, D. O'Regan and K. Balachandran, On recent developments in the theory of abstract differential equations with fractional derivatives, Nonlinear Anal.: TMA, 73(2010), 3462-3471.

R. Herrmann, Fractional Calculus: An Introduction for Physicists, World Scientific, Singapore, (2011).

R. Hilfer, Applications of Fractional Calculus in Physics, World Scientific, Singapore, (2000).

Y. Hino, S. Murakami and T. Naito, Functional Differential Equations with Infinite Delay, Lecture Notes in Math., vol. 1473, Springer-Verlag, Berlin, (1991).

L. Hu and G. Kallianpur, Schröodinger equations with fractional Laplacians, Appl. Math. Optim., 42(2000), 281-290.

L. Hu, Y. Ren and R. Sakthivel, Existence and uniqueness of mild solutions for semilinear integro-differential equations of fractional order with nonlocal initial conditions and delays, Semigroup Forum, 79(2009), 507-514.

S. Iqbal, J. Pečarić and Y. Zhou, Generalization of an inequality for integral transforms with kernel and related results, J. Inequalities Appl., (2010), Article ID 948430, 17 pages.

F. Isaia, On a nonlinear integral equation without compactness, Acta. Math. Univ. Comenianae., LXXV(2006), 233-240.

V. Isakov, Inverse Problems for Partial Differential Equations, Springer-Verlag, New York, (1998).

T. Iwabuchi and R. Takada, Global solutions for the Navier-Stokes equations in the rotational framework, Math. Ann., 357(2013), 727-741.

O. K. Jaradat, A. Al-Omari and S. Momani, Existence of the mild solution for fractional semilinear initial value problems, Nonlinear Anal.: TMA, 69(2008), 3153-3159.

W. Jiang, The controllability of fractional control systems with control delay, Comput. Math. Appl., 64(2012), 3153-3159.

H. Jiang, F. Liu, I. Turner *et al.*, Analytical solutions for the multi-term timespace Caputo-Riesz fractional advection-diffusion equations on a finite domain, J. Math. Anal. Appl., 389(2012), 1117-1127.

F. Jiao and Y. Zhou, Existence of solutions for a class of fractional boundary value problem via critical point theory, Comput. Math. Appl., 62(2011), 1181-1199.

F. Jiao and Y. Zhou, Existence results for fractional boundary value problem via critical point theory, Int. J. Bifurcation Chaos, 22(2012), 1-17.

S. Jimbo, The singularly perturbed domain and the characterization for the eigenfunctions with Neumann boundary conditions, J. Differential Equations, 77(1989), 322-350.

F. Kappel and W. Schappacher, Some considerations to the fundamental theory of infinite delay equations, J. Differential Equations, 37(1980), 141-183.

A. A. Kilbas, H. M. Srivastava and J. J. Trujillo, Theory and Applications of Fractional Differential Equations, in: North-Holland Mathematics Studies, vol. 204, Elsevier Science B.V., Amsterdam, (2006).

M. Kirane, Y. Laskri and N.-e. Tatar, Critical exponents of Fujita type for certain evolution equations and systems with spatiotemporal fractional derivatives, J. Math. Anal. Appl., 312(2005), 488-501.

V. Kiryakova, Generalized Fractional Calculus and Applications, in: Pitman Research Notes in Mathematics Series, vol. 301, Longman Scientific & Technical, Harlow, (1994).

J. Klafter, S. C. Lim and R. Metzler (Eds.), Fractional Dynamics in Physics, World Scientific, Singapore, (2011).

M. Klimek, Lagrangian and Hamiltonian fractional sequential mechanics, Czechoslovak J. Phys., 52(2002), 1247-1253.

M. Klimek, Existence-uniqueness result for a certain equation of motion in fractional mechanics, Bull. Polish Acad. Sci., 58(4)(2010), 573-581.

V. Kobolev and E. Romanov, Fractional Langevin equation to describe anomalous diffusion, Pro. Theor. Phys. Sup., 139(2000), 470-476.

G. Koch, N. Nadirashvili, G. A. Seregin et al., Liouville theorems for the Navier-Stokes equations and applications, Acta Math., 203(2009), 83-105.

A. N. Kochubei, Distributed order calculus and equations of ultraslow diffusion, J. Math. Anal. Appl., 340(2008), 252-281.

A. Kolmogorov, S. Fomine and V. M. Tihomirov, Eléments de la théorie des fonctions et de l'analyse fonctionnelle, Éditions Mir, Moscow, (1974), Avec uncomplément sur les algèbres de Banach, par V.M. Tikhomirov, Traduit du russe par Michel Dragnev.

L. Kong, Existence of solutions to boundary value problems arising from the fractional advection dispersion equation, Electronic J. Differential Equations, 2013(106), 1-15.

N. Kosmatov, Initial value problems of fractional order with fractional impulsive conditions, Results. Math., 63(2013), 1289-1310.

H. Kozono, $L^1$-solutions of the Navier-Stokes equations in exterior domains, Math. Ann., 312(1998), 319-340.

S. Kumar and N. Sukavanam, Approximate controllability of fractional order semilinear systems with bounded delay, J. Differential Equations, 252(2012), 6163-6174.

V. Lakshmikantham, Theory of fractional functional differential equations, Nonlinear Anal.: TMA, 69(2008), 3337-3343.

V. Lakshmikantham, D. D. Bainov and P. S. Simeonov, Theory of impulsive differential equations, World Scientific, Singapore-London, (1989).

V. Lakshmikantham and S. Leela, Nonlinear Differential Equations in Abstract Spaces, Pergamon Press, New York, (1969).

V. Lakshmikantham, S. Leela and J. V. Devi, Theory of Fractional Dynamic Systems, Cambridge Scientific Publishers, Cambridge, (2009).

V. Lakshmikantham, S. Leela, Z. Drici et al., Theory of Causal Differential Equations, Atlantis Studies in Mathematics for Engineering and Science, vol. 5, World Scientific, (2010).

V. Lakshmikantham and A. S. Vatsala, General uniqueness and monotone iterative technique for fractional differential equations, Appl. Math. Lett., 21(2008a), 828-834.

V. Lakshmikantham and A. S. Vatsala, Basic theory of fractional differential equations, Nonlinear Anal.: TMA, 69(2008b), 2677-2682.

V. Lakshmikantham, L. Wen and B. Zhang, Theory of Differential Equations with Unbounded Delay, Kluwer Academic Publishers, Dordrecht, (1994).

Y.-H. Lan, H. X. Huang and Y. Zhou, Observer-based robust control of $\alpha$ ($1 < \alpha < 2$) fractional-order uncertain systems: an LMI approach, IET Control Theory Appl., 6(2012), 229-234.

Y.-H. Lan and Y. Zhou, LMI-based robust control of fractional-order uncertain linear systems, Comput. Math. Appl., 62(2011), 1460-1471.

Y.-H. Lan and Y. Zhou, High-order $D^{\alpha}$-type iterative learning control for fractional-order nonlinear time-delay systems, J. Optim. Theory Appl., 156(2013a), 153-166.

Y.-H. Lan and Y. Zhou, $D^{\alpha}$-type iterative learning control for fractional-order linear time-delay systems, Asian J. Control, 15(2013b), 669-677.

P. G. Lemarié-Rieusset, Recent Developments in the Navier-Stokes Problem, Chapman CRC Press, (2002).

F. Li, Z. Liang and Q. Zhang, Existence of solutions to a class of nonlinear second order two-point boundary value problems, J. Math. Anal. Appl., 312(2005), 357-373.

C. F. Li, X. N. Luo and Y. Zhou, Existence of positive solutions of boundary value problem for fractional differential equations, Comput. Math. Appl., 59(2010), 1363-1375.

B. Li, S. Sun and Y. Sun, Existence of solutions for fractional Langevin equation with infinite-point boundary conditions, J. Comput. Appl. Math., (2016), 1-10.

Z. Li and M. Yamamoto, Initial-boundary value problems for linear diffusion equation with multiple time-fractional derivatives, preprint, arXiv:1306.2778, (2013).

K. Li, J. Peng and J. Jia, Cauchy problems for fractional differential equations with Riemann-Liouville fractional derivatives, J. Funct. Anal., 263(2012), 476-510.

J. Liang, J. H. Liu and T.-J. Xiao, Nonlocal impulsive problems for nonlinear differential equations in Banach spaces, Math. Comput. Model., 49(2009), 798-804.

S. C. Lim, M. Li and L. P. Teo, Langevin equation with two fractional orders, Phys. Lett., 372(2008), 6309-6320.

W. Lin, Global existence theory and chaos control of fractional differential equations, J. Math. Anal. Appl., 332(2007), 709-726.

J. L. Lions and E. Magenes, Non-homogeneous Boundary Value Problems and Applications, Springer-Verlag, (1972).

J. Liu, Nonlinear impulsive evolution equations, Dyn. Contin. Discrete Impuls. Syst., 6(1999), 77-85.

L. Liu, F. Guo, C. Wu et al., Existence theorems of global solutions for nonlinear Volterra type integral equations in Banach spaces, J. Math. Anal. Appl., 309(2005), 638-649.

Z. Liu and X. Li, On the controllability of impulsive fractional evolution inclusions in Banach spaces, J. Optim. Theory Appl., 156(2013), 167-182.

Z. Liu and X. Li, Approximate controllability of fractional evolution systems with Riemann-Liouville fractional derivatives, SIAM J. Control Optim., 53(2015), 1920-1933.

C. F. Lorenzo and T. T. Hartley, Variable order and distributed order fractional operators, Nonlinear Dynam., 29(2002), 57-98.

S. Lu, F. J. Molz and G. J. Fix, Possible problems of scale dependency in applications of the three-dimensional fractional advection-dispersion equation to natural porous media, Water Resources Res., 38(2002), 1165-1171.

Y. Luchko, Boundary value problems for the generalized time-fractional diffusion equation of distributed order, Fract. Calc. Appl. Anal., 12(2009a) 409-422.

Y. Luchko, Maximum principle for the generalized time-fractional diffusion equation. J. Math. Anal. Appl., 351(2009b), 218-223.

Y. Luchko, Some uniqueness and existence results for the initial-boundary-value problems for the generalized time-fractional diffusion equation, Comput. Math. Appl., 59(2010), 1766-1772.

Y. Luchko, Initial-boundary-value problems for the generalized multi-term time-fractional diffusion equation, J. Math. Anal. Appl., 374(2011), 538-548.

Y. Luchko and R. Gorenflo, An operational method for solving fractional differential equations with the Caputo derivatives, Acta Math. Vietnam., 24(1999), 207-233.

A. Lunardi, Analytic Semigroups and Optimal Regularity in Parabolic Problems, Birkhäuser, Basel, (1995).

A. C. J. Luo and V. S. Afraimovich (Eds.), Long-Range Interaction, Stochasticity and Fractional Dynamics, Springer, (2010).

V. Lupulescu, Causal functional differential equations in Banach spaces, Nonlinear Anal.: TMA, 69(2008), 4787-4795.

E. Lutz, Fractional Langevin equation, Phys. Rev., 64(2001), 051106.

J. T. Machado, V. Kiryakova and F. Mainardi, Recent history of fractional calculus, Commun. Nonlinear Sci. Numer. Simul., 16(2011), 1140-1153.

M. Magdziarz and T. Zorawik, Stochastic representation of a fractional subdiffusion equation, The case of infinitely divisible waiting times, Lévy noise and space-time-dependent coefficients, Proc. Amer. Math. Soc., (2015).

R. L. Magin, Fractional calculus models of complex dynamics in biological tissues, Comput. Math. Appl., 59(5)(2010), 1586C1593.

F. Mainardi, P. Paraddisi and R. Gorenflo, Probability Distributions Generated by Fractional Diffusion Equations, in: J. Kertesz, I. Kondor (Eds.), Econophysics: An Emerging Science, Kluwer, Dordrecht, (2000).

F. Mainardi and P. Pironi, The Fractional Langevin equation: Brownian motion revisited, Extracta Math., 10(1996), 140-154.

A. B. Malinowska and D. F. M. Torres, Introduction to the Fractional Calculus of Variations, Imp. Coll. Press, London, (2012).

H. Markus, The Functional Valculus for Sectorial Operators, Oper. Theory Adv. Appl., vol. 69, Birkhäuser-Verlag, Basel, (2006).

N. Masmoudi and T. K. Wong, Local-in-time existence and uniqueness of solutions to the Prandtl equations by energy methods, Comm. Pure Appl. Math., 68(2015), 1683-1741.

J. Mawhin, Topological Degree Methods in Nonlinear Boundary Value Problems, CMBS Regional Conference Series in Mathematics, 40, AMS, Providence, R.I., (1979).

J. Mawhin and M. Willem, Critical Point Theory and Hamiltonian Systems, Springer, New York, (1989).

A. McIntosh, Operators which have an $H^\infty$ functional calculus, Proc. Centre Math. Anal. Austral. Nat. Univ., 14(1986), 210-231.

M. M. Meerschaert, D. A. Benson, H. Scheffler et al., Stochastic solution of space-time fractional diffusion equations, Phys. Rev. E, 65(2002), 1103-1106.

M. M. Meerschaert, E. Nane and H. P. Scheffler, Stochastic model for ultraslow diffusion, Stochastic Process. Appl., 116(2006), 1215-1235.

M. M. Meerschaert, E. Nane and P. Vellaisamy, Distributed-order fractional Cauchy problems on bounded domains, preprint, arXiv:0912.2521v1, (2009).

M. M. Meerschaert and C. Tadjeran, Finite difference approximations for fractional advection-dispersion flow equations, J. Comput. Appl. Math., 172(2004), 65-77.

R. Metzler and J. Klafter, The random walk's guide to anomalous diffusion: a fractional dynamics approach, Phys. Rep., 339(2000a), 1-77.

R. Metzler and J. Klafter, Boundary value problems for fractional diffusion equations, Phys. A, 278(2000b), 107-125.

R. Metsler and J. Klafter, The restaurant at the end of the random walk: recent developments in the description of anomalous transport by fractional dynamics, J. Phys. A, 37(2004), 161-208.

M. W. Michalski, Derivatives of noninteger order and their applications, Inst. Math., Polish Acad. Sci., (1993).

R. Mikulevicius, H. Pragarauskas and N. Sonnadara, On the Cauchy-Dirichlet problem in the half space for parabolic SPDEs in weighted Hölder spaces, Acta Appl. Math., 97(2007), 129-149.

K. S. Miller and B. Ross, An Introduction to the Fractional Calculus and Differential Equations, John Wiley, New York, (1993).

S. Momani and O. Zaid, Analytical solution of a time-fractional Navier-Stokes equation by a domian decomposition method, Appl. Math. Comput., 177(2006), 488-494.

H. Mönch, Boundary value problems for nonlinear ordinary differential equations of second order in Banach spaces, Nonlinear Anal.: TMA, 4(1980), 985-999.

G. M. Mophou, Existence and uniqueness of mild solution to impulsive fractional differential equations, Nonlinear Anal.: TMA, 72(2010), 1604-1615.

G. M. Mophou and G. M. N'Guérékata, Existence of mild solutions of some semilinear neutral fractional functional evolution equations with infinite delay, Appl. Math. Comput., 216(2010), 61-69.

J. Mu, B. Ahmad and S. Huang, Existence and regularity of solutions to time-fractional diffusion equations, Special Issue on Time fractional PDEs, Comput. Math. Appl., (2016), http://dx.doi.org/10.1016/j.camwa.2016.04.039.

V. Mureşan, Existence, uniqueness and data dependence for the solutions of some integro-differential equations of mixed type in Banach space, J. Anal. Appl., 23(2004), 205-216.

I. P. Natanson, Theory of Functions of a Real Variable, vol. II, Frederick Ungar Publishing Co., New York, (1960).

J. M. A. M. van Neerven and B. Straub, On the existence and growth of mild solutions of the abstract Cauchy problem for operators with polynomially bounded resolvent, Houston J. Math., 24(1998),137-171.

G. M. N'Guerekata, A Cauchy problem for some fractional abstract differential equation with nonlocal conditions, Nonlinear Anal.: TMA, 70(2009), 1873-1876.

J. J. Nieto and D. O'Regan, Variational approach to impulsive differential equations, Nonlinear Anal.: RWA, 10(2009), 680-690.

T. Odzijewicz, A. B. Malinowska and D. F. M. Torres, Fractional variational calculus with classical and combined Caputo derivatives, Nonlinear Anal.: TMA, 75(2012a), 1507-1515.

T. Odzijewicz, A. B. Malinowska and D. F. M. Torres, Fractional calculus of variations in terms of a generalized fractional integral with applications to physics, Abstr. Appl. Anal., (2012b).

T. Odzijewicz, A. B. Malinowska and D. F. M. Torres, Generalized fractional calculus with applications to the calculus of variations, Comput. Math. Appl., (2012c).

D. O'Regan and R. Precup, Existence criteria for integral equations in Banach spaces, J. Inequal. & Appl., 6(2001), 77-97.

D. N. Pandey, A. Ujlayan and D. Bahuguna, On a solution to fractional order integro-differential equations with analytic semigroups, Nonlinear Anal.: TMA, 71(2009), 3690-3698.

A. Pazy, Semigroups of Linear Operators and Applications to Partial Differential Equations. Applied Mathematical Sciences, vol. 44. Springer-Verlag, New York, (1983).

F. Periago and B. Straub, A functional calculus for almost sectorial operators and applications to abstract evolution equations, J. Evol. Equs., 2(2002), 41-68.

S. Picozzi and B. West, Fractional Langevin model of memory in financial markets, Phys. Rev., 66(2002), 046118.

I. Podlubny, Fractional Differential Equations, Academic Press, San Diego, (1999).

R. Ponce, Hölder continuous solutions for fractional differential equations and maximal regularity, J. Differential Equations, 255(2013), 3284-3304.

T. R. Prabhakar, A singular integral equation with a generalized Mittag-Leffler function in the kernel, Yokohama Math. J., 19(1971), 7-15.

E. M. Rabei, K. I. Nawafleh, R. S. Hijjawi *et al.*, The Hamilton formalism with fractional derivatives, J. Math. Anal. Appl., 327(2007), 891-897.

P. H. Rabinowitz, Minimax Methods in Critical Point Theory with Applications to Differential Equations, CBMS, American Mathematical Society, vol. 65, (1986).

M. H. M. Rashid and Y. El-Qaderi, Semilinear fractional integro-differential equations with compact semigroup, Nonlinear Anal.: TMA, 71(2009), 6276-6282.

G. Raugel and G. R. Sell, Navier-Stokes equations on thin 3D domains, I, Global attractors and global regularity of solutions, J. Amer. Math. Soc., 6(1993), 503-568.

F. Riewe, Nonconservative Lagrangian and Hamiltonian mechanics, Phys. Rev. E, 53(1996), 1890-1899.

F. Riewe, Mechanics with fractional derivatives, Phys. Rev. E, 55(3, part B)(1997), 3581-3592.

H. E. Roman and P. A. Alemany, Continuous-time random walks and the fractional diffusion equation, J. Phys. A, 27(1994), 3407-3410.

A. Rontó and M. Rontó, Successive approximation method for some linear boundary value problems for differential equations with a special type of argument deviation, Miskolc Math. Notes, 10(2009), 69-95.

W. Rudin, Functional Analysis (2nd ed.), McGraw-Hill, (1991).

I. A. Rus, Metrical fixed point theorems, Univ. of Cluj-Napoca (Romania), (1979).

I. A. Rus, Picard mappings: results and problems, Babes-Bolyai Univ,, Cluj-Napoca, Seminar on Fixed Point Theory, preprint 6(1987), 55-64.

I. A. Rus, Weakly Picard mappings, Comment. Math. Univ. Carolin., 34(1993), 769-773.

I. A. Rus, Picard operators and applications, Sci. Math. Jpn., 58(2003), 191-219.

I. A. Rus and S. Mureşan, Data dependence of the fixed points set of some weakly Picard operators, Proc. Itinerant Seminar, Srima Publishing House, Cluj-Napoca (Romania), (2000), 201-207.

J. Sabatier, O. Agrawal and J. T. Machado, Advances in Fractional Calculus. Theoretical Developments and Applications in Physics and Engineering, Springer-Verlag, Berlin, (2007).

K. Sakamoto and M. Yamamoto, Initial value/boundary value problems for fractional diffusion wave equations and applications to some inverse problems, J. Math. Anal. Appl., 382(2011), 426-447.

H. A. H. Salem, On the nonlinear Hammerstein integral equations in Banach spaces and application to the boundary value problem of fractional order, Math. Comput. Modelling, 70(2009), 1873-1876.

S. G. Samko, A. A. Kilbas and O. I. Marichev, Fractional Integral and Derivatives: Theory and Applications, Gordon and Breach Science Publishers, Longhorne, PA, (1993).

T. Sandev, R. Metzler and Ž. Tomovski, Velocity and displacement correlation functions for fractional generalized Langevin equations, Fract. Calc. Appl. Anal., 15(2012), 426-450.

T. Sandev, Ž. Tomovski and J. L. A. Dubbeldam, Generalized Langevin equation with a three parameter Mittag-Leffler noise, Physica A: Stat. Mech. Appl., 390(2011), 3627-3636.

W. R. Schneider and W. Wayes, Fractional diffusion and wave equation, J. Math. Phys., 30(1989), 134-144.

J. B. Seaborn, Hypergeometric Functions and Their Applications, Springer-Verlag, New York, (1991).

M. A. Şerban, I. A. Rus and A. Petruşl, A class of abstract Volterra equations, via weakly Picard operators technique, Math. Inequal. Appl., 13(2010), 255-269.

M. F. Shlesinger, B. J. West and J. Klafter, Lévy dynamics of enhanced diffusion: applications to turbulence, Phys. Rev. Lett., 58(1987), 1100-1103.

X. B. Shu, Y. Z. Lai and Y. M. Chen, The existence of mild solutions for impulsive fractional partial differential equations, Nonlinear Anal.: TMA, 74(2011), 2003-2011.

L. M. Sokolov, A. V. Chechkin and J. Klafter, Distributed-order fractional kinetics, Acta Phys. Polon. B, 35(2004), 1323-1341.

C. A. Stuart, Bifurcation into Spectral Gaps, Société Mathématique de Belgique, (1995).

P.-L. Sulem and C. Sulem, The Nonlinear Schrödinger Equation: Self-focusing and Wave Collapse, Hardback, (1999).

H.-R. Sun and Q.-G. Zhang, Existence of solutions for a fractional boundary value problem via the Mountain Pass method and an iterative technique, Comput. Math. Appl., 64(2012), 3436-3443.

W. Sudsutad, B. Ahmad, S. K. Ntouyas et al., Impulsively hybrid fractional quantum Langevin equation with boundary conditions involving Caputo $q_k$-fractional derivatives, Chaos, Solitons & Fractals, 91(2016), 47-62.

W. Sudsutad, S. K. Ntouyas and J. Tariboon, Systems of fractional Langevin equations of Riemann-Liouville and Hadamard types, Adv. Differ. Equ., 2015(2015), 1-24.

C. L. Tang and X. P. Wu, Periodic solutions for second order systems with not uniformly coercive potential, J. Math. Anal. Appl., 259(2001), 386-397.

V. E. Tarasov, Fractional Dynamics: Application of Fractional Calculus to Dynamics of Particles, Fields and Media, Springer, (2010).

V. E. Tarasov, Fractional dynamics: Application of Fractional Calculus to Dynamics of Particles, Fields and Media, Springer Science and Business Media, (2011).

V. E. Tarasov, Theoretical physics models with integro-differentiation of fractional order, IKI, RCD, (2011). (in Russian)

A. Zain and A. M. Tazali, Local existence theorems for ordinary differential equations of fractional order, Ordinary and Partial Differential Equations, Lecture Notes in Math., Springer, Dundee, 964(1982), 652-665.

C. Thaiprayoon, S. K. Ntouyas and J. Tariboon, On the nonlocal Katugampola fractional integral conditions for fractional Langevin equation, Adv. Differ. Equ., 2015(2015), 1-16.

Y. Tian and Z. Bai, Existence results for the three-point impulsive boundary value problem involving fractional differential equations, Comput. Math. Appl., 59(2010), 2601-2609.

Y. Tian and Y. Zhou, Positive solutions for multipoint boundary value problem of fractional differential equations, J. Appl. Math. Comput., 38(2012), 417-427.

S. P. Toropova, Relation of semigroups with singularity to integrated semigroups, Russian Math., 47(2003), 66-74.

T.-P. Tsai, Asymptotic dynamics of nonlinear Schrödinger equations with many bound states, J. Differential Equations, 192(1995), 225-282.

S. Umarov and R. Gorenflo, Cauchy and nonlocal multi-point problems for distributed order pseudo-differential equations, Z. Anal. Anwend., 24(2005), 449-466.

W. Varnhorn, The Stokes Equations, Akademie Verlag, Berlin, (1994).

W. von Wahl, The Equations of Navier-Stokes and Abstract Parabolic Equations, Springer-Verlag, (2013).

W. von Wahl, Gebrochene potenzen eines elliptischen operators und parabolische differentialgleichungen in Räumen hölderstetiger Funktionen, Nachr. Akad. Wiss. Göttingen, Math.-Phys. Klasse, 11(1972), 231-258.

Y. Wang, Global existence and blow up of solutions for the inhomogeneous nonlinear Schrödinger equation in $\mathbb{R}^2$, J. Math. Anal. Appl., 338(2008), 1008-1019.

X. Wang, Impulsive boundary value problem for nonlinear differential equations of fractional order, Comput. Math. Appl., 62(2011), 2383-2391.

G. Wang, B. Ahmad and L. Zhang, Some existence results for impulsive nonlinear fractional differential equations with mixed boundary conditions, Comput. Math. Appl., 59(2010), 1389-1397.

G. Wang, B. Ahmad and L. Zhang, Impulsive anti-periodic boundary value problem for nonlinear differential equations of fractional order, Nonlinear Anal.: TMA, 74(2011), 792-804.

R. N. Wang, D. H. Chen and T. J. Xiao, Abstract fractional Cauchy problems with almost sectorial operators, J. Differential Equations, 252(2012), 202-235.

J. R. Wang, X. W. Dong and Y. Zhou, Existence, attractive and stability of solutions for quadratic Urysohon fractional integral equations, Commun. Nonlinear Sci. Numer. Simul., 17 (2012), 545-554.

J. R. Wang, X. W. Dong and Y. Zhou, Analysis of nonlinear integral equations with Erdelyi-Kober fractional operator, Commun. Nonlinear. Sci. Numer. Simul., 17(2012), 3129-3139.

J. R. Wang, Z. Fan and Y. Zhou, Nonlocal controllability of semilinear dynamic systems with fractional derivative in Banach spaces, J. Optim. Theory Appl., 154(2012), 292-302.

J. R. Wang, M. Fečkan and Y. Zhou, On the new concept of solutions and existence results for impulsive fractional evolution equations, Dyn. Partial Differ. Equ., 8(2011), 345-361.

J. R. Wang, M. Fečkan and Y. Zhou, Ulam's type stability of impulsive ordinary differential equations, J. Math. Anal. Appl., 395(2012), 258-264.

J. R. Wang, M. Fečkan and Y. Zhou, Nonexistence of periodic solutions and asymptotically periodic solutions for fractional differential equations, Commun. Nonlinear Sci. Numer. Simul., 18(2013a), 246-256.

J. R. Wang, M. Fečkan and Y. Zhou, Relaxed controls for nonlinear fractional impulsive evolution equations, J. Optim. Theory Appl., 156(1)(2013b), 13-32.

J. R. Wang, M. Fečkan and Y. Zhou, Fractional order iterative functional differential equations with parameter, Appl. Math. Modelling, 37(2013c), 6055-6067.

J. R. Wang, M. Fečkan and Y. Zhou, Presentation of solutions of impulsive fractional Langevin equations and existence results, Eur. Phys. J. Special Topics, 222(2013d), 1855-1872.

J. R. Wang, L. Lv and Y. Zhou, Ulam stability and data dependence for fractional differential equations with Caputo derivative, Electron. J. Qual. Theory Differ. Equ., 63(2011), 1-10.

J. R. Wang, L. Lv and Y. Zhou, Boundary value problems for fractional differential equations involving Caputo derivative in Banach spaces, J. Appl. Math. Comput., 38(2012a), 209-224.

J. R. Wang, L. Lv and Y. Zhou, New Concepts and results in stability of fractional differential equations, Commun. Nonlinear Sci. Numer. Simul., 17(2012b), 2530-2538.

J. R. Wang and W. Wei, A class of nonlocal impulsive problems for integrodifferential equations in Banach spaces, Results Math., 58(2010), 379-397.

J. R. Wang, W. Wei and Y. Zhou, Fractional finite time delay evolution systems and optimal controls in infinite dimensional spaces, J. Dyn. Control Syst., 17(2011), 515-535.

J. R. Wang, X. Xiang and Y. Peng, Periodic solutions of semilinear impulsive periodic system on Banach space, Nonlinear Anal.: TMA, 71(2009), e1344-e1353.

G. Wang, L. Zhang and G. Song, Systems of first order impulsive functional differential equations with deviating arguments and nonlinear boundary conditions, Nonlinear Anal.: TMA, 74(2011), 974-982.

J. R. Wang and Y. Zhou, A class of fractional evolution equations and optimal controls, Nonlinear Anal.: RWA, 12(2011a), 262-272.

J. R. Wang and Y. Zhou, Study of an approximation process of time optimal control for fractional evolution systems in Banach spaces, Adv. Difference Equ., vol. 2011b, Article ID 385324, 15 pages.

J. R. Wang and Y. Zhou, Analysis of nonlinear fractional control systems in Banach spaces, Nonlinear Anal.: TMA, 74(2011c), 5929-5942.

J. R. Wang and Y. Zhou, Time optimal controls problem of a class of fractional distributed system, Int. J. Dyn. Syst. Differ. Equs., 3(2011d), 363-382.

J. R. Wang and Y. Zhou, Existence of mild solutions for fractional delay evolution systems, Appl. Math. Comput., 218(2011e), 357-367.

J. R. Wang and Y. Zhou, Existence and controllability results for fractional semilinear differential inclusions, Nonlinear Anal.: RWA, 12(2011f), 3642-3653.

J. R. Wang and Y. Zhou, Mittag-Leffer-Ulam stabilities of fractional evolution equations, Appl. Math. Lett., 25(2012a), 723-728.

J. R. Wang and Y. Zhou, Complete controllability of fractional evolution systems, Commun. Nonlinear Sci. Numer. Simul., 17(2012b), 4346-4355.

J. R. Wang, Y. Zhou and M. Fečkan, Alternative results and robustness for fractional evolution equations with periodic boundary conditions, Electron. J. Qual. Theory Differ. Equ., 97(2011), 1-15.

J. R. Wang, Y. Zhou and M. Fečkan, On recent developments in the theory of boundary value problems for impulsive fractional differential equations, Comput. Math. Appl., 64(2012a), 3008-3020.

J. R. Wang, Y. Zhou and M. Fečkan, Nonlinear impulsive problems for fractional differential equations and Ulam stability, Comput. Math. Appl., 64(2012b), 3389-3405.

J. R. Wang, Y. Zhou and M. Fečkan, Abstract Cauchy problem for fractional differential equations, Nonlinear Dynam., 71(2013), 685-700.

J. R. Wang, Y. Zhou and M. Fečkan, On the nonlocal Cauchy problem for semilinear fractional order evolution equations, Cent. Eur. J. Math., online, (2014).

J. R. Wang, Y. Zhou and M. Medved, On the solvability and optimal controls of fractional integrodifferential evolution systems with infinite delay, J. Optim. Theory Appl., 152(2012a), 31-50.

J. R. Wang, Y. Zhou and M. Medved, Picard and weakly Picard operators technique for nonlinear differential equations in Banach spaces, J. Math. Anal. Appl., 389(2012b), 261-274.

J. R. Wang, Y. Zhou and M. Medved, Qualitative analysis for nonlinear fractional differential equations via topological degree method, Topol. Methods Nonlinear Anal., 40(2012c), 245-271.

J. R. Wang, Y. Zhou and M. Medved, Existence and stability of fractional differential equations with Hadamard derivative, Topol. Methods Nonlinear Anal., 41(2013), 113-133.

J. R. Wang, Y. Zhou and W. Wei, A class of fractional delay nonlinear integrodifferential controlled systems in Banach spaces, Commun. Nonlinear Sci. Numer. Simul., 16(2011a), 4049-4059.

J. R. Wang, Y. Zhou and W. Wei, Impulsive problems fractional evolution equations and optimal controls in infinite dimensional spaces, Topol. Methods Nonlinear Anal., 38(2011b), 17-43.

J. R. Wang, Y. Zhou and W. Wei, Study in fractional differential equations by means of topological degree method, Numer. Funct. Anal. Optim., 33(2012a), 216-238.

J. R. Wang, Y. Zhou and W. Wei, Optimal feedback control for semilinear fractional evolution equations in Banach spaces, Systems Control Lett., 61(2012b), 472-476.

J. R. Wang, Y. Zhou and W. Wei, Stabilization of solutions to nonlinear impulsive evolution equations, Kybernetika, 48(2012c), 1211-1228.

J. R. Wang, Y. Zhou and W. Wei, Fractional Schrödinger equations with potential and optimal controls, Nonlinear Anal.: RWA, 13(2012d), 2755-2766.

J. R. Wang, Y. Zhou and W. Wei, Cauchy problems for fractional differential equations via Picard and weakly Picard operators technique, Fixed Point Theory, 14(2013), 219-234.

J. R. Wang, Y. Zhou and W. Wei et al., Nonlocal problems for fractional integrodifferential equations via fractional operators and optimal controls, Comput. Math. Appl., 62(2011), 1427-1441.

W. Wei, S. H. Hou and K. L. Teo, On a class of strongly nonlinear impulsive differential equation with time delay, Nonlinear Dyn. Syst. Theory, 6(2006), 281-293.

W. Wei, X. Xiang and Y. Peng, Nonlinear impulsive integro-differential equation of mixed type and optimal controls, Optimization, 55(2006), 141-156.

F. B. Weissler, The Navier-Stokes initial value problem in $L^p$, Arch. Rational Mech. Anal., 74(1980), 219-230.

B. West, M. Bologna and P. Grigolini, Physics of Fractal Operators, Springer-Verlag, Berlin, (2003).

S. Westerlund, Dead matter has memory! Causal consulting, Kalmar, Sweden, (2002).

J. H. Wu, Unified treatment of local theory of NFDEs with infinite delay, Tamkang J. Math., 22(1991), 51-72.

X. Xu, J. Cheng and M. Yamamoto, Carleman estimate for a fractional diffusion equation with half order and application, Appl. Anal., 90(2011), 1355-1371.

T. J. Xiao and J. Liang, The Cauchy Problem for Higher Order Abstract Differential Equations, Lecture Notes in Math., vol. 1701, Springer, Berlin, New York, (1998).

A. Yagi, Abstract Parabolic Evolution Equations and their Applications, Springer-Verlag, Berlin Heidelberg, (2010).

T. Yang, Impulsive Control Theory, Springer, Berlin, (2001).

L. Yang and H. Chen, Nonlocal boundary value problem for impulsive differential equations of fractional order, Adv. Differ. Equ., 2011(2011), Article ID 404917, 16 pages.

H. Ye, J. Gao and Y. Ding, A generalized Gronwall inequality and its application to a fractional differential equation, J. Math. Anal. Appl., 328(2007), 1075-1081.

R. Zacher, A De Giorgi-Nash type theorem for time fractional diffusion equations, Math. Ann., 356(2013), 99-146.

G. M. Zaslavsky, Fractional kinetic equation for Hamiltonian chaos, chaotic advection, tracer dynamics and turbulent dispersion, Physica D, 76(1994), 110-122.

G. M. Zaslavsky, Chaos, fractional kinetics, and anomalous transport, Phys. Rep., 371(2002), 461-580.

G. M. Zaslavsky, D. Stevens and H. Weitzner, Self-similar transport in incomplete chaos, Phys. Rev. E., 48(1993), 1683-1694.

E. Zeidler, Nonlinear Functional Analysis and its Applications, vol. II A and B, Springer-Verlag, New York, (1990).

S. Zhang, Existence of positive solution for some class of nonlinear fractional differential equations, J. Math. Anal. Appl., 278(2003), 136-148.

S. Zhang, Existence of solution for a boundary value problem of fractional order, Acta Math. Scientia, 26(2006), 220-228.

S. Zhang, Positive solutions to singular boundary value problem for nonlinear fractional differential equation, Comput. Math. Appl., 59(2010), 1300-1309.

M. Zheng, F. Liu, I. Turner *et al.*, A novel high order space-time spectral method for the time fractional Fokker-Planck equation, SIAM J. Sci. Comput., 37(2015), A701-A724.

Y. Zhou (Editor), Advances in fractional differential equations, Comput. Math. Appl., 59(3)(2010), 1047-1376.

Y. Zhou (Editor), Advances in fractional differential equations (II), Comput. Math. Appl., 62(3)(2011), 821-1618.

Y. Zhou (Editor), Advances in fractional differential equations (III), Comput. Math. Appl., 64(10)(2012), 2965-3484.

Y. Zhou (Editor), Control and optimization of fractional systems, J. Optim. Theory Appl., 156(1)(2013), 1-182.

Y. Zhou, Existence and uniqueness of fractional functional differential equations with unbounded delay, Int. J. Dyn. Syst. Diff. Equs., 1(4)(2008), 239-244.

Y. Zhou, Existence and uniqueness of solutions for a system of fractional differential equations, Fract. Calc. Appl. Anal., 12(2009), 195-204.

Y. Zhou, Basic Theory of Fractional Differential Equations, World Scientific (1st ed.), Singapore, (2014).

Y. Zhou, Fractional Evolution Equations and Inclusions: Analysis and Control, Academic Press (An imprint of Elsevier), (2016).

Y. Zhou and F. Jiao, Existence of extremal solutions for discontinuous fractional functional differential equations, Int. J. Dyn. Syst. Diff. Equs., 2(2009), 237-252.

Y. Zhou and F. Jiao, Nonlocal Cauchy problem for fractional evolution equations, Nonlinear Anal.: RWA, 11(2010a), 4465-4475.

Y. Zhou and F. Jiao, Existence of mild solutions for fractional neutral evolution equations, Comput. Math. Appl., 59(2010b), 1063-1077.

Y. Zhou, F. Jiao and J. Li, Existence and uniqueness for *p*-type fractional neutral differential equations, Nonlinear Anal.: TMA, 71(2009a), 2724-2733.

Y. Zhou, F. Jiao and J. Li, Existence and uniqueness for fractional neutral differential equations with infinite delay, Nonlinear Anal.: TMA, 71(2009b), 3249-3256.

Y. Zhou, F. Jiao and J. Pečarić, Abstract Cauchy problem for fractional functional differential equations in Banach spaces, Topol. Methods Nonlinear Anal., 42(2013), 119-136.

Y. Zhou and L. Peng, On the time-fractional Navier-Stokes equations, Special Issue on Time fractional PDEs, Comput. Math. Appl., (2016), http://dx.doi.org/10.1016/j.camwa.2016.03.026.

Y. Zhou, X. H. Shen and L. Zhang, Cauchy problem for fractional evolution equations with Caputo derivative, Eur. Phys. J. Special Topics, 222(2013), 1747-1764.

Y. Zhou, V. E. Tarasov, J. J. Trujillo *et al.*, (Eds.), Dynamics of Fractional Partial Differential Equations, Eur. Phys. J. Special Topics, 222(2013), 1743-2011.

Y. Zhou, Y. Tian and Y.-Y. He, Floquet boundary value problem of fractional functional differential equations, Electron. J. Qual. Theory Differ. Equ., 50(2010), 1-13.

Y. Zhou, V. Vijayakumar and R. Murugesu, Controllability for fractional evolution inclusions without compactness, Evol. Equ. Control Theory, 4(2015), 507-524.

Y. Zhou, L. Zhang and X. H. Shen, Existence of mild solutions for fractional evolutions, J. Integr. Equs. Appl., 25(2013), 557-586.

L. P. Zhu and G. Li, Nonlocal differential equations with multivalued perturbations in Banach spaces, Nonlinear Anal.: TMA, 69(2008), 2843-2850.

A. Zoia, M.-C. Néel and A. Cortis, Continuous-time random-walk model of transport in variably saturated heterogeneous porous media, Phys. Rev. E, 81(2010), 031104.

A. Zoia, M.-C. Néel and M. Joelson, Mass transport subject to time-dependent flow with nonuniform sorption in porous media, Phys. Rev. E, 80(2009), 056301.

# Index

$p$-function, 42, 43, 57

Arzela-Ascoli theorem, 3, 41, 51, 65, 70, 74, 101, 115, 127–128, 141–142, 175, 212, 221, 231, 283, 313
  $PC$-type, 3, 188, 196
asymptotic expansion, 12, 159
atomic, 43, 44, 46, 54, 59–61, 65

Banach contraction mapping principle, 18, 51, 65–66, 106, 170, 186, 195, 211, 218, 342
Beta function, 170, 201–203, 293–295
Bochner
  integrable, 3, 15, 90
  theorem, 3, 90
bounded
  norm-bounded, 19
  totally, 19, 39, 41
  uniformly, 3, 32, 40, 51, 120, 128, 135, 140–142, 324

Caputo
  fractional derivative, 4, 6–8, 28, 42, 58, 71, 87, 89, 98, 102, 115, 134, 157, 177–178, 182, 184, 190, 206, 213, 228, 252, 286, 288, 321, 323, 337
  fractional derivative operator, 239, 286
Carathéodory, 33
  $L^{\frac{1}{\beta}}$-Carathéodory, 28, 34–35
  condition, 87, 89, 147
  function, 34
Cauchy sequence, 1
Chandrabhan, 34

$L^{\frac{1}{\gamma}}$-Chandrabhan, 28, 34–35
  function, 34
compact
  mapping, 142
  operator, 85, 120, 136, 175, 220–222, 254
compactness, 100, 127–128, 141–142, 164, 172, 196, 220, 268
condition
  $L^{\frac{1}{\beta}}$-Carathéodory, 28
  $L^{\frac{1}{\delta}}$-Lipschitz, 28
  $L^{\frac{1}{\gamma}}$-Chandrabhan, 28
  $(PS)_c$, 20–22
  $(PS)_c^*$, 22
  Ambrosetti-Rabinowitz, 238, 243
  boundary, 12, 225
  Carathéodory, 87, 89, 147
  Dirichlet type boundary, 251, 287
  Hölder, 323
  initial, 116, 117, 147, 323
  Neumann boundary, 154–155, 178
  nonlocal, 116–117, 145, 147
  Palais-Smale (PS), 20, 240, 243, 277
cone, 19, 20, 33, 41
continuous
  absolutely, 34, 37, 38, 59, 89, 91, 92, 97, 118, 250, 309
  completely, 18–19, 29, 31–33, 39, 49, 51, 63, 65, 73–74, 84–85, 130, 149, 151, 188, 189, 212
  equi-uniformly, 230–231
  Hölder, 72, 77, 153, 166, 176, 287, 288, 301–302, 304, 306, 310, 321, 323–324, 331
  Lipschitz, 178–179

strongly, 23, 120, 136, 165, 254, 326,
			337, 339–340, 342
	uniformly, 23, 68, 160
	uniformly Lipschtiz, 189
critical point, 22, 226
	theorem, 20, 226, 275
	theory, 223, 225–226

degree
	Brouwer, 98
	coincidence, 98
	Leray-Schauder, 98
	topological, 87
delay
	bounded, 27–28, 43
	infinite, 58–59
	unbounded, 28
Dirichlet integral, 320
Dunford-Riesz integral, 24

embedding
	compact, 311–312
	continuous, 312, 324
equation
	differential, 98
	diffusion, 323
	Euler, 275
	Euler-Lagrange, 225, 285, 309, 320–321
	evolution, 115
	functional differential, 43
	integral, 88, 117, 128, 134, 142, 164,
			184, 192, 199, 214, 216, 225,
			281–282
	Langevin, 198–199
	Navier-Stokes, 285–286
	neutral functional differential, 43, 59
	partial differential, 98, 116, 156, 223,
			225, 285
	Schrödinger, 336–337
equicontinuity, 115, 220
equicontinuous, 3, 15, 29, 32–33, 40, 51,
		65, 70, 93, 95, 100, 120–122, 124–127,
		131–132, 136–141, 143–144, 173, 188,
		196, 221, 283
	function, 3, 40–41
	set, 187, 212
	uniformly, 318
existence, 16–17, 20, 27, 28, 31, 44, 46, 51,
		60, 62, 65, 71–72, 74, 102, 115–117, 134,
		145, 147–149, 157–158, 164, 171,

181–182, 185, 190, 193, 195, 198–199,
		206–207, 212–214, 216, 223, 225–226,
		231, 238, 245, 250–251, 257, 259,
		271–272, 274–275, 279, 281, 283,
		285–287, 290, 297, 311, 317–318,
		320–323, 327, 331, 337, 340

Fatou lemma, 249
Fick's law, 224
fixed point theorem
	Darbo, 115
	Darbo-Sadovskii, 18
	Hybrid, 19, 28
	Krasnoselskii, 19, 33, 46, 60, 65, 115
	O'Regan, 19, 149
	Schaefer, 18, 147–148, 187–188, 222
	Schauder, 18, 71, 74, 84, 115, 221
Fountain theorem, 21
	Dual, 21
Fourier transform, 9, 225, 233, 338
Fréchet
	derivative, 20, 44, 46
	differentiable, 20
fractional
	*p*-type neutral differential equation, 27
	*p*-type neutral functional differential
			equation, 42
	abstract differential equation, 87
	abstract evolution equation, 115
	advection dispersion equation, 223–224,
			271–272
	boundary value problem, 20, 223
	calculus of variation, 225, 226
	Cauchy equation, 108
	Cauchy problem, 102–104, 106–107,
			109, 179, 182
	derivative, 4, 116, 198, 225, 232,
			271–272, 285, 309
	derivative operator, 9, 10, 225, 275
	differential equation, 27, 87, 102, 190,
			199, 206, 223, 225, 322
	differential operator, 223, 225
	diffusion equation, 116, 177, 201, 285,
			321–323
	evolution equation, 115–117, 134, 145,
			147
	functional differential equation, 27, 41,
			89
	impulsive boundary value equation,
			181

impulsive differential equation,
        181–182, 190
impulsive evolution equation, 182, 213
impulsive initial value equation, 181
impulsive Langevin equation, 182, 198,
        203
initial-boundary value problem, 178
integral equation, 98, 99, 103, 107–110,
        182–183, 199
integral operator, 9, 10, 225
iterative functional differential
        equation, 28, 72
neutral differential equation, 27–28
neutral functional differential equation,
        58
ordinary differential equation, 87, 201
partial differential equation, 130, 285
Schrödinger equation, 285, 336–337
variational principle, 225
variational problem, 225
Fredholm alternative theorem, 146
Fubini Theorem, 159

Gâteaux
    derivative, 22
    differentiable, 22, 314–315
Galerkin approximation, 225
Gronwall inequality, 176, 192, 301
    singular, 181

Hölder
    condition, 323
    continuous, 72, 77, 153, 166, 176, 287,
        288, 301–302, 304, 306, 310,
        321, 323–324, 331
    estimate, 323
    inequality, 2, 29, 35, 45, 48, 60, 62, 75,
        76, 90, 93–94, 104–105, 129,
        185, 189, 193, 195, 210–211,
        230, 268, 272–273, 310
Helmholtz projector, 287
Hille-Yosida theorem, 337
Hodge projection, 287

impulsive
    differential equation, 181
inequality
    Gronwall, 176, 192, 301
    Hölder, 2, 29, 35, 45, 48, 60, 62, 75–76,
        90, 93–94, 104–105, 129, 185,

189, 193, 195, 210–211, 230,
        268, 272–273, 310
Poincaré, 14
singular Gronwall, 181
Young, 233
infinitesimal generator, 23, 116, 120,
        134–135, 216, 337
iteration method, 287

Lagrange
    mean value theorem, 202
Lagrangian functional, 314
Laplace
    operator, 155–156, 178, 337
    transform, 119–120, 134–135, 164–165,
        215
Laplacian operator, 154, 178
Lebesgue
    dominated convergence theorem, 3, 39,
        71, 93, 100, 105, 124–125,
        138–139, 161, 173, 219, 238,
        275, 291, 293, 328–330, 334
    integrable, 3, 29, 35–36, 38, 45, 60, 90,
        92
    measurable, 29, 33–34, 44–45, 60
Liouville-Weyl
    fractional derivative, 8, 233
    fractional differential operator, 9
    fractional integral, 7
    fractional integral operator, 9, 224
Lipschitz
    $L^{\frac{1}{\delta}}$-Lipschitz, 28, 35

mean value theorem, 302, 307
measurable function, 3, 35–36, 45, 65, 90
measure of noncompactness, 14, 87, 89, 95
    Hausdorff, 14–15, 115, 117, 134
    Kuratowski, 14–15, 115
Mittag-Leffler function, 12, 157–158, 182,
        184, 198, 200–201, 288

Nemitskiĭ operator, 178–179
Neumann theorem, 146
nonlinear alternative of Leray-Schauder
        type, 19, 189
nonlocal term, 98
norm
    Bielecki, 102
    Chebyshev, 73, 102
    Euclidean, 309

graph, 25, 171
operator, 153
supremum, 46, 324

Peano theorem, 87
Picard operator, 17–18, 87, 102, 106–107
   weak, 17–18, 102, 109–110
Pompeiu-Hausdorff functional, 17–18,
   111–113
priori
   bound, 150, 188
   estimate, 17, 98, 340
problem
   bifurcation, 98
   boundary value, 98, 181, 190, 223
   Cauchy, 87, 89, 98, 102, 115–117, 120,
      126–128, 130–131, 133–135,
      140–145, 153, 157, 164, 169,
      181–182, 213–214
   impulsive, 183, 191
   initial value, 27–28, 58, 87, 89, 321, 323,
      337
   initial-boundary value, 322–323
   nonlocal, 98–99, 101, 115, 145–146,
      181–182, 216

regularity, 242, 278, 285–287, 301, 315,
   321, 323, 327, 331
relative compactness, 115, 128, 142
relatively compact, 2–3, 14, 18, 40–41,
   56–57, 96–97, 100–101, 127–128, 133,
   140–142, 196, 231
resolvent operator, 26, 153, 157–158
Riemann-Liouville
   fractional derivative, 4–6, 9, 12, 91,
      115–117, 271, 309, 323
   fractional integral, 4–5, 116, 223, 250,
      262
   fractional integral operator, 224, 226,
      309
   fractional partial derivative, 130
Riesz representation theorem, 44

sectorial operator, 25, 153, 155, 158, 325
   almost, 23, 115, 153, 156–157
semigroup, 25–26
   $C_0$-semigroup, 22–23, 115–117, 120,
      126, 130, 134–135, 140, 145,
      213, 338
   analytic, 25, 154, 288, 325, 336

integrated, 179
property, 10, 136
solution, 27–31, 34, 38, 43–46, 51, 53,
   55–56, 58–60, 62, 67–68, 70–75, 78,
   80–83, 87–89, 92–93, 97–99, 101–104,
   107–110, 122, 128, 131, 137, 142, 145,
   148–152, 157–158, 165–166, 171, 173,
   181–186, 188–193, 196–200, 203,
   206–207, 212, 214, 223–226, 231–232,
   234, 236, 238, 247, 250, 252–253,
   256–262, 266, 271–272, 274–275,
   280–281, 283, 287, 301, 309, 321–323,
   326, 328, 331
   $PC$-mild, 182, 214, 216, 218, 221–222
   classical, 116, 157–158, 164, 166, 169,
      176, 179, 225, 287, 301, 304,
      308, 322–323, 338
   extremal, 27–28, 33
   lower, 34–35, 42
   maximal, 38, 41–42
   mild, 115–118, 120, 122, 126–128, 130,
      133–135, 137, 140–149, 152,
      157–158, 164–166, 169–172,
      175–176, 178–179, 214–215,
      285–287, 289–290, 297–298, 301,
      304, 306, 337, 339–342
   minimal, 38, 41–42
   operator, 71, 115, 338–339
   positive, 82, 192, 197
   upper, 34–35, 42
   weak, 242, 272, 274–276, 278–281, 285,
      309, 311, 317, 320–321, 323
space
   Banach, 1–3, 7, 12, 14, 18–22, 25, 46,
      59, 61, 74, 84–85, 87, 89, 98,
      102, 116, 120, 134–135, 145,
      153, 156, 171, 213, 228, 243,
      252, 262, 288, 290, 311–312, 324
   Cartesian product, 228
   complete metric, 17–18, 74, 103, 170,
      292
   divergence-free function, 287
   fractional derivative, 225–226, 228
   fractional power, 288
   fractional Sobolev, 314
   Hölder continuous function, 324
   Hölder continuous functions, 324
   Hilbert, 14, 226, 232, 262, 272
   Lebesgue, 153
   metric, 17–18, 169

of *p*-integrable functions, 309
of absolutely continuous functions, 2
of continuous and bounded functions, 58
of continuous functions, 2, 28, 42, 59, 89, 98, 102, 116, 153, 309
of distributions, 13
of linear bounded operators, 22
of measurable functions, 2, 59, 109
phase, 58–59
Sobolev, 225, 287, 324
Stokes operator, 287
successive approximation, 102, 104, 107

topological degree, 15, 87
transform
Fourier, 9, 225, 233, 338
Laplace, 119–120, 134–135, 164–165, 215

uniform operator topology, 121, 136, 159–161, 289, 326
uniformly Lipschtiz continuous, 189
uniqueness, 27, 28, 44, 51, 53, 60, 65, 71, 77, 84, 102, 116, 157, 164–165, 171, 181–182, 185, 189–190, 193, 197–198, 206–207, 214, 216, 231, 286–287, 290, 313, 323, 337, 340

variational
functional, 225–226, 272
method, 223, 225, 262, 311, 318, 321
structure, 225–226, 228, 231, 234, 272–274, 311, 314

Wright function, 13, 288, 326, 338

Young
inequality, 233
theorem, 226